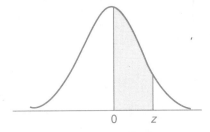

The entries in the table are the probabilities that a random variable having the standard normal distribution takes on a value between 0 and z; they are equal to the area under the curve shaded in the figure.

AREAS UNDER THE STANDARD NORMAL CURVE

z	.00	.01	.02	.03	.04	.05	.06	.07	.08	.09	z
.0	.0000	.0040	.0080	.0120	.0160	.0199	.0239	.0279	.0319	.0359	.0
.1	.0398	.0438	.0478	.0517	.0557	.0596	.0636	.0675	.0714	.0753	.1
.2	.0793	.0832	.0871	.0910	.0948	.0987	.1026	.1064	.1103	.1141	.2
.3	.1179	.1217	.1255	.1293	.1331	.1368	.1406	.1443	.1480	.1517	.3
.4	.1554	.1591	.1628	.1664	.1700	.1736	.1772	.1808	.1844	.1879	.4
.5	.1915	.1950	.1985	.2019	.2054	.2088	.2123	.2157	.2190	.2224	.5
.6	.2257	.2291	.2324	.2357	.2389	.2422	.2454	.2486	.2517	.2549	.6
.7	.2580	.2611	.2642	.2673	.2704	.2734	.2764	.2794	.2823	.2852	.7
.8	.2881	.2910	.2939	.2967	.2995	.3023	.3051	.3078	.3106	.3133	.8
.9	.3159	.3186	.3212	.3238	.3264	.3289	.3315	.3340	.3365	.3389	.9
1.0	.3413	.3438	.3461	.3485	.3508	.3531	.3554	.3577	.3599	.3621	1.0
1.1	.3643	.3665	.3686	.3708	.3729	.3749	.3770	.3790	.3810	.3830	1.1
1.2	.3849	.3869	.3888	.3907	.3925	.3944	.3962	.3980	.3997	.4015	1.2
1.3	.4032	.4049	.4066	.4082	.4090	.4115	.4131	.4147	.4162	.4177	1.3
1.4	.4192	.4207	.4222	.4236	.4251	.4265	.4279	.4292	.4306	.4319	1.4
1.5	.4332	.4345	.4357	.4370	.4382	.4394	.4406	.4418	.4429	.4441	1.5
1.6	.4452	.4463	.4474	.4484	.4495	.4505	.4515	.4525	.4535	.4545	1.6
1.7	.4554	.4564	.4573	.4582	.4591	.4599	.4608	.4616	.4625	.4633	1.7
1.8	.4641	.4649	.4656	.4664	.4671	.4678	.4686	.4693	.4699	.4706	1.8
1.9	.4713	.4719	.4726	.4732	.4738	.4744	.4750	.4756	.4761	.4767	1.9
2.0	.4772	.4778	.4783	.4788	.4793	.4798	.4803	.4808	.4812	.4817	2.0
2.1	.4821	.4826	.4830	.4834	.4838	.4842	.4846	.4850	.4854	.4857	2.1
2.2	.4861	.4864	.4868	.4871	.4875	.4878	.4881	.4884	.4887	.4890	2.2
2.3	.4893	.4896	.4898	.4901	.4904	.4906	.4909	.4911	.4913	.4916	2.3
2.4	.4918	.4920	.4922	.4925	.4927	.4929	.4931	.4032	.4934	.4936	2.4
2.5	.4938	.4940	.4941	.4943	.4945	.4946	.4948	.4949	.4951	.4952	2.5
2.6	.4953	.4955	.4956	.4957	.4959	.4960	.4961	.4962	.4963	.4964	2.6
2.7	.4965	.4966	.4967	.4968	.4969	.4970	.4971	.4972	.4973	.4974	2.7
2.8	.4974	.4975	.4976	.4977	.4977	.4978	.4979	.4979	.4980	.4981	2.8
2.9	.4981	.4982	.4982	.4983	.4984	.4984	.4985	.4985	.4986	.4986	2.9
3.0	.4987	.4987	.4987	.4988	.4988	.4989	.4989	.4989	.4990	.4990	3.0
3.1	.4990	.4991	.4991	.4991	.4992	.4992	.4992	.4992	.4993	.4993	3.1
3.2	.4993	.4993	.4994	.4994	.4994	.4994	.4994	.4995	.4995	.4995	3.2
3.3	.4995	.4995	.4995	.4996	.4996	.4996	.4996	.4996	.4996	.4997	3.3
3.4	.4997	.4997	.4997	.4997	.4997	.4997	.4997	.4997	.4997	.4998	3.4
3.5	.4998	.4998	.4998	.4998	.4998	.4998	.4998	.4998	.4998	.4998	3.5
3.6	.4998	.4998	.4999	.4999	.4999	.4999	.4999	.4999	.4999	.4999	3.6
3.7	.4999	.4999	.4999	.4999	.4999	.4999	.4999	.4999	.4999	.4999	3.7
3.8	.4999	.4999	.4999	.4999	.4999	.4999	.4999	.4999	.4999	.4999	3.8
3.9	.5000										

For $z \geq 3.90$, the areas are .5000 to four decimal places.

CONTEMPORARY STATISTICS

A COMPUTER APPROACH

Sheldon P. Gordon
Suffolk Community College

Florence S. Gordon
New York Institute of Technology

McGraw-Hill, Inc.
New York   St. Louis   San Francisco   Auckland   Bogotá   Caracas
Lisbon   London   Madrid   Mexico City   Milan   Montreal   New Delhi
San Juan   Singapore   Sydney   Tokyo   Toronto

CONTEMPORARY STATISTICS
A Computer Approach

Copyright © 1994 by McGraw-Hill, Inc. All rights reserved. Printed in the United States of America. Except as permitted under the United States Copyright Act of 1976, no part of this publication may be reproduced or distributed in any form or by any means, or stored in a data base or retrieval system, without the prior written permission of the publisher.

 This book is printed on recycled paper containing a minimum of 50% total recycled fiber with 10% postconsumer de-inked fiber.

1 2 3 4 5 6 7 8 9 0   DOC   DOC   9 0 9 8 7 6 5 4 3

ISBN 0-07-023901-0

This book was set in Century Old Style by CRWaldman Graphic Communications.
The editors were Jack Shira and Jack Maisel;
the designer was Irwin Hahn;
the production supervisor was Annette Mayeski.
R. R. Donnelley & Sons Company was printer and binder.

Library of Congress Cataloging-in-Publication Data

Gordon, Sheldon P.
    Contemporary statistics: a computer approach / Sheldon P. Gordon, Florence S. Gordon.
       p.   cm.
    Includes index.
    ISBN 0-07-023901-0
    1. Statistics.   2. Statistics—Data processing.   I. Gordon, Florence S.   II. Title.
QA276.12.G685   1994
519.5—dc20   93-20896

About the Authors

SHELDON P. GORDON is professor of mathematics at Suffolk Community College in New York. He received his Ph.D. in applied mathematics from McGill University. Dr. Gordon is coeditor of the MAA Notes volume, *Statistics for the Twenty First Century*, published by the Mathematical Association of America. He is the project director of the NSF-funded Math Modeling/PreCalculus Reform Project, which is developing an alternative to traditional precalculus/college algebra and trigonometry courses based on the applications of mathematics. He is also a member of the working group of the Harvard Calculus Reform Project. Dr. Gordon is the author of over 75 articles in mathematics and mathematical education, especially the use and implications of technology in the mathematics curriculum. He is the author of the textbook *Calculus and the Computer*, as well as the author of graphics software packages in all areas of the undergraduate mathematics curriculum.

FLORENCE S. GORDON is professor of mathematics at New York Institute of Technology. She received her Ph.D. in mathematical statistics from McGill University. She is the coeditor of the MAA Notes volume, *Statistics for the Twenty First Century*, a collection of articles on new approaches to teaching the first course in statistics written by the leading statistics educators in the country and abroad. Dr. Gordon is a member of the working group of the Math Modeling/PreCalculus Reform Project and is responsible for developing the project materials related to data analysis and probability applications. She is the author of over 30 articles in mathematical statistics and mathematical education. She has also presented numerous workshops and talks at major mathematics meetings on the use of technology in teaching statistics and related courses.

This book is dedicated to
　　Jean Shanfield and Gertrude Gordon

and the memories of
　　Morris Shanfield and Hyman Gordon

who instilled in us a love for learning

and to
　　Craig and Kenneth Gordon

who have filled us with love and pride

Contents

Preface   x

Chapter 1: Introduction to Statistics and the Computer   1

    1.1 Why Study Statistics?   2
    1.2 Introduction to the IBM PC   5
    1.3 Introduction to the Macintosh   10
    1.4 Introduction to Minitab   13

Chapter 2: Descriptive Statistics   18

    2.1 Collecting and Organizing Data   20
    2.2 Using Computers to Organize Data   29
    2.3 Graphical Displays of Data   36
    2.4 Stem-and-Leaf Plots   46
    2.5 Classification of Statistical Data   51
    2.6 Selecting a Random Sample   53
          Student Projects   60
          Chapter 2 Summary   62

Chapter 3: Numerical Description of Data   64

    3.1 Measures of Central Tendency: The Mean, Median and Mode   65
    3.2 Measuring the Variation in Data: The Standard Deviation   74
    3.3 Significance of the Standard Deviation   85
    3.4 z Values   90
    3.5 Percentiles, Quartiles and Box Plots   96
    3.6 Comparing Two Sets of Data   101
    3.7 Grouped Data   112
          Student Projects   119
          Chapter 3 Summary   120

Chapter 4: Probability — 123

    4.1 Introduction to Probability   125
    4.2 Fundamental Counting Principle   141
    4.3 Permutations and Combinations   144
    4.4 Mutually Exclusive and Independent Events   153
    4.5 Dependent Events and Conditional Probability   167
    4.6 Random Patterns in Chaos (Optional)   171
        Student Projects   *176*
        Chapter 4 Summary   *179*

Chapter 5: Discrete Probability Distributions — 181

    5.1 Random Variables and Probability Distributions   182
    5.2 Binomial Distribution   190
    5.3 Binomial Probabilities   199
    5.4 The Hypergeometric Distribution (Optional)   207
    5.5 The Poisson Distribution (Optional)   213
    5.6 The Trinomial Distribution (Optional)   221
        Student Projects   *227*
        Chapter 5 Summary   *228*

Chapter 6: The Normal Distribution — 231

    6.1 Introduction to the Normal Distribution   232
    6.2 Applications of the Normal Distribution   244
    6.3 The Normal Approximation to the Binomial Distribution   258
    6.4 When Is a Set of Data Normal?   276
        Chapter 6 Summary   *279*

Chapter 7: Sampling Distributions — 282

    7.1 Sampling and the Distribution of Sample Means   283
    7.2 Applications of the Central Limit Theorem   296
    7.3 The *t*-Distribution for Small Samples   304
    7.4 Overview of Probability Situations   308
    7.5 Investigating Other Sampling Distributions (Optional)   313
        Student Projects   *323*
        Chapter 7 Summary   *324*

Chapter 8: Estimation   328

    8.1 Confidence Intervals for Means   328
    8.2 Confidence Intervals for Means Based on Small Samples   345
    8.3 Confidence Intervals for Proportions   355
    8.4 Determining the Sample Size n 366
    8.5 Summary of Confidence Intervals   372
         Student Projects   *375*
         Chapter 8 Summary   *376*

Chapter 9: Hypothesis Testing   378

    9.1 Hypothesis Tests for Means   379
    9.2 Hypothesis Testing Using *P*-Values   397
    9.3 Hypothesis Tests for Means Using Small Samples   404
    9.4 Hypothesis Tests for Proportions   415
    9.5 Differences of Means: Hypothesis Tests and Estimation   424
    9.6 The Paired-Data Test   442
    9.7 Differences of Proportions: Hypothesis Tests and Confidence Intervals   449
    9.8 Summary of Hypothesis Testing   461
         Student Projects   *465*
         Chapter 9 Summary   *466*

Chapter 10: Correlation and Regression Analysis   468

    10.1 Introduction to Correlation and Linear Regression   469
    10.2 Correlation   473
    10.3 Linear Regression   489
    10.4 The Standard Error of the Estimate   502
    10.5 Applications of Statistical Inference in Regression Analysis   509
    10.6 Extensions of the Regression Concept   516
         Student Projects   *523*
         Chapter 10 Summary   *524*

Chapter 11: The Chi-Square Distribution   528

    11.1 Introduction to Chi-Square Analysis   529
    11.2 Contingency Tables and the Chi-Square Distribution   535

11.3 The Goodness-of-Fit Test 550
11.4 The Kolmogorov-Smirnov Test for Normality (Optional) 558
11.5 The Distribution of Sample Variances: Estimation and Hypothesis Tests (Optional) 566
Student Projects 578
Chapter 11 Summary 579

Chapter 12: Analysis of Variance 582

12.1 Introduction to One-Way Analysis of Variance 583
12.2 Applications of ANOVA 597
Student Projects 604
Chapter 12 Summary 605

Chapter 13: Nonparametric Statistical Tests 607

13.1 Introduction 608
13.2 The Sign Test 608
13.3 The Rank-Sum Test 619
13.4 The Spearman Rank Correlation Test 627
13.5 The Runs Test 633
Student Projects 642
Chapter 13 Summary 643

Appendix I Baseball Statistics Data Set 645

Appendix II Weather Data Set 651

Appendix III Stock Exchange Data Set 653

Table I Random Numbers 661

Table II Binomial Probabilities 662

Table III Areas under the Standard Normal Curve 665

Table IV Critical Values of the *t*-Distribution 666

Table V Critical Values of *r* 667

Table VI Values of χ_α^2 668

Table VII Critical Values for the Kolmogorov-Smirnov Test of Goodness of Fit 669

Table VIIIA Critical Values of the F-Distribution ($\alpha = 0.05$)   670

Table VIIIB Critical Values of the F-Distribution ($\alpha = 0.01$)   672

Table IX Critical Values of Spearman's Rank Correlation Coefficient   674

Table X Critical Values for Total Number of Runs   675

Solutions to Selected Problems   676

Index   692

Preface

Virtually every facet of daily life is affected by statistics. Students taking an introductory course in statistics should begin to understand and appreciate how statistics applies to the world around them. This new awareness can make a first course in statistics one of the most interesting courses in college. *Contemporary Statistics: A Computer Approach* is intended for this introductory course in statistics and is appropriate for students who have little experience in algebra or mathematics. This book focuses on the broad usefulness of statistics, while carefully helping students develop a clear understanding of statistical methods and statistical reasoning.

Our objective has been to write an introductory textbook which makes statistics come alive in several ways:

- By involving students in applying statistical concepts and methods to conduct a variety of statistical projects to study situations of interest to them;
- By integrating computer graphics software to help students understand and explore statistical concepts as well as visualize statistical ideas and methodology;
- By presenting "real" data in relevant examples and realistic situations to illustrate the concepts.

Philosophy of *Contemporary Statistics: A Computer Approach*

In writing a new textbook for introductory statistics, we have adopted the following basic principles advocated by many leaders in statistical education today.

LEARN BY DOING The best way for students to appreciate the power and usefulness of statistics is to apply it directly. *Contemporary Statistics* encourages students to "experience" the subject by conducting a variety of individual and group projects as well as exploring concepts through the use of the accompanying statistical software. This experience provides students exposure to the tasks of formulating statistical questions, collecting and analyzing data, developing conclusions based on the statistical analysis, and preparing statistical reports to present their results. These projects also help students organize their ideas and encourage development of their writing and communication skills. Invariably, these activities result in an increased level of enthusiasm for the subject and a better understanding of statistical ideas.

FOCUS ON IDEAS, NOT CALCULATIONS Given the premise that many routine computations can and should be done by machine, *Contemporary Statistics* focuses on the use of the computer to develop students' conceptual understanding of statistics. The accompanying software package,

GSTAT, is a simple easy-to-use tool for graphically displaying statistical ideas and gives students a dynamic perspective that cannot be achieved through a textbook alone. By the use of random simulations of virtually every topic in statistics and probability, the software provides an effective way to visualize concepts while allowing students to see the patterns of outcomes associated with probability experiments and verify the theoretical predictions. We presume that students will use a combination of statistical calculators and the graphics package for all complex calculations.

EXPLORATORY DATA ANALYSIS (EDA) Students should become accustomed to looking at data and should be comfortable working with different types of data displays such as histograms, box plots, stem-and-leaf plots, etc. By means of a variety of graphical representations within the text and through the use of the accompanying software, students are encouraged to recognize pattern and shape as they examine the effects of using different numbers of classes or different class widths for frequency distributions or histograms or the appearance of outliers and their effect on the statistical measures.

DISTINGUISH ROUTINE CALCULATIONS FROM APPLICATIONS Ideally, students should be able to distinguish between routine statistical calculations and the actual application of statistical ideas. To accomplish this, *Contemporary Statistics* uses carefully crafted exercise materials which not only develop routine calculation skills, but also engage students in a wide variety of problems relating to the application and exploration of data.

Special Features of the Book

GSTAT Statistical Software accompanies the text and is designed to provide students with a visual and exploratory framework for understanding virtually every concept encountered in introductory statistics. The use of GSTAT assumes no previous computer experience and offers maximum flexibility. GSTAT can be used in conjunction with the text to perform one or more of the following important functions:

1. Provide graphical simulations and demonstrations of concepts which allows students to visualize and actively explore statistics and probability concepts;
2. Serve as a tool to perform lengthy computations and complicated procedures, such as calculating descriptive measures of data (mean and standard deviation), regression equations, correlation coefficients, chi-square values, F-values associated with analysis of variance, and several others;
3. Furnish tutorials aimed at developing the skills needed to solve important problems in statistics, such as finding normal probabilities, constructing confidence intervals for means or proportions, and conducting hypothesis tests for mean or proportions.

The software package consists of 40 programs and will support either IBM or Macintosh computers. The IBM versions are available on both $3\frac{1}{2}$-inch or $5\frac{1}{4}$-inch formats. Detailed instructions and suggestions for the use of GSTAT are found throughout the text in sections called *Computer Corners*.

Student Projects, integrated throughout the text, offer suggestions for a wide variety of individual and small-group projects for students to conduct. These special sections at the end of each chapter have been carefully chosen to appeal to student interests and provide detailed descriptions of what constitutes a good statistical report.

Problem Sets are carefully crafted to provide three distinct levels of exercises: *Mastering the Techniques* (purely mechanical practice to develop basic statistical skills); *Applying the Concepts* (relating basic skills to real-world problems); and *Computer Applications* (more detailed real-world applications involving extensive computations or open-ended investigations).

Extensive Data is provided which gives students an understanding of the role of statistics in all aspects of modern life. The book contains numerous sets of timely "real" data including baseball statistics, weather information, and stock market results. These data sets provide students with the opportunity to compare results shown in the book with daily information while they are taking the course. These data sets are also integrated throughout the examples and problems sets.

Minitab Methods sections throughout the text introduce students to one of the most widely used commercial statistical packages. These sections provide a self-contained introduction to Minitab and illustrate its uses.

In the News sections begin each chapter and feature a variety of news items relating to statistics. These topics are used as a springboard for a series of thought-provoking questions designed to encourage students to think critically about statistics.

Enrichment Topics concerning subjects such as chaos, the Kolmogorov-Smirnov test for normality, and a variety of additional sampling distributions not normally covered in courses at this level are made possible by the use of the computer.

Suggested Course Outlines

Contemporary Statistics is designed to fit a variety of different courses. It can serve a two-semester sequence, covering all of the usual topics in probability and statistics, which relies heavily on coverage of data analysis and student completion of numerous statistical projects. However, it is more likely that the book will be used in a one-semester course, which increases the book's flexibility. As recommended by leaders in statistics education, our preference is to de-emphasize probability as much as possible, thus allowing more time to be devoted to statistical ideas and methods. To this end, we typically omit many sections in Chapters 4 and 5. A reasonably paced course could include

- Chapters 1 to 3
- Chapters 4 and 5 (selected sections)
- Chapters 6 to 9 (most sections)
- Chapters 10 or 11 (selected sections)

A one-semester course covering a considerable amount of probability material from Chapters 4 and 5 will likely have to scale back on the statistical coverage. Such a course might include

- Chapters 1 to 5
- Chapters 6 and 7 (most sections)
- Chapters 8 or 9 (beginning sections)
- Chapter 10 (brief introduction to regression)

More specific suggestions on course outlines based on the text are included in the Instructor's Manual.

Supplements

Instructor's Resource Manual contains an abundance of ideas and tools for effective use of the text and accompanying GSTAT software programs. Developed by the authors, this manual offers a general introduction, followed by chapter-by-chapter suggestions regarding topic coverage, use of the various software programs in GSTAT, and ways in which this approach can be tailored to address a variety of particular course and student requirements.

Instructor's Version of GSTAT supports actual classroom demonstrations by avoiding the tutorial type of prompts found in the student software, thus allowing an instructor to create quick and effective classroom presentations. These programs are formatted in large lettering to best support large lecture situations.

Student Solutions Manual contains detailed solutions for every different type of problem found in the text and has answers, including many partial solutions, for every odd-numbered question in the problem sets. The manual also includes numerous helpful study and problem solving hints.

Acknowledgments

A project such as this that combines text and software so intimately owes much to many different people. We are especially indebted to Eliot Silverman for first giving us a glimpse of the potential power of computer graphics simulations for conveying understanding of statistical concepts. We are also extremely grateful for the many helpful discussions with and advice provided by Dr. Paul Kaplan throughout the development of the text. We want to thank Jane-Marie Wright and Beverly Broomell for the lovely and thorough job they did in producing the Student Solutions Manual that accompanies the book. We also want to thank Marilyn Campo for her invaluable assistance in preparing the index.

Prototypes of many of the software programs included in the GSTAT package were developed under the State University of New York's *Faculty Grants for the Improvement of Undergraduate Instruction* program. Additional support was provided by New York Institute of Technology's *AAUP/NYIT* Faculty Research Grant Program. We gratefully acknowledge the support from both of these wonderful programs.

A great many people at McGraw-Hill have provided tremendous help and cooperation throughout the preparation of the book. Jack Shira, our mathematics editor, provided tremendous encouragement, assistance, and many helpful suggestions in the writing of this book. Jack Maisel did wonders in his role of editing supervisor to pull all the text and software pieces together in what was an extremely complicated project. We also want to thank Denise Schanck, Nancy Evans and Maggie Lanzillo for their help throughout. They are a great team! We also appreciate Robert Weinstein for sharing our vision for this project in its early stages and for his encouragement and support.

We want to express our sincere appreciation to our sons, Craig and Kenneth, for their assistance and support over many years while we worked on this project. It's hard enough when one parent is engrossed in writing a book; when both are involved, the demands on family life increase by several orders of magnitude. Without their patience and understanding, this project would not have been possible.

McGraw-Hill and the authors are grateful to the following reviewers for their helpful comments and suggestions: Jasper E. Adams, Stephen F. Austin State University; Paul Alper, College of St. Thomas; Chris Burditt, Napa Valley College; Daniel M. Cherwien, Cumberland County College; Charles E. Eyler, Johnson State College; Frank Gunnip, Oakland Community College; Shu-ping Hodgson, Central Michigan University; Carol Kublin, State University of New York–Cobleskill; Robert Lacher, South Dakota State University; Debra A. Landre, San Joaquin Delta College; Annette Noble, University of Maryland Eastern Shore; Albert E. Parish, Jr., College of Charleston; Robert Patenaude, College of the Canyons; John Reeder, American River College; Lawrence Riddle, Agnes Scott College; Bruce Sisko, Belleville Area College; Paul Speckman, University of Missouri; Ara B. Sullenberger, Tarrant County Junior College; Dorothy Sulock, University of North Carolina–Asheville; Kevin Vang, Minot State University; Joseph Walker, Georgia State University; and Deborah White, College of the Redwoods.

To the Student

Welcome to the world of Statistics! As you read this book, you will see that statistics is one of the most powerful and useful fields of mathematics. It provides us with the necessary tools to analyze information, make predictions, and arrive at decisions in virtually every area of human endeavor. Yet, for all their power, the ideas and methods of modern statistics are quite simple. The techniques you will learn here are the same ones used everyday by business, government, and research. They are the basis for the reports you can find daily in the news.

To appreciate statistics, you must experience statistics. We have written a book which you should find easy to read and understand. We have carefully crafted problem sets so you can first concentrate on developing the skills you need to perform basic calculations and then apply those skills to realistic situations. Active involvement in working these problems is an essential part of developing statistical understanding and mastering the subject. Try to keep current with the assignments and you will find the course enjoyable.

The computer package (GSTAT) that accompanies the book is also designed to help you master statistics. GSTAT software contains routines to perform computations, graphic simulations and demonstrations, and tutorials to help you achieve a far deeper understanding of statistics. If possible, you should run the programs as you read through the text, or at least try to run some of them as you get the chance. Certainly, if you are having difficulty with one of the topics, you should sit down at a computer and go through enough examples to help you understand and become proficient with the procedure at hand.

We have included a variety of statistical projects for you to conduct during the course of the semester and hope that your instructor will assign several of them to you. They will give you the opportunity to apply what you have learned to subjects that are of interest to you. With such projects, select a topic about which you want to find out something, not just something to get the work done. Statistics gives you the tools to analyze data, make predictions and decisions, or answer pertinent questions. Also, we include information and suggestions for creating a formal report or presentation of your findings. Our experience has consistently been that students who begin projects early and also complete them early almost always enjoy the process and rarely have any problems. Those who leave them for the night before invariably run into problems.

This course should be one of the most interesting and rewarding courses you will take in college. Good luck and enjoy!

Sheldon P. Gordon
Florence S. Gordon

CONTEMPORARY STATISTICS

A COMPUTER APPROACH

1 Introduction to Statistics and the Computer

The vast majority of college students picked Levi's 501 jeans as the most "in" clothing, says a study sponsored by Levi's. And in separate studies funded by the cloth-diaper and disposable-diaper industries, guess what: cloth diapers were shown to be better for the environment than paper—and vice versa.

In recent years, research studies like these have become one of America's most powerful and popular tools of persuasion. The business of studying public opinion and consumer habits has exploded in the past two decades. Today, studies have become vehicles for polishing corporate images, influencing juries, shaping debate on public policy, selling shoe polish and satisfying the media's—and the public's—voracious appetite for information.

Yet while studies promise a quest for truth, many today are little more than vehicles for pitching a product or opinion. An examination of hundreds of recent studies indicates that the business of research has become pervaded by bias and distortion. The result is a corruption of the information used every day by America's voters, consumers and leaders. A growing number of studies are actually sponsored by companies or groups with a real—usually financial—interest in the outcome. And often the study question is posed in such a way that the response is predictable:

- When Levi Strauss & Co. asked students which clothes would be most popular this year, 90% said Levi's 501 jeans. They were the only jeans on the list.
- A Gallop poll sponsored by the disposable-diaper industry asked; "It is estimated that disposable diapers account for less than 2% of the trash in today's landfills. In contrast, beverage containers, third-class mail and yard waste are estimated to account for about 21% of the trash in landfills. Given this, in your opinion, would it be fair to ban disposable diapers?" 84% said no.

The news media also play a role in disseminating sloppy or biased research to consumers. Journalists often publicize reports about a study without examining the study's methodology to see if it was conducted properly. Statistics are thrown around with abandon, even when sample sizes are so small they're meaningless.

Besides interviewing too few people, there are other ways a survey can be flawed: Those surveyed may not be representative of the population, the analysis of the data may be faulty, or the conclusions may be screened so only the best are reported. For example:

"There's good news for the 65 million Americans currently on a diet" trumpeted a news release for a diet-products company. Its study showed that people who lose weight can keep it off. The sample: 20 graduates of the company's program who endorse it in commercials.

There is still much good research being done, of course. In medicine and other physical sciences, research must be quantifiable and replicable to be taken seriously.

The Wall Street Journal, November 14, 1991

1.1 Why Study Statistics?

The time may not be very remote when it will be as essential for efficient citizenship that everyone be able to think statistically as it now is to be able to read or to write.

When H. G. Wells made this prediction early in the 20th century, it might well have been considered an overstatement, if not an outright foolish remark. From our viewpoint as the 20th century comes to a close, his statement is a prophecy which has come true.

Our society has developed into one where science and technology affect everything around us. Statistics is one of the most important of these scientific tools. Virtually all facets of our lives are affected by statistics. Statistics has become a necessary element in most academic fields including the sciences, engineering, business, political science, economics, psychology, sociology, education, medicine, nursing, and other health-related areas.

The effects of statistics are not limited to academic areas. We cannot pick up a newspaper or watch television without being exposed to statistical ideas in one form or another. To emphasize the extent to which statistics pervades our lives, we list just a few of the items which are statistical in nature that we found in two randomly chosen issues of *USA TODAY*. They are not arranged in any particular order.

The average hourly wage of workers is up 3.9% compared to last year.

Average grocery prices are up 1.4% over the previous month.

A table shows average teacher salaries in various states.

A table lists the average weekly pay for various professions.

Of law school graduates from one branch of a state university, 56.8% failed the bar exam compared to 36.4% of law graduates from another branch of the same university.

A state health department estimates that the infant mortality rate is 9.3 deaths per 1000 live births, compared to the nationwide average of 10.0 deaths.

The infant mortality rate of blacks is approximately double that of whites.

A survey was conducted to determine which athletes or singers teenagers want to see on TV commercials for a line of basketball sneakers.

The results of a Commerce Department report on durable-goods orders compared to prior years were revealed.

A new movie is taking in an average of $26,459 per screen on which it is shown.

A Proctor-Silex survey finds that 39% of all people watch their bread while it is toasting.

A tire advertisement claims the tires will last at least 50,000 miles before the tread wears out.

A Gallup poll finds that 90% of women are happy; 46% of women with children are satisfied with their lives compared to 39% of women without children.

The ratings of a new late-night TV talk show were compared to those of local programming.

Nielsen TV ratings were given for the week including the number of viewers, percentage of TV sets in use, and rankings.

The total number of miles driven last year is estimated to be 1.9 trillion, up 43% since 1970.

The District of Columbia has the highest annual average pay, $28,477, compared to the national average of $20,855.

The typical U.S. household produces an average of 6 pounds of garbage per day.

A study claims that 25% of all landfills in the country violate safety standards for groundwater.

The Labor Department projects a growth of 18.4 million new jobs over the next 12 years.

The results of a study of blacks show the number of families, number of one-parent families, average level of education, average family income, average life expectancy, average percentage who vote, etc.

Pepsi Cola is test-marketing a new cherry Pepsi product based on the market shares of different soft drinks.

The percentage of females and minorities working on newspapers is significantly lower than that in the total work force.

Predictions on the Dow-Jones average were shown.

> A study of the effects on children of exposure to low levels of lead indicates that it can significantly lower their intelligence.
>
> A study of infection rates for AIDS among different groups was reported.
>
> A medical study indicates that a compound found in orange peel reduces the growth of breast cancer cells.

In addition to such specific news items, we can always find the usual types of statistical data in the daily papers:

> Weather reports including temperatures, rainfall, snowfall, etc., by city and by region
>
> Business and stock reports including stock market averages, individual stock results, bond rates, gold and metal indices, grain indices, foreign exchange rates, bank rates, and so forth
>
> Sports reports including league standings and records, odds or betting line on the day's games, individual and team statistics, and predictions on forthcoming seasons

As you can see, statistics affects all aspects of our lives today. To assure yourself that the above list is fairly typical, we strongly suggest that you pick up a current issue of your local newspaper or *USA TODAY* and read it through carefully, while marking off every item that seems statistical in nature. You will find that we are not exaggerating the prevalence of statistical ideas.

However, as the article from the *Wall Street Journal* reprinted in our "In the News" section at the beginning of this chapter shows, you should not automatically accept on faith anything you read in the newspapers or in advertisements just because it is statistical. It is all too easy to distort a study to get the outcomes desired—just phrase the questions the right way or ask the right set of people. By the time you finish this course, you will be able to read such claims and reports more critically and be able to interpret them more intelligently.

As a consequence, we can paraphrase Wells and state unequivocally that no person can function intelligently in today's world who has not studied statistics. Moreover, it is not just a question of using statistics directly. Decisions are being made all around us based on statistical reasoning and statistical predictions. It is essential that all educated people have some understanding of what statistics is all about, how it is used, and perhaps just as importantly, how it can be misused.

Our approach in this book is to introduce the basic ideas of modern statistics and to demonstrate how they are applied. We hope to give you an appreciation of the nature of statistics and a mastery of the most important statistical methods. One especially nice thing about studying statistics is that the techniques you learn in this course are precisely the methods being applied in the real world.

As the book (and your course) progresses, you should be alert to newspaper and magazine articles and news reports that are statistical in nature. The ideas and methods developed here apply directly all around you.

1.2 Introduction to the IBM PC

In this section, we introduce you to the IBM PC (personal computer) and its use in conjunction with this book and the accompanying software package. Those of you using a Macintosh instead of the IBM PC (and its clones) should read Section 1.3 to learn about the Macintosh. Those using an Apple II should consult with your instructor.

As we said earlier, statistics and its applications affect almost all facets of our lives. From an historical perspective, however, statistics has achieved this preeminence only in the past forty years or so. The rapid growth in the importance of statistics can be attributed directly to the availability of high-speed computers. In turn, now that inexpensive microcomputers are widely available for educational and home uses, it makes sense to take this trend one step further and use the computer in the statistics classroom.

In practical situations, the applications of statistics involve heavy computations with data. Without computers, such applications could not be handled in a reasonable amount of time. As a result, we will use computers throughout this book to perform most of the heavier computational procedures.

However, the computer has a much more important role to play in a course such as this one. The primary goal of most students taking this course is to develop an *understanding* of and *appreciation* for the concepts and methods of modern statistics. Using computer graphics, the statistical ideas will come alive for you.

To begin, a microcomputer system has three components: a keyboard for entering data and commands to the computer, a disk drive, and a TV monitor for displaying the computer output. Many of you will also have a printer connected to the computer so that you can get printed copies of your results.

The computer keyboard is essentially identical to a typewriter keyboard in terms of location of the usual keys. There are several extra keys on the computer keyboard, but you will not be using many of them. In place of the Return key on a typewriter, the PC has a key marked ↵ or Enter or both located in the usual position to the right of the letters of the alphabet. This key plays the same role as the Return key. You must press it to enter any response, just as you would press the Return key to move to the next line on a typewriter.

The On-Off control for a computer is often located toward the rear on the right side of the disk drive as you face it. On the IBM PC, this control is a red switch which is On in the upper position and Off in the lower position. (In other PC compatibles, the On-Off switch is often located in some out-of-the-way place where there is no danger of turning it off accidentally.)

The programs you will be using in conjunction with this course are all contained on a computer *disk* that accompanies this book. The book will contain either two soft $5\frac{1}{4}$-inch disks or one rigid $3\frac{1}{2}$-inch disk. Open the protective

FIGURE 1.1

wrapping surrounding the disk, and remove it from its holder. See Figure 1.1. Be careful how you hold it, since disks can be easily damaged. In particular, don't touch the exposed portions of the disk that are visible through the holes of the soft disk, and don't bend the disk. Don't expose the disk to magnetic fields such as magnets, electric devices, or metal detectors in libraries and airports.

Before you can run a program, you must first turn on the monitor and then the computer. If your computer has a hard or C: drive, skip to the next paragraph. Otherwise, you will have to load DOS (disk operating system) as we describe here. If there is only one drive (one door), it is automatically your primary floppy drive. If there are two drives, then the primary floppy drive is usually either on the left-hand side or on top. (It is sometimes marked "A".) Insert the DOS disk into the primary drive, with the label facing up and toward you, as shown in Figure 1.2, and close the door if there is a "latch." Next, turn on the computer by moving the On-Off switch to the On position. Loading DOS may take as long as 30 seconds, during which time you may hear some noises coming from the disk drive unit. Don't be concerned about the noise or the delay. You may be asked questions about the current date and time; just press the Enter key in response. When DOS has been loaded, the "A>" prompt should appear on the screen.

If your computer has a hard disk, simply turn on the power to both the monitor and the computer. This will load DOS from the hard disk into your computer's memory. Until DOS has finished loading, you should not insert any floppy disk into the primary disk drive. As part of the start-up procedure, you might be asked a few questions about the current date and time; just

FIGURE 1.2

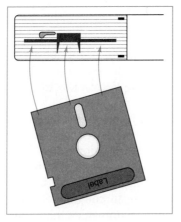

press the Enter key in response. After a few seconds, the screen should display the "C>" prompt.

Depending on the computer system you are using, there are several very simple steps that some of you may have to follow at this point to start the statistics program.

- If you are using an older system with a CGA monitor (on the IBM computer, this is labeled an "IBM Personal Computer Color Display"), you must run a special program to make the Greek characters display properly. With your DOS disk in drive A, or with your DOS directory as the active directory, type GRAFTABL and press Enter. The GRAFTABL program is distributed with all versions of DOS.
- If you have only one floppy drive and no hard drive, remove the DOS disk from the disk drive and insert either the $5\frac{1}{4}$-inch statistics disk labeled "Disk 1" or the $3\frac{1}{2}$-inch statistics disk into the drive. Now type STAT and press the Enter key to begin the program.
- If you have two floppy drives, leave your DOS disk in the primary drive and insert the appropriate statistics disk in the other drive. Type B: and press the Enter key to make the B drive active; then type STAT and press the Enter key to begin the program.
- If you have a hard disk, insert the appropriate statistics disk into your primary drive and type A:. Then type STAT and press the Enter key to begin. Incidentally, the statistics programs can be copied to and run from your hard disk, if you wish.

After you finish the above steps, you will see a display screen that reads as follows:

```
                    CONTEMPORARY STATISTICS

         1. STATISTICAL ANALYSIS OF DATA     15. THE POISSON DISTRIBUTION
         2. RANDOM NUMBER GENERATOR          16. TRINOMIAL DISTRIBUTION SIMULATION
         3. COMPARING TWO SETS OF DATA       17. NORMAL DISTRIBUTION PROBABILITY
         4. GROUPED DATA ANALYSIS            18. NORMAL DISTRIBUTION SIMULATION
         5. COIN FLIPPING SIMULATION         19. NORMAL APPROXIMATION TO BINOMIAL
         6. DICE ROLLING SIMULATION          20. CENTRAL LIMIT THEOREM SIMULATION
         7. LAW OF LARGE NUMBERS             21. t-DISTRIBUTIONS
         8. BIRTHDAY PROBLEM SIMULATION      22. DISTRIBUTION OF SAMPLE MEDIANS
         9. DRUNKARDS (RANDOM) WALK SIMULATION 23. DISTRIBUTION OF SAMPLE MODES
        10. PATTERNS IN CHAOS                24. DISTRIBUTION OF SAMPLE MIDRANGES
        11. THE BINOMIAL DISTRIBUTION        25. CONFIDENCE INTERVALS: MEANS/PROPS
        12. BINOMIAL SIMULATION              26. CONFIDENCE INTERVAL SIMULATION
        13. BINOMIAL PROBABILITY             27. DISTRIBUTION OF SAMPLE PROPORTIONS
        14. THE HYPERGEOMETRIC DISTRIBUTION  28. HYPOTHESIS TESTS

            Use the arrow keys to move to the item you want; then press Enter.
               Continue pressing the arrow keys to see additional items.
                         Press ESC to Quit the program.
```

From this display, you can select any of the listed programs, numbered 1 through 28, by using the four arrow keys on the keyboard to highlight the desired program and then pressing the Enter key. The menu program will then load and run the indicated program. A third column of program choices (numbers 29 through 40) is also available. Press the right-arrow key to move from the first to the second column and then press it again to obtain the following display. You can move back with the left-arrow key. As before, make your selection by highlighting the desired program and pressing Enter.

```
                    CONTEMPORARY STATISTICS

   15. THE POISSON DISTRIBUTION              29. HYPOTHESIS TESTING SIMULATION
   16. TRINOMIAL DISTRIBUTION SIMULATION     30. DIST'N OF DIFFERENCE OF MEANS
   17. NORMAL DISTRIBUTION PROBABILITY       31. DIST'N OF DIFF'CE OF PROPORTIONS
   18. NORMAL DISTRIBUTION SIMULATION        32. LINEAR REGRESSION ANALYSIS
   19. NORMAL APPROXIMATION TO BINOMIAL      33. CORRELATION SIMULATION
   20. CENTRAL LIMIT THEOREM SIMULATION      34. REGRESSION SIMULATION
   21. t-DISTRIBUTIONS                       35. CHI-SQUARE ANALYSIS
   22. DISTRIBUTION OF SAMPLE MEDIANS        36. CHI-SQUARE DISTRIBUTIONS
   23. DISTRIBUTION OF SAMPLE MODES          37. KOLMOGOROV-SMIRNOV TEST
   24. DISTRIBUTION OF SAMPLE MIDRANGES      38. DISTRIBUTION OF SAMPLE VARIANCES
   25. CONFIDENCE INTERVALS: MEANS/PROPS     39. ANALYSIS OF VARIANCE
   26. CONFIDENCE INTERVAL SIMULATION        40. SIMULATION OF THE RUNS TEST
   27. DISTRIBUTION OF SAMPLE PROPORTIONS
   28. HYPOTHESIS TESTS


        Use the arrow keys to move to the item you want; then press Enter.
           Continue pressing the arrow keys to see additional items.
                       Press ESC to Quit the program.
```

We suggest that you move back and forth several times through these columns of options before you make any selection.

We do not discuss the particular details of any of the programs in the package at this time. Each is described in detail as needed throughout the book. However, most of the programs have certain common characteristics which we do discuss here. Each program starts with a title and a short description of what it does. You are then asked to supply a few entries, such as how many repetitions you want. For each such entry, simply type in the number you want and then press the Enter key. Occasionally, you will see a prompt asking you to <PRESS ANY KEY TO BEGIN> or <PRESS ANY KEY TO CONTINUE>. In each case, simply press any key on the keyboard.

Furthermore, each program uses the Esc key (for *escape*, usually located at the upper left or lower left of the keyboard) as a means of stopping a program. When you press Esc, depending on the status of the program, you may be shown a menu screen with a variety of choices or return to the main menu.

If you are running the statistics program from $5\frac{1}{4}$-inch floppy disks, you will need to change disks to run some of the programs. When prompted by the program, simply remove the statistics disk from the floppy drive, place the other disk in the drive, and press Enter.

For now, we suggest that you get a feel for working with the main menu as follows:

Select any program.

Have the computer start that program.

Press the Esc key once or twice to stop execution of the program.

If you are not returned to the main menu, then press the space bar once or twice to go to the options menu and then select and enter the appropriate number to return to the main menu.

Repeat this procedure with other choices.

Almost all the programs on your disk involve a computer graphics display of the statistical concept or method under discussion. After the initial entries are made, the program will transfer to a graphics screen and draw the appropriate display. When the display is completed, there is usually a beep, and the message <PRESS ANY KEY TO CONTINUE> appears. As mentioned above, simply press any key. This will take you to an accompanying text screen where more detailed information is displayed.

From this text screen, most of the programs are designed so that pressing the space bar will take you to an options page where you will be given choices for continuing or ending the program. Also, from the text page, if you press any key other than the space bar, the program will take you directly back to the graphical display. Thus, you can switch back and forth between the graph and the text by pressing any key other than the space bar. However, from the text page, the space bar will take you to the options page.

With each program, the last option on the options screen will always be to return to the main menu. If you select that choice, the program will bring you back to the original menu and give you the choice of any other program. To quit from the program altogether, you must return to the main menu and press Esc.

If you run into trouble, you can start the computer system over again at any time either by turning it off, waiting at least 10 seconds, and turning it on again, or by pressing three keys simultaneously: Ctrl, Alt, and Del. Remove your floppy disk and repeat the procedures you used to start the program. As we indicated above, you can always stop a program during its operation by pressing the Esc key.

There is one other feature that can be very useful if you have a printer attached to the computer. You can print a copy of whatever is on the screen, either text or graphics. First make sure that the printer is attached, that it is turned on, and that the On-Line key or button has been set. Then whenever you press the Prt/Scr key (usually located in the upper right of the keyboard), the program will temporarily stop and whatever shows on the screen is printed for you. As soon as the printout is complete, the program resumes execution. Some older models require that you press either of the shift keys and Prt/Scr simultaneously.

We encourage you to make full use of the programs on the disk. They are intended to serve a variety of purposes:

1. They will perform most of the heavy computations for you.
2. They will display the concepts and methods of statistics to enhance your understanding.
3. Some serve as tutorials to assist you in mastering some of the essential procedures that you will need to apply the statistical methods.

You should turn to the computer often as the course progresses for any of the above purposes. Further, the more you use the computer and these programs, the more confident you will be in computer usage and, we hope, the more you will appreciate the statistical ideas and methods. Use them frequently and enjoy!

But we do want to emphasize that simply using these programs is not enough to master the necessary statistical procedures. You must work through the various problems and procedures, using pencil, paper, and a handheld calculator, so that you will develop full facility with these methods as well as a thorough understanding of the statistical concepts and ideas.

1.3 Introduction to the Macintosh

In this section, we introduce the Macintosh family of computers and their use in conjunction with this book and the accompanying software package. Those of you using an IBM PC or its clones instead of the Macintosh should read Section 1.2 instead to learn about the PC.

As we indicated in Section 1.1, statistics and its applications affect almost all facets of our lives. From an historical perspective, however, statistics has achieved this preeminence only in the past forty years or so. The rapid growth in the importance of statistics can be attributed directly to the availability of high-speed computers. In turn, now that inexpensive microcomputers are widely available for educational and home use, it makes sense to take this trend one step further and use the computer in the statistics classroom.

In practical situations, the applications of statistics involve heavy computations with data. Without computers, such applications could not be handled in a reasonable amount of time. As a result, we will use computers throughout this book to do most of the heavier computational procedures.

However, the computer has a much more important role to play in a course such as this one. The primary goal of most students taking this course is to develop an *understanding* of and *appreciation* for the concepts and methods of modern statistics. Using computer graphics, the statistical ideas will come alive for you.

To begin, a microcomputer system has three components: a keyboard for entering data and commands to the computer, a disk drive, and a TV monitor for displaying the computer output. Many of you will also have a printer connected to the computer so that you can get printed copies of your results.

The keyboard of the Macintosh is essentially identical to that of a typewriter in terms of the placement of the usual keys. There are several extra keys on the keyboard, but you will not be using very many of them. The Macintosh has a Return key which is equivalent to the Return key on a typewriter. You must press it to enter any response, just as you would press the Return key to move to the next line on a typewriter.

On older Macintoshes the On-Off control is located in the rear around the left side of the computer's base as you face it. This control is a switch which is On in the upper position and Off in the lower position. On other models, you must turn on the monitor and the computer separately. The monitor uses a push button arm in back on the right. The computer uses a switch just below the monitor button. Recent models have an On-Off key on the keyboard.

The programs you will be using in conjunction with this course are all contained on a computer *disk* that accompanies this book. Open the protective wrapping surrounding the disk, and remove it from its holder. See Figure 1.3. Be careful how you handle the disk. In particular, don't expose it to magnetic fields such as magnets, electric devices, or metal detectors in libraries and airports.

You control the computer with the keyboard or the mouse device connected to it. As you move the mouse around on a smooth surface, the cursor on the screen moves likewise. When you highlight a symbol (called an *icon*) or an item you want, you select it by pressing the button on the mouse—this is called *clicking*. Before you can run any of the statistics programs, you must first start up your Macintosh computer. If your computer does not have a hard drive, insert your start-up disk in the disk drive as shown in Figure 1.4 and turn on the power.

When the original screen, called the *desktop*, is displayed, use the mouse to move to the start-up disk icon and select it by clicking the mouse; then press the Apple key (bottom row of the keyboard) and the letter "e" to eject

FIGURE 1.3

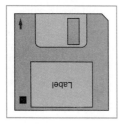

FIGURE 1.4

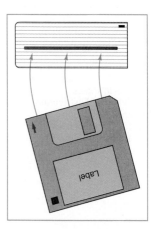

the start-up disk. Then insert your statistics disk. Highlight the icon that represents a computer disk and double click (press the button twice fairly rapidly) to "open" the statistics disk and then double click again on the stat icon to start the program.

If your computer has a hard drive, simply turn on the power. When the desktop screen is displayed, insert your statistics disk into the disk drive, as shown in Figure 1.4. Use the mouse to point to the icon that represents a computer disk and double click to "open" the statistics disk and then double click again on the stat icon to start the program.

After you finish the above steps, you will see a display screen that reads as follows:

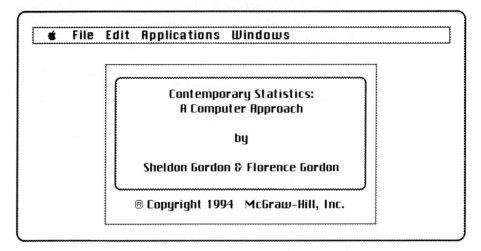

The Stat program works like most Macintosh applications. To see a list of programs, move the mouse to point to the Applications option at the top of the screen and press the mouse button. You will then see the following list of programs:

Statistical Analysis of Data
Random Number Generator
Comparing Two Sets of Data
Grouped Data Analysis
Coin Flipping Simulation
Dice Rolling Simulation
Law of Large Numbers
Birthday Problem Simulation
Drunkards (Random) Walk
Patterns in Chaos
The Binomial Distribution
Binomial Simulation
Binomial Probability
The Hypergeometric Distribution

The Poisson Distribution
Trinomial Distribution Simulation
Normal Distribution Probability
Normal Distribution Simulation
Normal Approximation to Binomial
Central Limit Theorem Simulation
t-Distributions
Distribution of Sample Medians
Distribution of Sample Modes
Distribution of Sample Midranges
Confidence Intervals: Means/Props
Confidence Interval Simulation
Distribution of Sample Proportions
Hypothesis Tests

Hypothesis Testing Simulation
Distribution of Difference of Means
Distribution of Difference of Proportions
Linear Regression Analysis
Correlation Simulation
Regression Simulation
Chi-Square Analysis
Chi-Square Distributions
Kolmogorov-Smirnov Test
Distribution of Sample Variances
Analysis of Variance
Simulation of the Runs Test

While holding the mouse button down, move the cursor to a specific program and then release the mouse button to run the program you have chosen. (This process is called *dragging* the mouse.)

We do not discuss the particular details of the programs in the package at this time. Each is described in detail as it is needed throughout the book. However, most programs have certain common characteristics which we do discuss here. Each program starts with a title and a short description of what it does. You are then asked to supply a few entries, such as how many repetitions you want.

There is one other feature that can be very useful if you have a printer attached to the computer. You can have a copy of whatever is on the screen printed on the printer. First, make sure that the printer is attached, that it is turned on, and that the On-Line key or button on the printer has been set. If the printer is a dot-matrix or ImageWriter printer, simply press the ⌘ key and then press Shift and 4 together. If the printer is a laser printer, then there is an extra step involved: you must press ⌘ and then press Shift and 3 together to create a printable MacPaint or Teachtext document. You then must load MacPaint and select the print option to have the screen contents printed.

We encourage you to make full use of the programs on the disk. They are intended to serve a variety of purposes:

1. They will perform most of the heavy computations for you.
2. They will display the concepts and methods of statistics to enhance your understanding.
3. Some serve as tutorials to assist you in mastering some of the essential procedures that you will need to apply the statistical methods.

You should turn to the computer often as the course progresses for any of the above purposes. Further, the more you use the computer and these programs, the more confident you will be in computer use and, we hope, the more you will appreciate the statistical ideas and methods. Use them frequently and enjoy!

But we do want to emphasize that simply using the programs is not enough to master the necessary statistical procedures. You must work through the various problems and procedures, using pencil, paper, and a handheld calculator, so that you will develop full facility with these methods as well as a thorough understanding of the statistical concepts and ideas.

1.4 Introduction to Minitab

The software package developed for this book is designed primarily to give you a deeper understanding of statistical ideas and methods by using the graphical capabilities of the computer to demonstrate the concepts. At the same time, much of the growth in modern statistics is due to the availability of powerful computer packages which perform statistical calculations. One of the most common of these computational packages is Minitab. (Others are SPSS, BMD, and SAS.) In this section, we give a brief overview of the Minitab system. We will provide additional details and examples throughout the book as particular capabilities of Minitab apply to the topic being discussed.

To use Minitab, you must think of data as being organized and stored in columns. The columns are designated C1, C2, C3, and so forth. You then issue certain commands to tell Minitab what to do. All Minitab commands refer to the particular column of data on which you want an operation performed.

Further, we note that Minitab is a *command-driven* system in the sense that it requires the user to type in a particular command for each operation desired. (By way of comparison, the graphics package that accompanies the book is *menu-driven* so that you only have to respond to questions as to what operations you want performed.)

In our discussions of Minitab, we follow these conventions: The commands that you must use are printed in **boldface**. Each command has a minimal *required* part, but can be augmented with additional *optional* parts to make the commands more intelligible to you. The output from Minitab is displayed in a computer-style font. For instance, the command

> **MEAN** OF THE ENTRIES IN COLUMN **C1**

causes Minitab to calculate and print the average of a set of numbers stored in column C1. This command can be shortened to read

> **MEAN C1**

Minitab simply ignores the additional optional words. Either way, the Minitab response will be, for instance,

> MEAN = 37.29

Furthermore, Minitab uses the symbol > as a prompt. When the system is waiting for you to enter a command, it prompts you with

> MTB>

and waits for your response. If you have indicated that you will be entering data values, it prompts you with

> DATA>

and then waits for you to enter your data values.

To enter a set of data into a particular column, you can use several different approaches. The Minitab command

> SET C1

allows you to enter a set of data typed on a single line, and Minitab assigns the entries to the indicated column, in this case C1. For example,

> SET C1
> 24 29 33 46 36 53 40 END

assigns the seven values to column C1. Notice that the entries do not require commas between them. Further, the word END is used to indicate that no more data are being entered.

A second approach can be used to enter two or more columns of data simultaneously. The command

> READ C1 C4

allows you to enter pairs of data values, two at a time, and store them in columns C1 and C4. For example,

> READ C1 C4
> 12   58
> 15   73
> 18   90
> 20   82
> END

enters four pairs of data values into these two columns.

In addition, it is possible to give a name to each column of data as follows:

> NAME C1 'AGE'
> NAME C4 'WEIGHT IN KG'

Notice that the name used for a column must be enclosed in a pair of apostrophes; do not use quotation marks for this.

You can have the data in a particular column or group of columns printed out by using the command

> PRINT C1 C4

and Minitab will respond with

ROW	AGE	WEIGHT IN KG
1	12	58
2	15	73
3	18	90
4	20	82

You can also indicate a particular column by referring to it by its name:

> PRINT 'AGE'

You can also save any set of data on a disk. If you type

> SAVE 'FILENAME'

where FILENAME is the name of the file, say PEOPLEDATA, then all the entries will be saved for subsequent use. You can then retrieve such an existing file by typing

> RETRIEVE 'PEOPLEDATA'

and continue with your statistical investigation. For instance, you could add data values to the columns you previously entered, change entries to see the effects, apply different statistical procedures to the same data, and so forth.

Also some commands in Minitab have optional subcommands. When these cases arise, you indicate to the Minitab system that you will be providing a subcommand by including a semicolon (;) at the end of the command line. The system will then prompt you for the desired subcommand, and you must enter the required statement and end it with a period. For instance, suppose you want to generate a set of 50 random numbers between 1 and 100. The appropriate command is

> RANDOM 50 C8;

which stores these 50 numbers in column 8. However, before executing this command, Minitab first responds with

> SUBC>

and is asking for more detail on the type of numbers you want. Therefore, you might respond with

> SUBC> **INTEGERS 1 100.**

to indicate that you want the random numbers to be integers between 1 and 100. (Note the semicolon at the end of the primary command and the period at the end of the subcommand.)

Having produced such a set of random numbers or having entered a set of desired values, you can request different operations to be performed on the data in that column. For example,

MEAN C1	gives the average of the entries.
SUM C1	gives the sum of the entries.
MAXIMUM C1	gives the largest value of the entries.
MINIMUM C1	gives the smallest value of the entries.
DESCRIBE C1	gives a large set of statistical measures of the entries in the column, including the mean, median, standard deviation, minimum and maximum, number of scores, and other quantities studied in this course.
ERASE C1	erases all data stored in Column 1 so that you can enter a new set of values for that column.
STOP	indicates that you are ending the Minitab session. Before stopping, remember to save any set of data that you want to use again since otherwise it will be lost.

We introduce other Minitab features as the need arises.

Descriptive Statistics

According to the U.S. Census Bureau, the population of the United States reached 249,632,972 in 1990. Of these, 51.25 percent, or 127,936,898, were women and 48.75 percent, or 121,696,074, were men. There were 63,604,432 people, or 25.6 percent under age 18. The breakdown of the populations was approximately

White	200 million
Black	30 million
American Indian, Eskimo or Aleut	2 million
Asian or Pacific Islander	7 million
Other	10 million

The population (in millions) of each of the 50 states is shown in the following table, along with the number of seats each state will get in the House of Representatives based on its current population.

The cost of the 1990 census was $2.6 billion, or an average of $10.42 for each man, woman and child counted. Have you ever wondered why the government spends so much money and expends so much effort just to count how many of us there are? Major political, economic and social decisions that actually affect all of us are based on the census statistics.

Possibly the most important outcome of the census is the distribution of the 435 seats in the House of Representatives. Based on the census, the number of representatives for various states will change, as shown in the accompanying table. States which have experienced increases in population over the preceding 10 years will have additional representatives while those that have lost population, or have experienced slower rates of population growth, will lose seats. In turn, states with more seats have more influence in Congress and so we can expect to see more federal money allocated for those states. This means more jobs, more public assistance programs, more public works projects, and so forth. On the other hand, states that have lost seats in the House can expect many of these federal expenditures in their areas to be diminished.

Is it any wonder that many states and municipalities are challenging the results of the census by claiming that their populations were grossly undercounted?

In this chapter, we will study how data are collected, organized and displayed in tables and graphs. This is essential so that we can make sense out of a large collection of values and consequently be able to interpret the data and make intelligent use of the results.

State	Population (millions)	Number of House Seats	Change from Previous House
California	29.84	52	7
New York	18.04	31	−3
Texas	17.06	30	3
Florida	13.00	23	4
Pennsylvania	11.92	21	−2
Illinois	11.47	20	−2
Ohio	10.89	19	−2
Michigan	9.33	16	−2
New Jersey	7.75	13	−1
North Carolina	6.66	12	1
Georgia	6.51	11	1
Virginia	6.22	11	1
Massachusetts	6.03	10	−1
Indiana	5.56	10	None
Missouri	5.14	9	None
Wisconsin	4.91	9	None
Tennessee	4.90	9	None
Washington	4.89	9	1
Maryland	4.80	8	None
Minnesota	4.39	8	None
Louisiana	4.24	7	−1
Alabama	4.06	7	None
Kentucky	3.70	6	−1
Arizona	3.68	6	1
South Carolina	3.51	6	None
Colorado	3.31	6	None
Connecticut	3.30	6	None
Oklahoma	3.16	6	None
Oregon	2.85	5	None
Iowa	2.79	5	−1
Mississippi	2.59	5	None
Kansas	2.49	4	−1
Arkansas	2.36	4	None
West Virginia	1.80	3	−1
Utah	1.73	3	None
Nebraska	1.58	3	None
New Mexico	1.52	3	None
Maine	1.23	2	None
Nevada	1.21	2	None
Hawaii	1.12	2	None
New Hampshire	1.11	2	None
Idaho	1.01	2	None
Rhode Island	1.01	2	None

State	Population (millions)	Number of House Seats	Change from Previous House
Montana	.80	1	−1
South Dakota	.70	1	None
Delaware	.67	1	None
North Dakota	.64	1	None
Vermont	.56	1	None
Alaska	.55	1	None
Wyoming	.46	1	None
District of Columbia	.61		None
Total	249.6	435	

2.1 Collecting and Organizing Data

In Section 1.1 we discussed numerous examples of how statistics affects virtually all aspects of our lives. We now introduce some of the basic terminology of statistics which is fundamental to all subsequent statistical ideas.

> **Statistics** is the study of the collection, organization, display and analysis of data, and the methods by which inferences and conclusions can be drawn from the data.

We discuss the collection, organization, and display of data in this chapter. The analysis of data is treated in Chapter 3, while the many applications of statistics are interspersed throughout the rest of the book.

It is important to realize that there are two ways to interpret data. We can consider all available data relating to some particular situation, such as the heights of all students in a college, the results of a national census, or the results in a national or local election. Such a set of values is known as a *population*. However, it can be extremely expensive, time-consuming, or simply impractical, if not impossible, to collect and analyze all these data when the population is large. Thus, in practice, it is usually desirable to study a relatively small fraction of a population which is representative of the entire population. Such a set is called a *sample*. For example, a political poll involves surveying a sample of 1000 voters, say, out of a much larger population.

> A **population** is the set of all possible data values for a subject under consideration.
>
> A **sample** is a set of data values drawn from the much larger population.

In a given situation, there is just one population, but there are many different possible samples.

There are two main branches of modern statistics:

> **Descriptive statistics** involves the collection, organization, and analysis of all data relating to some population or sample under study.
>
> **Inferential statistics** involves making predictions or decisions about an entire population based on results from the data in an appropriately chosen sample drawn from that population.

For example, the U.S. Census, weather records, sports records, and health records are all illustrations of descriptive statistics. The goal is to amass all available data on the subject and have them available for various uses. On the other hand, political polls, TV ratings, consumer taste tests of new products, test marketing of new products, tests of the effectiveness of new drugs, and production quality control are all examples of the use of inferential statistics. Only a relatively small fraction of the population (the sample) is studied in each case. The results from this sample are then used to *infer* something about the entire population. It might be to determine what percentage of the population prefers a certain candidate who is running for office, what percentage of families watch certain television programs, what levels of sales might be expected for a new product, or what percentage of items produced by a factory are expected to be defective.

Historically, descriptive statistics was used in most statistical applications. During the last 50 years, however, inferential statistics has come to dominate statistical applications. The cost savings and the ease of collecting and analyzing sample data are responsible for this change in focus. Thus, the emphasis in this text is on modern statistical inference and its applications.

However, descriptive statistics is the foundation for inferential statistics. We therefore start with a discussion of some important ideas from descriptive statistics as the basis for our later study of inferential statistics.

We begin by discussing the *collecting* of data. Suppose we are studying the height of students in a statistics class. Our approach might be to have each student call out her or his height. Measured in inches, the heights for this group of 36 students are as follows:

> 68, 63, 67, 66, 69, 72, 62, 64, 66, 68, 66, 62, 60, 70, 71, 63,
> 67, 63, 66, 65, 69, 67, 72, 68, 74, 65, 66, 61, 64, 61, 62, 64,
> 65, 65, 71, 64

If we examine these numbers, it is very difficult to determine any patterns or relationships. They appear to be haphazard, which is not surprising since it is unlikely that seating arrangements for students are based on heights. To discover if a pattern exists, our next step should be to *organize* these data in numerical order as follows:

60, 61, 61, 62, 62, 62, 63, 63, 63, 64, 64, 64, 64, 65, 65, 65, 65, 65, 66, 66, 66, 66, 67, 67, 68, 68, 68, 68, 69, 69, 70, 71, 71, 72, 72, 74

We can now make a few meaningful observations about the data. The heights clearly go from a minimum of 60 inches to a maximum of 74 inches. Thus, the spread from the shortest to the tallest is 14 inches. This is known as the *range* for the data. In general, for any set of data,

Range = largest − smallest

Moreover, there are relatively few heights at the lower end (60 or 61) and relatively few at the upper end (70 and over). In fact, most heights seem to be clustered near the center, from about 62 to 68 inches.

When we are describing such a set of data, it is usually important to indicate how many times each measurement occurs. This is a simple task, now that our data are in numerical order. The number of times that an item occurs is called its *frequency*.

The frequency is the number of occurrences of a measurement.

We display the frequency associated with each height in Table 2.1.

TABLE 2.1

Height (inches)	Frequency
60	1
61	2
62	3
63	3
64	4
65	5
66	4
67	2
68	4
69	2
70	1
71	2
72	2
73	0
74	1

One disadvantage of this table is that there are many different entries and several have rather low frequencies. In fact, for completeness, we even included 73 with a corresponding frequency of 0. It often makes sense to combine the data in groups or *classes*. For instance, we might set up the following classes: 60–62, 63–65, 66–68, 69–71, and 72–74. These numbers represent the beginning and end of each class and so are known as the *class limits* for that class.

> The **class limits** are the lowest and highest data values for a class.

We can now create a summary table which tells us how many entries are in each class.

Class	Class Limits	Frequency
1	60–62	6
2	63–65	12
3	66–68	10
4	69–71	5
5	72–74	3
		36

Since such a table shows us how the measurements are spread out or *distributed*, we call this a *frequency distribution table* or simply a *frequency distribution*. Note that such a frequency distribution can be produced from any set of data without first ordering them.

When a frequency distribution is constructed, the following guidelines apply:

1. A frequency distribution should have a minimum of 5 classes and a maximum of 20. For small data sets, use between 5 and 10 classes. For large data sets, use up to 20 classes.
2. Each data entry must fall into one and only one class.
3. There should be no gaps. The largest value in a class should be 1 less than the smallest value in the next class. Moreover, if there are no entries for a particular class, that class must still be included with a frequency of 0.
4. Each class should have the same width. The width can be found as

$$\text{Class width} = \frac{\text{largest entry} - \text{smallest entry}}{\text{number of classes}}$$

We note that if this is not an integer, it is usually rounded *up* to the next integer. In our example, the width of each class is 3.

5. It is sometimes desirable to use some type of tail-end designation for the first and last classes, such as "under 60" and "over 74."
6. It is sometimes desirable to use *class boundaries* between successive classes, so that the above frequency distribution might appear as follows:

Class	Class Boundaries	Frequency
1	59.5–62.5	6
2	62.5–65.5	12
3	65.5–68.5	10
4	68.5–71.5	5
5	71.5–74.5	3
		36

Note that the values used for the class boundaries are created by using one more decimal place than appears in the data. For instance, if 65.5 were a data value, we could not use the above version of the frequency distribution since it would be unclear whether 65.5 fell into the second or third class. In such a case, the boundaries must be set up by using an additional decimal place.

> The **class boundaries** are the average of the upper limit of one class and the lower limit of the next class.

Using this approach, we see that each class width is now clearly equal to 3; it is not possible to misinterpret the width as 2 by calculating $62 - 60 = 2$.

There are several other useful variations on a frequency distribution. In addition to considering just the frequency of each occurrence or each class, we are also interested in the percentage of data values that fall into each class. Thus, in our height example, the first class, 60 to 62, contains 6 of the 36 members of the group. This corresponds to

$$\frac{6}{36} = \frac{1}{6} = .167 = 16.7\% \approx 17\%$$

of the data. (The symbol $\approx$ indicates "approximately equal.") This value of .167, or 16.7%, is known as the *relative frequency* for this class.

> Relative frequency for a class $= \dfrac{\text{number of entries in class}}{\text{total number of entries}}$

If we repeat this process for the remaining data, then we can extend the above frequency distribution to produce a *relative frequency distribution* for our data:

Class	Class Limits	Frequency	Relative Frequency
1	60–62	6	$\frac{6}{36} = 16.7\%$
2	63–65	12	$\frac{12}{36} = 33.3\%$
3	66–68	10	$\frac{10}{36} = 27.8\%$
4	69–71	5	$\frac{5}{36} = 13.9\%$
5	72–74	3	$\frac{3}{36} = 8.3\%$
		36	100.0%

Since we round individual percentages, it is possible that the sum of the relative frequencies or percentages might not always precisely equal 100%. Thus, we should expect approximate values in relative frequency distributions.

The next type of tabular display is known as a *cumulative frequency distribution*, which (as its name suggests) contains a column for the running cumulative total of frequencies for all classes.

> The **cumulative frequency** of a class is the total of all class frequencies up to and including the present class.

The cumulative frequency distribution for our example is as follows:

Class	Class Limits	Frequency	Cumulative Frequency
1	60–62	6	6
2	63–65	12	18
3	66–68	10	28
4	69–71	5	33
5	72–74	3	36
		36	

Notice that the final entry for the cumulative frequencies (36) is precisely equal to the total number of entries in the original set of data.

We illustrate the above ideas in the following example.

EXAMPLE 2.1 Suppose we collect data on the dollar amount that each student in a class spent on textbooks this semester. The 36 amounts are as follows:

205, 233, 195, 214, 225, 247, 198, 186, 202, 236, 227, 214, 226, 231, 257, 207, 221, 188, 218, 225, 245, 208, 197, 232, 190, 186, 204, 162, 215, 226, 186, 207, 236, 275, 220, 205

We first organize these entries in numerical order:

162, 186, 186, 186, 188, 190, 195, 197, 198, 202, 204, 205, 205, 207, 207, 208, 214, 214, 215, 218, 220, 221, 225, 225, 226, 226, 227, 231, 232, 233, 236, 236, 245, 247, 257, 275

The spread in the costs is therefore from $162 up to $275. Thus, the range is

$$275 - 162 = 113$$

To set up a frequency distribution (or a relative frequency distribution), first we have to decide which classes to use. For our data, an appropriate choice might be 160–169, 170–179, ..., 270–279. The resulting frequency distribution is shown in Table 2.2.

Note that this frequency distribution has too many classes for our rather small data set. Moreover, there are few entries in the classes at both ends. As

TABLE 2.2

Class	Class Limits	Frequency
1	160–169	1
2	170–179	0
3	180–189	4
4	190–199	4
5	200–209	7
6	210–219	4
7	220–229	7
8	230–239	5
9	240–249	2
10	250–259	1
11	260–269	0
12	270–279	1
		36

a result, it is preferable to combine the entries at the ends. We then use tail-end classes of "under 180" and "250 and over." The modified frequency distribution for these data is shown in Table 2.3. The corresponding relative frequency distribution is shown in Table 2.4.

By using our modified relative frequency distribution, it is now easy to discover useful information about the data. For instance, clearly most of the students spent between $180 and $239 on books. Further, about 64% spent between $200 and $239 on books. Similarly, we can conclude that 75% spent over $200 on books for the semester.

The corresponding cumulative frequency distribution for this set of data is shown in Table 2.5.

TABLE 2.3

Class	Class Limits	Frequency
1	Under 180	1
2	180–189	4
3	190–199	4
4	200–209	7
5	210–219	4
6	220–229	7
7	230–239	5
8	240–249	2
9	250 and over	2
		36

TABLE 2.4

Class	Class Limits	Frequency	Relative Frequency
1	Under 180	1	2.8%
2	180–189	4	11.1%
3	190–199	4	11.1%
4	200–209	7	19.4%
5	210–219	4	11.1%
6	220–229	7	19.4%
7	230–239	5	13.9%
8	240–249	2	5.6%
9	250 and over	2	5.6%
		36	100.0%

It is important to realize that there is no single correct way to set up a frequency distribution, a relative frequency distribution, or a cumulative frequency distribution for a set of data. There is always a choice as to the number of classes used. In fact, if you take the same data and organize them into frequency distributions using different numbers of classes, the results can be dramatically different, as we illustrate below.

Further, there are no restrictions regarding what class limits to use, but uniform width is usually desirable. Again, this can lead to major differences in the final tabulation, especially if many entries fall at or close to the boundaries between classes.

For instance, suppose we want to restructure the data in Example 2.1, using 6 classes instead of 9 classes. Since the range is 113, we first find the class width as follows:

$$\frac{113}{6} \approx 18.8$$

TABLE 2.5

Class	Class Limits	Frequency	Cumulative Frequency
1	Under 180	1	1
2	180–189	4	5
3	190–199	4	9
4	200–209	7	16
5	210–219	4	20
6	220–229	7	27
7	230–239	5	32
8	240–249	2	34
9	250 and over	2	36
		36	

Since the result is *not* a whole number, we round *up* to use a class width of 19. Therefore, the new frequency distribution is as follows:

Class	Class Limits	Frequency
1	162–180	1
2	181–199	8
3	200–218	11
4	219–237	12
5	238–256	2
6	257–275	2
		36

We note that these results are quite different from our earlier Table 2.2.

Exercise Set 2.1

Mastering the Techniques

In Exercises 1 to 4, construct frequency distributions with the given number of classes. (*Note:* There may be many different "right" answers depending on how you construct the classes.)

1. 54, 75, 121, 142, 154, 159, 171, 189, 203, 211, 225, 247, 251, 259, 264, 278, 290, 305, 315, 322, 355, 367, 388, 450, 490.
   Use 6 classes; then use 9 classes.
2. 210, 216, 224, 235, 248, 255, 260, 266, 270, 275, 278, 283, 285, 287, 288, 290, 293, 298, 305, 316.
   Use 5 classes; then use 7 classes.
3. 53, 47, 59, 66, 36, 69, 84, 77, 42, 57, 51, 60, 78, 63, 46, 63, 42, 55, 63, 48, 75, 60, 58, 80, 44, 59, 60, 75, 49, 63.
   Use 6 classes; then use 10 classes.
4. 2.8, 3.5, 7.2, 5.8, 6.3, 4.1, 5.7, 8.2, 2.3, 4.4, 7.1, 8.0, 6.8, 5.2, 4.3, 3.0, 3.6, 5.4, 6.3, 6.6, 5.7, 8.2, 4.9, 6.0, 7.2.
   Use 7 classes; then use 6 classes.
5. Convert the frequency distribution in Exercise 1 to a relative frequency distribution and a cumulative frequency distribution.
6. Convert the frequency distribution in Exercise 2 to a relative frequency distribution and a cumulative frequency distribution.
7. Convert the frequency distribution in Exercise 3 to a relative frequency distribution and a cumulative frequency distribution.
8. Convert the frequency distribution in Exercise 4 to a relative frequency distribution and a cumulative frequency distribution.
9. Using the relative frequency distribution for Exercise 5, what percentage of the entries are **(a)** above 150; **(b)** between 100 and 350?
10. Using the relative frequency distribution for Exercise 6, what percentage of the entries are **(a)** below 250; **(b)** between 250 and 299?

Applying the Concepts

11. A group of 25 students in a high school class has the following scores on the Scholastic Aptitude Test (SAT): 860, 940, 1120, 900, 840, 980, 1050, 1220, 860, 770, 1010, 870, 890, 910, 930, 1040, 1280, 1020, 970, 1330, 890, 980, 1260, 980, 760. Construct a frequency distribution for these data with 6 classes.
12. Some "1-pound" bags of carrots in a supermarket are weighed, and the following results, in ounces, are found: 17, 19, 16, 20, 17, 19, 22, 16, 17, 17, 19, 16, 15, 20, 18, 22, 16, 17, 18, 17, 16, 19, 18, 16, 16, 19, 18, 15, 20, 18, 16, 18, 16, 19, 15, 16, 18, 16, 15, 18. Construct a frequency distribution for these data using 8 classes. What percentage of the bags are underweight?
13. The weights in pounds of a group of people signing up at a health club are 135, 175, 166, 148, 183, 206, 190, 128, 147, 156, 166, 174, 158, 196, 120, 165, 189, 174, 148, 225, 192, 177, 154, 140, 180, and 172. Construct the relative and cumulative frequency distributions for these data using 6 classes; using 7 classes. What percentage of the group weighs over 180 pounds?
14. The following are the speeds, in miles per hour, of a group of cars on a highway as measured with a radar gun: 58, 62, 59, 53, 61, 55, 57, 54, 59, 53, 66, 60, 58, 60, 61, 58, 56, 60, 58, 62, 57, 55, 53, 55, 61, 57, 52, 58, 49, 54, 52, 55, 57, 60, 64. Construct the relative and cumulative frequency distributions for these data using 5 classes. What percentage of the cars are traveling over the legal speed limit of 55 miles per hour?
15. The following are the heights in meters achieved by the athletes competing in the pole vault event in a track and field competition: 5.2, 5.6, 4.9, 5.3, 5.8, 4.8, 5.0, 5.2, 5.4, 4.8, 4.4, 5.1, 5.5, 4.9, 5.2, 5.7, 5.0, 5.3, 4.9, 4.8. Construct the relative frequency distribution for these data using 5 classes. What percentage of the vaulters cleared 5.5 meters? What percentage did not clear 5 meters?

2.2 Using Computers to Organize Data

In the last section, we discussed ways to organize data which help simplify the data analysis. While the ideas involved in organizing data are simple in principle, they can become quite tedious to implement for large data sets. Although most of our work in this book involves relatively small data sets (say, at most 50 items), in practice, sets of data often contain tens of thousands of entries. Obviously, for such large data sets, even the simplest procedure becomes overwhelming if it must be done by hand. Thus, the use of computers has become an essential element in the study of statistics. Even with the relatively small data sets we will encounter, our work can usually be simplified when a computer program performs the required calculations.

We now indicate how the computer can be used to perform the statistical calculations described in Section 2.1. We strongly recommend that you review the computer instructions from Chapter 1 and that you keep them handy whenever you use the computer.

Begin by loading your statistics disk. When the main menu appears, select Program 1: Statistical Analysis of Data. (On IBM compatibles just use the arrow keys to highlight the desired program and then press the Enter key. On the Macintosh, use the mouse to highlight the desired program name and then press or click the mouse control.) This program will perform a wide variety of statistical operations. In fact, it will do almost everything we discuss in this and the following chapter on any set of data with up to 500 entries.

To use this program, you must first enter your data. You can either enter the data directly from the keyboard by typing in the entries or load a set of data which was previously saved on the disk. Since we have not saved any data previously, indicate that you will enter the data directly. That is, on an IBM compatible, in response to the display

```
YOU MAY ENTER YOUR DATA

1  FROM THE KEYBOARD NOW
2  FROM A DISK FILE

WHAT IS YOUR CHOICE ▶
```

type the numeral 1 and press the Enter key. On the Macintosh, the data entry screen will appear after you highlight and click on the OK button. To load an existing data file that you saved, select the file option from the menu box that appears at the top of the screen.

We demonstrate the use of this program for the following set of data, which represents the number of hours per week that employees of a fast-food restaurant work:

20, 25, 35, 30, 50, 30, 16, 22, 45, 40, 30, 24, 36, 44, 52, 33, 42, 80, 38, 28, 18, 20, 45, 33, 28

To enter the data, simply type each value and then press the Enter key after each. (Do not worry about any errors you may make in typing these entries; we will describe how to correct mistakes below.) After you have entered all

FIGURE 2.1

```
YOU MAY:

 1  EDIT THIS SET OF DATA
 2  SEE MEAN AND STANDARD DEVIATION
 3  SEE AN ORDERED LIST AND MEDIAN
 4  SEE A FREQUENCY DISTRIBUTION
 5  CONSTRUCT A HISTOGRAM
 6  CONSTRUCT A STEM AND LEAF PLOT
 7  CONSTRUCT A BOX AND WHISKER PLOT
 8  SAVE RESULTS TO DISK FILE
 9  ENTER A NEW SET OF DATA
10  RETURN TO THE MAIN MENU

WHAT IS YOUR CHOICE ▶
```

the data, type the number 9999 followed by Enter. The number 9999 indicates the end of the data.

At this point, the program displays an options menu which provides you with a choice of available options for dealing with the data. See Figure 2.1 for a display of options on an IBM compatible.

For now, we discuss only those options that relate to topics already considered. The other options are discussed in later sections.

Option 1 gives you the opportunity to edit the data. This allows you to correct any entries you may have typed in by mistake. More importantly, you can also change any of the data values, add new values, or delete existing values to see the effects of changes in the data on different statistical procedures.

Option 3 organizes your data in numerical order. You can select either ascending or descending order. The program will also print values for the median and other quantities discussed in Chapter 3.

Option 4 constructs a relative frequency distribution for the data.

Option 8 allows you to save the present set of data as a disk file for future use.

Option 9 allows you to enter a new set of data.

Option 10 ends the program by returning you to the main menu for additional program choices or to exit the statistics package altogether.

Although the above descriptions are intended to suggest that this program has many different capabilities and is very powerful for statistical computations, it is very simple to use. We illustrate its use with the data set you just entered:

20, 25, 35, 30, 50, 30, 16, 22, 45, 40, 30, 24, 36, 44, 52, 33, 42, 80, 38, 28, 18, 20, 45, 33, 28

First, suppose that the number 80 listed above is actually wrong—it should be 60. On the Macintosh, use the mouse to highlight the incorrect number. Then type the correct number and press Return. On an IBM compatible, select Option 1 from the option menu to edit the data. You will then be asked

```
YOU CAN DO THE FOLLOWING:

1 ADD EXTRA DATA POINTS
2 CHANGE DATA POINTS
3 DELETE DATA POINTS

WHAT IS YOUR CHOICE <1 − 3> ▶
```

Since we want to change 80 to 60, we respond by choosing Option 2. The program then lists the first 20 numbers in the given order and asks:

```
DO YOU WANT A CHANGE IN THIS GROUP ▶
```

Since the entry we want to change, 80, is item number 18 in this group, the appropriate response is Y or y. The program then asks

> WHICH ITEM NUMBER SHOULD CHANGE ▶

You should respond with 18 followed by the Enter key. The program then asks

> WHAT IS THE NEW VALUE ▶

You should respond by entering 60 followed by the Enter key. The program then shows the revised list and again asks if you want a change. Since you have made the desired change, simply type N or n and press Enter. The program will return to the options menu. Of course, if you want to make other changes in the data, your response would be Y or y.

Incidentally, if you type Y by mistake when you intend N, so that the program asks which entry to change, the easiest solution is to select any entry whatsoever and simply enter it again unchanged.

The other editing features—adding a new data point or deleting a data point—are treated in a very similar fashion. Feel free to experiment with the program, making any changes you want.

Suppose that you have made the desired change in the data set and have returned to the options menu. Select Option 3 now to have the data put into numerical order. The program will then ask

> YOU CAN HAVE THE DATA IN:
> 1. ASCENDING ORDER
> 2. DESCENDING ORDER
>
> WHAT IS YOUR CHOICE ▶

Respond by choosing Option 1. The computer displays the data in ascending order in a single column until the screen is filled, followed by a message

> <PRESS ANY KEY TO CONTINUE>

as shown in Figure 2.2. When you press any key, the program will display the next group of entries. The process continues until all values have been listed on the screen. A final screen displays the median and other information about the data to be discussed in Chapter 3.

IBM PC or compatible users can get hard copy of any page by pressing the PrtSc key or the Shift and PrtSc keys together. Macintosh users can get hard copy on an ImageWriter printer by pressing and holding the ⌘ key

FIGURE 2.2

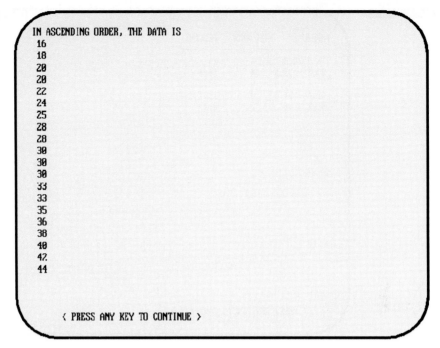

and then pressing Shift and 4 together. A printable MacPaint document can be created for a laser printer by simultaneously pressing three keys: the ⌘, Shift, and 3.

After all these displays, you can return to the options menu by pressing any key. To construct a frequency distribution for your data, select Option 4. You will be asked how many classes you want (between 5 and 20). Enter 5, say. You will then be asked for the lowest and highest values you want to use. With the above set of data, we might use the smallest and largest values, a minimum of 16 to a maximum of 60. The corresponding display is shown in Figure 2.3. When you are ready, simply press any key on an IBM PC compatible or click on the appropriate window on the Macintosh to return to the options menu. You may want to experiment with this set of data to see the effects of using different numbers of classes or different values for the minimum and maximum. To change the way in which the frequency distribution is constructed, choose Option 4 again. This time, for instance, request six classes ranging from a minimum of 10 to a maximum of 69. The result is shown in Figure 2.4.

Suppose that you have temporarily finished with this set of data, but you want to save it for future use as a file on your disk. (In this way, you will not have to type in all the entries again.) From the options menu, select Option 8. The program will ask you for the name of this data set. On an IBM compatible, the name can be any combination of up to 8 letters or numbers, provided the first character is a letter. For instance, you might choose to call this data file HOURS or HOURS93. However, the name cannot start with a numeral

FIGURE 2.3

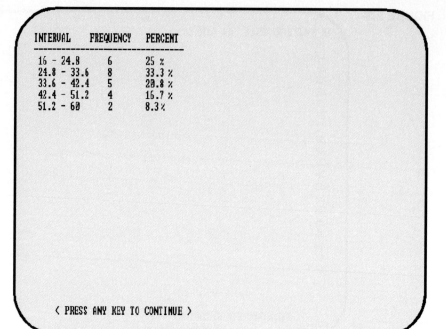

FIGURE 2.4

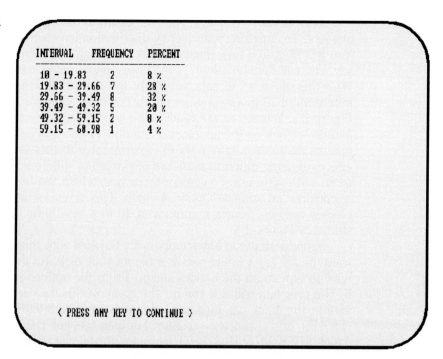

or a symbol. On the Macintosh, names can be as long as you like and can include any character.

CAUTION Do not use the same name for different data files, or else the computer will destroy the old set in the process of saving the new set.

After the program saves the data file consisting of the actual data entries as well as some additional statistical facts about the data set, you will return to the options menu.

A computer disk can store only a finite number of files and programs. If you try to store too many data files, the program may display the error message THERE IS NO MORE SPACE ON DISK or DISK FULL. The options menu is then displayed. You will have to put another formatted disk into the disk drive before you can store your data file.

If you have finished, return to the main menu by selecting Option 10. However, if you want to work with a new set of data, select Option 9. Again, you can either enter the new set of data from the keyboard, as discussed above, or load the data from an existing data file. For example, if you want to load the data file HOURS that you previously saved, then type the name HOURS in response to a prompt and press the Enter key. The program will load the data file and you can choose other options to use with that set of data again.

This program and many others on your disk are intended to make life much easier for you in terms of statistical calculations. Since the programs do all the computations, you have the opportunity to sit back and think about the significance of the statistical concepts and methods. Hopefully, they will help bring you to a deeper understanding of statistics and its uses.

However, you should not become so dependent on these programs that you don't bother to develop the ability to perform these procedures by hand with paper, pencil, and a handheld calculator. It is important to learn not only when to use the computer as a very useful statistical tool, but also when not to rely on it exclusively, preventing you from mastering the necessary statistical skills.

Exercise Set 2.2

Computer Applications

For each of the following sets of data, use Program 1: Statistical Analysis of Data to construct a relative frequency distribution for the given numbers of classes. In each case, experiment with the effects of using different minimum and maximum values. In those cases where the data are not arranged in order, have the program order them as well.

1. 54, 75, 121, 142, 154, 159, 171, 189, 203, 211, 225, 247, 251, 259, 264, 278, 290, 305, 315, 322, 355, 367, 388, 450, 490.
Use 6 classes; then use 9 classes.

2. 210, 216, 224, 235, 248, 255, 260, 266, 270, 275, 278, 283, 285, 287, 288, 290, 293, 298, 305, 316.

   Use 5 classes; then use 7 classes.

3. 53, 47, 59, 66, 36, 69, 84, 77, 42, 57, 51, 60, 78, 63, 46, 63, 42, 55, 63, 48, 75, 60, 58, 80, 44, 59, 60, 75, 49, 63.

   Use 6 classes; then use 10 classes.

4. 2.8, 3.5, 7.2, 5.8, 6.3, 4.1, 5.7, 8.2, 2.3, 4.4, 7.1, 8.0, 6.8, 5.2, 4.3, 3.0, 3.6, 5.4, 6.3, 6.6, 5.7, 8.2, 4.9, 6.0, 7.2. Use 7 classes; then use 6 classes.

5. 860, 940, 1120, 900, 840, 980, 1050, 1220, 860, 770, 1010, 870, 890, 910, 930, 1040, 1280, 1020, 970, 1330, 890, 980, 1260, 980, 760.

   Use 6 classes; then use 8 classes.

6. 17, 19, 16, 20, 17, 19, 22, 16, 17, 17, 19, 16, 15, 20, 18, 22, 16, 17, 18, 17, 16, 19, 18, 16, 16, 19, 18, 15, 20, 18, 16, 18, 16, 19, 15, 16, 18, 16, 15, 18.

   Use 8 classes.

7. 135, 175, 166, 148, 183, 206, 190, 128, 147, 156, 166, 174, 158, 196, 120, 165, 189, 174, 148, 225, 192, 177, 154, 140, 180, 172.

   Use 6 classes; then use 7 classes.

In Exercises 8 to 10, refer to Appendix I, which lists the leading batters in the National League at the end of the 1992 baseball season.

8. Construct a relative frequency distribution for the batting averages of these players, using different numbers of classes and appropriate minimum and maximum values. Then save this set of data as a data file under the name BATAVG for later use.

9. Consider the number of home runs hit by each player. Use Program 1: Statistical Analysis of Data to put the data in numerical order, and then construct a variety of relative frequency distributions using different numbers of classes. Save this set of data as a data file under the name HOMERUN.

10. Consider the number of hits by each player. Use the program to put the data in numerical order, and then construct a variety of relative frequency distributions using different numbers of classes. Save this set of data as a data file under the name HITS.

11. Appendix II gives the maximum temperature in major cities around the world on a given day. Use Program 1: Statistical Analysis of Data to put these values in order, and then display the data in several different relative frequency distributions. Save these data as a data file named TEMPS.

12. Appendix III gives data on the prices of stocks in a variety of major corporations on one recent trading day on the New York Stock Exchange. The second-to-the-last column in the table gives the closing price of each stock. Select a sample consisting of every tenth closing price in the list. Use Program 1: Statistical Analysis of Data to put these closing prices in order, and then display the data in a relative frequency distribution. Save these data as a data file named STOCKS.

2.3 Graphical Displays of Data

In Sections 2.1 and 2.2, we saw how important it is to organize a set of data and to display it in a form that simplifies our analysis. In those sections, we ordered our data set and summarized it in tabular form as a frequency distribution, a relative frequency distribution, or a cumulative frequency distribution. Another approach is to use a graphical display to give a visual dimension to the data. We discuss some of the most common and useful types of graphs in this section.

We begin with the *bar chart*, or *bar graph*. Suppose we use the data on student heights (rounded to the nearest inch) from Section 2.1. The relative frequency distribution for these data is as follows:

Class	Class Limits	Frequency	Relative Frequency
1	60–62	6	$\frac{6}{36}$ = 16.7%
2	63–65	12	$\frac{12}{36}$ = 33.3%
3	66–68	10	$\frac{10}{36}$ = 27.8%
4	69–71	5	$\frac{5}{36}$ = 13.9%
5	72–74	3	$\frac{3}{36}$ = 8.3%

To construct a bar chart, we start with horizontal and vertical axes. We label the quantity being studied (the height of each student) horizontally from left to right. The markings along the horizontal axis should correspond to the limits of the classes in the above frequency distribution. The corresponding frequency in each class is measured vertically upward. A vertical bar is then drawn across each class interval with height equal to the frequency for that class. The bar graph corresponding to the data in the above frequency distribution is shown in Figure 2.5.

We could also have drawn a bar chart by using the relative frequencies instead of the frequencies for each class. The relative frequencies are measured along the vertical axis as percentages, as shown in Figure 2.6.

It is also possible to construct such bar charts with horizontal rather than vertical bars. (See Figure 2.7.)

FIGURE 2.5

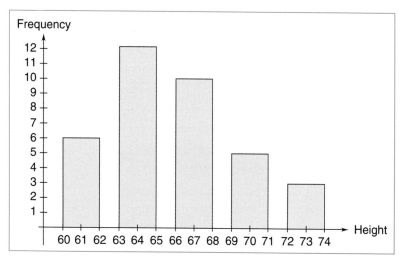

FIGURE 2.6

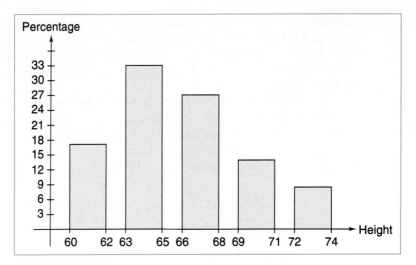

Any of these displays can be used to analyze our data. For instance, we can quickly determine that the most common heights for the people surveyed lie between 63 and 68 inches.

We note that a bar chart is especially useful for displaying nonnumeric data such as a company's sales on a monthly basis.

Another important graphical display is the *histogram*, as shown in Figure 2.8. In the bar graph (Figure 2.5), the first bar represents heights in the class between 60 and 62 inches while the second bar represents heights from 63 to 65 inches. In the histogram (Figure 2.8), the gap between 62 and 63 is replaced by a single vertical line between the first two boxes. However, it is not clear what the vertical line should represent—is it 62 or 63 or what? To

FIGURE 2.7

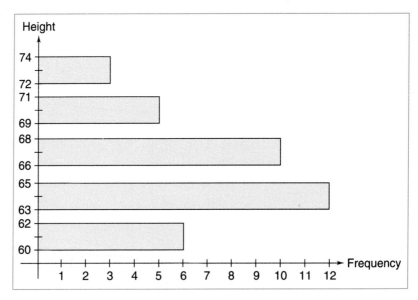

FIGURE 2.8

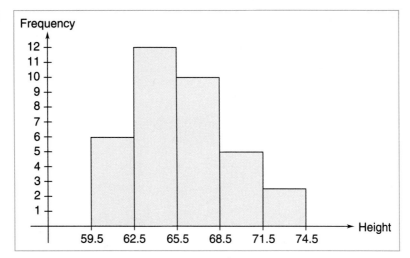

resolve this ambiguity, we agree that the vertical line represents $62\frac{1}{2} = 62.5$, which is the class boundary between the two classes. In the same way, the next vertical line represents 65.5, the one after that is at 68.5, and so forth. To complete this arrangement, we agree that the left-hand line represents 59.5 and that the right-hand line is at 74.5. In this way, it is perfectly clear which data values fall into which box of the histogram. Moreover, the width of each box represents the same 3-inch spread.

If the data involve decimals rather than integers, we modify this procedure slightly. For example, suppose that the data range from 4.13 to 4.79 and we want to display them in 6 classes. The width of each class is found from

$$\frac{4.79 - 4.13}{6} = \frac{.66}{6} = .11$$

We therefore use class limits of 4.13 to 4.23, then 4.24 to 4.34, and so forth. In this case, the resulting class boundaries are 4.125, 4.235, 4.345, and so on. Notice that the class boundaries (say 4.235) are determined by averaging successive pairs of upper and lower class limits (4.23 and 4.24).

Another type of graphical display is the *frequency polygon*. To construct this type of graph, we first determine the measurement corresponding to the midpoint of each class. This value is called the *class mark*, or *class midpoint*, and is given by

$$\text{Class mark} = \frac{\text{lower limit} + \text{upper limit}}{2}$$

Thus, in the class 59.5 to 62.5, the class mark is

$$\text{Class mark} = \frac{59.5 + 62.5}{2} = \frac{122}{2} = 61$$

FIGURE 2.9

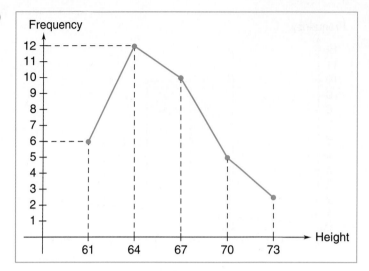

Similarly, the class mark for the next class, 62.5 to 65.5, is 64, and so forth. We now plot these points and connect them with a series of straight-line segments, as shown in Figure 2.9.

Finally, it is common practice to "tie down" the two ends of this graph to the horizontal axis. We pretend there are two additional classes—one higher, the other lower—both with frequencies of 0. When we connect these final two points, we obtain the frequency polygon for the data, as shown in Figure 2.10.

As an alternate method of constructing a frequency polygon, you could start with a histogram for the data and mark off the center point at the top of each box, as shown in Figure 2.11. If you connect these points with a series of line segments and "tie down" the two ends, the resulting graph is the

FIGURE 2.10

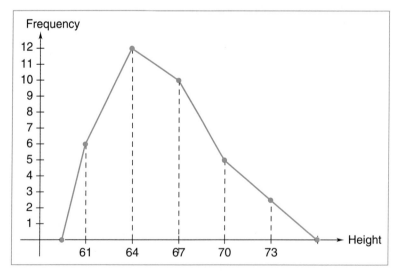

FIGURE 2.11

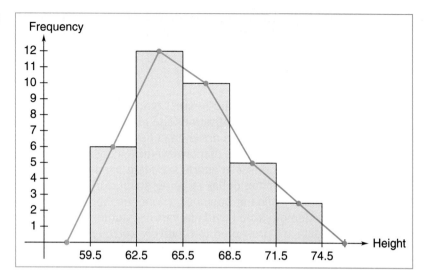

frequency polygon. However, it should then be redrawn without showing the underlying histogram so that it looks like that in Figure 2.10.

The final graphical display that we consider is the *pie chart* or *circle graph*. It is used to display relative frequencies rather than actual frequencies for the data. We draw a circle and then divide it into a series of wedges or slices to represent each class in the relative frequency distribution. The size of each slice is proportional to the percentage of the data that fall into the corresponding class. For the data on student heights, there are five classes. Since the first class (60 to 62) contains 16.7% of the data, the corresponding slice contains 16.7% of the circle. Similarly, the second class (63 to 65) contains 33.3% of the data, and the corresponding slice contains 33.3% of the circle, and so forth. The pie chart for these data is shown in Figure 2.12.

Although the pie chart is easy to interpret, it is without a doubt the most complicated of the graphical displays to construct accurately. We have to determine how large each slice is corresponding to predetermined percentages

FIGURE 2.12

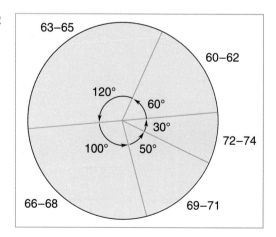

of the circle. Thus, since our first slice corresponds to 16.7% of the data, we must find 16.7% of the full 360° in the circle. This is

$$16.7\% \text{ of } 360° = .167 \times 360°$$
$$= 60.12° \approx 60°$$

which is rounded to the nearest whole degree. To draw an angle of 60° requires the use of a protractor. The angles for the other slices are calculated in a similar fashion and are then constructed, also with the aid of a protractor.

The four graphical displays we have considered are used widely for many different situations. Pie charts are often used to show the breakdown in budgets (how your tax dollar is being spent, for instance). Frequency polygons are routinely used in business, economic, and scientific applications, particularly to demonstrate trends in various studies. As such, they tend to be used when data are presented to relatively knowledgeable readers. Bar charts and histograms are used in a wide variety of applications and are often found in newspaper and magazine articles. For our purposes, the histogram proves to be the most useful.

EXAMPLE 2.2 Construct a bar chart, histogram, frequency polygon, and pie chart for the following set of data on the number of videotapes owned by 80 families who have video cassette recorders (VCRs).

Class	Class Limits	Frequency	Relative Frequency
1	1–10	5	6%
2	11–20	29	36%
3	21–30	21	26%
4	31–40	16	20%
5	41–50	7	9%
6	51–60	2	3%

FIGURE 2.13

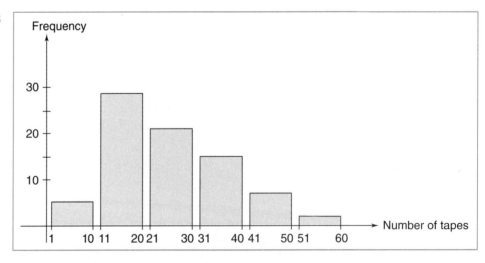

FIGURE 2.14

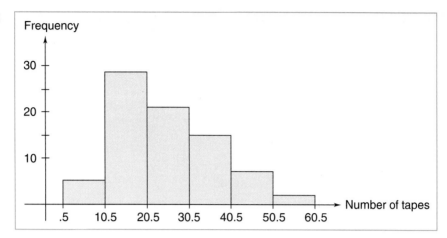

SOLUTION The bar chart for these data is particularly easy to construct and is shown in Figure 2.13.

To construct the histogram, we first determine the class boundaries which locate the vertical lines in each box. These are .5, 10.5, 20.5, 30.5, and so forth. The histogram for these data is shown in Figure 2.14.

To construct the frequency polygon, we need to determine the class marks for each class. These are just the averages of the class boundaries of each class interval. We therefore obtain for the first class mark

$$\frac{.5 + 10.5}{2} = \frac{11}{2} = 5.5$$

and so forth for the others. We now plot these values with the corresponding heights, connect them with line segments, and tie down the two ends to obtain the frequency polygon shown in Figure 2.15.

Finally, to construct the pie chart, we must determine what percentage of the total circle corresponds to each class. Since the first class (1 to 10)

FIGURE 2.15

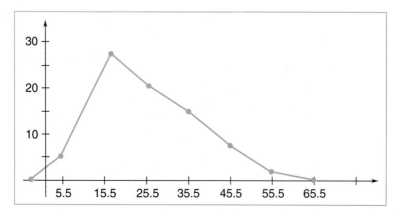

FIGURE 2.16

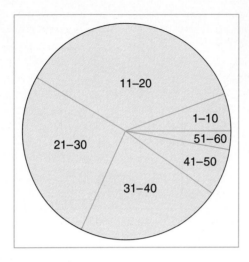

consists of 6% of the total data, we obtain

$$6\% \text{ of } 360° = .06 \times 360° = 21.6° \approx 22°$$

Similarly, for the second class,

$$36\% \text{ of } 360° = .36 \times 360° = 129.6° \approx 130°$$

and so on for the other classes. The six angles we need are then approximately 22°, 130°, 94°, 72°, 32°, and 11°. Notice that, because of the individual roundings, the sum of these six angles is actually 361°. The resulting pie chart is shown in Figure 2.16. ∎

Furthermore, we constructed this pie chart by drawing the slices in the same order as the classes in the relative frequency distribution. This is not necessary, and often pie charts are drawn with the slices displayed in order of increasing or decreasing size.

Computer Corner

Since the histogram is the most important graphical display in this course, we have provided a routine for constructing it as part of Program 1: Statistical Analysis of Data. Select this program from the main menu and enter your data, either from the keyboard or from a previously saved data file. From the options menu, select Option 5, CONSTRUCT A HISTOGRAM. Its structure is analogous to the option for constructing a frequency distribution for the data. The program asks you how many classes you want for your data—this can be anything from 5 up to 20. It then asks for the minimum and maximum values you want to use in constructing the histogram. The resulting histogram is then drawn on the screen.

At this point, you can transfer to an accompanying text screen which shows the relative frequency distribution for the data. In this way, you will see not only the frequencies, but also the class limits used for the histogram. On an IBM compatible, from the text screen, if you press the space bar, you transfer to the options menu to make another choice if you want. If you press any key other than the space bar, the histogram is displayed again. You can switch back and forth indefinitely between the histogram and the table by successively pressing any key other than the space bar. On the Macintosh, you click on the appropriate window to move back or forth.

Exercise Set 2.3

Mastering the Techniques

In Exercises 1 to 7, refer to the corresponding exercises in Exercise Set 2.1 to see the relative frequency distribution you constructed. Use these results to construct (a) a bar graph, (b) a histogram, (c) a frequency polygon, and (d) a pie chart.

1. 54, 75, 121, 142, 154, 159, 171, 189, 203, 211, 225, 247, 251, 259, 264, 278, 290, 305, 315, 322, 355, 367, 388, 450, 490.
   Use 6 classes; then use 9 classes.
2. 210, 216, 224, 235, 248, 255, 260, 266, 270, 275, 278, 283, 285, 287, 288, 290, 293, 298, 305, 316.
   Use 5 classes; then use 7 classes.
3. 53, 47, 59, 66, 36, 69, 84, 77, 42, 57, 51, 60, 78, 63, 46, 63, 42, 55, 63, 48, 75, 60, 58, 80, 44, 59, 60, 75, 49, 63.
   Use 6 classes; then use 10 classes.
4. 2.8, 3.5, 7.2, 5.8, 6.3, 4.1, 5.7, 8.2, 2.3, 4.4, 7.1, 8.0, 6.8, 5.2, 4.3, 3.0, 3.6, 5.4, 6.3, 6.6, 5.7, 8.2, 4.9, 6.0, 7.2.
   Use 7 classes; then use 6 classes.
5. 860, 940, 1120, 900, 840, 980, 1050, 1220, 860, 770, 1010, 870, 890, 910, 930, 1040, 1280, 1020, 970, 1330, 890, 980, 1260, 980, 760.
   Use 7 classes; then use 8 classes.
6. 17, 19, 16, 20, 17, 19, 22, 16, 17, 17, 19, 16, 15, 20, 18, 22, 16, 17, 18, 17, 16, 19, 18, 16, 16, 19, 18, 15, 20, 18, 16, 18, 16, 19, 15, 16, 18, 16, 15, 18.
   Use 8 classes.
7. 135, 175, 166, 148, 183, 206, 190, 128, 147, 156, 166, 174, 158, 196, 120, 165, 189, 174, 148, 225, 192, 177, 154, 140, 180, 172.
   Use 6 classes; then use 7 classes.

Applying the Concepts

8. In a high school class, 25 students have the following scores on the SAT: 860, 940, 1120, 900, 840, 980, 1040, 1220, 860, 770, 1010, 870, 890, 910, 930, 1040, 1280, 1020, 970, 1330, 890, 980, 1260, 980, 760. Use a histogram, frequency polygon,

or pie chart for this data set based on 6 classes to determine the percentage who scored over 1000.

9. A group of "1-pound" bags of carrots in a supermarket are weighed, and the following results, in ounces, are found: 17, 19, 16, 20, 17, 19, 22, 16, 17, 17, 19, 16, 15, 20, 18, 22, 16, 17, 18, 17, 16, 19, 18, 16, 16, 19, 18, 15, 20, 18, 16, 18, 16, 19, 15, 16, 18, 16, 15, 18. Use a histogram for this set of data using 8 classes to determine the percentage of bags that are underweight.

10. The weights, in pounds, of a group of people signing up at a health club are as follows: 135, 175, 166, 148, 183, 206, 190, 128, 147, 156, 166, 174, 158, 196, 120, 165, 189, 174, 148, 225, 192, 177, 154, 140, 180, 172. Use the histogram or pie chart for this data set using 7 classes to find the percentage who weigh between 150 and 195 pounds.

11. The following are the speeds, in miles per hour, of a group of cars on a highway as measured with a radar gun: 58, 62, 59, 53, 61, 55, 57, 54, 59, 53, 66, 60, 58, 60, 61, 58, 56, 60, 58, 62, 57, 55, 53, 55, 61, 57, 52, 58, 49, 54, 52, 55, 57, 60, 64. Construct the histogram and frequency polygon for this set of data using 5 classes. Use them to find the percentage who exceed the 55 miles per hour speed limit.

Computer Applications

12. Use the set of data on batting averages from Exercise 8 of Section 2.2 that you saved as a disk file named BATAVG to construct a histogram for the data. In particular, select Program 1: Statistical Analysis of Data, load the data file, and then select Option 5 (the histogram). Use different numbers of classes and appropriate minimum and maximum values.

13. Repeat Exercise 12 using the number of home runs hit by each player in the data file HOMERUN that you created in Exercise 9 of Section 2.2.

14. Repeat Exercise 12 for the data set on the number of hits by each player that you saved under the name HITS.

15. Repeat Exercise 12 for the data set on the maximum temperature in major cities around the world that you saved under the name TEMPS.

16. Repeat Exercise 12 for the data on the prices of stocks of major corporations on the New York Stock Exchange that you saved under the name STOCKS.

2.4 Stem-and-Leaf Plots

The techniques for displaying data that we considered in Section 2.3 all have one major disadvantage. Although they are all excellent methods for summarizing and displaying sets of data, they lose the individual data entries. That is, while we might know that there are 17 items in a certain class from 39.5 to 49.5 in a frequency distribution, we have absolutely no idea what values those 17 items actually represent. Consequently, we do not know how many 40s there are, how many 41s, etc., just by looking at the frequency distribution.

There is another technique for displaying data, the *stem-and-leaf plot*, which combines the visual impact of the histogram or bar chart with the detail of the original list of data entries. We illustrate a stem-and-leaf plot with an example. Suppose the students in a certain class receive the following test scores on an examination:

52, 57, 59, 61, 64, 64, 66, 67, 70, 72, 75, 77, 78, 78, 81, 83,
84, 84, 85, 86, 88, 89, 90, 90, 92, 94, 95, 95, 98, 99

We could organize this data set into a frequency distribution:

Class	Frequency
50–59	3
60–69	5
70–79	6
80–89	8
90–99	8

or a (horizontal) bar chart, shown in Figure 2.17. In either case, the original data are not explicitly shown. (The vertical entries are given in an unusual orientation intentionally for reasons you will see below.)

In a stem-and-leaf plot, we consider all entries in each of the classes. Let's look at the group of entries in the 50s: 52, 57, and 59. We separate the last

FIGURE 2.17

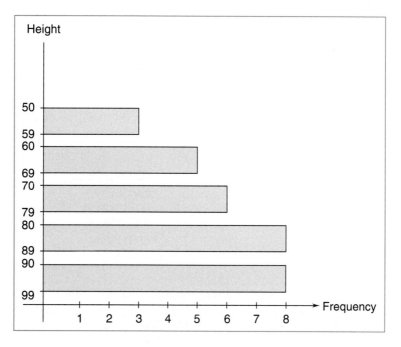

digit of each entry (the 2, 7, and 9) from the primary number, 50. We do the same for the group in the 60s: 61, 64, 64, 66, and 67; the last digits are 1, 4, 4, 6, and 7. We continue this process and display the results as follows:

50s:	2, 7, 9
60s:	1, 4, 4, 6, 7
70s:	0, 2, 5, 7, 8, 8
80s:	1, 3, 4, 4, 5, 6, 8, 9
90s:	0, 0, 2, 4, 5, 5, 8, 9

To streamline this display, we replace the 0s in 50, 60, etc., with an asterisk as a placeholder and so obtain the stem-and-leaf plot

5*	279
6*	14467
7*	025788
8*	13445689
9*	00245589

We note that the leading digits (5, 6, 7, 8, and 9) are called the *stems* and the trailing digits to the right are the *leaves*.

The length of each row in the stem-and-leaf plot is equal to the frequency of the class. Thus, the result combines the visual effect of a horizontal bar chart (Figure 2.17) with the detail of a list of the individual entries. However, the bar chart or the frequency distribution would tell us only that there are three entries in the 50s. The stem-and-leaf plot allows us to observe precisely what these three entries are.

Note that while the leaves in each stem do not have to be arranged in numerical order, it is certainly preferable to do that first.

Furthermore, if there are a reasonably large number of entries in each class, then it is sometimes convenient to break down the data into groups of 5 rather than groups of 10. For example, we might want to classify a set of data into groups of 50 to 54, 55 to 59, 60 to 64, 65 to 69, etc. To do this with placeholders, we introduce a second symbol in addition to the asterisk. Thus, we might use a dot as follows:

5*	000111223334
5.	555667799
6*	0011223444
6.	5566778889
7*	000113344
⋮	⋮

Furthermore, if the numbers are larger, then the stems can contain several digits while the leaves consist of just the trailing digit. We illustrate this in the following example.

EXAMPLE 2.3 Construct the stem-and-leaf plot based on the following entries: 110, 112, 114, 114, 117, 119, 122, 125, 126, 128, 129, 133, 134, . . . , 190, 193, 197.

SOLUTION The result is

$$
\begin{array}{r|l}
11* & 024479 \\
12* & 25689 \\
\vdots & \vdots \\
19* & 037
\end{array}
$$

■

The above sets of data were particularly suitable to a stem-and-leaf plot because it was easy to group the data in multiples of 10. For data that contain decimal values such as 4.7, we construct the stem-and-leaf plot as shown in the next example.

EXAMPLE 2.4 The reading scores on a standardized test for a group of third-grade students are as follows:

1.6, 1.9, 2.1, 2.4, 2.6, 2.7, 2.7, 2.8, 2.8, 2.8, 2.9, 2.9, 2.9, 3.0, 3.0, 3.0, 3.0, 3.1, 3.1, 3.2, 3.2, 3.3, 3.3, 3.3, 3.4, 3.4, 3.5, 3.6, 3.7, 3.7, 3.8, 4.0, 4.2, 4.5, 4.6, 5.1, 5.5, 5.7, 6.2, 6.3

Construct the stem-and-leaf plot for these data.

SOLUTION For this data set, the stem is the whole number including the decimal point while the leaf is the trailing decimal digit. Therefore, the stem-and-leaf plot is

$$
\begin{array}{r|l}
1.* & 69 \\
2.* & 14677888999 \\
3.* & 000011223334456778 \\
4.* & 0256 \\
5.* & 157 \\
6.* & 23
\end{array}
$$

■

Unfortunately, not all sets of data can be organized into a stem-and-leaf plot. First, there should not be too much spread in the data. For example, if the data values ranged from .5 to 1122, then it would be difficult to organize them appropriately in such a display. Similarly, if there is very little spread, such as a set of integers between 40 and 46, then the stem-and-leaf plot is inappropriate. Further, the numbers in the data should not be extremely large. For instance, if the entries were in the hundreds of thousands, such as 235,024 and 482,163, then just separating the last digit would be meaningless. Finally, the set of data should be relatively small. For instance, if there were several thousand entries, the resulting stem-and-leaf plot would be impractically large and therefore useless as a display.

Computer Corner

You have probably noticed that a stem-and-leaf plot is one of the options in Program 1: Statistical Analysis of Data. It is very simple to use. Enter a set of data, either from the keyboard or from a disk file. From the options menu, select Option 6, CONSTRUCT A STEM-AND-LEAF PLOT, and the program displays the data as a stem-and-leaf plot. When the display is complete, you can return to the options menu by pressing any key or by clicking on the appropriate window.

Exercise Set 2.4

Mastering the Techniques

Construct the stem-and-leaf plot for each of the following sets of data.

1. 53, 47, 59, 66, 36, 69, 84, 77, 42, 57, 51, 60, 78, 63, 46, 63, 42, 55, 63, 48, 75, 60, 58, 80, 44, 59, 60, 75, 49, 63
2. 17, 39, 26, 20, 17, 49, 22, 36, 17, 17, 39, 26, 15, 20, 18, 22, 46, 17, 28, 17, 36, 19, 18, 46, 16, 19, 28, 15, 20, 48, 16, 18, 36, 29, 15, 16, 28, 36, 15, 18
3. 210, 216, 224, 235, 248, 255, 260, 266, 270, 275, 278, 283, 285, 287, 288, 290, 293, 298, 305, 316
4. 960, 944, 1026, 905, 940, 980, 1044, 1020, 960, 975, 1005, 975, 990, 905, 930, 1040, 1080, 1025, 975, 1030, 990, 980, 1060, 980, 960
5. 304, 221, 242, 254, 259, 271, 289, 253, 311, 225, 247, 251, 209, 314, 278, 290, 295, 315, 222, 285, 317, 288, 250, 290
6. 135, 175, 166, 148, 183, 206, 190, 128, 147, 156, 166, 174, 158, 196, 120, 165, 189, 174, 148, 225, 192, 177, 154, 140, 180, 172
7. 2.8, 3.5, 7.2, 5.8, 6.3, 4.1, 5.7, 8.2, 2.3, 4.4, 7.1, 8.0, 6.8, 5.2, 4.3, 3.0, 3.6, 5.4, 6.3, 6.6, 5.7, 8.2, 4.9, 6.0, 7.2

Applying the Concepts

8. A video rental store selects a group of members to see how many tapes they rented over the course of a year. These are the results:

   17, 41, 24, 21, 16, 39, 62, 19, 14, 37, 59, 46, 33, 50, 28, 12,
   26, 37, 48, 27, 56, 39, 28, 16, 26, 19, 38, 25, 26, 17, 26, 38,
   16, 49, 25, 46, 38, 26, 15, 28

   Construct the stem-and-leaf plot for these data.

9. The weights, in pounds, of a group of people signing up at a health club are

   135, 175, 166, 148, 183, 206, 190, 128, 147, 156, 166, 174, 158, 196, 120, 165, 189, 174, 148, 195, 192, 177, 154, 140, 180, 172

   Construct the stem-and-leaf plot for this data set.

10. A group of 40 students in a high school class have the following scores on the SAT:

> 860, 940, 1120, 900, 840, 980, 1040, 1220, 860, 770, 1000, 870, 890, 920, 930, 1040, 1280, 1020, 970, 1330, 890, 980, 1260, 980, 760, 770, 980, 1280, 830, 1110, 750, 810, 850, 1140, 1090, 850, 1270, 770, 920, 1030

Construct the stem-and-leaf plot for this set of data. (*Hint:* Ignore the final 0 in each entry.)

Computer Applications

11. Repeat Exercise 1 using the stem-and-leaf plot option of Program 1: Statistical Analysis of Data.
12. Repeat Exercise 3 using Program 1.

2.5 Classification of Statistical Data

By its very nature, statistics is concerned with numerical data. We might be interested in the scores of children on a standardized test, the weights of frozen chickens in a supermarket, the weekly profit for a business, the number of people who see a particular movie, the number of people who prefer a certain Presidential candidate in a group of 500 voters, and so forth. In each instance, the quantity we study—a measurement or an observation—is a number.

There are many cases, however, in which the data we study are not numerical. For example, suppose an attitudinal poll is conducted on the issue of capital punishment based on this statement: "Capital punishment is an effective deterrent to major crimes." The possible responses allowed might include *strongly agree*, *agree*, *disagree*, and *strongly disagree*. Furthermore, suppose that the person conducting the poll is concerned with the effects of the respondent's gender or religion on the reply, so that he or she provides the following categories: Female and Male for gender and Protestant, Catholic, Jewish, Other and None for religion.

To work with such nonnumerical data using the statistical tools we will develop in later chapters, we need to impose a numerical scheme on the data. For example, with gender, 0 might be assigned to Male and 1 to Female. With religion, the scheme might be to use 1 for Protestant, 2 for Catholic, 3 for Jewish, 4 for Other, and 5 for None.

In each of these cases, the numerical data have been artificially created, but none of the numbers have any numerical meaning. We call such data *nominal data* because they are numerical in name only.

Unfortunately, many of the usual operations we perform on numbers have no meaning in such a situation. For example, one of the most important ideas underlying arithmetic is that of *order* in a numerical sense. We use this idea whenever we work with inequalities, $>$ or $<$. Thus, if Alison weighs more than Barry, we might write $A > B$ to compare their weights. However, this notion of order or inequality has no meaning in terms of the gender or religion categories above. We cannot numerically compare Catholic (2) to Protestant (1) in the sense of order.

Data for which numerical order is meaningful are known as *ordinal data*. For example, suppose you are asked to rate a series of movies that you have seen on a scale of 1 (dreadful) to 10 (fantastic). Such ratings have no real significance in the sense of the usual arithmetic operations, but they certainly represent a way to introduce an ordering relation.

Furthermore, let us reconsider the previous illustration on the attitudinal survey regarding capital punishment. The responses to the question might also be digitized by, for instance, using 4 for strongly agree, 3 for agree, 2 for disagree, and 1 for strongly disagree. With this scheme, we see that these data are ordinal.

In addition to order, there are several other underlying ideas in arithmetic involving numbers. These are the basic arithmetic operations—addition, subtraction, multiplication and division. These, in turn, give rise to further classifications of statistical data. Consider temperature readings on the Fahrenheit scale. It is clear that there is a very distinct ordering involved: a temperature of 45°F is clearly lower than a temperature of 57°F.

Moreover, in addition to the ordering of these temperature readings, we must recognize the fact that the *difference* between readings is also significant. When we compare temperatures of 45°F, 57°F, and 69°F, say, then the jump from 45°F to 57°F is precisely the same as the jump from 57°F to 69°F. In particular, 69°F − 57°F = 57°F − 45°F = 12°F, and the 12-degree temperature difference is fixed.

Let's consider one other aspect of temperature data. Suppose we have two temperature readings, 32°F and 64°F. You might be tempted to say that a temperature of 64°F is twice as warm as a temperature of 32°F. However, a little thought will convince you that this is not really a valid conclusion. The temperature readings used are based on the Fahrenheit measurement scale. Consider what happens if we look at them in terms of their Celsius equivalents. The Celsius equivalent of 32°F (the freezing point of water) is 0°C while the equivalent of 64°F is

$$\left(\tfrac{5}{9}\right)(64 - 32) = \left(\tfrac{5}{9}\right)(32) = 17.8°C$$

Obviously, 17.8°C is not twice as warm as 0°C. Clearly, the notion that one temperature reading is double (or any other multiple of) another reading depends on the scale in use. As a result, we see that multiplication (or division) of temperature readings carries little meaning.

A set of statistical data in which we can form differences of measurements, but cannot multiply or divide, is known as a set of *interval data*. Therefore, temperature readings are interval data.

Alternatively, let us consider weights. Suppose two children weigh 30 and 60 pounds, respectively. We would certainly be tempted to say that the second child weighs twice as much as the first. In view of the above example, some of you may be a little leery of such a conclusion. Suppose we convert these measurements to different scales and see what happens. First, we convert each to ounces and find that the respective weights are 480 and 960 ounces; the heavier child is still twice as heavy as the lighter child. Similarly, if we convert both weights to their kilogram equivalents, we find that the children weigh 13.64 kg and 27.27 kilograms; again, one weighs twice as much as the other. We therefore see that the choice of unit does not make a difference in terms of the multiple. The reason for this is that when we deal with weights, there is a fixed lower limit of 0 used in all scales. This is unlike the case of temperatures in which a variety of different reference values are used.

Statistical data in which we can form multiples and quotients, as well as differences, are known as *ratio data*. Data on such things as weights, heights, lengths, elapsed times, numbers of people, amounts of monetary transactions, and so forth all fall into this category. In fact, most types of data that we encounter are ratio data.

Exercise Set 2.5

Mastering the Techniques

Classify each of the following sets of data as nominal, ordinal, interval, or ratio. Remember to use the most specific category which describes the data.

1. Batting averages for major league players
2. Scores on the SAT
3. Responses to a voter's preferences in an election poll asking if the voter prefers candidate A, B, or C
4. Rating a person's 10 favorite TV programs
5. The prices of different brands of cereal in a store
6. The telephone numbers of the students in your class
7. The starting salaries offered to new graduates
8. Demographic information regarding race on a survey
9. The number of cigarettes smoked per day on a survey
10. The ratings given to a professor on a course evaluation
11. The degree of difficulty of the problems on the SAT
12. The boiling points of different water-alcohol mixtures
13. The number of movies that a person sees each month
14. The social security numbers of the members of your family

2.6 Selecting a Random Sample

As mentioned before, the major thrust in modern statistics is in the direction of *inferential statistics*. Essentially, this involves collecting data from a sample drawn from a much larger population and using the sample data to make predictions or come to decisions about the entire population. The key to doing this effectively lies in selecting an appropriate random sample from the population in question. In this section, we discuss ways in which such a sample can be obtained.

To begin, let's conduct a simple experiment. Close this book, then open it to a randomly selected page, and select a random word somewhere on that page. Please do not continue reading until you have done this simple experiment.

We will now make some predictions based on our performance of this experiment in many different classes over a good number of years.

Prediction 1: The page you selected is numbered between 200 and 400. (Typically, about 80% of all students in a class will select a page from the middle third of any book.)

Prediction 2: The page you selected is a right-hand page. (About 75 to 80% of all students pick the right-hand side.)

Prediction 3: The word you selected is in the middle third of the page. (Again, this is where some 75% of all students find their words.)

Prediction 4: The word is in the middle third of the line. (About two-thirds of all students so select it.)

While you may be in the minority and these predictions did not apply to your choice, we assure you that these are fairly typical of what will happen in a classroom.

Obviously, if we can make such predictions with reasonably good accuracy, then the "random" selection process most people use is not truly random. If you are selecting a page at random from a book, there is no reason why page 1 has less of a chance of being picked than page 250. In fact, every page should have an equal chance of being selected. In a fully comparable way, if you were to pick a name at random from a telephone directory, Aaron Aarons should have just as great a chance of being selected as does Fred L. Mullins somewhere in the middle of the book, and neither should have a better chance of being picked than Zelda Zminski at the far end. Consequently, we need a way to guarantee that, when we select a sample, *all* members of the population have the same chance of being selected. Otherwise, the sample is not truly random.

> The choice of a single item from a group is called **random** if every item in the group has the same chance of being selected as any other item.

To satisfy this requirement, statisticians have developed a variety of approaches. The most common is to work with a *table of random digits* or *table of random numbers*, as shown in Table I at the back of the book. This table contains thousands of individual one-digit numbers that are grouped five at a time for convenience. However, they should not be interpreted as being five-digit numbers. In fact, this table is just a small portion of a much larger table of random numbers which was produced for professional statistical purposes.

To use this table, we randomly select any entry in the table and work either horizontally or vertically from that number. To illustrate, we start at the upper left entry of the table and work horizontally across the top row. The numbers shown there begin with

04433   80674   24520   18222   10610   05794   37515

We use these random numbers to select a series of randomly chosen words from this book in the following way. To locate a particular word, it is necessary to pick first a random page, then a random line on the page, and finally a random word on that line. Since there are approximately 700 pages in this book, the choice of a page number requires three digits. Therefore, we select the first three random digits and interpret them as a three-digit number, 044, so that the randomly selected page is page 44. Further, there are approximately 50 lines of text per page. Therefore, we use the *next* two digits, 3 and 3, to represent a two-digit number and so consider line 33 on page 44. Finally, suppose there are 10 words to the line. We therefore use the next single digit, 8, to locate the 8th word on the 33rd line of the 44th page as the first random selection.

We note that an entry of 0 for the random word on a line should be interpreted as word number 10 on that line.

If we now want to select a second random word in the book, we simply repeat this procedure by continuing with the random digits in the table. Thus, the next three digits, 067, give us page 67. The following two digits, 42, indicate line 42 on that page. The following single digit, 4, indicates the 4th word of that line.

The next random selection will occur on page 520, line 18, word 2, and so forth. Of course, if there were no page 520, then the process would not produce a viable result. In such a case, we simply move on to the next set of digits and turn to page 221, line 6, word 1.

This process continues until we have made as many random selections as we need. If we finish a line (or column) in the table, then we simply pick up with the first entry in the next line (or column). In the unlikely event that a random selection actually duplicates a previous choice, we must disregard it and choose again.

EXAMPLE 2.5 Pick a random sample of three batters in the American League using the list given in Appendix I.

SOLUTION We begin by noting that there are 95 names in the list, so we need three different random numbers, all between 01 and 95. For convenience, we again start in the upper left corner of the table, but this time we work down instead of across. The entries are therefore

06683   95358   38018   62488 ...

Thus, our first choice is the batter in position 6 in the list, Carlos Baerga. Our second selection is in position 68, Mark Whiten. Our third pick is in position 39, Randy Velarde.

Of course, if one of the two-digit random numbers we selected from the table were larger than 95, we would discard it and take the next two digits. Further, do not be misled by the fact that, in Example 2.5, the players picked were numbered 6, 68, and 39 to suggest that a truly random sample *requires* one selection from the beginning, another from the middle, and a third from the end of a list. This is certainly not the case. It could happen that all three picks are among the top 10 or the bottom 12. The key is simply that *all* players in the list have the same chance of being selected. There is absolutely no preference for any one over any other in the list. In fact, if there were such a preference, then the selection process would not be random.

As you will see, the notion of *randomness* pervades all of statistics. For example, any statistical study requires selecting truly random samples that are representative of the population being studied. In Figure 2.18, we reproduce an item from *The New York Times* describing how a political opinion survey was conducted for the newspaper. If you read the highlighted portion, you will notice that such polls use tables of random digits in just the manner we discussed in this section. We will return to this item in a later chapter when we discuss how to analyze and interpret the results of such surveys. We will consider random events in much greater detail in Chapter 4. We will also develop statistical tests to determine if a set of numbers or other items is truly random in later chapters.

Before leaving the topic of generating random samples from a population, we briefly mention several other aspects of sampling that arise in practical cases. Suppose you are planning to conduct a telephone survey of randomly selected individuals in a certain area. One approach is to select random names from a local phone book to call. The problem with this is that you will miss all people with unlisted numbers and so potentially skew the results of the study. There are several ways to avoid this. In practice, pollsters often select the random listings from the book, then add 1 to the last digit of the phone number, and call that number. Alternatively, you can identify the telephone exchanges in the region, generate four-digit random numbers, and call those phone numbers.

Next, consider a statewide political poll to determine candidate preference. To achieve a high degree of accuracy in predicting the percentage or proportion of the state's adult population that will vote for a certain candidate, the pollsters must select not just a random sample from the population, but a random sample which mirrors the different demographic groups present in the population. Thus, an appropriate random sample should consist of approximately 50% women and 50% men. In addition, if 20% of the state's population is black, then the sample should contain approximately 20% blacks. Similarly, the sample should reflect the various ethnic, religious and economic groups present in the population. This type of sampling is known as *stratified sampling*, although it is somewhat more sophisticated than we will consider in this book.

Another approach to sampling sometimes used is *cluster sampling*. Suppose a study is to be conducted in some large community on a door-to-door basis. If a random sample of 300 households, say, were selected, then the

FIGURE 2.18

How the Poll Was Conducted

The latest New York Times Poll is based on telephone interviews conducted Jan. 30, and 31 with 1,187 adults around the United States, excluding Alaska and Hawaii.

The sample of telephone exchanges called was selected by a computer from a complete list of exchanges in the country. The exchanges were chosen to insure that each region of the country was represented in proportion to its population. For each exchange, the telephone numbers were formed by random digits, thus permitting access to both listed and unlisted residential numbers. The numbers were then screened to limit calls to residences.

The results have been weighted to take account of household size and number of residential telephones and to adjust for variations in the sample relating to region, race, sex, age and education.

In theory, in 19 cases out of 20 the results based on such samples will differ by no more than three percentage points in either direction from what would have been obtained by interviewing all adult Americans. The potential error for smaller subgroups is larger. For example, for Republican primary or caucus voters, it is plus or minus five percentage points.

In addition to sampling error, the practical difficulties of conducting any survey of public opinion may introduce other sources of error into the poll.

pollsters would spend an inordinate amount of time traveling from one site to the next. In the interest of saving time and money, the company conducting the poll may divide the entire community into a series of small neighborhoods or even individual blocks, called *clusters*. The pollsters then select, using a random process such as we described above, one or several particular clusters in the community, and the pollsters approach everyone in these randomly selected clusters. It is certainly possible that the results of such a sample are not typical of the entire community. Nonetheless, cluster sampling is often used for convenience in situations such as this. However, it is not a technique you should use naively when you conduct your own statistical studies, say by selecting your own friends and neighbors and considering them a cluster.

Computer Corner

Let's consider one alternative approach to generating a random sample. Rather than working from the random number table directly, we use a program on your disk which has been developed expressly to generate random samples. Program 2: Random Number Generator generates sets of random numbers in which each set consists of your choice of one, two, three or four random numbers within any desired set of ranges. You also have to enter the number of random samples you want.

To illustrate the program, suppose you want to generate 10 different dates between January 1, 1940, and December 30, 1993. When you select the program, it first asks you how many sets of random numbers you want. Since you want 10, that should be your response. It then asks you how many random numbers per set you desire. Your response should be 3, since you want one random number to represent the month, another to represent the day, and still another to represent the year. The program then asks you for the lower and upper limits on each of the random numbers. Thus, for the first random number, the minimum value you want is 1 and the maximum is 12 (for January through December). For the second random number, you might request a lower limit of 1 and an upper limit of 31. For the third random number, the limits should be 1940 and 1993. The program then randomly generates 10 sets of such random numbers, as shown in the typical output in Figure 2.19. You will be prompted to decide whether or not you want the results to be printed out.

FIGURE 2.19

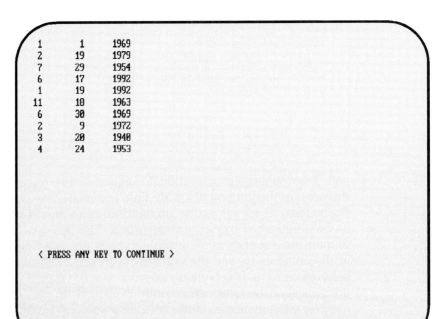

You should be careful to ignore dates such as February 31, 1955, if they arise. In fact, it might be a good idea to anticipate such potential complications and ask for a slightly larger set of random numbers than you actually need. In that way, you can discard any inappropriate sets (meaningless dates or possible duplications).

After the display is complete, you can transfer to the options menu by pressing any key or by clicking on the appropriate window. One of the options allows you to generate another set of random numbers based on the same set of requirements. Another choice allows you to start over and construct other sets based on other values. A third choice allows you to have a set of results printed out on a printer.

However, if you select this option, make sure that a printer is attached to your computer and that it is turned on. If not, it is possible for the computer to "lock up," and you will be unable to do anything without rebooting it (starting over, literally). Also, when you use the printer, be sure that it is set to "ON LINE" instead of "OFF LINE." Otherwise, the printer will not respond to the program's instructions to print.

Incidentally, every time you run this program, even with the same set of choices, it will generate a different set of random numbers for your use.

Exercise Set 2.6

Mastering the Techniques

Use the table of random digits (Table I) to generate the following sets of random numbers.

1. A set of 10 random numbers between 0 and 20
2. A set of 10 random numbers between 5 and 30
3. A set of 12 random numbers between 200 and 600
4. A set of 12 random numbers between 500 and 999
5. Two sets of 10 random numbers, one between 0 and 9, the other between 0 and 99
6. Two sets of 10 random numbers, one between 100 and 200, the other between 3000 and 3999

Applying the Concepts

7. Select a random sample consisting of the high temperatures in five randomly selected cities based on the data given in Appendix II.
8. Select a random sample consisting of the number of home runs hit by eight randomly selected hitters in the National League using the baseball statistics in Appendix I.
9. Select a random sample consisting of six randomly selected lines on randomly selected pages in this book. In each case, count the number of words appearing in the selected line and record the results.

10. Select a random sample consisting of eight randomly chosen sections in this book. In particular, select first a random chapter and then a random section in that chapter. In each case, make a notation regarding the page in the book on which the selected section begins.
11. Select a random sample consisting of 10 topics listed in the Index at the end of this book. In each case, note the (first) page number associated with that topic.
12. Select a random sample consisting of 10 people listed in your local telephone directory. In each case, select a random page, a random column on that page, and a random position in that column for your choice.

Computer Applications

13. Use Program 2: Random Number Generator to generate a list of 20 random numbers between 0 and 99.
14. Use the program to generate a list of 20 randomly selected birth dates between January 1, 1900 and December 31, 1993.
15. Use the program to generate a list of 20 randomly selected stocks based on the data in Appendix III.
16. Use the program to generate a list of 20 randomly selected names from your local telephone directory. (See Exercise 12.)

Student Projects

We have tried to indicate throughout this book that applications of statistics affect all aspects of our lives. However, if all that you, as a student, do is to concentrate on statistics as a classroom exercise, then you will be missing much of the significance of the subject. But, if you participate actively in collecting, organizing, and analyzing data from the world about you, then you will achieve a much deeper understanding of what statistics is and how it is used.

To encourage this type of active role, we include a series of suggested statistical projects at the end of each chapter. These projects will give you the opportunity to get personally involved by working with real-world data while reinforcing the statistical ideas in the chapter.

In conducting any such statistical study, you must be very careful to define as specifically as possible the population from which you are selecting the sample. Moreover, you must be certain that the process you use for drawing the sample is indeed random. At the same time, you should realize that sometimes there may be practical limitations to getting a truly random sample and, in such a case, you should carefully indicate the limitations. We illustrate these concerns in the following example.

EXAMPLE Suppose you wish to study the total bills paid by customers in a fast-food restaurant. A very simplistic approach would be to go in, pick a group of 25 random customers (using the table of random digits or Program 3), and record the total paid by each. However, does your selection process key on the customers as they enter the restaurant? Alternatively, does it take into account a random selection of the employee in front of whom they line up?

Even when the selection process is random, you must still carefully identify the population you are studying. For instance, does it make sense, in your study, to consider bills for individuals as well as those for a family of eight people, or should you impose some limitation? If you do impose such a limitation, you are identifying a particular population. Similarly, when are you conducting this study? Certainly the time of day affects the amount of the purchase—total bills during dinner hours will certainly be much higher than those during the midmorning hours, let alone in the middle of the night. If you collect your sample over only a few hours, then you are sampling the population of customers who come in during that period. Moreover, note that the total bills in one restaurant in a chain can be quite different from those in other branches in other locations.

Moreover, the day of the week may be a significant factor in terms of the bill. It is possible that customers purchase more during the weekend than on a weekday. Similarly, the weather may be a factor. In addition, special promotions run by the restaurant can affect the totals on the bills.

Thus, for instance, if you conduct your study during the early afternoon hours on several different days during the week, then your population might be the total customer bills in one particular Burger Heaven restaurant between the hours of 2 p.m. and 4 p.m. on weekdays only. As such, any conclusions you reach will apply to only this population. They do not necessarily apply to any other set of circumstances.

Collect a *random sample* consisting of at least 25 numbers on one of the following topics:

1. Go to the produce department in a grocery store or supermarket and randomly select a group of "1-pound" bags of carrots (or other prepackaged vegetable). Weigh each bag in your sample, and record the weight in ounces.
2. Select a random set of stock prices (probably the closing prices) for companies listed in the current stock market reports in your local newspaper.
3. Choose a random set of current sports results (batting averages in baseball, field goal percentages in basketball, total yards gained by a team in football, number of goals scored by players in hockey, etc.).
4. Collect the prices on a random set of items in a specialty shop or department in a department store. (For instance, you might want to consider clothing, automotive supplies, appliances or electronics.)
5. Select a random set of values representing the weekly gross for the major movies playing around the United States this week as listed in a publication such as *Variety*.

> 6. Collect data from a random selection of students at your school on a single quantity such as height, weight, SAT score, GPA, shoe size (men only or women only), number of credits completed, etc.
> 7. Measure the length of time that a group of randomly chosen cars stand at a red light. You should also record 0s for any cars that you select that arrive during the green cycle and so have no wait.
>
> Once you have collected this set of data, you should then
>
> (a) Construct an appropriate relative frequency distribution.
> (b) Construct the associated histogram, bar chart, and/or frequency polygon.
> (c) Construct the stem-and-leaf plot.
>
> Finally, you should incorporate all these items into a formal statistical project report. The report should include
>
> - A statement of the topic being studied
> - A statement of the specific population being investigated
> - The source of the data
> - A discussion of how the data were collected and why you do or do not believe it is a random sample
> - The tables and diagrams displaying the data
> - A discussion of any surprises you may have noted in connection with collecting the data or organizing them
>
> You should keep the actual or raw data you collect for this project for use in related statistical projects later in the course.

CHAPTER 2 SUMMARY

Statistics involves the collection, organization, display, and analysis of data as well as the methods by which conclusions can be drawn by using the data. Data can be interpreted as representing either a **population** or a **sample** drawn from a population.

Descriptive statistics consists of methods used to organize, display and analyze data from some population or sample. **Inferential statistics** uses the data from a sample to either make predictions or draw conclusions about the underlying population from which the sample was drawn.

You can organize data in a variety of ways. It is often desirable to organize a set of data in numerical order. You also can organize it into a **frequency distribution**, a **relative frequency distribution**, or a **cumulative frequency distribution**. If all classes have the same class width, then

$$\text{Width} = \frac{\text{largest value} - \text{smallest value}}{\text{number of items in data}}$$

The **class boundaries** separate the different classes. The **class marks** are the midpoints of each class.

You can display a set of data by using a variety of graphical methods including **bar charts**, **histograms**, **frequency polygons**, **pie charts**, and **stem-and-leaf plots**.

You can classify data as **nominal data, ordinal data, interval data**, and **ratio data**.

Whenever you conduct any statistical study involving a sample drawn from a population, you must obtain a **random sample**. To do so, you can use either the table of **random numbers**, Table I, or a computer program to produce a list of appropriate random numbers.

Review Exercises

Refer to the table on the racial makeup of the U.S. population at the beginning of this chapter for Exercises 1 to 5.

1. Construct a bar chart for this data.
2. Construct a pie chart for the data.
3. Explain why it would not make sense to construct a cumulative frequency distribution for this data.
4. What percentage of the population is black?
5. What percentage of the population is non-white?

Refer to the table on the population of each state at the beginining of this chapter for Exercises 6 to 12.

6. Construct a frequency distribution for this data based on classes under 2 million, 2 to 4 million, 4 to 6 million, ..., 10 million and over.
7. Construct the histogram corresponding to the frequency distribution in Exercise 6.
8. Construct the histogram for this data based on 5 classes, 0 to 4 million, 4 to 8 million, etc. How does it compare to the one in Exercise 7?
9. What percentage of states have populations under 6 million?
10. What percentage of the U.S. population lives in the northeast (Maine, Vermont, New Hampshire, Massachusetts, Rhode Island, Connecticut, New York and New Jersey)?
11. What percentage of the U.S. population lives in states beginning with an M?
12. What percentage of the states begin with an M? Does your answer to Exercise 11 seem reasonably compatible with this? Would you expect it to be?

3 Numerical Description of Data

According to an international study, the average American high school student's mathematical skills are the poorest when compared with skills of high school students in other major industrialized nations. On a test administered to high school students in each country, the average score for American students was lower than the average scores for students in Japan, Germany, Great Britain, Canada, Korea and elsewhere. Leading mathematics educators attribute this poor performance to a variety of causes including outmoded approaches to teaching mathematics that stress unnecessary skills and provide little understanding of the mathematics and its uses, poor teaching on the part of many teachers, poor motivation on the part of many students, poor attitudes among parents toward mathematics and a poor perception about mathematics in society in general.

No one would dare brag that she or he is unable to read for fear of being ridiculed. However, how often have you heard people bragging that they are unable to do mathematics? It is so common, in one form or another, that it may not even sound strange. Yet in today's technologically oriented society, mathematical knowledge and skills are almost as vital as reading and writing skills. Inability to do mathematical reasoning and problem solving is a major handicap in almost any career.

One of the major directions for mathematics reform calls for virtually all students to have a far greater exposure to statistics via a course such as this one. It is essential that all informed citizens learn to understand data and to know how to interpret, use and make decisions based on data.

Notice how the very reports that call our attention to the problems in mathematics education make use of statistical information and statistical arguments. For example, the report referred to above uses the mean scores on mathematics tests to make a point. A follow-up study subsequently suggested that the initial comparison between different groups of students is badly flawed because of the vast differences in culture and income among countries.

The key element that makes the comparison between student performance in the different countries so compelling is the use of the average or mean score that different groups achieved. The average provides a simple way to summarize all the data using a single number. Moreover, the different mean scores provide us with a way to compare two or more sets of data and to decide how significant a difference in averages is between two or more groups, if all other factors are comparable.

In this chapter, we will consider ways to summarize a set of data numerically as well as methods to measure how much variation there is in the data.

3.1 Measures of Central Tendency: The Mean, Median and Mode

Our primary goal in statistics is to use data for prediction and decision making. Although all the techniques discussed in Chapter 2 are very effective for displaying data, they are not particularly suitable for achieving this objective. In addition, we must develop some ways to describe a set of data *numerically* which will provide us with the necessary tools for statistical inference.

Let us again consider the set of data on the heights (in inches) of students in a statistics class. The individual values, in numerical order, are

60, 61, 61, 62, 62, 62, 63, 63, 63, 64, 64, 64, 64, 65, 65, 65, 65, 66, 66, 66, 66, 66, 67, 67, 67, 68, 68, 68, 69, 69, 70, 71, 71, 72, 72, 74

We have already seen two numerical ways to describe this set of data:

Number of measurements $n = 36$

Range $= 74 - 60 = 14$ inches

We can summarize this set of data in a relative frequency distribution as follows:

Class	Class Limits	Frequency	Relative Frequency
1	60–62	6	16.7%
2	63–65	12	33.3%
3	66–68	10	27.8%
4	69–71	5	13.9%
5	72–74	3	8.3%
		36	

From this type of table, we can conclude that most of the heights are between 63 and 68 inches.

This type of observation regarding the location of most of the data entries is extremely important in any statistical analysis. We must be able to determine where the majority or "center" of the values occurs. Methods for measuring this center are known as *measures of central tendency*.

The most common and most effective numerical measure of the center of a set of data is the *mean*. It is the arithmetic average of the values. For the set of data on student heights, we find the mean by adding all the measurements and then dividing the total by the number of entries $n = 36$. Thus, the mean of these data is

$$\frac{60 + 61 + 61 + \cdots + 74}{36} = \frac{2376}{36} = 66 \text{ inches}$$

In general, suppose we have a set of n values or measurements

$$x_1, x_2, x_3, \ldots, x_n$$

To find their mean, first we sum these values

$$x_1 + x_2 + x_3 + \cdots + x_n$$

and then we divide by the number of entries, n, so that

$$\text{Mean} = \frac{x_1 + x_2 + x_3 + \cdots + x_n}{n}$$

This type of summation of a set of n terms occurs often in statistics, and so a special symbol is used to abbreviate it. We introduce the Greek letter Σ (capital sigma) to stand for "summation", so that Σx represents the sum of all the x values. Thus, we rewrite the above formula for the mean as

$$\text{Mean} = \frac{\Sigma x}{n}$$

In later chapters, we will use a random sample drawn from a larger population and want to compare the mean of the sample to the mean of the population. For that reason, it is necessary to use different notations to immediately identify whether a mean or other statistical measure refers to a sample or to a population. Quantities which describe a population are called *population parameters* or simply *parameters*. They are often denoted by Greek letters. Quantities which describe a sample are called *sample statistics* or simply *statistics*. They are denoted by the lowercase letters of the alphabet.

For example, we denote the mean of a population of size N by the Greek letter μ (mu). Thus, the above formula for the mean of a population can now be written as

$$\mu = \frac{\Sigma x}{N}$$

When we have a sample drawn from a population, the mean of the sample is denoted by the symbol $\bar{x}$ (read "x bar"). However, the process of calculating the mean is identical, so that the mean of a sample of size n is given by the equivalent formula

$$\bar{x} = \frac{\Sigma x}{n}$$

Notice that N represents the size of a population while n represents the number of measurements in a sample.

Although the mean is the single most useful quantity we will encounter to describe a set of data, it is not the only, or even always the best, way of measuring the center of a set of data. Consider the following situation. A business employs 100 people, each of whom is paid a total of $10,000 per year. Therefore, the total of the salaries of all 100 workers is $1,000,000 per year. In addition, the owner takes home $10,000,000 each year, after expenses. Thus, the total income of the $N = 101$ people involved in this operation is $11,000,000, so that the mean income for these 101 people is

$$\mu = \frac{\$11,000,000}{101} = \$108,911$$

Obviously, this figure is totally meaningless, if not intentionally misleading. There are 100 workers who barely come within $100,000 of the mean and the owner whose income is incredibly far above the mean. Certainly, in this case, the mean is nowhere near the center of the data set.

In such situations, a better measure of the center of data is the *median*. The median corresponds to the *middle* of a set of data when the numbers are in numerical order. For example, if we have the five numbers

10, 15, 25, 30, 45

then the middle entry is the third one, and so the median is 25. On the other hand, if there are six numbers, say,

6, 10, 15, 25, 30, 45

then there is no precise middle value. In such cases, the median is defined to be the *average* of the *two* middle entries. In our example, the median is the average of the third and the fourth entries, or

$$\text{Median} = \frac{15 + 25}{2} = \frac{40}{2} = 20$$

If the number of measurements n is an odd number, the **median** is the middle value.
If the number of measurements n is an even number, the **median** is the average of the middle two values.

68 CHAPTER 3 • Numerical Description of Data

CAUTION The median makes sense only when the data are arranged in either ascending or descending order. If you inadvertently try to calculate a median by taking the center of a nonordered set of data, the result is meaningless.

Thus, in the business salary example, the median is the 51st value, or $10,000. On the other hand, for the student height example, there are 36 measurements, and so we must average the middle two. Therefore, the median height is the average of the 18th and 19th values, 65 and 66. Consequently, the median is equal to 66.5 inches.

In practice, the median is used for situations where one or a few data values lie very far from the center of the data. In such cases, the mean is greatly distorted because of the effects of these few entries. However, the median is not affected at all by such extreme values. Thus, the median is often used in economic applications such as family income or the price of homes in a community. According to the 1990 census, the median value of one-family homes in the United States was $79,100; the median age of the population was 32.9 years.

EXAMPLE 3.1 From the newspaper clipping in Figure 3.1, we see that the median price for a used home in western Suffolk County, New York, is around $150,000. However, notice that one of the individual homes mentioned sold for $477,000. Several extremely expensive homes such as this one (or several particularly inexpensive homes) would distort the mean selling price of homes significantly. The median selling price is not unduly affected by such extreme values. ∎

FIGURE 3.1

The High Cost of Shelter

Median Used-Home Prices

Month	Western Suffolk	Nassau	Queens
December	$145,000	$175,000	$156,200
January	136,000	174,000	146,000
February	145,000	173,000	142,000
March	140,000	176,000	145,000
April	136,900	184,500	145,000
May	148,000	188,500	144,500
June	152,000	191,200	160,000
July	154,000	192,800	159,000
August	155,000	191,000	165,000
September	154,000	193,000	157,000
October	146,000	191,500	161,000
November	150,000	187,500	160,000
December	149,000	189,000	159,000

SOURCE: Multiple Listing Service Inc., of Long Island Board of Realtors

The next situation illustrates the need for yet another measure of central tendency. A video rental business has only limited physical space and only a fixed amount of money to spend. The owner must decide very judiciously which tapes to purchase. In such a situation, the mean and the median are not particularly helpful to her. Instead, what the owner needs to know is which movies are most popular and which, therefore, are likely to be rented with the greatest frequency.

> The **mode** for a set of data is the value that occurs most frequently.

Thus, in the student height example, the value 65 inches occurs most frequently (5 times) and so it is the mode.

It is certainly possible for the greatest frequency to correspond to several different values, which results in more than one mode. For instance, there might be five 68s in this same set of data. In that case, both 65 and 68 would be considered modes for the data. A data set with two modes is called *bimodal*; a data set with three modes is called *trimodal*; if there are more than three modes, it is called *multimodal*. At the other extreme, if each data value occurs only once, then there is no mode.

There is one last measure of central tendency that we consider for a set of data.

> The **midrange** is the average of the largest and smallest values for the data.

For the student height example, the minimum and maximum values are 60 and 74 inches. Thus, the midrange is

$$\tfrac{1}{2}(60 + 74) = \tfrac{1}{2}(134) = 67 \text{ inches}$$

In this book, we work primarily with the mean since it is the most useful measure for inferential statistics. We will encounter a few applications that involve the median, particularly when the data set contains some very extreme values. We note that some statisticians are emphasizing the role of the median because it is less sensitive to the extremes in data. Only occasionally do we discuss the mode and the midrange.

EXAMPLE 3.2 A study is made of class sizes in a certain elementary school. The following data were collected:

　　　　　　Kindergarten:　　20, 20, 21, 21
　　　　　　First grade:　　　22, 23, 22, 23
　　　　　　Second grade:　　20, 19, 21, 20
　　　　　　Third grade:　　　22, 23, 23, 23

Fourth grade: 27, 27, 26
Fifth grade: 20, 19, 22, 20
Sixth grade: 23, 24, 24, 24

Find the mean, median, mode and midrange for the number of children in each class in this building.

SOLUTION We have the number of children in each of $N = 27$ different classrooms. We begin by putting these data in numerical order. (Although this is not necessary for calculating the mean, it is required for finding the median.) We therefore obtain

$$19, 19, 20, 20, 20, 20, 20, 20, 21, 21, 21, 22, 22, 22,$$
$$22, 23, 23, 23, 23, 23, 23, 24, 24, 24, 26, 27, 27$$

Consequently, the mean for this population is

$$\mu = \frac{\Sigma x}{N} = \frac{599}{27} = 22.2$$

and so the mean number of children per class is 22.2 students. (Notice that the value we use for the mean μ contains at least one more decimal place than the original data.)

Since there are $N = 27$ values in this set of data, the median is the 14th entry and so is 22 children per class.

Further, there are six classes with 20 students and six classes with 23 students. Consequently, this data set is bimodal, and 20 and 23 are the two modes.

Finally, the minimum and maximum values are 19 and 27, and hence the midrange is $\frac{1}{2}(19 + 27) = \frac{1}{2}(46) = 23$. ■

EXAMPLE 3.3 The school in Example 3.2 also has four Special Education classes containing 10, 11, 11 and 8 students. Find the mean, median, mode and midrange for the number of students per class in all 31 rooms.

SOLUTION If we include the four additional classes in the above data set, the data list becomes

$$8, 10, 11, 11, 19, 19, 20, 20, 20, 20, 20, 20, 21, 21, 21, 22,$$
$$22, 22, 22, 23, 23, 23, 23, 23, 23, 24, 24, 24, 26, 27, 27$$

As a result, the mean of this set of $N = 31$ values is

$$\mu = \frac{639}{31} = 20.6$$

while the median is 22; the two modes are still 20 and 23, and the midrange is now $\frac{1}{2}(8 + 27) = 17.5$. ■

Notice that the mean has dropped considerably (from 22.2 to 20.6) because of the additional, comparatively low values. Thus, it may no longer be as representative as it was for Example 3.2. The median, however, is unaffected.

A common complaint about statistics is that it is possible to prove virtually anything at all by using (or maybe misusing) statistics. It is certainly true that unscrupulous people can lie by using statistics and that statistically ignorant people can misinterpret or misuse statistical data. However, it is also possible to make different points, in entirely valid ways, by just emphasizing selected statistical information that suits one's position. Example 3.3 provides insight into how this can be done. Although the values for the mean and the median are fairly close, it is still possible to come to different conclusions based on using each one. For example, the school board responsible for this particular school could use the mean of 20.6 to demonstrate to parents that small class sizes are maintained. On the other hand, school board members could use the median of 22 instead to demonstrate fiscal responsibility. At the same time, the teachers' union would probably use the median, since it is the largest and so would allow them to argue that class sizes are too large. In each case, nobody is really lying, but each is correctly using the statistics to make a desired point. Unless a person knows the difference and is paying careful attention, he or she could easily be confused or misled by such arguments.

Computer Corner

Many of the above calculations can be performed by Program 1: Statistical Analysis of Data. Option 2 produces the mean for the data. This option also calculates another important measure for a set of data known as the standard deviation, which is the subject of the next section. The value for the median is calculated as part of Option 3 for an ordered list of the data. It also calculates the quartiles, which are discussed later in this chapter.

Minitab Methods

Minitab is especially effective in calculating numerical measures associated with sets of data. Before reading on, we recommend that you review Section 1.4 on the primary features of Minitab.

Suppose you enter a set of data into column C1 using the **SET** command, and you name it with **NAME C1 'WEIGHT'**. As shown in the sample Minitab session in Figure 3.2, you can obtain a list of the data by using **PRINT C1** and various measures such as the mean and the median. The appropriate commands are:

MEAN C1

MEDIAN C1

SUM C1

MINIMUM C1

MAXIMUM C1

FIGURE 3.2

```
MTB > SET C1
DATA> 136 137 157 144 190 164 147 150 136 163 148 174 211 169 148 184
DATA> 163 144 130 181 156 147 170 148 182 159 140 137 122 158
DATA> END
MTB > NAME C1 'WEIGHT'
MTB >
MTB > PRINT C1
WEIGHT
    136    137    157    144    190    164    147    150    136    163    148
    174    211    169    148    184    163    144    130    181    156    147
    170    148    182    159    140    137    122    158
MTB > MEAN C1
   MEAN    =    156.50
MTB > MEDIAN C1
   MEDIAN  =    153.00
MTB > SUM C1
   SUM     =    4695.0
MTB > MINIMUM C1
   MINIMUM =    122.00
MTB > MAXIMUM C1
   MAXIMUM =    211.00
MTB > DESCRIBE C1
              N       MEAN     MEDIAN     TRMEAN      STDEV     SEMEAN
WEIGHT       30     156.50     153.00     155.46      19.84       3.62
            MIN        MAX         Q1         Q3
WEIGHT   122.00     211.00     143.00     169.25

MTB > SAVE 'WEIGHTS'
Worksheet saved into file : WEIGHTS.MTW
MTB > STOP
```

In addition, the Minitab command **DESCRIBE** produces a table of all these measures for any desired column of data. This command also calculates several other measures, including the standard deviation (STDEV), the standard error of the mean (SEMEAN), and the first and third quartiles (Q1 and Q3), which we will study in later sections. You can save this set of data for future use with **SAVE 'WEIGHTS'** and exit Minitab with **STOP**.

Exercise Set 3.1

Mastering the Techniques

Calculate by hand or hand-held calculator the mean, median, mode (if there is one), and midrange for the following data sets.

1. 11, 12, 15, 20, 23, 25, 30, 33, 37, 44
2. 21, 21, 22, 23, 24, 25, 26, 26, 27, 27, 27, 28, 28, 28, 28, 29, 29, 29, 30, 31
3. 53, 47, 59, 66, 36, 69, 84, 77, 42, 57, 51, 60, 78, 63, 46, 63, 42, 55, 63, 48, 75, 60, 58, 80, 44, 59, 60, 75, 49, 63
4. 2.8, 3.5, 7.2, 5.8, 6.3, 4.1, 5.7, 8.2, 2.3, 4.4, 7.1, 8.0, 6.8, 5.2, 4.3, 3.0, 3.6, 5.4, 6.3, 6.6, 5.7, 8.2, 4.9, 6.0, 7.2
5. 860, 940, 1120, 900, 840, 980, 1050, 1220, 860, 770, 1010, 870, 890, 910, 930, 1040, 1280, 1020, 970, 1330, 890, 980, 1260, 980, 760

SECTION 3.1 • Measures of Central Tendency: The Mean, Median and Mode

6. 17, 39, 26, 20, 17, 49, 22, 36, 17, 17, 39, 26, 15, 20, 18, 22, 46, 17, 28, 17, 36, 19, 18, 46, 16, 19, 28, 15, 20, 48, 16, 18, 36, 29, 15, 16, 28, 36, 15, 18
7. 304, 221, 242, 154, 259, 271, 189, 253, 311, 225, 247, 151, 209, 314, 278, 190, 195, 315, 222, 185, 317, 288, 250, 190
8. 135, 175, 166, 148, 183, 206, 190, 128, 147, 156, 166, 174, 158, 196, 120, 165, 189, 174, 148, 225, 192, 177, 154, 140, 180, 172

Applying the Concepts

9. In a high school class, 25 students have the following scores on the SAT:

   860, 940, 1120, 900, 840, 980, 1050, 1220, 860, 770, 1010, 870, 890, 910, 930, 1040, 1280, 1020, 970, 1330, 890, 980, 1260, 980, 760

   Of the mean, median, mode and midrange of the SAT scores for these students, which should be used to indicate how well the teacher is preparing the students? Which one would suggest the greatest need for improvement?

10. A video arcade has 40 different video game machines, and a study is made to see how many times each game is used in a given time period. The results are as follows:

    17, 19, 16, 20, 17, 19, 22, 16, 17, 17, 19, 16, 15, 20, 18, 22,
    16, 17, 18, 17, 16, 19, 18, 16, 16, 19, 18, 15, 20, 18, 16, 18,
    16, 19, 15, 16, 18, 16, 15, 18

    Find the mean, median, mode and midrange of the number of uses per machine.

11. The weights, in pounds, of a group of people signing up at a health club are

    135, 175, 166, 148, 183, 206, 190, 128, 147, 156, 166, 174, 158, 196, 120, 165, 189, 174, 148, 225, 192, 177, 154, 140, 180, 172

    Find the mean, median and mode of the weights.

12. A group of people at the above health club sign up for a special weight reduction program. After 1 month, their weight losses, in pounds, are

    8, 12, 5, 21, 17, 10, 2, 8, 16, 20, 11, 8, 6, 14, 3, 9, 12, 17, 14, 8, 10, 6, 15

    Find the mean, median and mode for these data. Which value should be used to advertise the success of the program?

13. Consider the set of data in Exercise 1. Suppose that each number is doubled. Calculate the mean, median, mode and midrange for the new set of numbers, and compare each to the corresponding value for the original set.

14. Consider the set of numbers in Exercise 1. Now suppose that each number has 200 added to it. Find the mean, median, mode and midrange; compare each to the corresponding value for the original set.

15. Consider the set of numbers in Exercise 3. Suppose each number is multiplied by 10. Before doing the work, anticipate what values you should get for the mean, median and mode. Verify your predictions by performing the calculations.

16. Consider the set of numbers in Exercise 4. Suppose 750 is subtracted from each of the numbers. Before you do the work, what do you predict the results will be for the mean, median and mode? Verify your predictions.

(Note: Exercises 13 to 16 indicate methods that can be used to simplify numerical calculations when either large numbers or decimal values are involved. The topic is known as *coding*, but we do not discuss it formally.)

Computer Applications

17. Repeat Exercise 2, using Option 2 of Program 1: Statistical Analysis of Data.
18. Repeat Exercise 3, using Option 2 of Program 1.
19. Repeat Exercise 4, using Option 2 of Program 1.
20. Repeat Exercise 5, using Option 2 of Program 1.
21. Use the data on baseball records in Appendix I (or in the computer data file HOMERUN you created) to find the mean and median number of home runs for the players in the list.
22. Find the mean and median for the temperatures in the cities listed in Appendix II using the data file TEMPS you saved.
23. Find the mean and median for the prices of stocks of the set of random companies listed in Appendix III that you previously selected and saved as the data file STOCKS.

3.2 Measuring the Variation in Data: The Standard Deviation

As we saw in Section 3.1, the mean of a set of data is the single most useful statistical measure of the data. However, the mean alone is usually inadequate to fully describe the data. In this section, we consider some additional measures which are used in conjunction with the mean.

Suppose there are two patients in a hospital room whose pulse readings are recorded 3 times a day. The readings for the first patient are 72, 75 and 78, while those for the second patient are 48, 64 and 113. For both patients, the mean pulse reading is 75 beats per minute. The readings for the first patient are very consistent and do not indicate any abnormal condition, since they are all close to a normal pulse rate. The readings for the second patient, however, vary considerably. This set of readings probably indicates a medical condition that requires attention. Thus, the fact that the two means are the same can be very misleading. The variation in the measurements for each patient is also a significant factor and one which we must be able to assess for any set of data.

Consider a situation where two students in a class have the following sets of grades:

$$\text{Student A:} \quad 80, 76, 78, 83, 83$$

$$\text{Student B:} \quad 85, 60, 95, 65, 95$$

SECTION 3.2 • Measuring the Variation in Data: The Standard Deviation

Both students have the identical mean of 80. However, student A is extremely consistent, and an average grade of 80 obviously is appropriate for her. Student B's grades, however, are very inconsistent. The low grades may indicate that he was not applying himself during several units in the course, or the high grades might indicate that he really has exceptional ability if he takes the work seriously. Clearly, the fact that the means are the same does not indicate that the two students' performances are necessarily equivalent. The amount of variation in a set of data is extremely important and cannot be ignored.

We have already encountered one simple way to measure the variation in a set of data—the range. Thus, for student A, the range in grades is $83 - 76 = 7$ points while for student B, the range in grades is $95 - 60 = 35$. Unfortunately, the range does not give sufficient information since it depends exclusively on two values in a set of data—the largest and the smallest. It does not take any other values into account.

We need a more accurate method for measuring the amount of *variation*, *spread*, *dispersion* or *consistency* of a set of data which takes into account the effects of all the data entries. First, we must agree on what we mean by "variation in data". We interpret variation as measuring how far each data value is from the mean (the center) of the data. Thus, if the data have mean μ, then the *deviation* for each value x is $x - \mu$. That is, if the mean of a set of numbers is 80 and one of the values is 83, then the deviation for that one value is $83 - 80 = 3$. Let's apply this idea to the five test scores for student A. The best way to do this is via a tabular array. The first column represents the actual x or data values, and the second column represents the corresponding individual deviations, $x - \mu = x - 80$.

x	$x - \mu$
80	$80 - 80 = 0$
76	$76 - 80 = -4$
78	$78 - 80 = -2$
83	$83 - 80 = 3$
83	$83 - 80 = 3$

Since we are more interested in the total deviation for all the data in the set, we sum the individual deviations and find that

$$\text{Total deviation} = 0 + (-4) + (-2) + 3 + 3 = 0$$

In fact, this total will always be zero for any data set whatsoever because the positive and negative values always cancel. Obviously, this approach to finding a way of measuring the deviation in a set of data is totally useless!

To prevent the positive and negative deviations from always canceling, we must avoid the negative values. This can be done either by using the absolute value of each deviation or by squaring each deviation. At one time, statisticians used the average of the absolute values of the individual deviations to measure the spread in a set of data. However, this quantity, called the *average deviation* or the *mean absolute deviation*, is not particularly effective for the type of work we wish to do in inferential statistics.

An alternate approach to eliminating negative quantities from the individual deviations is to square each value. We again perform the needed calculations in a table:

x	$x - \mu$	$(x - \mu)^2$
80	$80 - 80 = 0$	$0^2 = 0$
76	$76 - 80 = -4$	$(-4)^2 = 16$
78	$78 - 80 = -2$	$(-2)^2 = 4$
83	$83 - 80 = 3$	$3^2 = 9$
83	$83 - 80 = 3$	$3^2 = 9$
		38

We now sum the squares of the individual deviations and so obtain a total of 38. Since this is based on the contributions from five different terms, we must average it over all the data values to get

$$\frac{38}{5} = 7.6$$

This number is known as the *variance* of the data, and it is a useful measure of the spread of the original values about the mean. When we are concerned with a population, the variance is written in terms of the Greek letter σ (lowercase sigma) and is denoted by σ^2. Thus, we can summarize the above calculations with the following formula:

$$\text{Population variance } \sigma^2 = \frac{\Sigma(x - \mu)^2}{N}$$

where N is the size of the population.

However, a far more useful measure of the spread or variability in a set of data is the *standard deviation*, which is defined as the square root of the variance:

$$\text{Standard deviation} = \sqrt{\text{variance}}$$

Since the standard deviation is the square root of the variance σ^2, the standard deviation is denoted by σ and is found from the formula

$$\text{Population standard deviation } \sigma = \sqrt{\frac{\Sigma(x - \mu)^2}{N}}$$

It is this quantity that we use throughout our study of statistics to measure the spread in data.

In the previous example, the standard deviation is given by

$$\sigma = \sqrt{\text{variance}}$$
$$= \sqrt{\sigma^2} = \sqrt{7.6} = 2.757$$

You will probably find it much easier to calculate the standard deviation for a set of data as a procedure using a tabular array, as we did above, rather than by using this formula. As you will see at the end of this section, it is easier still to use computer programs to carry out the calculations for you. Many calculators also have statistical functions which include calculating the standard deviation.

One special advantage of working with the standard deviation is that it is measured in the same units as the original data. Thus, if the original set of numbers represents costs of a certain type of item in dollars, then both the mean and the standard deviation are measured in dollars. If the numbers represent lengths of time measured in seconds, then the mean and standard deviation also are measured in seconds.

EXAMPLE 3.4 Find the variance and standard deviation of the grades for Student B.

SOLUTION We perform the necessary calculations in the following table:

x	$x - \mu = x - 80$	$(x - \mu)^2$
85	85 − 80 = 5	25
60	60 − 80 = −20	400
95	95 − 80 = 15	225
65	65 − 80 = −15	225
95	95 − 80 = 15	225
		1100

Since the sum of the squares of the individual deviations is 1100, the population variance for this set of data is given by

$$\sigma^2 = \frac{1100}{5} = 220$$

and the standard deviation is

$$\sigma = \sqrt{220} = 14.83$$

Since this value is approximately 5 times the standard deviation for Student A's grades (2.757), we conclude that the spread in Student B's results is about 5 times that of Student A.

CHAPTER 3 · Numerical Description of Data

The standard deviation is the way in which we measure the amount of variation in a set of data.

> The larger that σ is for a set of numbers, the greater the spread or variability among those numbers.
>
> The smaller the value for σ, the smaller the amount of variation in the data.

It is important to realize that, even though the standard deviation is affected by the largest and smallest values in the data, it also measures the spread due to each and every value of x.

EXAMPLE 3.5 The eight tellers working in a bank serve the following numbers of customers in a 1-hour period:

$$24, 31, 20, 16, 35, 28, 25, 29$$

Find the mean and the standard deviation for the number of customers served per teller.

SOLUTION We can work with the data in the order given, but it is usually preferable to rewrite them in numerical order. Thus, we set up these data values in a table, as shown in Table 3.1, where we put the values in ascending order.

TABLE 3.1

x	$x - \mu = x - 26$	$(x - \mu)^2$
16	$16 - 26 = -10$	100
20	$20 - 26 = -6$	36
24	$24 - 26 = -2$	4
25	$25 - 26 = -1$	1
28	$28 - 26 = 2$	4
29	$29 - 26 = 3$	9
31	$31 - 26 = 5$	25
35	$35 - 26 = 9$	81
$\Sigma x = 208$		$\Sigma(x - \mu)^2 = 260$

$$\mu = \frac{208}{8} = 26$$

Since the sum of the eight original measurements is 208, the mean is $\mu = 26$ customers served per teller. The sum of the squared deviations is

260, so that the variance is

$$\sigma^2 = \frac{260}{8} = 32.5$$

and so the standard deviation is

$$\sigma = \sqrt{32.5} = 5.70 \text{ customers served per teller} \blacksquare$$

Incidentally, it is not necessary to put the data in order to calculate the mean and the standard deviation, as we did above. However, there are certain advantages to doing so. Notice the middle column in Table 3.1 under $x - \mu$. There is an obvious pattern to the entries: they begin with the most negative value, increase up through 0, and continue to the most positive value. Similarly, the last column begins with a large positive quantity, decreases to a relatively small value, and then increases again to a large quantity. These patterns will repeat whenever the data are in numerical order and so represent an easy way to check that you have not made certain types of errors in the computation.

All the above ideas for the variance and the standard deviation were developed in the context of a population. Very similar ideas exist for the variance and standard deviation of a sample drawn from a population, with one significant difference. When we deal with a sample, we cannot average the sum of the squared deviations, $(x - \bar{x})^2$, over the entire set of data. Instead, it is necessary to make the following modification:

$$\text{Sample variance} = \frac{\Sigma(x - \bar{x})^2}{n - 1}$$

— Deviations measured from center at sample mean $\bar{x}$.

and

$$\text{Sample standard deviation} = \sqrt{\frac{\Sigma(x - \bar{x})^2}{n - 1}}$$

That is, instead of dividing by all n data points, it is necessary to divide by $n - 1$ of them. Although the reasons for this change are beyond the scope of this book, we do want to point out that using $n - 1$ for a sample instead of N for a population gives a better estimate of the population standard deviation. Other than this, the procedure involved in calculating the variance and the standard deviation for a sample is totally parallel to what we did for a population. Thus, first we find the mean $\bar{x}$ for the sample, we calculate the individual deviations $x - \bar{x}$, we square each deviation, we sum the squares, and finally we divide the total by $n - 1$ instead of N to get the variance.

Moreover, just as σ^2 and σ represent the variance and standard deviation of a population, respectively, we use the symbols s^2 and s to stand for the variance and standard deviation, respectively, of a sample. Thus,

and

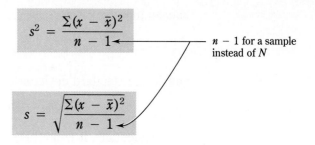

EXAMPLE 3.6 A sample of 10 drivers is selected, and each is asked how many years he or she has been driving. The results, in years, are

$$8, 15, 22, 11, 18, 4, 10, 16, 30, 16$$

Find the mean and standard deviation for the number of years this group has been driving.

SOLUTION Since we are told that the data represent a sample of drivers rather than a population, we have to determine $\bar{x}$ and s. Obviously, $n = 10$. We thus set up Table 3.2 with the data in numerical order.

TABLE 3.2

x	$x - \bar{x}$	$(x - \bar{x})^2$
4	$4 - 15 = -11$	121
8	$8 - 15 = -7$	49
10	$10 - 15 = -5$	25
11	$11 - 15 = -4$	16
15	$15 - 15 = 0$	0
16	$16 - 15 = 1$	1
16	$16 - 15 = 1$	1
18	$18 - 15 = 3$	9
22	$22 - 15 = 7$	49
30	$30 - 15 = 15$	225
$\Sigma x = 150$		$\Sigma(x - \bar{x})^2 = 496$

$$\bar{x} = \frac{150}{10} = 15 \text{ years}$$

From this table, we find that the sample mean $\bar{x} = 15$. Notice that we still divide by $n = 10$ here; it is only in the formula for the standard deviation that we divide by $n - 1$ instead of N. The sum of the individual squared deviations is 496, so that the sample variance is

$$s^2 = \frac{496}{n-1} = \frac{496}{10-1} = \frac{496}{9} = 55.11$$

and so the sample standard deviation is
$$s = \sqrt{55.11} = 7.42$$

When we calculated the standard deviation, either as a population parameter σ or as a sample statistic s, we essentially did it as a *procedure* rather than by emphasizing the formulas for these quantities. It is also possible to convert the previous formulas for σ and s to alternate (but equivalent) versions, known as *calculating* or *shortcut formulas*, which often involve somewhat less work if you are doing the calculations by hand. We do not go into the details of the algebraic conversion here. We simply quote the final results, namely,

$$\sigma = \sqrt{\frac{N(\Sigma x^2) - (\Sigma x)^2}{N^2}}$$

and

$$s = \sqrt{\frac{n(\Sigma x^2) - (\Sigma x)^2}{n(n-1)}}$$

In both formulas, $(\Sigma x)^2$ indicates that the sum of all the x values is squared whereas (Σx^2) indicates that each of the x values is first squared and then summed. If you use this calculating formula for s, be very careful that you divide by both n and $n-1$.

We illustrate the calculating-formula approach by repeating the calculations done in Example 3.6.

EXAMPLE 3.7 Use the calculating formula for s to find the sample standard deviation for the data in Example 3.6 on how long a group of drivers have been driving.

SOLUTION To apply the calculating formula, we also arrange the data in a table, Table 3.3.

TABLE 3.3

x	x^2
4	16
8	64
10	100
11	121
15	225
16	256
16	256
18	324
22	484
30	900
$\Sigma x = 150$	$\Sigma x^2 = 2746$

As a consequence, we find that

$$s = \sqrt{\frac{n(\Sigma x^2) - (\Sigma x)^2}{n(n-1)}}$$

$$= \sqrt{\frac{10(2746) - 150^2}{10(10-1)}}$$

$$= \sqrt{\frac{27{,}460 - 22{,}500}{10(9)}}$$

$$= \sqrt{\frac{4960}{90}}$$

$$= \sqrt{55.11}$$

$$= 7.42$$

which is the same value we obtained previously. ∎

In many cases, the calculating-formula approach involves considerably less work to obtain the same answer for σ or s. This is particularly true if the value for the mean, either μ or $\bar{x}$, is not an integer. On the other hand, if the mean is an integer, then our original approach is easier to apply since the numbers do not become so large. Further, with the wide availability of statistical software and handheld calculators that perform statistical calculations, many people feel that the calculating formulas have become obsolete.

By now, you should be able to calculate the standard deviation for a set of data, whether it is a population or a sample. You will usually be able to tell the difference either by the notation (either $\bar{x}$ or s is used to indicate a sample or μ or σ to indicate a population) or by the phrasing (the use of the word *sample* is an obvious giveaway; its absence may sometimes suggest that the data in question are a population). It is essential that you read all problems extremely carefully to distinguish between a sample and a population from the context.

At this point, most of you probably have at best a fuzzy idea of what a standard deviation is all about. This is typical at this stage in this course. The mean is a very obvious quantity, and you have worked with the notion of averages for many years. The standard deviation is a far less obvious concept; yet, as you will discover, it is almost as important as the mean. For now, you should be content with the descriptions we gave above in introducing the need for the standard deviation—it is a way of measuring the amount of spread or variability in a set of numbers. In subsequent sections, as we develop interpretations of what the standard deviation is and how it is used, you will find yourself gaining a greater understanding and mastery of it.

There is a general rule of thumb that allows you to verify that the value calculated for the standard deviation, either σ or s, is reasonable. For most sets of data, the standard deviation turns out to be very roughly equal to one-sixth of the range:

$$\sigma \approx \frac{\text{range}}{6} \quad \text{or} \quad s \approx \frac{\text{range}}{6}$$

This estimate should never be used as the value for the standard deviation since it is only a rough approximation. In fact, some statisticians use range/4 or range/5 as the approximation. As such, it is used only to make certain that you are in the right ballpark.

Computer Corner

Program 1: Statistical Analysis of Data also calculates the standard deviation for any set of data. Option 2 gives both sample standard deviation s and the population standard deviation σ, in addition to the mean, for any set of data entered, either from the keyboard or from a disk file. Although it does not specifically give the value for the variance, you can find it easily—just square the standard deviation.

Minitab Methods

Minitab also performs the calculation of the standard deviation for any set of data in any desired column, say C1. Simply enter the command

STDEV C1

and Minitab will print the standard deviation for column C1. This value is also printed as one portion of the results given in response to the command

DESCRIBE C1

Exercise Set 3.2

Mastering the Techniques

For the following data sets, calculate the mean, variance, and standard deviation, using both the definition and the calculating formula. Assume that the data represent a population in each case.

1. 11, 12, 15, 20, 23, 25, 30, 33, 37, 44
2. 6, 9, 10, 10, 13, 17, 18, 21
3. 122, 125, 128, 130, 132, 132, 134, 136, 138, 143
4. 305, 355, 327, 376, 388, 347, 360, 321, 372, 334
5. 5.4, 5.6, 5.7, 5.9, 6.0
6. 8.0, 8.2, 8.5, 8.7, 8.8, 9.0, 9.2, 9.6
7. Repeat Exercise 1, assuming the data represent a sample.
8. Repeat Exercise 2, assuming the data represent a sample.

9. Repeat Exercise 3, assuming the data represent a sample.
10. Repeat Exercise 4, assuming the data represent a sample.
11. Repeat Exercise 5, assuming the data represent a sample.
12. Repeat Exercise 6, assuming the data represent a sample.

Applying the Concepts

13. During one week, a hospital emergency room treats the following numbers of patients each day:

    160, 244, 126, 205, 140, 180, 144

    Find the mean, variance and standard deviation for the number of patients treated per day.

14. A supermarket has eight checkout lines, and the manager does a study to see how many people are waiting for each register during a particularly busy Saturday afternoon. The results, in minutes, are

    7, 9, 6, 2, 5, 8, 7, 6

    Find the mean, variance and standard deviation for the number of customers waiting on line.

15. Suppose the manager of the supermarket in Exercise 14 does a different study to measure the time that customers wait on line to check out. He selects a random group of 10 customers and finds the following times, in minutes:

    4, 7, 3, 4, 5, 2, 11, 7, 6, 8

    Find the mean and standard deviation for the waiting time based on these data.

16. A sample of 12 people enrolled in a special weight reduction program are studied. After 1 month, their weight losses, in pounds, are

    8, 12, 5, 21, 17, 10, 2, 8, 16, 20, 11, 8

    Find the mean, variance and standard deviation for the weight lost.

17. Consider the set of data in Exercise 1. Suppose that each entry is doubled. Calculate the mean, variance and standard deviation for the new set of numbers, and compare each to the corresponding value for the original set.

18. Consider the set of numbers in Exercise 1. Suppose now that each number has 10 subtracted from it. Find the mean, variance and standard deviation; compare each to the corresponding value for the original set.

19. Consider the set of numbers in Exercise 2. Suppose each number is multiplied by 10. Before doing the work, anticipate what values you should get for the mean, variance and standard deviation. Verify your predictions by performing the calculations.

20. Consider the set of numbers in Exercise 4. Suppose 300 is subtracted from each number. Before you do the work, what do you predict the results will be for the mean, variance and standard deviation? Verify your predictions.

21. Samples of two brands of pens are selected and tested to see how many hours they can be used before running out of ink. The results are as follows:

    Brand A: 23, 25, 27, 28, 30, 35

    Brand B: 16, 22, 28, 33, 46

Calculate the mean and standard deviation for each. Assume that both brands cost the same. If you consider only the mean in making a decision, which brand should you purchase? Would your decision change if you consider the standard deviation? If so, why?

22. Two typists are competing for a job which requires a high degree of accuracy. Samples of their work are studied to measure the number of typing errors they make per page. The results are

$$\text{Typist A:} \quad 0, 0, 2, 3, 5, 8, 10$$

$$\text{Typist B:} \quad 2, 3, 3, 4, 5, 5, 6, 6$$

Based on these data, which typist should be hired?

Computer Applications

Use Program 1: Statistical Analysis of Data to find the mean and standard deviation for the following data sets.

23. 21, 21, 22, 23, 24, 25, 26, 26, 27, 27, 27, 28, 28, 28, 28, 29, 29, 29, 30, 31
24. 53, 47, 59, 66, 36, 69, 84, 77, 42, 57, 51, 60, 78, 63, 46, 63, 42, 55, 63, 48, 75, 60, 58, 80, 44, 59, 60, 75, 49, 63
25. 2.8, 3.5, 7.2, 5.8, 6.3, 4.1, 5.7, 8.2, 2.3, 4.4, 7.1, 8.0, 6.8, 5.2, 4.3, 3.0, 3.6, 5.4, 6.3, 6.6, 5.7, 8.2, 4.9, 6.0, 7.2
26. 860, 940, 1120, 900, 840, 980, 1050, 1220, 860, 770, 1010, 870, 890, 910, 930, 1040, 1280, 1020, 970, 1330, 890, 980, 1260, 980, 760
27. 17, 39, 26, 20, 17, 49, 22, 36, 17, 17, 39, 26, 15, 20, 18, 22, 46, 17, 28, 17, 36, 19, 18, 46, 16, 19, 28, 15, 20, 48, 16, 18, 36, 29, 15, 16, 28, 36, 15, 18
28. Consider the set of data from Exercise 23. Replace the final entry (31) with 41, and use the program to calculate the new values of the mean and standard deviation. How do they compare with the original values? Repeat this process by replacing 41 with 51. How do the mean and standard deviation change? Finally, replace 51 with 28. How do the mean and the standard deviation compare to the original values?
29. Use the data on baseball records in Appendix I (or in the computer data file HOMERUN) to find the mean and standard deviation for the number of home runs for each player on the list.
30. Find the mean and standard deviation for the prices of the sample set of stocks you saved as the data file STOCKS.

3.3 Significance of the Standard Deviation

In Section 3.2, we described how to calculate the standard deviation as a measure of the spread or variability in a set of data. We now consider the question of *how* the standard deviation measures this spread and, equally important, how it is used.

Consider a population having a mean μ and a standard deviation σ. Since the mean can be thought of as the center of the data, all the entries are therefore spread out around the mean. We can use the standard deviation to

FIGURE 3.3

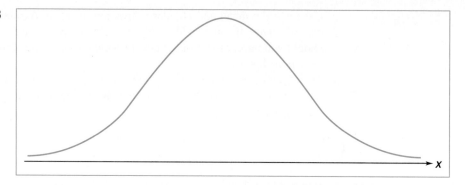

describe this spread in a very precise way. The following *empirical rule* specifies how close the data values are to the mean. The rule applies to bell-shaped sets of data such as that shown in Figure 3.3. Fortunately, many sets of data fall into such a pattern.

Empirical Rule

1. Approximately 68% of all members of the population lie within 1 standard deviation of the mean μ.
2. Approximately 95% of all members of the population lie within 2 standard deviations of the mean μ.
3. Approximately 99.7% of all members of the population lie within 3 standard deviations of the mean μ.

See Figure 3.4.

Suppose we have a bell-shaped population with mean $\mu = 100$ and standard deviation $\sigma = 20$. According to this rule, we can predict that approximately 68% of the population lies within 1 standard deviation (20 points)

FIGURE 3.4

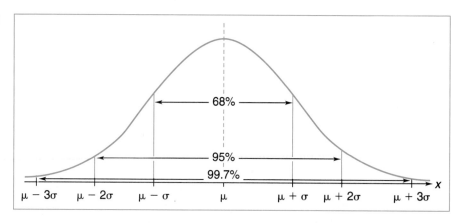

FIGURE 3.5

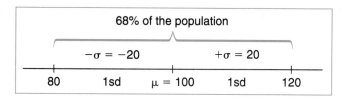

of the mean $\mu = 100$. That is, approximately 68% of the numbers in this population lie between $\mu - \sigma = 100 - 20 = 80$ and $\mu + \sigma = 100 + 20 = 120$. This is an interval centered at the mean μ that extends left and right 1 standard deviation or 20 points. See Figure 3.5. Similarly, a spread of 2 standard deviations corresponds to moving left and right 40 points from the mean. This produces the interval from 60 to 140 which, according to the rule, should contain approximately 95% of the population. See Figure 3.6.

FIGURE 3.6

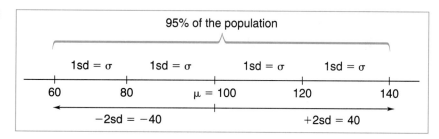

Further, approximately 99.7% of the population lies within 3 standard deviations of μ. This represents a spread of 60 points on each side of the mean $\mu = 100$ and so produces an interval from 40 to 160. Thus, virtually the entire population lies in this interval. See Figure 3.7.

Now consider a different population, one having the same mean $\mu = 100$, but with a standard deviation σ of 6 instead of 20. According to the empirical rule, we should expect that approximately 68% of the members of this new population lie within 1 standard deviation (6 points) of the mean, or from 94 to 106. Similarly, approximately 95% of this population should fall within 2 standard deviations (12 points) of the mean, or within the interval from 88 to 112. Finally, virtually all members of this population lie within 3 standard deviations (18 points) of the mean, or between 82 and 118.

FIGURE 3.7

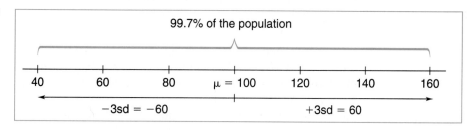

These two examples show that when the standard deviation increases, the spread of the data also increases. That is, it is necessary to go farther from the mean to encompass 68% of the entries. When the standard deviation is small, then 68% of the population lies in a considerably smaller range. An analogous argument applies when we want to characterize the spread in the data to include 95% (or 99.7%) of the population. The larger the standard deviation, the greater the spread needed to include 95% (or 99.7%) of the data.

Suppose instead that the population has a mean $\mu = 100$ and a standard deviation $\sigma = 2$. We then conclude that approximately 68% of the population should lie in the interval from 98 to 102, that about 95% of the population should lie in the interval from 96 to 104, and that about 99.7% should lie in the interval from 94 to 106.

It is important to realize that these three values—68%, 95%, and 99.7%—are just approximations. For certain types of populations, they will be highly accurate; for others, they will be only rough estimates. To see how accurate they can be, consider the set of numbers discussed in Chapter 2 representing the heights of students in a class. The values used, in inches, were

60, 61, 61, 62, 62, 62, 63, 63, 63, 64, 64, 64, 64, 65, 65, 65, 65, 65,
66, 66, 66, 66, 67, 67, 67, 68, 68, 68, 69, 69, 70, 71, 71, 72, 72, 74

The mean of these data is $\mu = 66$ inches. Further, if we apply Program 1 to these data or perform the calculations by hand, the standard deviation $\sigma = 3.416$. Therefore, the interval corresponding to 1 standard deviation is centered at $\mu = 66$ and goes up and down 3.416, so that it extends from 62.584 to 69.416 inches. Consequently, all the heights from 63 up through 69 inches fall into this interval. There are precisely 24 such heights out of the 36, and this corresponds to $\frac{24}{36} = 66.7\%$ which is very close to the predicted value of 68%.

Moreover, a range of 2 standard deviations involves a spread of 6.831 each way from the mean, which produces the interval from 59.169 to 72.831. Thus, all heights from 60 up through 72 inches will lie in this interval, and there are 35 of them. This corresponds to $\frac{35}{36} = 97\%$. Finally, a spread of 3 standard deviations produces the interval from 55.753 to 76.247, which includes all members of the group.

EXAMPLE 3.8 The mean score of all students taking the verbal portion of the SAT during a recent year was 421. If the standard deviation of their scores was 80, use the empirical rule to describe the distribution of the scores, assuming they are bell-shaped.

SOLUTION According to the empirical rule, approximately 68% of all students taking the SAT verbal section had scores falling within 1 standard deviation of the mean $\mu = 421$, or from 341 to 501. Moreover, approximately 95% fell within 2 standard deviations of μ and so were between 261 and 581. Furthermore, if this interval contains approximately 95% of the population, then only about 5% of the population falls outside the interval. That is, only about 5% of all people who took the SAT verbal section had scores below 261 or above 581. Finally, about 99.7% of all students had scores between 181 and 661. (Of course, the minimum possible score is actually 200.) ∎

Although the empirical rule we have been using is accurate for bell-shaped sets of data, statisticians use a rule for populations which have other shapes. *Chebyshev's Inequality* applies to all sets of data, bell-shaped or otherwise.

> **Chebyshev's Theorem**
>
> Let k be any number greater than or equal to 1. The proportion of measurements in any set of data that lie within k standard deviations of the mean μ is at least
>
> $$1 - \frac{1}{k^2}$$

For $k = 2$, this formula tells us that at least $1 - \frac{1}{4} = \frac{3}{4} = 75\%$ of *any* population lies within 2 standard deviations of the mean. For $k = 3$, at least $1 - \frac{1}{9} = \frac{8}{9} = 88.9\%$ of any population lies within 3 standard deviations of the mean. For $k = 4$, at least $1 - \frac{1}{16} = \frac{15}{16} = 93.8\%$ of the data lies within 4 standard deviations of the mean.

Let us compare these results to the ones given by the empirical rule for bell-shaped data sets. Chebyshev's Inequality guarantees that at least 75% of any population lies within 2 standard deviations of μ while the empirical rule tells us that approximately 95% of the data from a bell-shaped population fall within the same 2 standard deviations of μ. Thus, Chebyshev's Inequality is less precise for bell-shaped populations. However, it is the best result available for non-bell-shaped populations.

Exercise Set 3.3

Mastering the Techniques

Given the following mean μ and standard deviation σ for a set of bell-shaped data, determine the intervals which will contain approximately 68% and 95% of all members of the population.

1. $\mu = 30, \sigma = 5$
2. $\mu = 30, \sigma = 9$
3. $\mu = 160, \sigma = 10$
4. $\mu = 160, \sigma = 25$
5. $\mu = 400, \sigma = 120$
6. $\mu = 400, \sigma = 20$
7. $\mu = 7, \sigma = 1.5$
8. $\mu = 7, \sigma = 2.4$
9. $\mu = 11.3, \sigma = 2.6$
10. $\mu = 11.3, \sigma = .8$

Applying the Concepts

In the following exercises, assume that the underlying population is bell-shaped.

11. The mean score on a recent SAT math section among college-bound students was 475. If the standard deviation was 75, within what interval of scores would you expect 68% of all such students to fall? 95%? 99.7%?
12. The mean SAT verbal score among college-bound students was 431 with a standard deviation of 92. Find intervals of scores which would encompass approximately 68% and 95% of the scores for all such students. What scores were achieved by approximately the top $2\frac{1}{2}$% of all such students?
13. A certain region records an average of 45 inches of rainfall per year with a standard deviation of 11 inches. Within what range of values of rainfall, in inches, would you expect approximately 68% of all years to receive?
14. A dairy fills its 1-quart milk containers to an average of 33 ounces with a standard deviation of .5 ounce. Within what range of values (in ounces) would you expect approximately 68% of all such containers to fall? What range of values would cover 95% of all containers? Approximately what percentage of all containers will be underfilled? (Assume the population is bell-shaped.)
15. Test records indicate that a certain group of cars have a mean gas mileage of 28 miles per gallon on the highway with a standard deviation of 2.5 miles per gallon. Approximately what percentage of all such cars would you expect to get between 25.5 and 30.5 miles per gallon? If government regulations specify that a minimum of 23 miles per gallon is required for this type of car, approximately what percentage of the cars would not satisfy the law? (Assume the population is roughly bell-shaped.)

Computer Applications

16. Using the data file BATAVG for the batting averages of baseball players and the values you previously found for the mean and standard deviation of these data, determine intervals of batting averages that should contain approximately 68% and 95% of all players listed. Compare these results to the actual list of batting averages given in Appendix I. What percentages of players actually fall within the two intervals you calculated?
17. Repeat Exercise 16 for the data file TEMPS listing temperatures in a variety of cities. It will probably be easier for you to use Option 3 of Program 1 to put the data in numerical order before you attempt to compare the predicted intervals to the actual results.
18. Repeat Exercise 16 for the data file STOCKS listing stock prices for a variety of stocks.

3.4 z Values

People are often interested in comparing individuals in different fields to see who is the best in some sense. It might be the best athlete of all time, the greatest man-made wonder in the world or one of many other possible categories. The problem in trying to make such comparisons is that there are always factors in each case which make it very hard to compare. We now illustrate one application of the standard deviation for a set of data which allows us to make such comparisons in a fairly simple manner.

Consider the problem of determining who is the greatest athlete in history. While there are many different ways that this question can be asked, we restrict our attention to one particular category in each of three sports. In baseball, we consider the number of home runs over a career; in basketball, the total number of points scored; and in football, the total number of yards rushing for a running back. The following table lists the top 10 players in each category:

Leading Home Run Hitters		Leading Rushers		All-Time Leading NBA Scorers	
Hank Aaron	755	Walter Payton	16,726 yds	Kareem Abdul-Jabbar	38,387 points
Babe Ruth	714	Tony Dorsett	12,739	Wilt Chamberlain	31,419
Willie Mays	680	Jim Brown	12,312	Elvin Hayes	27,313
Frank Robinson	586	Franco Harris	12,120	Oscar Robertson	26,710
Harmon Killebrew	573	Eric Dickerson	11,903	John Havlicek	26,395
Reggie Jackson	563	John Riggins	11,352	Moses Malone	25,737
Mike Schmidt	548	O. J. Simpson	11,236	Alex English	25,613
Mickey Mantle	536	Ottis Anderson	10,101	Jerry West	25,192
Jimmy Foxx	534	Earl Campbell	9,407	Adrian Dantley	23,177
Ted Williams	521	Jim Taylor	8,597	Elgin Baylor	23,149

Clearly, each sport is measuring a totally different achievement. Consequently, there is no direct way to select the single highest-rated player among the three groups. Nevertheless, it is important to be able to compare entries from different groups such as these.

Before we do anything formally, let's first decide what we mean by the phrase *single best athlete*. Within each of these three groups, it is obvious which player is the most outstanding. The top-rated player is better than any of his direct competitors on the list of the top 10 of all time, let alone compared to all the other players in each sport. However, what we want to determine is which of these top three players is the most exceptional—the one who is rated farthest above all the others in his own group. The key to doing this is the standard deviation.

Consider the top 10 home run hitters. First we find the mean and standard deviation for the number of home runs in this select group. To distinguish these parameter values for home runs from the comparable parameters for the other sports, we use subscripts. Thus, μ_A and σ_A stand for the mean and standard deviations of number of home runs in this list. We find

$$\mu_A = 601.1 \qquad \sigma_A = 79.63$$

Thus, we see that Hank Aaron with 755 home runs has 153.9 home runs above the mean $\mu_A = 601.1$. However, by itself, this is not particularly useful information since it depends on only the number of home runs, and so we cannot compare this value to the results for a different sport. On the other hand, using the standard deviation $\sigma_A = 79.63$, we see that Aaron's 755 home runs are approximately 2 standard deviations above the mean. It is this type of comparison in terms of the *number* of standard deviations above or below the mean which will allow us to determine which player is farthest above the mean of his own group.

To formalize these ideas, we introduce the notion of the *z value* or *z score* or *standard score* associated with a measurement x. The *z* value is given by

$$z = \frac{x - \mu}{\sigma} \tag{3.1}$$

> The *z* value associated with a measurement x represents the number of standard deviations that x lies away from the mean μ.

That is, if x were 2 standard deviations above the mean, then the corresponding *z* value would be $z = 2$, regardless of the values for μ and σ. For Hank Aaron with $x = 755$ home runs, $\mu_A = 601.1$ and $\sigma_A = 79.63$, so his *z* value is

$$z = \frac{755 - 601.1}{79.63} = \frac{153.9}{79.63} = 1.93$$

In a similar way, Babe Ruth, the second all-time home run hitter with $x = 714$ home runs, has a corresponding *z* value of

$$z = \frac{714 - 601.1}{79.63} = \frac{112.9}{79.63} = 1.42$$

This means that Ruth's achievement falls just about one and a half standard deviations above the mean of the 10 best home run hitters. The worst hitter on the list, Ted Williams with $x = 521$, has a *z* value of

$$z = \frac{521 - 601.1}{79.63} = \frac{-80.1}{79.63} = -1.01$$

The fact that this comes out negative simply indicates that Williams' value of $x = 521$ is approximately 1 standard deviation *below* the mean.

> If a number x is **above** the mean μ, the corresponding *z* value will be **positive**.
>
> If x is **below** the mean μ, the corresponding *z* value will be **negative**.

The mean itself, $x = \mu$, corresponds to a z value of

$$z = \frac{x - \mu}{\sigma} = \frac{\mu - \mu}{\sigma} = \frac{0}{\sigma} = 0$$

Having found the z value for the best all-time home run hitter, we now repeat this procedure for best running back and best basketball scorer. Among the 10 best rushers, the mean and standard deviation are

$$\mu_B = 11{,}649 \quad \text{and} \quad \sigma_B = 2114$$

Consequently, using formula (3.1) for the z value, we find that corresponding to Walter Payton's $x = 16{,}726$ yards,

$$z = \frac{16{,}726 - 11{,}649}{2114} = \frac{5077}{2114} = 2.40$$

This tells us that Payton's achievement is more than two and a third standard deviations above the mean $\mu_B = 11{,}649$ of his competition in the 10-best list. However, since the z value for Hank Aaron is $z = 1.93$, Payton's achievement is more outstanding relative to his competition than is Aaron's achievement compared to his competition.

Finally, for leaders in basketball scoring, the mean and standard deviation are

$$\mu_C = 27{,}309 \quad \text{and} \quad \sigma_C = 4301.4$$

Therefore, the z value for Kareem Abdul-Jabbar with $x = 38{,}387$ is

$$z = \frac{38{,}387 - 27{,}309}{4301.4} = \frac{11{,}078}{4301.4} = 2.58$$

That is, Abdul-Jabbar's total is approximately two and a half standard deviations above the mean of his group. Thus, his is the most outstanding achievement, compared to his competition, of all three groups. In other words, the z value, 2.58 for Abdul-Jabbar, is larger than the value for either Hank Aaron ($z = 1.93$) or Walter Payton ($z = 2.40$).

The notion of z values as a way of measuring the number of standard deviations that a measurement x lies either above or below the mean μ arises throughout our study of statistics. It is one of the most important and useful concepts we have at our disposal.

Incidentally, as you will see later, z values are usually written to two decimal places. If a calculation involves more than two decimals, then it should be rounded appropriately. Thus, if $z = 5/3 = 1.666667$, we round to $z = 1.67$ while $z = 4/3 = 1.33333$ is rounded to $z = 1.33$.

Most of the calculations involving z values that we will encounter in this book are similar to what we did above. We will start with the mean μ and the standard deviation σ for a set of data and will have to determine the value of z corresponding to a particular measurement x by using formula (3.1).

We will also occasionally have cases in which this procedure is reversed. That is, rather than knowing the value for the measurement x and having to find the z value, we will be given the z value and have to find the measurement x that corresponds to it. The simplest way to do this is by using the following formula, which comes from formula (3.1) by solving algebraically for x:

$$x = \mu + z\sigma \tag{3.2}$$

To derive this, we start with the formula

$$z = \frac{x - \mu}{\sigma}$$

Upon multiplying both sides by σ, we obtain

$$x - \mu = z\sigma$$

When we solve this equation for x, we obtain Equation (3.2).

EXAMPLE 3.9 The mean and standard deviation of a population are 2500 and 400, respectively. Find

(a) The z value corresponding to $x = 3000$
(b) The z value corresponding to $x = 1993$
(c) The x value corresponding to $z = 2$
(d) The x value corresponding to $z = 1.42$
(e) The x value corresponding to $z = -.35$

SOLUTION To determine the z values in parts a and b, we use formula (3.1) with $\mu = 2500$ and $\sigma = 400$. Therefore, if $x = 3000$,

$$z = \frac{3000 - 2500}{400} = \frac{500}{400} = 1.25$$

and if $x = 1993$,

$$z = \frac{1993 - 2500}{400} = \frac{-507}{400} = -1.27$$

To determine the x measurement based on a given z value, we use formula (3.2). Thus, if $z = 2$,

$$x = 2500 + 2(400) = 2500 + 800 = 3300$$

That is, $x = 3300$ is 2 full standard deviations above the mean $\mu = 2500$. If $z = 1.42$, then

$$x = 2500 + 1.42(400) = 2500 + 568 = 3068$$

and if $z = -.35$, then

$$x = 2500 + (-.35)(400) = 2500 - 140 = 2360$$

Notice in this last case that the value of x came out below the mean $\mu = 2500$ since the z value is negative. ∎

Exercise Set 3.4

Mastering the Techniques

1. Given $\mu = 40$ and $\sigma = 10$, find the z value corresponding to each of the following x values: $x = $ 50, 55, 63, 77, 47, 31, 40.
2. Given $\mu = 130$ and $\sigma = 12$, find the z value corresponding to each of the following x values: $x = $ 150, 145, 161, 122, 115, 100.
3. Given $\mu = 40$ and $\sigma = 10$, find the x measurement corresponding to each of the following z values: $z = $ 1.5, 2.25, .47, 2.06, 1.94, -2, -1.28.
4. Given $\mu = 130$ and $\sigma = 12$, find the x measurement corresponding to each of the following z values: $z = $ 1.5, 2.25, .47, 2.06, 1.94, -2, -1.28.

Applying the Concepts

5. The heights (in inches) of the starting players on two basketball teams are as follows:

    A: 84, 80, 77, 75, 74

    B: 81, 76, 76, 74, 73

    (a) Which player is the tallest relative to the rest of his team?
    (b) Which player is the shortest relative to the rest of his team?
6. The gross receipts (in millions of dollars) for the top five movies during Christmas week in two different years are

    Year 1: 12.2, 10.6, 9.4, 9.0, 8.8

    Year 2: 13.5, 11.6, 11.0, 10.1, 8.8

    (a) On the average, which is the better year?
    (b) Which film is the most successful compared to its competition?
    (c) Which film is the least successful compared to its competition?
7. The comparison made in this section among home run hitters, running backs and basketball scorers applies as of the time this book was written. Some of these listings will have certainly changed by the time you read this, since some of the players listed are still active. Find the current lists for the top players in each of these three categories, and see who is the current best of the best.

Computer Applications

8. Depending on the season, compare the relative performance of players in two different leagues. For instance, in baseball, you should compare the leading hitter in the American League to the leading hitter in the National League to see which one is relatively farther above the mean of his group. Use Program 1: Statistical Analysis of Data to calculate the mean and standard deviation for the two sets of data.
9. Analyze the relative performance of the best player on a team compared to the rest of his team during one year to the performance of the best player in a previous year by researching statistics in a back issue of a newspaper or in a sports encyclopedia.

10. Analyze the relative performance of the best-rated car in a category such as time to go from 0 to 50 miles per hour compared to the time needed for the best-rated car several years ago. Consult current and back issues of most car and road magazines to find the data.
11. Analyze the relative heights of the tallest man and woman in your class; the shortest man and woman in your class.

3.5 Percentiles, Quartiles and Box Plots

We now consider another way to describe a data set which is an extension of the notion of median. As we saw in Section 3.1, the median is determined in such a way that when the data are arranged in numerical order, 50% of the data entries lie below the median and 50% lie above it.

> The **kth percentile** for a set of data in numerical order is that value x having the property that k percent of the data entries lie at or below x.

We use P_k to represent the kth percentile. For example, the 90th percentile, P_{90}, of a data set is the value of x such that 90% of the data fall at or below x (and hence only 10% of the data are above x). The 50th percentile is that value of x for which 50% of the data are below x. Thus, the 50th percentile is precisely the same as the median.

We also say that an individual, a score, or a measurement is *in the kth percentile* to indicate that k percent of the population lies below the value in question.

The rules for determining a given percentile are analogous to the rules we developed to calculate the median. Suppose, for example, we have 40 entries in a set of *ordered data* and, for convenience, we label them

$$x_1, x_2, x_3, \ldots, x_{40}$$

The median or 50th percentile is between the 20th and 21st entries, and so we average them to produce

$$P_{50} = \frac{x_{20} + x_{21}}{2}$$

In a similar way, the 20th percentile falls between x_8 and x_9 since there are 8 entries $x_1, x_2, \ldots, x_8$ (20% of the total number of entries) below it and 32 entries $x_9, x_{10}, \ldots, x_{40}$ (80% of the total) above it. We therefore locate the 20th percentile at the average of x_8 and x_9:

$$P_{20} = \frac{x_8 + x_9}{2}$$

Further, the 75th percentile falls between x_{30} and x_{31}, since 75% of the total number of entries (30 values) lie below it and 25% of the total entries (10 values) lie above it. Therefore,

$$P_{75} = \frac{x_{30} + x_{31}}{2}$$

We note that the method for finding a given percentile is not as simple as merely observing whether the number of entries is odd or even, as we did for the median. Since the value for a given percentile depends on *both* the number of entries in the list and the percentile value k desired, each case must be considered and analyzed on its own.

EXAMPLE 3.10 For the data given on the American League baseball records in Appendix I, find the 50th, 20th, and 75th percentiles for batting averages.

SOLUTION The list includes the top 95 full-time hitters. The 50th percentile, or the median, is precisely equal to the middle or 48th entry, and this corresponds to Stankiewicz's average of .268.

To locate the 20th percentile, first we notice the data are listed from highest to lowest. However, to find the 20th percentile (that value below which 20 percent of the data lie), we must think of the batting averages as ranging from the lower end to the higher end. We therefore find that 20% of 95 is 19, and so we take the 19th entry from the bottom of the list. This is Reynold's .247. Thus, the 20th percentile is $P_{20} = .247$.

To determine the 75th percentile, first we find that 75% of 95 entries is 71.25. Hence, we average the 71st and 72nd entries from the bottom of the list, or equivalently the 24th and 23rd entries from the top of the list. These are Lofton's .285 and Olerud's .284, and so we find that the 75th percentile is just

$$P_{75} = \frac{.285 + .284}{2} = .2845$$

■

Percentiles divide a set of data into 100 equal parts. A related concept is that of the *quartile*, which divides a set of data into four equal parts. The first quartile locates the bottom 25% of the data; the second quartile identifies the second 25% or middle of the data; the third quartile locates the third 25% or three-quarters mark of the data. These three quartiles are denoted, respectively, by Q_1, Q_2, and Q_3.

> Quartile Q_1 is that value at or below which 25% of all data entries fall.
>
> Quartile Q_2 is that value at or below which 50% of all data entries fall.
>
> Quartile Q_3 is that value at or below which 75% of all data entries fall.

FIGURE 3.8

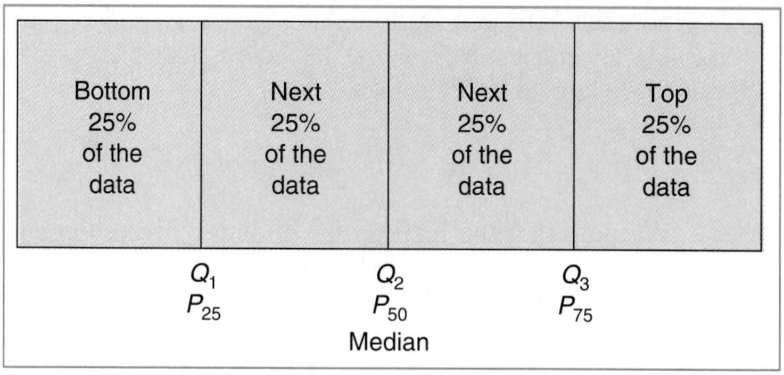

| Bottom 25% of the data | Next 25% of the data | Next 25% of the data | Top 25% of the data |

Q_1 $\quad\quad\quad$ Q_2 $\quad\quad\quad$ Q_3
P_{25} $\quad\quad\quad$ P_{50} $\quad\quad\quad$ P_{75}
$\quad\quad\quad\quad$ Median

Thus, the first quartile is precisely equivalent to the 25th percentile. The second quartile is equivalent to the 50th percentile, or the median. The third quartile is equivalent to the 75th percentile. We illustrate these ideas in Figure 3.8.

Furthermore, we say that a given value of x from a data set lies *in* a given quartile if it lies between that quartile value and the preceding one. Thus, a value of x is in the second quartile if it falls within the second quartile of the data set.

EXAMPLE 3.11 Find the first, second and third quartiles for the American League batting averages.

SOLUTION We have already found that the median or 50th percentile is .268, and this is also the second quartile. Further, we already found the 75th percentile, .2845, and this is identical to the third quartile. Thus, all that remains is to determine the first quartile. Since 25% of 95 is 23.75, we must average the 23rd and 24th entries from the bottom of the list, and these are Leius' .249 and White's .248, so that the first quartile Q_1 is .2485. ∎

Another variation on the percentile is the *decile*, which divides a set of data into 10 equal parts. Thus, the first decile locates the bottom 10% of the data and is equivalent to the 10th percentile, the second decile locates the next tenth of the data and is equivalent to the 20th percentile, and so forth.

EXAMPLE 3.12 Find the first and fourth deciles for the data on the American League batting averages.

SOLUTION Since the first decile is equivalent to the 10th percentile and 10% of 95 is 9.5, the first decile is given by the average of the 9th and 10th entries from the bottom of the table. These are Milligan's .240 and Tettleton's .238, and so the first decile is .239. Similarly, since 40% of 95 is 38, the fourth decile

FIGURE 3.9

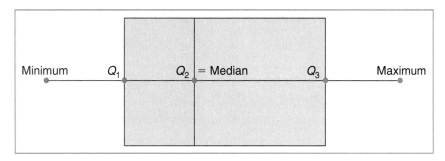

is the 38th entry from the bottom of the list. Therefore,

$$\text{Fourth decile} = .262 \qquad \blacksquare$$

Statisticians have recently developed an effective graphical means for displaying the information given by the quartiles of a data set. It is known as a *box-and-whisker plot* or more simply as a *box plot*. To construct it, first we organize a data set into four groups based on the lowest value, the first quartile Q_1, the second quartile (or median) Q_2, the third quartile Q_3, and the largest value. Typically, these values are set up along a horizontal line running from left to right. Furthermore, the first and third quartiles are used to construct a box centered on this line, as shown in Figure 3.9. Finally, a third vertical line is drawn which corresponds to the median. The three values used—Q_1, Q_2, and Q_3—are sometimes called *hinges*.

A box plot provides a useful picture of the spread of the data. In particular, the box portion corresponds to the middle 50% of the data entries (those between the 25th and 75th percentiles). For typical bell-shaped data sets, the middle 50% of the data within the box are much more closely clustered about the center than the outlying 25% at each end. Therefore, for such sets of data, the length of the box is usually considerably less than half the distance between the minimum and maximum values. The box plot displays both the extremes in the data set and the location of the middle 50% of the values about the median.

EXAMPLE 3.13 Construct the box plot for the American League batting averages.

SOLUTION We note that the minimum batting average listed is .222 while the maximum is .343. We have already determined that the median is .268 and that the first quartile is .2485 while the third quartile is .2845. Consequently, the box plot for these data is as shown in Figure 3.10. ∎

FIGURE 3.10

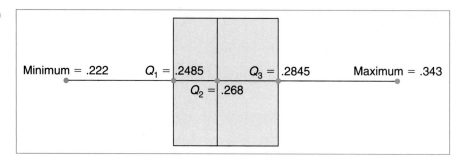

Notice that in the box plot in Figure 3.10, the box portion, which represents the middle 50% of all batting averages, actually covers a disproportionately small part of the range from .222 to .343. Thus, we see that the bulk of the players have averages that are clustered fairly tightly together between .248 and .285. Moreover, the 50% of the batting averages that are outside this interval tend to be more spread out than the ones inside the interval.

Computer Corner

You can use Program 1: Statistical Analysis of Data to calculate the quartiles for any data set and to construct the box plot associated with the data. Once you have entered your data, either from the keyboard or from a disk file, Option 3 arranges the data in order, either ascending or descending, and then shows the median and the first and third quartiles. Option 7 constructs the box-and-whisker plot, and this displays the minimum and maximum entries in the data as well as the median and the first and third quartiles.

Minitab Methods

Minitab is very useful for producing the box plot for any set of data you have entered, say in column C1. The command

BOXPLOT C1

produces the box plot of the data in column C1, along with values for the various quartiles. In addition, the command **DESCRIBE C1** that we discussed previously will produce a variety of statistical measures for the data in column C1, including the first and third quartiles. Figure 3.11 shows a typical Minitab session.

FIGURE 3.11

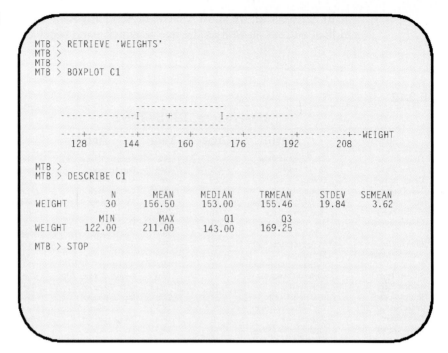

Exercise Set 3.5

Applying the Concepts

1. Use the data set in Appendix I on batting averages in the National League to find **(a)** the three quartiles; **(b)** the 2nd, 3rd, and 7th deciles; and **(c)** the 95th percentile. Draw the box plot for this set of data.
2. Use the data set on home runs in the National League to determine **(a)** the three quartiles; **(b)** the 2nd, 4th, and 6th deciles; and **(c)** the 98th percentile. Draw the box plot for this set of data.
3. Use the data set on temperature readings in Appendix II to determine **(a)** the three quartiles; **(b)** the 2nd, 3rd, and 8th deciles; and **(c)** the 95th percentile. Draw the box plot for this data set.

3.6 Comparing Two Sets of Data

Some of the most important applications of statistics involve situations where we have two related sets of data and wish to compare them. For instance, we might want to compare women and men professors to see if they are paid comparable salaries. We might want to compare the prices of homes in two cities. We might want to compare the batting averages of players in the American League and the National League. Typically, the objective of such a comparison is to see if there is a significant difference in some measure of the data, say between the means of the two groups or between the medians of the groups.

In Figure 3.12, we show a list of all Academy Award winners for best actor and best actress as well as the age at which each received the award.

FIGURE 3.12*

Best Actors, 1928–1991

Year	Actor	Movie	Age	Year	Actor	Movie	Age
1928	Emil Jannings	*The Way of All Flesh*	44	1959	Charlton Heston	*Ben Hur*	35
1929	Warner Baxter	*In Old Arizona*	38	1960	Burt Lancaster	*Elmer Gantry*	47
1930	George Arliss	*Disraeli*	46	1961	Maximilian Schell	*Judgment at Nuremburg*	31
1931	Lionel Barrymore	*A Free Soul*	53	1962	Gregory Peck	*To Kill a Mockingbird*	46
1932	Fredric March	*Dr. Jekyll and Mr. Hyde*	35	1963	Sidney Poitier	*Lilies of the Field*	39
1932	Wallace Beery	*The Champ*	47	1964	Rex Harrison	*My Fair Lady*	56
1933	Charles Laughton	*The Private Life of Henry VIII*	34	1965	Lee Marvin	*Cat Ballou*	41
				1966	Paul Scofield	*A Man for All Seasons*	44
1934	Clark Gable	*It Happened One Night*	33	1967	Rod Steiger	*In the Heat of the Night*	42
1935	Victor McLaglen	*The Informer*	49	1968	Cliff Robertson	*Charly*	43
1936	Paul Muni	*The Story of Louis Pasteur*	41	1969	John Wayne	*True Grit*	62
1937	Spencer Tracy	*Captains Courageous*	37	1970	George C. Scott	*Patton*	43
1938	Spencer Tracy	*Boys' Town*	38	1971	Gene Hackman	*The French Connection*	40
1939	Robert Donat	*Goodbye Mr. Chips*	34	1972	Marlon Brando	*The Godfather*	48
1940	James Stewart	*The Philadelphia Story*	32	1973	Jack Lemmon	*Save the Tiger*	48
1941	Gary Cooper	*Sergeant York*	40	1974	Art Carney	*Harry and Tonto*	56
1942	James Cagney	*Yankee Doodle Dandy*	43	1975	Jack Nicholson	*One Flew over the Cuckoo's Nest*	38
1943	Paul Lukas	*Watch On the Rhine*	48				
1944	Bing Crosby	*Going My Way*	43	1976	Peter Finch	*Network*	60
1945	Ray Milland	*The Lost Weekend*	40	1977	Richard Dreyfuss	*The Goodbye Girl*	32
1946	Fredric March	*The Best Years of Our Lives*	49	1978	Jon Voight	*Coming Home*	40
				1979	Dustin Hoffman	*Kramer vs. Kramer*	42
1947	Ronald Colman	*A Double Life*	56	1980	Robert De Niro	*Raging Bull*	37
1948	Laurence Olivier	*Hamlet*	41	1981	Henry Fonda	*On Golden Pond*	76
1949	Broderick Crawford	*All the King's Men*	38	1982	Ben Kingsley	*Gandhi*	39
1950	Jose Ferrer	*Cyrano de Bergerac*	38	1983	Robert Duvall	*Tender Mercies*	55
1951	Humphrey Bogart	*The African Queen*	52	1984	F. Murray Abraham	*Amadeus*	45
1952	Gary Cooper	*High Noon*	51	1985	William Hurt	*Kiss of the Spider Woman*	35
1953	William Holden	*Stalag 17*	35	1986	Paul Newman	*Color of Money*	61
1954	Marlon Brando	*On the Waterfront*	30	1987	Michael Douglas	*Wall Street*	33
1955	Ernest Borgnine	*Marty*	38	1988	Dustin Hoffman	*Rain Man*	51
1956	Yul Brynner	*The King and I*	41	1989	Daniel Day-Lewis	*My Left Foot*	58
1957	Alec Guinness	*The Bridge on the River Kwai*	43	1990	Jeremy Irons	*Reversal of Fortune*	43
				1991	Anthony Hopkins	*The Silence of the Lambs*	55
1958	David Niven	*Separate Tables*	49				

*This illustration was suggested by an article by Richard Brown and Gretchen Davis in *The Mathematics Teacher*, vol. 83, February 1990.

FIGURE 3.12 (Continued)

Best Actresses, 1928–1991

Year	Actress	Movie	Age	Year	Actress	Movie	Age
1928	Janet Gaynor	Seventh Heaven	22	1963	Patricia Neal	Hud	37
1929	Mary Pickford	Coquette	36	1964	Julie Andrews	Mary Poppins	30
1930	Norma Shearer	The Divorcee	26	1965	Julie Christie	Darling	24
1931	Marie Dressler	Min and Bill	62	1966	Elizabeth Taylor	Who's Afraid of Virginia Woolf?	34
1932	Helen Hayes	The Sin of Madelon Claudet	32	1967	Katharine Hepburn	Guess Who's Coming to Dinner	60
1933	Katharine Hepburn	Morning Glory	26				
1934	Claudette Colbert	It Happened One Night	29	1968	Katharine Hepburn	The Lion in Winter	61
1935	Bette Davis	Dangerous	27	1968	Barbra Streisand	Funny Girl	26
1936	Luise Rainer	The Great Ziegfeld	24	1969	Maggie Smith	The Prime of Miss Jean Brodie	35
1937	Luise Rainer	The Good Earth	25				
1938	Bette Davis	Jezebel	30	1970	Glenda Jackson	Women in Love	34
1939	Vivien Leigh	Gone with the Wind	26	1971	Jane Fonda	Klute	34
1940	Ginger Rogers	Kitty Foyle	29	1972	Liza Minnelli	Cabaret	26
1941	Joan Fontaine	Suspicion	24	1973	Glenda Jackson	A Touch of Class	37
1942	Greer Garson	Mrs. Minever	34	1974	Ellen Burstyn	Alice Doesn't Live Here Anymore	42
1943	Jennifer Jones	The Song of Bernadette	24				
1944	Ingrid Bergman	Gaslight	29	1975	Louise Fletcher	One Flew over the Cuckoo's Nest	41
1945	Joan Crawford	Mildred Pierce	41				
1946	Olivia de Havilland	To Each His Own	30	1976	Faye Dunaway	Network	35
1947	Loretta Young	The Farmer's Daughter	34	1977	Diane Keaton	Annie Hall	31
1948	Jane Wyman	Johnny Belinda	34	1978	Jane Fonda	Coming Home	41
1949	Olivia de Havilland	The Heiress	33	1979	Sally Field	Norma Rae	33
1950	Judy Holliday	Born Yesterday	28	1980	Sissy Spacek	Coal Miner's Daughter	30
1951	Vivien Leigh	A Streetcar Named Desire	38	1981	Katharine Hepburn	On Golden Pond	74
1952	Shirley Booth	Come Back, Little Sheba	45	1982	Meryl Streep	Sophie's Choice	33
1953	Audrey Hepburn	Roman Holiday	24	1983	Shirley MacLaine	Terms of Endearment	49
1954	Grace Kelly	The Country Girl	26	1984	Sally Field	Places in the Heart	38
1955	Anna Magnani	The Rose Tattoo	48	1985	Geraldine Page	Trip to Bountiful	61
1956	Ingrid Bergman	Anastasia	41	1986	Marlee Matlin	Children of a Lesser God	21
1957	Joanne Woodward	The Three Faces of Eve	27	1987	Cher	Moonstruck	41
1958	Susan Hayward	I Want to Live	40	1988	Jodie Foster	The Accused	26
1959	Simone Signoret	Room at the Top	38	1989	Jessica Tandy	Driving Miss Daisy	80
1960	Elizabeth Taylor	Butterfield 8	28	1990	Kathy Bates	Misery	43
1961	Sophia Loren	Two Women	27	1991	Jodie Foster	The Silence of the Lambs	29
1962	Anne Bancroft	The Miracle Worker	31				

Suppose that we are interested in studying whether there is a significant difference in the ages of the winners based on their gender. As you will see, there are many different ways to make such a comparison, and most are based on the ideas we have already discussed.

We begin by organizing the data into tabular form as a pair of frequency distributions. As we saw in Chapter 2, whenever we construct a frequency distribution, we can obtain different results depending on the number of classes and on the class limits used. If we are to construct two separate frequency distributions and compare them intelligently, they both must contain the identical classes.

Thus, for the given sets of data, we might consider classes 20 to 29, 30 to 39, ..., 70 to 79. Moreover, if you examine the data closely, you will notice that there is a tie for best actress in 1968 between Katharine Hepburn (age 61) and Barbra Streisand (age 26). The easiest way to account for such a tie is to average the two ages, so we use 43 as the winner's age. The corresponding relative frequency distributions for actors and actresses are displayed in Table 3.4.

Comparing the two sets of entries shows that the best-actress winners tend to be younger than the best-actor winners. It is precisely this type of conjecture, that a difference exists between two groups, that we want to study.

We can investigate such a suspicion by using a graphical comparison of the two sets of data. Of the four major graphical displays we studied in Chapter 2—the bar chart, histogram, frequency polygon, and pie chart—the first two are the most effective tools for such a comparison. We work with the histogram here. Moreover, as with the frequency distribution, whenever we use a histogram, very different results can occur depending on the number of classes and the corresponding class limits. Again, we must be careful to use the same classes to make an intelligent comparison. The resulting histograms are shown in Figure 3.13.

This display provides even more evidence that there seems to be a definite tendency for Oscars to be awarded to actresses at a younger age than to actors.

Notice, incidentally, that the two histograms are arranged one above the other. In this way, it is much easier to compare the two sets of data than if the displays were side by side. But it is easier to compare two frequency distribution tables if they are arranged side by side. In fact, it is usually preferable to list both sets of data in the same frequency distribution table, as we did in Table 3.4.

The tabular and graphical displays can only suggest a significant difference between two sets of data. To come to any meaningful conclusions about whether such a difference actually exists, we must compare some of the numerical measures for the sets of data. Since we suspect that there is a difference in the ages at which actors and actresses win Academy Awards, we now turn to a comparison of the means and standard deviations. For the 64 actors, the mean age $\mu_M = 44$ years with standard deviation $\sigma_M = 8.75$ years, while

FIGURE 3.13

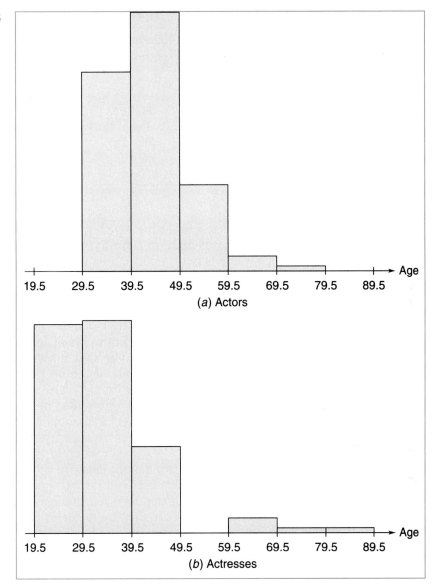

for the actresses $\mu_W = 35.3$ years with standard deviation $\sigma_W = 11.5$ years. (Since the data sets consist of all Academy Award winners, we are studying populations rather than samples.) Based on these calculations, we see that the mean age of actresses who have won an Academy Award is considerably lower than the mean age of actors.

In Sections 9.5 and 9.6, we develop some extremely powerful statistical

TABLE 3.4

Classes	Actors	Relative Frequency	Actresses	Relative Frequency
20–29	0	0	23	.36
30–39	21	.33	24	.38
40–49	29	.45	12	.19
50–59	10	.16	0	0
60–69	3	.05	3	.05
70–79	1	.02	1	.02
80–89	0	0	1	.02

techniques which allow us to decide whether an apparent difference in the means of two samples is actually *statistically significant* or is only a relatively minor difference which is not significant in a statistical sense.

When we are comparing two such sets of data, it is useful to compare other measures, particularly the medians. We begin by constructing the box-and-whisker plot for each set of data, as shown in Figure 3.14. Again, the two displays are positioned one above the other, and the horizontal scales are the same to aid in the visual comparison. From this diagram, again the ages of the majority of the best actors appear to be shifted more to the right (toward higher ages) than those for the majority of the best actresses. In fact, the median age for actors is 42.5 while that for actresses is 33.

Statisticians have also developed sophisticated methods to compare the medians of two sets of data to see whether the difference between them is statistically significant. However, the techniques are outside the scope of this book.

In addition to *comparing* two sets of data, we often want to determine whether there is any *relationship* between the two quantities being studied. For instance, is there a relationship between the ages of best actresses (or best actors) over time in the sense that the Academy tends to award Oscars to younger or older stars in any given year? Just by looking at the set of data

FIGURE 3.14

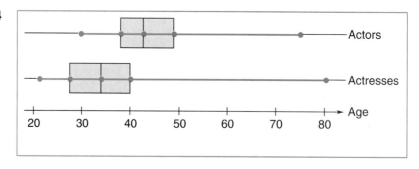

on the ages of best actresses in Figure 3.12, we might observe a trend of younger actresses receiving Oscars in the early years and older actresses receiving them in later years.

One of the most effective graphical methods for determining if there appears to be a relationship or pattern or trend between two variables is the *scatterplot*. To construct it, we interpret the pairs of values as pairs of coordinates in an algebraic sense and plot them as points in the plane. Thus, we have a pair of axes with the years on the horizontal and the ages of the actresses on the vertical. For 1928, with $x = 28$, we have Janet Gaynor at age $y = 22$ to produce the point $(x,y) = (28,22)$. The next pairing is $x = 29$ for 1929 with Mary Pickford at age $y = 36$ to produce the point $(29,36)$. The complete scatterplot for this set of paired data is shown in Figure 3.15. From the display, we observe that, overall, the points seem to fall into a rising pattern and so there does appear to be some relationship between the ages of the best actress and the year of the award. In Chapter 10, we develop a method for determining precisely what this relationship is.

Further, we will also develop a measure, known as the *correlation coefficient*, which is used to determine the degree of relationship or correlation between the two quantities. In the present case, it turns out that there is a statistically significant degree of correlation between the ages of the best actresses and the year in which the Oscar was awarded.

FIGURE 3.15

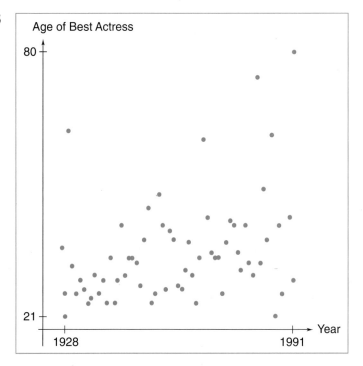

**FIGURE 3.15
(Continued)**

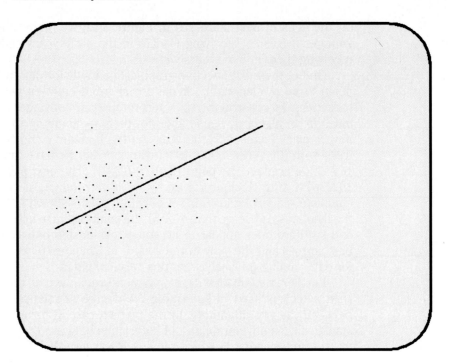

Computer Corner

We have included Program 3: Comparing Two Sets of Data to allow you to investigate the relationships between two related sets of data. The program is just an expansion of Program 1: Statistical Analysis of Data, but now you must enter two sets of data, one after the other. After you enter each of the items in the first set (the x's), type 9999 to indicate that the data are complete; then enter the second set of entries, one at a time, and again use 9999 to indicate that you have finished. You will then be faced with basically the same set of options as in Program 1. You may see the values for the mean and standard deviation (both for a population and for a sample) for the two sets of data; you can see ordered lists of the two sets (in ascending order only) along with the medians and quartiles of both sets; you can construct a single frequency distribution table for the two sets of data with any number of classes from 5 to 20; you can construct histograms for the two sets of data; you can see the joint box-and-whisker plots for the two sets of data. Moreover, you can edit either one or both sets of data to see the effects of changes on entries in either or both, and you can save both sets of data as a disk file for future investigations.

Minitab Methods

You can use Minitab to compare two sets of data either numerically or graphically. Suppose the two data sets are stored in columns C1 and C2. You can have Minitab print both columns simultaneously by using the command

> PRINT C1-C2

In a similar way, you can have Minitab produce the associated numerical summaries of the two sets of data with the command

> DESCRIBE C1-C2

This includes such information as the mean, standard deviation, median, and quartiles for the two sets of data.

You can also have Minitab produce the various graphical displays for the two sets of data by using the commands

> HISTOGRAM C1-C2
> STEM-AND-LEAF C1-C2
> BOXPLOT C1-C2

From such displays, you can try to see, by eye, if there appears to be any difference between the two sets of data.

Figure 3.16 shows a typical Minitab session.

Exercise Set 3.6

Applying the Concepts

1. Select a random sample of 25 of the stocks listed in Appendix III. Then select a second random sample of 25 stocks (not the same ones) from a current stock list in your newspaper. For simplicity, you may want to discard any stock whose price is under $5 or over $150. Construct relative frequency distributions, histograms, and box plots for the two sets of data. In addition, calculate the mean and the median for the two sets. Based on these, is there any apparent difference in the two sets of values? (For reference, the stock data shown in Appendix III correspond to a time when the Dow-Jones industrial average was approximately 3300.)

FIGURE 3.16

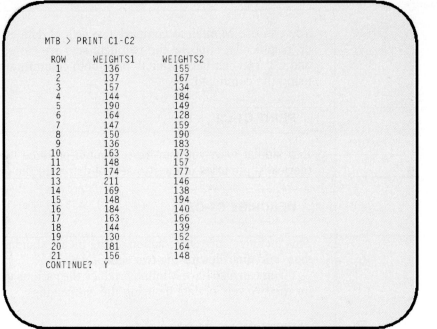

```
MTB > HISTOGRAM C1-C2
Histogram of WEIGHTS1   N=30
Midpoint    Count
    120       1    *
    130       1    *
    140       7    *******
    150       6    ******
    160       7    *******
    170       3    ***
    180       3    ***
    190       1    *
    200       0
    210       1    *
CONTINUE?
Histogram of WEIGHTS2   N=30
Midpoint    Count
    120       1    *
    130       2    **
    140       5    *****
    150       5    *****
    160       7    *******
    170       3    ***
    180       3    ***
    190       4    ****
```

```
MTB > STEM-AND-LEAF C1-C2
Stem-and-leaf of WEIGHTS1   N=30
Leaf Unit = 1.0
    1     12 2
    6     13 06677
   14     14 04477888
   (5)    15 06789
   11     16 3349
    7     17 04
    5     18 124
    2     19 0
    1     20
    1     21 1
CONTINUE?
Stem-and-leaf of WEIGHTS2   N=30
Leaf Unit = 1.0
    1     12 2
    2     12 2
    3     13 4
    6     13 889
    8     14 04
   12     14 6689
   13     15 2
   (5)    15 57769
   12     16 04
   10     16 67
    8     17 3
    7     17 7
    6     18 34
    4     18 56
    2     19 04
```

2. Select a random sample of 25 National League pitchers from the data shown in Appendix I. Draw the scatterplot of the number of walks (BB for bases on balls) versus the number of hits (H) allowed by each pitcher. Does there appear to be any pattern? That is, does it appear that, in general, the number of walks increases as the number of hits allowed increases? Does it appear that the number of walks given decreases as the number of hits increases? Or does the scatterplot seem to indicate that there is no general pattern?

3.7 Grouped Data

We have seen the importance of quantities such as the mean, median and standard deviation for measuring essential characteristics of any set of data. In all the situations we have encountered so far, we had the original set of raw data on which to carry out these calculations.

In many situations, however, we are given only a set of data previously summarized in a frequency distribution or graphical display. This might arise with tables of data developed by a government agency, a corporation or business report or a research study in many fields. In such instances, we do not have access to the original data values. We must be able to find the mean, standard deviation and median of the data from the table. We illustrate how to estimate these quantities in the following situation.

A survey of the speeds (measured to the nearest mile per hour) of 400 cars on a road is conducted. The results are summarized in the following Table 3.5:

TABLE 3.5

Class Boundaries	Frequency
27.5–32.5	23
32.5–37.5	70
37.5–42.5	199
42.5–47.5	100
47.5–52.5	8
	400

We now want to find the appropriate statistical measures of central tendency (the mean and median) and the measures of variability (the variance and standard deviation). The problem is, How can we calculate such quantities when all we have is a set of *grouped data*? We cannot calculate these quantities exactly; that requires knowing the actual raw data. The following methods allow us to approximate these values based on the grouped data.

We begin by considering the first class, consisting of the speeds of 23 vehicles traveling between 27.5 and 32.5 miles per hour. Unfortunately, we have no idea what the precise speed of any of these 23 vehicles is. We therefore assume that all 23 of these cars are traveling at the center value or class mark for this class, which is 30 miles per hour.

We note that this assumption that all 23 values are 30 miles per hour will not produce an exact answer. However, the errors that occur as we overestimate some values and underestimate others tend to balance each other out.

Similarly, we assume that the 70 cars in the second class traveling between 32.5 and 37.5 miles per hour are all moving at the class mark for this class, namely, 35 miles per hour. In an analogous way, we assume that the 199 cars in the third class, between 37.5 and 42.5 miles per hour, are all moving at a speed of 40 miles per hour, and so forth. In this way, we are creating a hypothetical list of 400 assumed car speeds. The first 23 entries in this list are 30s, the next 70 entries are 35s, the next 199 are 40s, and so on, as shown in Figure 3.17.

To calculate the mean for this set of "data," we could sum the 400 entries in this hypothetical list. However, since the first 23 entries are all equal to 30, their total contribution is precisely $23 \times 30 = 690$. Similarly, the next 70 entries are all equal to 35, so their total contribution is $70 \times 35 = 2450$, and

FIGURE 3.17

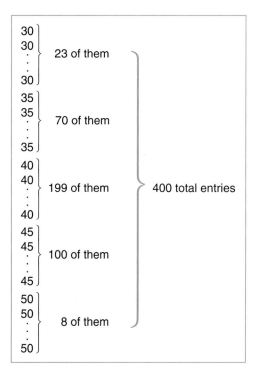

so forth. As a result, the grand total of all 400 entries is

$$690 + 2450 + 7960 + 4500 + 400 = 16,000$$

and the hypothetical mean is therefore

$$\bar{x} = \frac{16,000}{400}$$

$$= 40 \text{ miles per hour}$$

It is possible to do this calculation much more easily just by extending the original frequency distribution table to reflect what we did. Specifically, we found the class mark x_{cm} for each class, multiplied it by the corresponding frequency f, summed the results, and finally calculated the estimated mean $\bar{x}$ by dividing by the total number of data entries, $n = 400$. We indicate this in the following table:

Class Boundaries	f	Class Mark x_{cm}	$f \cdot x_{cm}$
27.5–32.5	23	30	$23 \times 30 = 690$
32.5–37.5	70	35	$70 \times 35 = 2,450$
37.5–42.5	199	40	$199 \times 40 = 7,960$
42.5–47.5	100	45	$100 \times 45 = 4,500$
47.5–52.5	8	50	$8 \times 50 = 400$
	$\Sigma f = 400$		$\Sigma f \cdot x = 16,000$

$$\bar{x} = \frac{16,000}{400} = 40$$

Notice that the total, 16,000, is divided by 400, the number of cars in the original data, and not by 5, the number of classes in the frequency distribution. In fact, were you to divide by 5 by mistake, the answer you would get for the mean speed would be the totally unreasonable value

$$\frac{16,000}{5} = 3200 \text{ miles per hour}$$

> The **mean** for a set of grouped data is approximately
>
> $$\bar{x} = \frac{\Sigma f \cdot x_{cm}}{n}$$

It is also possible to estimate the median for a set of grouped data, and we consider this later in this section. We first extend the above ideas to find a value for the standard deviation. We apply the usual process for calculating the standard deviation to the hypothetical list in Figure 3.17. That is, we subtract the mean $\bar{x} = 40$ from the group of twenty-three 30s, then from the group of seventy 35s, and so forth. We then square each of these deviations, sum them, divide this total by $n - 1 = 399$ to estimate the variance s^2, and finally take the square root to obtain the standard deviation s.

Finding the standard deviation is also simpler if we extend the above table. Consider the first class of 23 entries, each assumed equal to the class mark $x_{cm} = 30$. In Table 3.6 we have a column for $x_{cm} - \bar{x}$ starting with $30 - 40 = -10$ corresponding to each of the 23 items in this class. The square of this is 100, under the heading $(x_{cm} - \bar{x})^2$. Since there are 23 entries in this class, each with the same contribution of $(x_{cm} - \bar{x})^2 = 100$, they produce a final value of $23 \times 100 = 2300$ under a heading of $f \cdot (x_{cm} - \bar{x})^2$. There are similar results for the other four classes. We summarize these ideas by extending the above table as in Table 3.6. As a result, the sample variance is approximately equal to

$$s^2 = \frac{7350}{399}$$
$$= 18.421$$

and the sample standard deviation is approximately

$$s = \sqrt{18.421} = 4.29 \text{ miles per hour}$$

Incidentally, if the above data were for a population rather than a sample, then the only changes would be in the notation (using μ, σ^2, and σ instead of $\bar{x}$, s^2, and s, respectively) and the expressions for the variance and the standard deviation would be divided by the population size N instead of by $n - 1$.

TABLE 3.6

Class Boundaries	f	x_{cm}	$f \cdot x_{cm}$	$x_{cm} - \bar{x}$	$(x_{cm} - \bar{x})^2$	$f \cdot (x_{cm} - \bar{x})^2$
27.5–32.5	23	30	690	−10	100	2300
32.5–37.5	70	35	2450	−5	25	1750
37.5–42.5	199	40	7960	0	0	0
42.5–47.5	100	45	4500	5	25	2500
47.5–52.5	8	50	400	10	100	800
	400					7350

The **standard deviation** for a set of grouped data is estimated as

$$s = \sqrt{\frac{\Sigma f \cdot (x_{cm} - \bar{x})^2}{n - 1}}$$

for a sample and as

$$\sigma = \sqrt{\frac{\Sigma f \cdot (x_{cm} - \mu)^2}{N}}$$

for a population.

We now show how to estimate the median for a set of grouped data. We again consider the previous example involving the speeds of 400 cars, as shown in Table 3.5. Since the number of entries $n = 400$ is even, the median corresponds to the average of the 200th and the 201st entries. We must locate approximately where these would be. Referring to the original frequency distribution, we see that the first two classes together account for $23 + 70 = 93$ entries. Moreover, the third class contains the next 199 data values. Therefore, the 200th entry must occur somewhere in the third class between 37.5 and 42.5. Although we could simply use the class mark of 40, this would be highly inaccurate if the median were to fall near either end of the class. Instead, we can get a much more accurate approximation for the median by locating the desired entry more carefully, as we now illustrate.

There are 199 entries in the third class from 37.5 to 42.5. If they were individual numbers, they would be item 94 up through item 292. We want to estimate the value for item 200. Clearly, it is the 106th entry out of the 199 items in this class. We therefore want to predict approximately where it should fall between 37.5 and 42.5. Specifically, this number should be about $\frac{106}{199}$ of the way from 37.5 to 42.5. Since the interval from 37.5 to 42.5 has length 5, we want to go

$$\frac{106}{199} \times 5 = .5327 \times 5$$

$$= 2.6633$$

above 37.5. That is, the 200th entry should be approximately

$$37.5 + 2.6633 = 40.1633$$

In a similar way, the 201st entry should be $\frac{107}{199}$ of the way from 37.5 to 42.5, or at

$$37.5 + \left(\frac{107}{199}\right)(5) = 37.5 + (.5377)(5)$$

$$= 37.5 + 2.6884$$

$$= 40.1884$$

Finally, since the median is the average of the 200th and 201st entries, it should be at $(40.1633 + 40.1884)/2 = 40.1759$.

The process used here is known as *interpolation*. It can be summarized by the following formula:

$$V = L + \frac{c}{f} W$$

where V is the value being calculated, L is the lower boundary of the class, f is the frequency of the class, c is the count or position within the class, and W is the width of the class. For the above situation where we found the 200th entry, $L = 37.5$, $f = 199$, $c = 106$, and $W = 5$, so that

$$V = 37.5 + \left(\frac{106}{199}\right)(5)$$

$$= 40.1633$$

Computer Corner

The mean and standard deviation for grouped data are particularly simple to obtain by using Program 4: Grouped Data Analysis. To use it, you have to respond to several simple questions. You must indicate how many classes you have (any value between 4 and 20), the lower limit for the first class, the upper limit for the first class, and the lower limit for the second class. For example, if the lower limit for the first class is 0, the upper limit for the first class is 99, and the lower limit for the second class is 100, then the program automatically sets up the classes as

0– 99
100–199
200–299
300–399
etc.

As an alternative to using class limits, the program also accepts class boundaries. Thus, if the upper boundary of the first class is equal to the lower boundary of the second class (say both are 99.5), then the program sets up the classes boundaries as

−.5– 99.5
99.5–199.5
199.5–299.5
299.5–399.5
etc.

Remember that the class boundaries must have one more decimal place than the original data.

Finally, you must enter the frequency for each of the classes for your data set.

The program shows the estimated values for the mean and standard deviation for your set of grouped data.

Problem Set 3.7

Mastering the Techniques

Estimate the mean, standard deviation and median for the following sets of data:

1.

Class Limits	f
1– 7	10
8–14	30
15–21	50
22–28	40
29–35	10

2.

Class Boundaries	f
121–170	10
171–220	50
221–270	120
271–320	80
321–370	40

Applying the Concepts

3. The speeds, to the nearest mile per hour, of a group of 200 cars on a road are as follows:

Class Boundaries	f
27.5–32.5	18
32.5–37.5	76
37.5–42.5	200
42.5–47.5	100
47.5–52.5	6

Estimate the mean, standard deviation and median speed of these cars.

4. A fast-food restaurant collects the following data on the amounts spent by customers on meals:

Class Limits, dollars	f
0–0.99	20
1.00–1.99	80
2.00–2.99	70
3.00–3.99	30
4.00–4.99	10

Estimate the mean, standard deviation and median amounts spent per customer.

Computer Applications

Estimate the mean and standard deviation for the following sets of data using Program 4: Grouped Data Analysis.

5.

Class Limits	f
100–199	47
200–299	139
300–399	328
400–499	555
500–599	419
600–699	280
700–799	311
800–899	107

6.

Class Limits	f
10.0–19.9	73
20.0–29.9	128
30.0–39.9	185
40.0–49.9	157
50.0–59.9	205
60.0–69.9	132
70.0–79.9	148
80.0–89.9	92
90.0–99.9	15

Continue the investigation you began in the Student Project at the end of Chapter 2 to calculate the various numerical descriptions for your set of data. For the data you previously collected,

1. Calculate the mean, median and mode.
2. Calculate the sample variance and standard deviation.
3. Calculate the z values corresponding to each data entry.
4. Locate the quartiles for the data.
5. Construct the box plot for the data.

Finally, incorporate all these items into a formal statistical project report including

- A statement of the topic being studied and a specific statement of the population being investigated
- The source of the data
- A discussion of how the data were collected and why you believe that it is a random sample
- A list of the data

> - All the statistical results you calculated and the box plot
> - Some of the tables and diagrams you constructed for the first project
> - A discussion of any surprises you may have noted in connection with collecting the data or organizing them

CHAPTER 3 SUMMARY

To perform any statistical analysis on a set of data, you must use numerical measures which describe those data. The **range** is

$$\text{Range} = \text{largest entry} - \text{smallest entry}$$

The measures of central tendency are the **mean, median** and **mode**.

For a population,

$$\text{Mean } \mu = \frac{\Sigma x}{N}$$

For a sample,

$$\text{Mean } \bar{x} = \frac{\Sigma x}{n}$$

The measures of spread or variation in a set of data are the **variance** and the **standard deviation**.

For a population, the standard deviation

$$\sigma = \sqrt{\frac{\Sigma(x - \mu)^2}{N}}$$

For a sample, the standard deviation is

$$s = \sqrt{\frac{\Sigma(x - \bar{x})^2}{n - 1}}$$

Shortcut formulas are available that are sometimes simpler to apply to calculate these quantities.

The **empirical rule** applies to many sets of data:

About 68% of the data fall within 1 standard deviation of the mean.

About 95% of the data fall within 2 standard deviations of the mean.

About 99.7% of the data fall within 3 standard deviations of the mean.

You use **z values**, given by

$$z = \frac{x - \mu}{\sigma} \quad \text{or} \quad z = \frac{x - \bar{x}}{s}$$

to measure the number of standard deviations that x lies above or below the mean. To find the value of x corresponding to a given z value, use

$$x = \mu + z\sigma$$

Chebyshev's Inequality, which applies to all sets of data, is less precise than the empirical rule.

You can describe a set of data by using **percentiles, quartiles** and **deciles**. Much of this information can be displayed in a **box plot**.

You can display a set of paired data by using a **scattergram**.

Review Exercises

1. The department manager in a company is analyzing the number of sick days her staff has used during the past year. She finds the following number of sick days: 3, 8, 0, 5, 7, 3, 8, 0, 12, 6, 2, 8, 7, 4, 9, 7, 4, 10. Find the mean, median, mode, variance and standard deviation for this set of data.

2. A study is conducted on the number of hours that freshmen sleep each night at a college. The times for a random sample of 20 students are: 6, 7, 5, 6, 8, 4, 5, 6.5, 7, 3.5, 5, 6, 4.5, 6, 6, 7.5, 3, 6, 5, 7. Find the mean, median, mode, variance and standard deviation for this set of data.

3. The daily caloric intake by students on a meal plan at a university has mean 1450 calories and standard deviation 300 calories. **(a)** What percentage of students would you expect to consume between 850 and 2050 calories each day? **(b)** Within what range would you expect 68% of the students' calorie intake to be?

4. For the data on the number of hours that students sleep, determine: **(a)** the first quartile; **(b)** the second quartile; **(c)** the 20th percentile; **(d)** the 90th percentile; **(e)** the sixth decile.

5. Construct the box plot for the data on the number of hours that students sleep in Review Exercise 2.

6. A state college publishes the following table giving the numbers of students receiving different levels of student aid:

Aid	Number
<999	300
1000–1999	1800
2000–2999	2100
3000–3999	1150
4000–4999	600
5000–5999	50

Estimate the mean, the standard deviation and the median for the student aid received by the students.

7. Some sports writers have claimed that the most outstanding record in professional sports is Nolan Ryan's strikeout record. Use the following data on the top

10 career strikeout leaders to compare Ryan's total to Kareem Abdul-Jabbar's point scoring total. Whose accomplishment is the most outstanding?

Nolan Ryan	5511
Steve Carlton	4136
Tom Seaver	3640
Burt Blyleven	3631
Don Sutton	3574
Gaylord Perry	3534
Walter Johnson	3508
Phil Nieckro	3342
Ferguson Jenkins	3192
Bob Gibson	3117

Probability

According to the FBI, the following cities had the highest rates of major crime, as measured by the number of reported cases of murder, rape, robbery, aggravated assault, burglary, larceny and motor vehicle theft per 1000 residents during 1990.

Crime in US cities

Here is a list of the nation's 52 largest cities in order of their overall rate of crime—murder, rape, robbery, aggravated assault, burglary, larceny-theft and motor vehicle theft. The list includes only cities with 300,000 or more people in 1990 and the rate is based on the number of crimes per 1,000 people. Arson is not included.

RANK		1990 RATE	1989 RANK	RATE
1.	Atlanta	192.4	2	206.9
2.	Miami	190.2	1	183.8
3.	Dallas	155.2	3	167.3
4.	Fort Worth	149.8	4	156.9
5.	St. Louis	146.7	5	153.3
6.	Kansas City, MO	129.5	9	127.2
7.	Seattle	126.0	7	129.1
8.	Charlotte, NC	125.9	6	132.4
9.	San Antonio	124.8	10	127.2
10.	New Orleans	124.4	17	112.6
11.	Detroit	121.9	14	120.9
12.	Tucson	118.8	11	125.5
13.	Boston	118.5	15	120.7
14.	Austin, TX	117.1	21	106.7
15.	Minneapolis	114.4	13	121.0
16.	Houston	113.4	20	108.2
17.	El Paso, TX	112.4	22	106.2
18.	Portland, OR	111.0	8	127.5
19.	Chicago	110.6	28	99.6

(continued)

Crime in US cities
(continued)

RANK		1990 RATE	1989 RANK	RATE
20.	Oakland, CA	109.1	12	125.3
21.	Washington	107.7	26	102.8
22.	Phoenix	107.6	19	108.7
23.	Oklahoma City	106.1	18	111.9
24.	Baltimore	106.0	33	93.5
25.	Fresno, CA	105.3	16	116.9
26.	Jacksonville, FL	104.6	24	103.3
27.	Memphis	102.0	37	88.8
28.	Albuquerque	100.6	27	99.7
29.	Columbus, OH	99.1	23	103.9
30.	New York City	97.0	29	96.7
31.	San Francisco	96.6	36	90.2
32.	Toledo, OH	96.1	30	95.4
33.	Long Beach, CA	95.7	31	94.1
34.	Tulsa, OK	95.3	35	91.8
35.	Milwaukee	93.0	39	87.6
36.	Los Angeles	92.3	34	92.7
37.	San Diego	91.5	32	93.7
38.	Sacramento, CA	91.3	25	103.2
39.	Cleveland	91.1	42	83.5
40.	Wichita, KS	89.3	40	87.4
41.	Buffalo	88.9	41	85.3
42.	Pittsburgh	87.6	38	88.8
43.	Nashville	78.8	47	69.7
44.	Denver	77.6	43	76.1
45.	Cincinnati	75.6	44	74.7
46.	Philadelphia	71.9	46	70.0
47.	Las Vegas	71.3	45	73.9
48.	Omaha	70.5	48	65.7
49.	Indianapolis	67.5	49	65.1
50.	Honolulu	61.0	50	62.3
51.	Virginia Beach	57.8	51	56.2
52.	San Jose, CA	48.7	52	51.4

NOTE: The Chicago figures exclude rapes because its criminal sexual assault statistics include crimes other than rapes. If those numbers were included, Chicago would have moved up only one place in the 1989 and 1990 rankings.

What do such statistics actually mean? For example, consider Atlanta, the worst city with respect to crime, with a crime rate of 192.4 per 1000 people, or 19.24%. Does this mean that approximately one out of every five residents of Atlanta was a victim of a crime during 1990? Does it mean that the chance of *any* single resident being a crime victim was about 20%? Does it mean that approximately 20% of the people in Atlanta are criminals? Does it say anything about the chance of being a crime victim there in 1991 or any other year? Or, for that matter, how accurate is the 20% figure?

Rather than answering these questions directly, we will pose several other questions for you to think about that bear on the above points. Do you think that any residents of Atlanta were victims of more than one crime? Do you think that all residents are *equally likely* to be victimized or are residents of some neighborhoods more likely and residents of other neighborhoods less likely to be subjected to criminal acts? How do visitors to Atlanta count in terms of their possibly being crime victims? Do the results from any single year necessarily carry over unchanged into following years or might a high crime rate one year reflect economic conditions, cutbacks in the police force, etc? Will a high crime rate one year cause stepped-up police activity the following year?

Finally, how are crime statistics collected? Consider the following scenario: a burglar breaks into a home with an illegal gun, encounters the resident, shoots her in the shoulder and makes off with some jewelry and the victim's car. Is that a single crime or is the criminal eventually charged with breaking and entering, possession of an illegal firearm, assault with a deadly weapon, attempted murder, burglary, theft of motor vehicle and fleeing the scene of the crime?

Note that all these issues involve the notion of chance and the likelihood of events occurring. In this chapter, we will see how the concepts of chance and likelihood are assessed mathematically using the theory of probability.

4.1 Introduction to Probability

In most areas of human endeavor, there is always an element of uncertainty. If we consider the weather, a sporting event, a stock transaction, an election result, or a matter relating to health, we are always faced with a certain degree of risk. In fact, according to the old adage, the only things in life that are certain are death and taxes. Therefore, we must be able to assess the degree of uncertainty in any given situation, and this is done mathematically by using *probability*. In particular, our primary goal in this course is to understand and use inferential statistics which will allow us to make predictions and decisions based on sample data. However, any statistical prediction or decision also involves an element of uncertainty, and we need the ideas of probability to assess its accuracy.

Inferential statistics stands on two supporting columns. The first consists of the ideas on descriptive statistics we have already discussed. The second supporting leg is the notion of probability, which we introduce in this chapter.

We use probability to measure how likely it is for events to occur. For example,

How likely is it for a coin to come up heads when it is flipped?

How likely is it that it will rain tomorrow?

How likely is it that a particular candidate will be elected?

How likely is it that a product coming off a production line is defective?

How likely is it that a person has heart disease?

To answer such questions, we must develop some basic ideas about the theory of probability. We begin with some terminology that we introduce in the process of discussing the act of flipping a fair coin and recording the outcomes.

> 1. Such an act is called an **experiment** in probability.
> 2. In this experiment, there are two possible **outcomes**: getting a head and getting a tail.
> 3. The **sample space** for an experiment is the set of all possible outcomes from that experiment.

In the coin flipping experiment, the sample space consists of a head (H) and a tail (T), and we write this as

$$\text{Sample space} = \{H, T\}$$

> 4. In any experiment, we are interested in some specific collection of outcomes. We call such a collection an **event**.

For instance, when flipping a coin, we may be interested in the event "obtaining a head."

> **5.** The **probability** of an event is a measure of the likelihood of that event occurring.

Suppose the experiment of flipping a coin is conducted repeatedly while we keep track of the number of heads and tails that occur. For example, we might obtain the following results:

Flips	Heads	Tails
10	6	4
100	54	46
1,000	485	515
10,000	5,038	4,962

On the basis of such results, we would conclude that the coin comes up heads approximately 50% of the time and so the probability of getting a head is $\frac{1}{2}$, or .5. This approach to probability is known as the *relative frequency concept* of probability since we can think of the probability of an event as the relative frequency of occurrence of that event in the long run.

In general, the probability of an event is always some number between 0 and 1. We can express it either as a decimal (such as .5) or as a fraction (such as $\frac{1}{2}$).

> The numerical value of the probability is an assessment of how likely the event is to occur.
>
> It is always between 0 and 1.
>
> The closer the probability is to 1, the more likely the event is to occur.
>
> The closer the probability is to 0, the less likely the event is to occur.

We denote events by letters such as A and B, and the associated probabilities are written as $P(A)$ and $P(B)$.

Thus, in the coin flipping experiment, the only possible events of interest are either

$$H: \text{getting a head}$$

or

$$T: \text{getting a tail}$$

The probability of getting a head is $\frac{1}{2}$, and we write

$$P(H) = \frac{1}{2}$$

while the probability of getting a tail is also $\frac{1}{2}$, so that

$$P(T) = \frac{1}{2}$$

Therefore, these two events are equally likely to occur.

It is essential to realize that probability is only an assessment of the likelihood of an event. When we say that the probability of heads on a fair coin is $\frac{1}{2}$, we mean that the chance of a head coming up is 50%. It does not offer any guarantee of what *will* happen on any individual flip of the coin. Thus, if we flip a fair coin twice, it is unreasonable to expect precisely one head and one tail. For that matter, it is conceivable (although extremely unlikely) that a fair coin comes up tails 20 times in a row. Nevertheless, intuitively, if we conduct such an experiment repeatedly, then approximately one-half of all the outcomes will be heads and one-half will be tails. That is, any discrepancies or variations tend to average out in the long run. This concept is known in statistics as the *Law of Large Numbers*. It is discussed in the Computer Corner at the end of this section.

We now apply the above ideas in a variety of other relatively simple contexts.

EXAMPLE 4.1 Consider the experiment of rolling a fair die, and record the outcomes. The sample space consists of the set of six possible outcomes

$$\text{Sample space} = \{1, 2, 3, 4, 5, 6\}$$

Consider the following three events: getting a 5, getting a 4, getting an even number on the die. The probability of obtaining a 5 is $P(5) = \frac{1}{6}$. Similarly, the probability of obtaining a 4 is $P(4) = \frac{1}{6}$. Finally, the probability of obtaining an even number is

$$P(\text{even}) = P(2 \text{ or } 4 \text{ or } 6) = \frac{3}{6} = \frac{1}{2}$$ ∎

EXAMPLE 4.2 Consider the experiment of picking a single card from a standard deck of cards. The sample space consists of the 52 possible outcomes—each of the 52 cards in the deck. The probability of picking the jack of hearts is therefore $\frac{1}{52}$. The probability of picking an ace is

$$P(\text{ace}) = \frac{4}{52}$$

The probability of picking a heart is

$$P(\text{heart}) = \frac{13}{52}$$ ∎

In each of these examples, we have intuitively used a basic principle of probability which allows us to determine the probability of an event when all the possible outcomes are *equally likely*.

> The outcomes in an experiment are *equally likely* when any one of the outcomes has the same likelihood of occurring as any other.

For instance, when a card is picked from a deck, there are 52 possible outcomes, and the probability that any one of the 52 cards will be picked is $\frac{1}{52}$. In such a case, the probability that an event A will occur is found from the following rule, known as the *classical interpretation of probability*:

> If an experiment has a finite number of outcomes which are equally likely, then the probability that an event A will occur is given by
>
> $$P(A) = \frac{\text{number of ways } A \text{ can occur}}{\text{total number of possible outcomes}} \quad (4.1)$$

Thus, in calculating the probability of getting an even number on the roll of a die, there are 6 possible equally likely outcomes of which exactly 3 are even. The probability is therefore $\frac{3}{6}$.

Of course, the problem in applying this principle is the assumption that the outcomes are *equally likely*. If they are not, then the calculation of the probability of an event becomes considerably more complicated. For example, suppose a die is loaded in such a way that 5 and 6 come up more often than the other four faces. Then the six outcomes are no longer equally likely, and the above principle and the intuitive approach we used before do not apply.

Furthermore, probability results often violate what our intuition suggests, and so we should not depend on our intuitive beliefs in attempting to assess probabilities. Yet, the results of probability theory always have to be matched up against how well they actually measure the likelihood of an event.

EXAMPLE 4.3 In the experiment of flipping two fair coins, consider the three events of obtaining 2 heads, obtaining 1 head (and 1 tail), and obtaining 0 heads (and 2 tails). We write these events as 2H, 1H, and 0H, and our objective is to find the associated probabilities $P(2H)$, $P(1H)$, and $P(0H)$. Since there are three outcomes, one might be tempted to conclude that each of the probabilities is equal to $\frac{1}{3}$. This is true only if the outcomes listed are equally likely. We could actually perform the experiment of flipping a pair of coins repeatedly and record the number of each occurrence. In fact, we heartily recommend that you do so. However, that can be a very time-consuming process since it may be necessary to repeat the experiment at least 100 times to obtain an appropriate pattern.

Instead, we will use Program 5: Coin Flipping Simulation to perform such a coin flipping experiment for us. When you select this program from the main menu, you will be asked how many coins you wish to flip. Type 2 and press Enter. The program then simulates the flipping of two fair coins repeatedly. The results are displayed graphically, as in Figure 4.1. There are three vertical columns, the left for 0 heads, the center for 1 head, and the right for 2 heads. Each time the program performs this experiment, it displays the occurrence in the appropriate column as a horizontal line segment. As you can see from Figure 4.1, the central column is considerably taller than

FIGURE 4.1

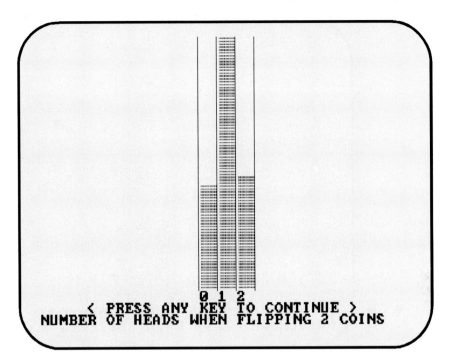

the other two. Thus, it seems that far more outcomes result in 1 head and 1 tail than either 2 heads or 2 tails.

When the graphic display is complete (there is a beep and you are told to PRESS ANY KEY TO CONTINUE), you can see the numerical results of this simulation. For instance, the corresponding results for this particular run are:

```
OUT OF 164 FLIPS OF 2 COINS
% WITH 0 HEADS = 22.56%
% WITH 1 HEADS = 53.04%
% WITH 2 HEADS = 24.39%
```

The experiment was performed 164 times, and the percentage of times that 1 head and 1 tail occurred was somewhat more than twice the frequency that either of the other two possibilities occurred. Certainly, this evidence indicates that the initial choice of $P(2H) = P(1H) = P(0H) = \frac{1}{3}$ is not correct.

We can verify this further by going to the options menu and selecting the first choice to run the program again. The program repeats with a new series of random experiments. A second set of results is displayed in Figure 4.2. This second run is different from the first—each time the program is used, you obtain a different series of results, just as you would obtain a different series of results if you performed the experiment of flipping the coins by hand. Again, the results clearly indicate that the split involving 1 head and 1 tail

FIGURE 4.2

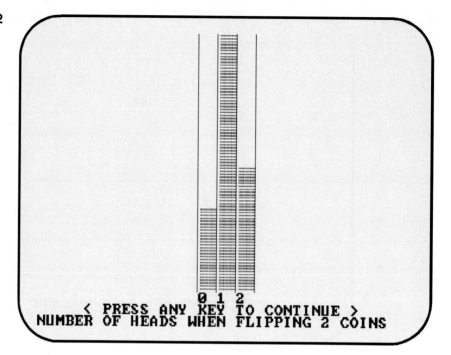

occurs approximately twice as often as either 2 heads or 0 heads. Thus, the three outcomes of 2 heads, 1 head, and 0 heads are seemingly *not* equally likely, and so the basic principle for calculating probabilities (which applies *only* when the various outcomes are equally likely) does not apply.

To account for this, let's consider precisely what happens when we flip two coins. Suppose that the coins we use are a penny and a nickel. Getting 2 heads involves both coins coming up heads; getting 0 heads involves both coming up tails. However, there are two different ways to obtain 1 head: The penny could come up a head and the nickel could be a tail, *or* the penny could be a tail and the nickel a head. Thus, there are really *four* different equally likely outcomes to this experiment, and so the sample space actually consists of

$$\text{Sample space} = \{HH, HT, TH, TT\}$$

Therefore, according to the basic principle (4.1),

$$P(2H) = \tfrac{1}{4}$$
$$P(1H) = \tfrac{2}{4} = \tfrac{1}{2}$$
$$P(0H) = \tfrac{1}{4}$$

These results certainly agree with the results from the computer simulation. They should also agree with the results of your performing the experiment by hand. ∎

There is an effective graphical way to display the various outcomes of many probability experiments known as a *tree diagram*. We illustrate its use in the same coin flipping experiment. Start with an initial point at the left and consider the first coin. Since it can come up either heads or tails, there are two branches, one corresponding to a head and the other to a tail. See Figure 4.3. In turn, each of these splits into two further branches depending on the two possible outcomes for the second coin, again either a head or a tail. Thus, if we examine Figure 4.3, we can count a total of four branches, which gives the size of the sample space. Moreover, by following each of these branches, we obtain the four equally likely outcomes, namely, HH, HT, TH, and TT. Thus, we obtain the same probabilities we found in Example 4.3.

We summarize this information in the following table, where we use x to represent the number of heads.

x	$P(x)$
0	$\tfrac{1}{4}$
1	$\tfrac{2}{4}$
2	$\tfrac{1}{4}$

There is one other important principle of probability that we must develop. In any experiment, suppose we restrict our attention to a set of events that are distinct or nonoverlapping and which together contain all the possible

FIGURE 4.3

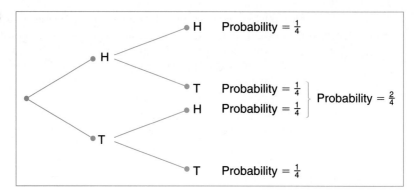

outcomes. In any such experiment, the sum of the probabilities of all the distinct events is always one. Thus, when we roll a die, the six events of getting a 1, getting a 2, ..., getting a 6 each have probability $\frac{1}{6}$, and their sum is 1. Similarly, when we flip two coins and the possible events are 0H, 1H and 2H, the sum of the corresponding probabilities is

$$\tfrac{1}{4} + \tfrac{2}{4} + \tfrac{1}{4} = 1$$

If we interpret this principle in terms of a tree diagram, then we conclude that the sum of the probabilities of all the end branches is 1.

EXAMPLE 4.4 Consider the experiment of flipping a set of three fair coins. The events of interest to us are 3 heads (and 0 tails), 2 heads (and 1 tail), 1 head (and 2 tails), and 0 heads (and 3 tails). Based on our experiences in Example 4.3, we might be tempted to experiment first with the computer to see what types of outcomes actually occur. From the options menu, select choice 2 to Change the Number of Coins and then, in response to the question of how many coins you want, type 3. A typical result of this program is shown in Figure 4.4.

From the display, it is clear that the probability of obtaining either 3 heads or 0 heads is considerably smaller than the probability of obtaining either type of split. Further, the probabilities of obtaining either 2 heads (and 1 tail) or 1 head (and 2 tails) seem approximately the same. Similarly, the probabilities of obtaining either 3 heads (and 0 tails) or 0 heads (and 3 tails) are also approximately the same. The corresponding numerical data support these tentative conclusions.

To obtain precise values for the associated probabilities, we must determine exactly what outcomes are possible. Think of the three coins as a penny, nickel and dime. It is obvious what must happen to obtain 3 heads or 0 heads. However, let's see how we would get a split involving 2 heads and 1 tail or 1 head and 2 tails. The easiest way to do this is to construct a tree diagram as shown in Figure 4.5. From it, we see that there are eight distinct possible outcomes, all equally likely, and so

Sample space = {HHH, HHT, HTH, HTT, THH, THT, TTH, TTT}

FIGURE 4.4

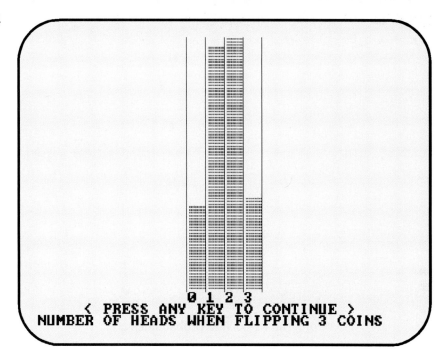

FIGURE 4.5

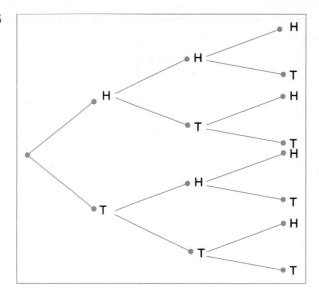

Therefore, the associated probabilities are

$$P(3H) = \tfrac{1}{8}$$
$$P(2H) = \tfrac{3}{8}$$
$$P(1H) = \tfrac{3}{8}$$
$$P(0H) = \tfrac{1}{8}$$

Notice that the sum of the probabilities is 1. Further, these theoretical predictions certainly agree with the results of the computer simulation. ■

We can also deduce these values from the following line of reasoning. Consider the event of getting 2 heads and 1 tail. We concentrate on the tail. The possible outcomes involve having the penny come up tails, the nickel come up tails, or the dime come up tails. Thus, there are actually three different outcomes that result in 2 heads and 1 tail. By an identical line of reasoning, there are three different outcomes that result in 1 head and 2 tails. Thus, of the eight possible outcomes, we again find that 3 heads and 0 heads can each occur in just one way while each of the two splits can occur in three different ways. We therefore obtain the same set of probabilities

x	$P(x)$
0	$\tfrac{1}{8}$
1	$\tfrac{3}{8}$
2	$\tfrac{3}{8}$
3	$\tfrac{1}{8}$

We encourage you to experiment with this program while using larger numbers of coins. The program works with up to 16 coins. As you will see in later sections, we will be interested in both the probabilities involved in such cases and the shape of the distribution of the results.

EXAMPLE 4.5 Consider the experiment of rolling two fair dice. In many games of chance, the item of interest is the sum of the faces on the two dice. Find

$$P(\text{sum} = 2), P(\text{sum} = 3), P(\text{sum} = 4), \ldots, P(\text{sum} = 12)$$

SOLUTION We first determine the appropriate sample space consisting of equally likely outcomes. Suppose that the dice are different colors, one red and the other green. Suppose that the red die lands with face 1 showing. This can be paired with any of six outcomes on the green die—a 1, 2, 3, 4, 5 or 6—and these six outcomes can be written

$$(1, 1), (1, 2), (1, 3), (1, 4), (1, 5), (1, 6)$$

In the same way, the red die could land with face 2 showing, and this could be paired with any of the six outcomes from the green die. These six outcomes are

$$(2, 1), (2, 2), (2, 3), (2, 4), (2, 5), (2, 6)$$

This argument can be extended to the red die landing with the 3, 4, 5, or 6 showing. Therefore, the entire sample space consists of the 36 possible outcomes which are listed in Figure 4.6. Since these 36 possible outcomes are equally likely, it follows that

$$P(\text{sum} = 2) = \tfrac{1}{36}$$
$$P(\text{sum} = 3) = \tfrac{2}{36} \quad [\text{either } (1, 2) \text{ or } (2, 1)]$$
$$P(\text{sum} = 4) = \tfrac{3}{36} \quad [(1, 3), (2, 2), \text{ or } (3, 1)]$$
$$P(\text{sum} = 5) = \tfrac{4}{36}$$
$$P(\text{sum} = 6) = \tfrac{5}{36}$$
$$P(\text{sum} = 7) = \tfrac{6}{36}$$
$$P(\text{sum} = 8) = \tfrac{5}{36}$$
$$P(\text{sum} = 9) = \tfrac{4}{36}$$
$$P(\text{sum} = 10) = \tfrac{3}{36}$$
$$P(\text{sum} = 11) = \tfrac{2}{36}$$
$$P(\text{sum} = 12) = \tfrac{1}{36}$$

■

We can check the accuracy of these probabilities by using Program 6: Dice Rolling Simulation. The program simulates rolling a pair of fair dice a total of 360 times and displays the results graphically using a series of 11 vertical columns corresponding to sums of 2, 3, 4, ..., 12. The number 360 is used since the probabilities associated with rolling a pair of dice all involve 36 in the denominator. As you see from the typical result shown in Figure 4.7,

FIGURE 4.6

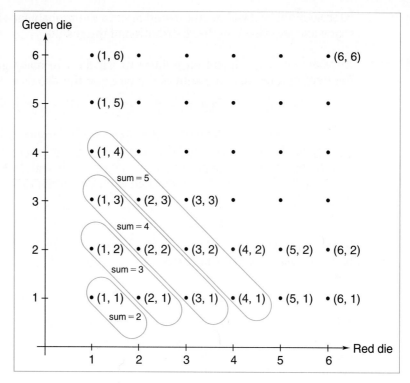

FIGURE 4.7

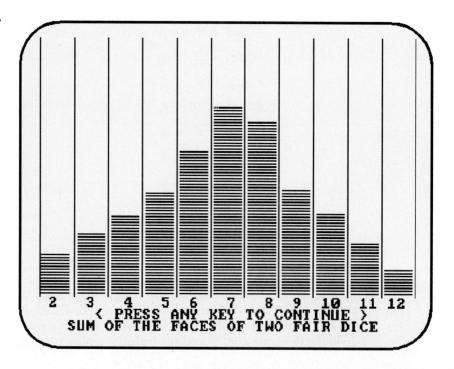

the number of times of each occurrence grows from a minimum for a sum of 2 up to a maximum near the center where the sum is 7 and then decays to a minimum for a sum of 12.

The accompanying text page shown in Table 4.1 clearly indicates that the simulated results match up fairly well with the theoretical predictions. In particular, we would expect that, out of the 360 repetitions of this experiment, a sum of 2 should occur approximately $\frac{1}{36}$ of the time, or $\frac{1}{36}$ of 360 = 10 times. Similarly, a sum of 3 should occur approximately $\frac{2}{36}$ of 360 = 20 times, and so forth. Thus theoretically, we would expect the following set of outcomes out of 360:

Sum	Expected Number of Outcomes
2	10
3	20
4	30
5	40
6	50
7	60
8	50
9	40
10	30
11	20
12	10
	360

TABLE 4.1

```
OUT OF 360 ROLLS OF TWO FAIR DICE:
    SUM OF  2 CAME UP 14 TIMES
    SUM OF  3 CAME UP 21 TIMES
    SUM OF  4 CAME UP 27 TIMES
    SUM OF  5 CAME UP 35 TIMES
    SUM OF  6 CAME UP 49 TIMES
    SUM OF  7 CAME UP 64 TIMES
    SUM OF  8 CAME UP 59 TIMES
    SUM OF  9 CAME UP 36 TIMES
    SUM OF 10 CAME UP 28 TIMES
    SUM OF 11 CAME UP 18 TIMES
    SUM OF 12 CAME UP  9 TIMES
    < SPACE BAR => MENU; OTHER => GRAPH >
```

When we compare these theoretical predictions to the 360 simulated repetitions of the experiment performed by the computer and summarized in Table 4.1, we see that the agreement is quite close.

Incidentally, while we could construct a tree diagram for this problem also, it would be considerably more complicated since each successive roll involves six different branches.

Computer Corner

We have already discussed two programs relating to the topics covered in this section, Program 5: Coin Flipping Simulation and Program 6: Dice Rolling Simulation. Another program you might want to use in an experimental mode at this time is Program 7: Law of Large Numbers. The program is designed to demonstrate this important theoretical principle via simulation.

The Law of Large Numbers guarantees that random variations tend to average out if a statistical experiment is conducted often enough. More specifically, it says that if an event having a certain probability is repeated indefinitely, then the proportion of favorable outcomes gets closer and closer to that probability as the number of repetitions increases.

We denote the probability of such an event by π since, as we will see in Chapter 5, it is a parameter for a population. This use of π for a probability has nothing to do with the mathematical constant ($\pi = 3.14159\ldots$).

For example, when we flip a fair coin, the probability of getting a head is $\pi = \frac{1}{2}$. Obviously, there is no guarantee of what will happen on any single flip. However, if the coin is flipped repeatedly, then in the long run, the number of times that a head comes up will be approximately one-half of the total number of flips. Thus, the proportion of favorable outcomes gets closer to $\frac{1}{2}$ as the number of flips increases. To see how this works, suppose the first few outcomes are H, H, T, H, H, etc. After the first flip, the proportion of heads is 1 (1 head out of 1 flip). After the first two flips, the proportion of successes is still 1 (2 heads out of 2 flips). After the first three flips, the proportion of successes is .667 (2 heads out of 3 flips). After the first four flips, the proportion is .75 (3 heads out of 4 flips). As this process continues, the proportion of heads will eventually approach $\pi = .5$, the theoretical probability of getting a head on any single flip.

In the program, you enter a value for the probability π as a decimal between 0 and 1 and the number of repetitions, say 1000. The program repeatedly simulates a random process with probability π of success, keeps track of the number of successes, and graphs the proportion of successes as the number of repetitions increases. The display demonstrates how the proportion of successes approaches the horizontal line representing the probability of success π. We show the results of one run of this program, using $\pi = .4$ with $n = 2000$ rep-

FIGURE 4.8

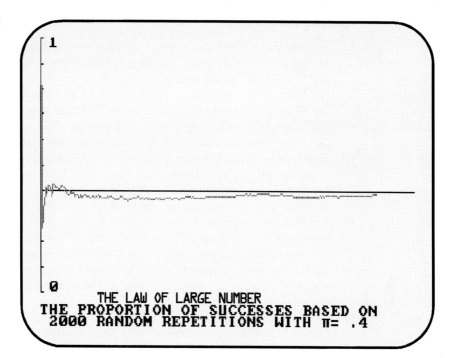

etitions, in Figure 4.8. Notice that initially the proportion of successes is not close to .4. (In the figure, it is considerably higher at the start.) Eventually, as the number of repetitions increases toward 2000, the overall proportion of successes definitely approaches $\pi = .4$.

We suggest that you experiment with this program using different values for π and n.

Exercise Set 4.1

Mastering the Techniques

In Exercises 1 to 10, a number from 1 to 25 is randomly selected.

1. What is the probability that the number is 10?
2. What is the probability that it is even?
3. What is the probability that it is a multiple of 5?
4. What is the probability that it is a prime number? (*Note*: 1 is not considered prime.)
5. What is the probability that it is greater than 18?
6. What is the probability that it is less than 7?
7. What is the probability that it is at least 11?
8. What is the probability that it is 11 or less?
9. What is the probability that it is no more than 6?
10. What is the probability that it is no less than 20?

Applying the Concepts

11. Suppose that the probability that a woman gives birth to a girl is .5. Consider a family of three children with regard to the sex of the children.
    (a) What is the sample space?
    (b) What is the probability of having 0 girls? 1 girl? 2 girls? 3 girls?
    (c) What is the probability that there are 2 or more girls?
    (d) What is the probability that a least one of the children is a boy?
    (e) What is the probability that there are at most two girls?

12. Suppose a set of four fair coins is flipped and the events of interest are the number of heads—0H, 1H, 2H, 3H, 4H.
    (a) What is the sample space for this experiment?
    (b) What is the probability associated with each event?

13. A car rental agency has 15 cars of a certain model. Four are gray, two are red, three are yellow, and six are blue. If the particular car you obtain when you rent one is selected randomly, what is the probability that (a) it is red or (b) it is blue? Suppose you adamantly refuse to accept a gray car, but any other color is acceptable. Under that condition, what is the probability that the car you get (c) is red or (d) is blue?

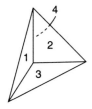

14. Certain role-playing games such as Dungeons and Dragons involve the use of a four-sided (tetrahedral) die whose sides are numbered 1, 2, 3 and 4 (the side not shown). Suppose you roll two such dice and are concerned with the sum of the two "down" faces.
    (a) What is the sample space for this experiment?
    (b) What is the probability associated with each event?
    (c) What is the probability of obtaining a total of at least 5?
    (d) What is the probability of obtaining a total less than 4?

The following questions all concern the student body at a college whose enrollment consists of 1200 white males, 1100 white females, 400 black males, 500 black females, 300 Asian males, and 100 Asian females.

15. Suppose one student is selected at random. What is the probability that he or she is (a) a white male, (b) a black female, (c) a male, (d) a female, (e) an Asian (f) not white?

16. Suppose the student selected is a female. What is the probability that she is (a) black, (b) Asian

17. Suppose the student selected is a male. What is the probability that he is (a) white, (b) not black?

18. Suppose the student selected is an Asian. What is the probability that he or she is female?

19. What is the probability that the student selected is either a white female or an Asian?

20. What is the probability that the student selected is either a black male or an Asian?

21. What is the probability that the student selected is neither black nor an Asian male?

4.2 Fundamental Counting Principle

Many of the probability ideas introduced in Section 4.1 require us to *count* the number of ways that various outcomes in a probability experiment can occur. The examples we considered previously were all quite straightforward. For more complicated situations, we need a more sophisticated way to count the number of outcomes that can occur.

Recall that the probability of an event A can be found from the formula

$$P(A) = \frac{\text{number of ways that } A \text{ can occur}}{\text{total number of possible outcomes}}$$

provided that all outcomes in the experiment are equally likely and there are only a finite number of possible outcomes. The denominator in this formula is equal to the size of the sample space for the experiment. To apply this result, we must be able to count the number of ways that the event A can occur and to count the number of outcomes in the sample space. Although we could draw a tree diagram or list all possible outcomes, these methods quickly become unworkable whenever the probability experiment becomes complicated. As a result, we must develop more effective methods for counting outcomes in an experiment.

Consider the experiment of flipping a coin and rolling a die. The sample space is

{H1, H2, H3, H4, H5, H6, T1, T2, T3, T4, T5, T6}

The total number of possible outcomes is 12, which is precisely equal to the product of 2 (the number of possible outcomes with the coin alone) and 6 (the number of possible outcomes with the die alone). This example provides the motivation for the following result.

> **Fundamental Counting Principle**
>
> If an event A can occur in a total of m ways and, independent of it, a second event B can occur in a total of n ways, then the combined event of A followed by B can occur in a total of $m \times n$ ways.

That is, we simply multiply the number of ways that A alone can occur and the number of ways that B alone can occur.

EXAMPLE 4.6 Consider the experiment involving rolling a die and picking a card from a standard deck. How many possible outcomes are there in the sample space?

SOLUTION Since there are 6 outcomes with the die and 52 with the card, the Fundamental Counting Principle tells us that the total number of possible outcomes in the experiment is $6 \times 52 = 312$.

The Fundamental Counting Principle can be extended to cover situations involving more than two events A and B. Thus, if we have three events A, B and C, then the total number of possible outcomes is the product of the three individual numbers of outcomes.

EXAMPLE 4.7 A student owns 3 pairs of sneakers, 5 pairs of jeans, and 20 shirts. If she randomly selects one of each in the morning without regard to color or matching, how many different outfits does she have?

SOLUTION According to the Fundamental Counting Principle, the total number of outfits possible is

$$3 \times 5 \times 20 = 300$$

EXAMPLE 4.8 A fast-food restaurant offers a choice of 7 main meals, 3 side orders, 5 different soft drinks, and 2 types of dessert. How many different meals consisting of one item from each category are possible?

SOLUTION The total number of possible meals is

$$7 \times 3 \times 5 \times 2 = 210$$

Often, the number of possible outcomes involves the choice of whether or not to include a particular item. We illustrate this situation in the next example.

EXAMPLE 4.9 An automobile manufacturer offers a certain car in 12 different colors, 3 different models (two-door, four-door and wagon), and 4 different engines. In addition, air conditioning is available. How many different car packages can be put together?

SOLUTION Since the buyer has the choice of either taking air conditioning or not, there are 2 ways to make this decision. Therefore, the total number of possible car packages is

$$12 \times 3 \times 4 \times 2 = 288$$

Now that we have a method for counting the number of possible outcomes in an experiment, we can extend the scope of probability problems we can solve.

EXAMPLE 4.10 Consider the experiment of rolling a die and picking a card from a deck. Find the probability that the sum of the face on the die and the number of spots on the card is 15. (Assume that the face cards—jack, queen and king—have 0 spots.)

SOLUTION We have already seen that the total number of possible outcomes in this experiment is $6 \times 52 = 312$. Of these, we can obtain a total of 15 either by having a 5 on the die and a 10 on the card (4 different ways) or by having a 6 on the die and a 9 on the card (another 4 different ways). Therefore, since the 312 individual outcomes are equally likely, the desired probability is

$$P(\text{total is 15}) = \frac{8}{312} = .056$$

EXAMPLE 4.11 A supermarket stocks 9 brands of peanut butter, 5 brands of grape jelly, and 15 varieties of loaves of bread. Of these, there are 2 store brands of peanut butter, 1 store brand of jelly, and 4 varieties of store-brand bread. If a customer takes one item from each category at random, what is the probability that he will select only store brands?

SOLUTION By the Fundamental Counting Principle, the total number of possible selections available is

$$9 \times 5 \times 15 = 675$$

To find the number of selections consisting of only store brands, we use the fact that there are 2 different varieties of store brand peanut butter, 1 variety of jelly, and 4 varieties of bread. As a result, the total number of ways that the customer can select only store brands is

$$2 \times 1 \times 4 = 8$$

Therefore, the probability of choosing only store brands is

$$\frac{8}{675} = .0118$$

We continue our investigation of counting problems in the next section where we will consider still more complicated situations.

Exercise Set 4.2

Applying the Concepts

1. A state's license plate is made up of three letters followed by three numbers. If there is no restriction on the letters that can be used, how many different license plates are possible?
2. A state's license plate is made up of two letters followed by four numbers. If there is no restriction on the letters that can be used, how many different license plates are possible?
3. How many different telephone numbers are possible consisting of three digits followed by another four digits?
4. In practice, no telephone number can have either a 0 or a 1 as the first digit. How many different telephone numbers are possible?
5. Suppose that a telephone number cannot have a 0 or a 1 as the first or second digit. How many different telephone numbers are possible?
6. Suppose that a telephone number cannot end in four 0s and cannot have a 0 or a 1 in the first two positions. How many different telephone numbers are possible?

7. In the United States, the call letters for a radio station must start with a K or a W (depending on whether the station is west or east of the Mississippi). If the station name must consist of exactly four letters, how many different station names are possible if no repetitions of a letter are allowed?
8. Repeat Exercise 7 if repetitions of a letter are permitted.
9. Twelve basketball players are on a team. Two of them are centers, five are guards, and five are forwards. In how many ways can a team (consisting of one center, two guards and two forwards) be formed?
10. Fifteen hockey players are on a team. Two are goalies, three are centers, six are wings, and four are defensemen. How many different teams of six players (consisting of one goalie, one center, two wings and two defensemen) can be formed?
11. A car seats six. In how many ways can five people be seated if only three of them can drive?
12. A housing contractor will build a house having two, three or four bedrooms, two or three bathrooms, and a one- or two-car garage. How many different houses are possible?
13. A fisherman knows there are four streams connecting his boat ramp to Trout Lake. There are five streams connecting Trout Lake to Bass Lake.
    (a) How many different routes can he travel from the dock to Bass Lake via Trout Lake?
    (b) How many different routes can he travel from Bass Lake to the dock via Trout Lake?
    (c) How many different round trips can he take if he refuses to go the same way back and forth?
14. There are four routes connecting Hicktown to Smallville. There are six further roads connecting Smallville to Riverburg. The county sheriff is responsible for patrolling these roads.
    (a) In how many ways can the sheriff make the trip from Hicktown to Riverburg via Smallville?
    (b) In how many ways can she make the round trip without going over the same road in both directions?
    (c) In how many ways can she make the round trip if she does not object to going over the same route twice?

4.3 Permutations and Combinations

In this section, we extend the ideas on counting to a wide variety of situations that require somewhat more sophisticated techniques.

Suppose a group of nine students decide to play a "pickup" baseball game. To avoid arguments, they agree to set up a batting order by picking names at random. How many different batting orders are possible if there are no other restrictions?

The question can be answered by applying the Fundamental Counting Principle. The leadoff batter can be any one of the nine players and so can be selected in nine ways. Once he or she has been picked, there are eight players

FIGURE 4.9

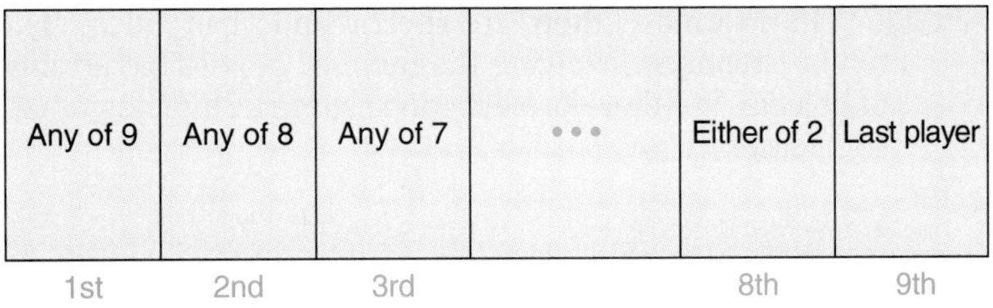

Any of 9	Any of 8	Any of 7	...	Either of 2	Last player
1st	2nd	3rd		8th	9th

left. The second hitter therefore can be selected in eight different ways. Consequently, the third hitter can be any one of the remaining seven; and so forth. By the time the ninth hitter is selected, the other eight have all been assigned, so there is only one player left to bat in the last spot. See Figure 4.9. By the Fundamental Counting Principle, the total number of different batting orders possible is

$$9 \times 8 \times 7 \times 6 \times 5 \times 4 \times 3 \times 2 \times 1 = 362{,}880$$

As a variation on the above problem, suppose there are 20 players at the field the following day. The coach will select nine to start the game without regard to position on the field. How many different batting orders are now possible? Consider the leadoff position. The coach can pick any one of the 20 players to bat first, leaving 19 others. He can then pick any of the remaining 19 players to bat second, any of the then remaining 18 players to bat third, and so forth. By the time the coach gets to the ninth position, there are 12 players remaining and any one of them can be selected. See Figure 4.10. Therefore, the total number of batting orders the coach can construct is given by

$$20 \times 19 \times 18 \times 17 \times \cdots \times 12$$

and this is approximately 60,949,000,000.

Before we proceed, several observations are in order. The Fundamental Counting Principle looks very simple and innocuous. However, as the two illustrations above demonstrate, the numbers involved can become incredibly large very quickly. The number of possible arrangements of a set of ballplayers (or many other sets of objects) can be astronomical.

FIGURE 4.10

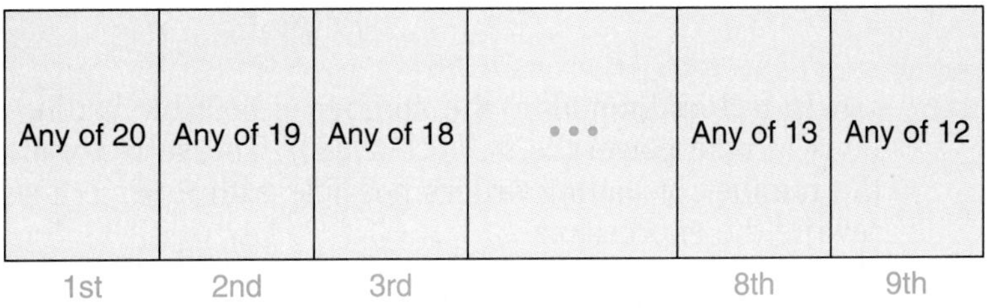

Any of 20	Any of 19	Any of 18	...	Any of 13	Any of 12
1st	2nd	3rd		8th	9th

Furthermore, there are several underlying ideas that apply to many different situations involving the number of possible arrangements for a group of objects. We therefore introduce some terminology that will be useful in all such cases.

> A **permutation** is a specific arrangement of a group of objects in some order.

Any other arrangement of the same objects is a different permutation. Therefore, rather than speaking of the number of possible batting orders or the number of possible arrangements, we speak of the "number of possible permutations of the 20 players" if we take groups of 9 at a time. In such cases, the key words are *order* or *arrangement*.

Further, the number of permutations possible in a group of objects typically involves a product of integers such as we obtained above. As a result, we introduce several mathematical abbreviations for writing such expressions. We first introduce *factorial notation:*

$$1! = 1$$
$$2! = 2 \times 1 = 2$$
$$3! = 3 \times 2 \times 1 = 6$$
$$4! = 4 \times 3 \times 2 \times 1 = 24$$
$$5! = 5 \times 4 \times 3 \times 2 \times 1 = 120$$

and so forth. These expressions are read "1 factorial", "2 factorial," and so on. In general, if n is any positive integer,

$$n! = n \times (n-1) \times (n-2) \times (n-3) \times \cdots \times 2 \times 1 \qquad n \geq 1$$

For example, if $n = 12$, then this becomes

$$12! = 12 \times (12-1) \times (12-2) \times (12-3) \times \cdots \times 3 \times 2 \times 1$$
$$= 12 \times 11 \times 10 \times \cdots \times 3 \times 2 \times 1$$

We note that this formula does not apply when $n = 0$. However, by convention, we define

$$0! = 1$$

In factorial notation, the number of possible batting orders or permutations of nine players is $9!$ ($= 362{,}880$). To use factorial notation to express the number of batting orders possible with 9 players selected from 20, we rewrite the expression

$$20 \times 19 \times 18 \times \cdots \times 12$$

as

$$(20 \times 19 \times 18 \times \cdots \times 12) \times \frac{11 \times 10 \times 9 \times \cdots \times 2 \times 1}{11 \times 10 \times 9 \times \cdots \times 2 \times 1}$$

since the quotient is simply equal to 1 and does not affect the value of the expression when we multiply by it. In turn, this is equal to

$$\frac{(20 \times 19 \times 18 \times \cdots \times 12) \times (11 \times 10 \times 9 \times \cdots \times 2 \times 1)}{11 \times 10 \times 9 \times \cdots \times 3 \times 2 \times 1} = \frac{20!}{11!}$$

$$= \frac{20!}{(20-9)!}$$

In general, suppose we have a group of n objects and wish to find the number of possible orders or permutations when we select r of the objects at a time. Based on the above illustration with $n = 20$ and $r = 9$, we see that this is

$$\frac{n!}{(n-r)!}$$

Further, we denote this quantity by $_nP_r$, so that

$$_nP_r = \frac{n!}{(n-r)!}$$

is the **number of permutations of n objects taken r at a time**.

EXAMPLE 4.12 Find the number of permutations possible for a group of 15 objects taken 3 at a time.

SOLUTION The number of possible permutations is $_{15}P_3$ and is equal to

$$_{15}P_3 = \frac{15!}{(15-3)!} = \frac{15!}{12!}$$

When we multiply out the factorials, we get

$$\frac{15!}{12!} = \frac{15 \times 14 \times 13 \times 12 \times 11 \times \cdots \times 2 \times 1}{12 \times 11 \times \cdots \times 2 \times 1}$$

$$= 15 \times 14 \times 13 = 2730$$

since all the other factors (12, 11, 10, ..., 3, 2, 1) cancel. ∎

The type of cancellation that occurred in this example is typical of what happens in general. You should first cancel all possible factors before multiplying the numbers in both the numerator and the denominator. Moreover, the value for *any* permutation must always be a positive integer. If you work out any problem and your result is not an integer, then you have made an error somewhere.

EXAMPLE 4.13 A student has 30 textbooks and has a bookshelf that will hold only 20 of them. How many different arrangements are possible for the 20 books?

SOLUTION The question is asking, How many permutations are possible for 30 objects taken 20 at a time? Using $n = 30$ and $r = 20$, we find that

$$_{30}P_{20} = \frac{30!}{(30-20)!}$$
$$= \frac{30!}{10!}$$
$$= \frac{30 \times 29 \times 28 \times \cdots \times 2 \times 1}{10 \times 9 \times \cdots \times 2 \times 1}$$

We note that all numbers from 1 up to 10 will cancel; however, that still leaves the product of all integers from 11 up to 30. Because of its magnitude, the result is typically expressed in scientific or exponential notation as 7.3097×10^{25} or approximately 73,097,000,000,000,000,000,000,000 arrangements. ∎

As we mentioned at the beginning of this section, the results of many counting problems are truly astronomical. Consequently, we do not normally multiply out such expressions unless the numbers involved are relatively small. We are usually content to leave such answers in factorial form.

EXAMPLE 4.14 An employee in a flower and plant shop wishes to select six plants out of the 250 in the store to make a window display. In how many different ways can he select and arrange these six plants?

SOLUTION We are looking for the number of arrangements possible for 6 plants chosen from 250. Thus, $n = 250$ and $r = 6$, so that the answer is

$$_{250}P_{6} = \frac{250!}{(250-6)!} = \frac{250!}{244!}$$ ∎

While the above permutation examples all involve the *order* or *arrangement* of a set of objects, similar problems occur where order is totally irrelevant. Instead, we want to count the number of possible groups without any regard to the order within the group.

> A **combination** is any collection of a group of objects without regard to order.

Problems involving combinations, where order *is not* relevant, are very similar to problems involving permutations, where order *is* critical. The only difference between permutations and combinations is whether order matters. Thus, we could alter the above problems slightly to eliminate the notion of order:

We could modify Example 4.13 to ask, Out of the student's 30 books, how many different groups of 20 books can be selected to fit on the bookshelf?

We could modify Example 4.14 to ask, Out of the 250 plants, how many different groups of 6 plants could be selected to be displayed in the window?

We now consider how to calculate the number of combinations possible when a group of r objects is selected from a larger set of n objects where order is irrelevant. Suppose we have seven people, Al, Barbara, Carlos, Don, Earl, Fred and Ginny or simply, A, B, C, D, E, F and G. Suppose further that we are going to select a group of two of them as a random sample drawn from this group. How many different random samples are possible? The order of selection is certainly not important here; we just want to know the number of possible groups consisting of two people selected out of the original seven. Thus, we have a combination problem.

To solve this problem, we first enumerate all possible samples of size 2:

AB, AC, AD, AE, AF, AG, BC, BD, BE, BF, BG, CD, CE, CF, CG, DE, DF, DG, EF, EG, FG

There are 21 such samples. Notice that we did not list BA, for instance; that sample is equivalent to AB since order is not important.

As with permutations, we must find a much better way to proceed than just listing and counting the possibilities. To see how to do this, consider the comparable problem of picking groups of size 2 from this set of 7 people where order *is* important. Thus, we find that the number of permutations is

$$_7P_2 = \frac{7!}{(7-2)!} = \frac{7!}{5!} = 7 \times 6 = 42$$

Notice that the number of permutations, 42, is precisely twice the number of combinations, 21. (Alternatively, the number of combinations, 21, is one-half the number of permutations, 42.) The reason for this is that in permutations all pairings such as AB and BA are counted separately while in combinations such pairings are equivalent and so are counted only once.

We now consider what happens if we select samples of size 3 instead of 2 from the group of 7 people. The number of permutations is

$$\begin{aligned} _7P_3 &= \frac{7!}{(7-3)!} \\ &= \frac{7!}{4!} \\ &= \frac{7 \times 6 \times 5 \times 4 \times 3 \times 2 \times 1}{4 \times 3 \times 2 \times 1} \\ &= 7 \times 6 \times 5 \\ &= 210 \end{aligned}$$

We now list all possible combinations:

ABC, ABD, ABE, ABF, ABG, ACD, ACE, ACF, ACG, ADE, ADF, ADG, AEF, AEG, AFG, BCD, BCE, BCF, BCG, BDE, BDF, BDG, BEF, BEG, BFG, CDE, CDF, CDG, CEF, CEG, CFG, DEF, DEG, DFG, EFG

There are 35 such combinations, and this is precisely one-sixth of the number of permutations.

To see why this is the case, consider the sample ABC. There are six possible rearrangements: ABC, ACB, BAC, BCA, CAB and CBA. All are counted when we find the number of permutations. However, with combinations, order is irrelevant and so we count this group only once. The same applies to any other group of three people. Therefore, there will be 6 times as many permutations as combinations with samples of size 3.

In general, suppose we have a group of n objects and select r of them at a time without regard to order. The number of permutations, $_nP_r$ includes the $r!$ possible rearrangements of the r objects. Since for combinations the order is irrelevant, it is necessary to divide by this $r!$ term, and so the number of combinations possible is given by

$$\frac{_nP_r}{r!} = \frac{\left[\frac{n!}{(n-r)!}\right]}{r!}$$

$$= \frac{n!}{(n-r)!r!}$$

Just as we introduced a special notation, $_nP_r$, for the number of permutations possible for n objects taken r at a time, we use a comparable notation for the number of combinations $_nC_r$. Thus,

$$_nC_r = \frac{n!}{r!(n-r)!}$$

is the number of possible combinations of n objects taken r at a time.

An alternative notation often used to represent the number of combinations possible for n objects taken r at a time is

$$\binom{n}{r} = {}_nC_r = \frac{n!}{r!(n-r)!}$$

The numbers of combinations of n objects taken r at a time are also known as *binomial coefficients*. They arise in many different areas of mathematics, including statistics. We will encounter them again in Chapter 5.

EXAMPLE 4.15

Samples of size 5 are to be selected from a group of 100 people. How many possible samples are there?

SOLUTION The question is equivalent to asking: How many combinations of 100 people taken 5 at a time are possible? The answer is given by

$$_{100}C_5 = \frac{100!}{5!(100-5)!}$$

$$= \frac{100!}{5!\,95!}$$

This reduces to

$$_{100}C_5 = \frac{100 \times 99 \times 98 \times 97 \times 96 \times 95 \times \cdots \times 2 \times 1}{(95 \times 94 \times \cdots \times 2 \times 1)(5 \times 4 \times 3 \times 2 \times 1)}$$

$$= \frac{100 \times 99 \times 98 \times 97 \times 96}{5 \times 4 \times 3 \times 2 \times 1}$$

$$= 6{,}145{,}920$$

As with permutations, you should cancel all possible terms before multiplying out the numerator and the denominator. Moreover, if the numbers are extremely large, we usually leave the answers to combination problems in terms of factorials. Furthermore, as with permutations, the answer to any combination problem must be a positive integer.

EXAMPLE 4.16

A donut shop sells 30 varieties of donuts. A customer asks for a box of a dozen different donuts, but does not specify any varieties. How many different boxes are possible?

SOLUTION Since order is not important (just the particular collection of different types of donuts), this is a combination problem. There are

$$_{30}C_{12} = \frac{30!}{12!\,18!}$$

possible boxes of donuts.

EXAMPLE 4.17

The professor in a communications course with 25 students randomly selects 7 of them to make presentations before the class. How many such groups of seven students are possible?

SOLUTION The use of the word *groups* suggests that order is irrelevant. Further, the fact that there is no indication of the order in which the presentations will be made also suggests that order does not matter. Therefore, the number of combinations of the 25 students taken 7 at a time is

$$_{25}C_7 = \frac{25!}{7!\,18!}$$

EXAMPLE 4.18 A telephone poll is to be conducted by randomly selecting six names from each page of a local telephone book. If the book contains 300 names per page, how many different groups of people are possible from each page of the directory?

SOLUTION Since we are asked for the number of different "groups," order is not important, and so this is a combination problem. Therefore, the number of possible groups that can be selected from each page of the directory is

$$_{300}C_6 = \frac{300!}{6!\,294!}$$

∎

Note that this answer just gives the number of possible samples based on using only one page of the phone directory. If we want the number of samples possible by using the first two pages, then the Fundamental Counting Principle gives

$$_{300}C_6 \times {}_{300}C_6 = ({}_{300}C_6)^2$$

It we want the number of samples based on three pages, the result is

$$_{300}C_6 \times {}_{300}C_6 \times {}_{300}C_6 = ({}_{300}C_6)^3$$

If the directory has 120 pages, then the total number of such samples possible is

$$({}_{300}C_6)^{120}$$

Exercise Set 4.3

Mastering the Techniques

Calculate each of the following quantities:

1. $_5P_2$
2. $_8P_4$
3. $_8P_3$
4. $_8P_5$
5. $_7P_1$
6. $_7P_0$
7. $_7P_7$
8. $_5C_2$
9. $_8C_4$
10. $_8C_3$
11. $_8C_5$
12. $_7C_1$
13. $_7C_0$
14. $_7C_7$

Applying the Concepts

15. A student is given a reading list consisting of 10 books and must select 6 of them to read and report on during the course of a semester. In how many ways can she pick the books to read?
16. A college offers 12 courses in literature. If a student intends to take one of these courses each of the eight semesters he attends the college, in how many ways can he pick the eight courses he will take?

17. The disk jockey at a campus radio station has received listener requests for nine different selections. If there is only time to play five of them, in how many ways can the DJ schedule the programming?
18. Find the number of different signals that can be formed by raising 3 flags on a ship's flagpole, one above the other, if 10 different flags are available.
19. NASA is selecting the flight crew of 3 astronauts for a forthcoming mission from a group of 16 astronauts. In how many ways can the crew be selected?
20. NASA is selecting the flight crew for a forthcoming mission from a group of 16 astronauts; 12 are military personnel and the remaining 4 are civilians. The crew will consist of a flight commander, a pilot and a mission specialist. The commander and the pilot are to be military personnel and the mission specialist is to be a civilian. In how many ways can the three crew members be selected for the three positions?
21. A group of nine girls is camping out. They have a tent that can shelter five people and one that can shelter four people. In how many ways can they arrange themselves?
22. In how many ways can a pile of eight books—five novels and three nonfiction—be arranged if the novels are to lie on top?
23. A final examination consists of two parts, and each part contains six questions. A student has to answer a total of eight questions, four from each part. In how many ways can the student choose the eight questions?
24. In how many ways can 6 fraternity pledges be selected from a group of 20 applicants?
25. A school maintains a list of 16 substitute teachers. On a given day, five teachers are absent. In how many ways can a set of substitutes be selected?
26. In a so-called combination lock, there are 50 different settings. To open the lock, you move to a certain number in one direction, then to a second number in the opposite direction, and finally to a third number in the original direction.
    (a) What is the total number of possible combinations if the first turn must be in a specific direction?
    (b) Explain why a combination lock should really be called a permutation lock.
27. Show that $_nP_0 = 1$ for any positive integer n.
28. Show that $_nP_n = n!$ for any positive integer n.
29. Show that $_nC_0 = 1$ for any positive integer n.
30. Show that $_nC_n = 1$ for any positive integer n.
31. Show that $_nC_r = {_nC_{n-r}}$ for any positive integer n and any positive integer $r \leq n$.

4.4 Mutually Exclusive and Independent Events

We now introduce three additional ideas about the probability of an event in a statistical experiment. These concepts will allow us to consider a wide variety of more complicated problems and will provide the groundwork for the probabilistic ideas used throughout statistics.

Suppose we begin by picking one card from a standard deck. Consider the three events

A: picking an ace

B: picking a king

C: picking a heart

We first consider events *A* and *B*. If the card we select is an ace, then it certainly is not a king, and so event *B* did not occur. Alternatively, if *B* occurs so that our card is a king, then we did not pick an ace and so event *A* does not occur.

On the other hand, consider events *A* and *C*. Suppose that event *A* occurs and we pick an ace. This does not automatically prevent the card selected from being a heart (we could have picked the ace of hearts). Therefore, it is still possible for event *C* to occur also. Similarly, suppose the card selected is a heart, so that event *C* occurs. It is still possible for the card to be the ace of hearts, so that event *A* could occur as well. Thus, it is possible that events *A* and *C* occur simultaneously, even though we saw above that it is impossible for events *A* and *B* to occur simultaneously. Clearly, these two cases represent two different types of situations.

We formalize these ideas with the following terminology.

> Two events *A* and *B* are **mutually exclusive** if the occurrence of one precludes the occurrence of the other.

Thus, we see that events *A* and *B* above (picking an ace and picking a king on the same card) are mutually exclusive while events *A* and *C* (picking an ace and picking a heart) are not mutually exclusive.

We can visualize the notions of mutually exclusive and not-mutually-exclusive events by introducing a geometric tool known as a *Venn diagram* for displaying probability. In such a display, we use a rectangle to represent the sample space for any experiment. Events of interest are then represented by circles within the sample space. We show this in Figure 4.11 for events *A*, *B*, *C* and *D*.

FIGURE 4.11

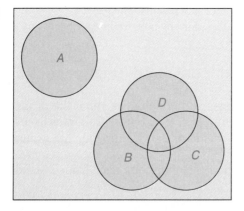

FIGURE 4.12

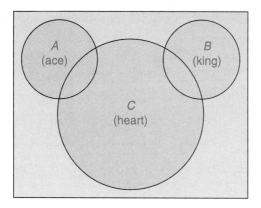

We now consider the specific Venn diagram in Figure 4.12 that corresponds to the above card example. The events labeled A and B (picking an ace and picking a king) are mutually exclusive, and so the circles representing them do not overlap. The circle representing event C (picking a heart) overlaps the other two since A and C are not mutually exclusive and B and C are not mutually exclusive. Thus, we see from this display that if the events are mutually exclusive, there is no overlap between the two circles. However, if they are not mutually exclusive, there will be some overlap or duplication between them.

We next consider the probabilities associated with these various events. We know that

$$P(A) = P(\text{ace}) = \tfrac{4}{52}$$
$$P(B) = P(\text{king}) = \tfrac{4}{52}$$
$$P(C) = P(\text{heart}) = \tfrac{13}{52}$$

Based on these probabilities, we are now in a position to consider some more interesting events associated with this experiment. For instance, what is the probability that the card selected is either an ace or a king? Since there are eight ways of picking either an ace or a king, we immediately see that

$$P(A \text{ or } B) = P(\text{ace or king}) = \tfrac{8}{52}$$

Alternatively, what is the probability that the card selected is either an ace or a heart? Since there are 4 aces and 13 hearts, we might be similarly tempted to conclude that this probability is $\tfrac{17}{52}$. However, this is incorrect since the ace of hearts is being counted twice—once as an ace and again as a heart. Therefore, the correct answer is

$$P(A \text{ or } C) = P(\text{ace or heart}) = \tfrac{16}{52}$$

Obviously, these two questions had to be handled carefully. The primary difference between them is that the first two events, A and B, are mutually exclusive and hence have no overlap while the second two events, A and C, are not mutually exclusive and so the overlap must be taken into account.

Based on these two cases, we are led to the following general principle, known as the *Addition Rule*:

Addition Rule

If events A and B are mutually exclusive, then

$$P(A \text{ or } B) = P(A) + P(B) \qquad (4.2)$$

If events A and B are not mutually exclusive, then

$$P(A \text{ or } B) = P(A) + P(B) - P(\text{both } A \text{ and } B) \qquad (4.3)$$

In other words, if two events A and B are mutually exclusive, then we can find the probability that either A or B will occur by simply adding the individual probabilities. If they are not mutually exclusive, then there is some overlap between them. Thus, to find the probability that either A or B will occur, we must subtract the probability of the duplication $P(\text{both } A \text{ and } B)$ from the sum $P(A) + P(B)$.

We can also interpret these ideas in the following way. If A and B are mutually exclusive, then $P(\text{both } A \text{ and } B) = 0$ since they cannot occur simultaneously. Therefore, we could actually consider only formula (4.3) since it reduces to formula (4.2) when the events are mutually exclusive.

EXAMPLE 4.19 In a group of 100 students, 60 are taking mathematics and 12 are taking culinary arts, but no one is taking both. What is the probability that a randomly selected student from the group is taking either course?

SOLUTION We display the given information in a Venn diagram shown in Figure 4.13. We are told that

$$P(\text{mathematics}) = \tfrac{60}{100}$$

$$P(\text{culinary arts}) = \tfrac{12}{100}$$

However, since no student is taking both courses, the events are mutually exclusive. Therefore,

$$P(\text{mathematics or culinary arts}) = \tfrac{60}{100} + \tfrac{12}{100} = \tfrac{72}{100} = .72 \qquad \blacksquare$$

FIGURE 4.13

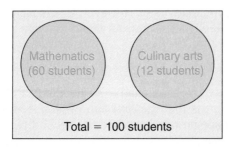

SECTION 4.4 • Mutually Exclusive and Independent Events

EXAMPLE 4.20 In a group of 120 students, 60 are taking mathematics, 40 are taking psychology, and 15 are taking both. A randomly chosen student from this group is selected. What is the probability that he or she is taking either mathematics or psychology? What is the probability that he or she is taking neither one?

SOLUTION We first display the given data pictorially, as shown in Figure 4.14. Since 15 students are in both groups, it follows that there are $60 - 15 = 45$ students taking mathematics but not psychology and $40 - 15 = 25$ students who are taking psychology but not mathematics. In the Venn diagram, the numbers in parentheses represent the number of students in each group.

Further, it is clear that

$$P(\text{mathematics}) = \tfrac{60}{120}$$

$$P(\text{psychology}) = \tfrac{40}{120}$$

$$P(\text{both mathematics and psychology}) = \tfrac{15}{120}$$

Since the events are not mutually exclusive, we use formula (4.3) to find that

$$P(\text{mathematics or psychology}) = P(\text{mathematics}) + P(\text{psychology}) - P(\text{both})$$

$$= \tfrac{60}{120} + \tfrac{40}{120} - \tfrac{15}{120}$$

$$= \tfrac{85}{120} = .708$$

Further, since this represents the probability of taking either mathematics or psychology or both, there must be $120 - 85 = 35$ remaining students from the group who are taking neither. Thus, the probability that the student selected is not taking either mathematics or psychology is

$$P(\text{neither mathematics nor psychology}) = \tfrac{35}{120} = .292 \qquad \blacksquare$$

When probabilities are given as fractions, as in Example 4.20, it is usually a good idea to leave the fractions unreduced in the anticipation that they will combine with a common denominator.

We now introduce a further notion regarding probability. Suppose we have a single event A based on a statistical experiment.

FIGURE 4.14

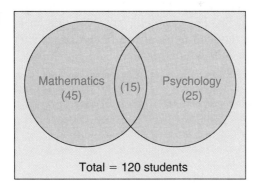

The **complementary event** A' is the event

A': A does not occur

For example, if A represents obtaining 5 on a roll of a single die, then A' represents obtaining any face other than 5. Thus, since $P(A) = \frac{1}{6}$, we see that $P(A') = \frac{5}{6}$, by counting the number of ways that we can get anything other than 5.

In general, if A is any event, it follows that

$$P(A') = 1 - P(A)$$

Alternatively,

$$P(A) + P(A') = 1$$

We can visualize these ideas with the Venn diagram shown in Figure 4.15. The single circle represents event A, and therefore the complementary event A' is represented by everything outside the circle and inside the rectangle for the sample space. The total probability associated with the experiment is always 1. Therefore, the probability of the complement $P(A')$ must be $1 - P(A)$.

As we will see, it is occasionally far easier to calculate $P(A')$ than it is to determine $P(A)$ itself. In such a case, we can use the above formula to find that

$$P(A) = 1 - P(A')$$

and so calculate $P(A)$ once we have found the probability of the complementary event A'. This approach is taken in several upcoming examples.

We now turn to another concept of probability involving two events A and B. Suppose we conduct the following experiment. A card is selected from a standard deck and put aside. A second card is then selected from the remaining deck. What is the probability that the second card is an ace?

Having picked and discarded the first card, we are selecting the second card from a reduced deck of 51 cards. Therefore, the total number of possible outcomes is 51, not 52. The problem, however, lies in knowing how many aces are left in the reduced deck. If the first card selected was an ace, then there are only three aces left and

$$P(\text{second card is an ace}) = \tfrac{3}{51}$$

FIGURE 4.15

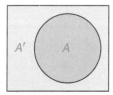

On the other hand, if the first card selected was not an ace, then all four aces are left in the deck and

$$P(\text{second card is an ace}) = \tfrac{4}{51}$$

As a result, we see that the probability for the second event *depends* on the outcome of the first event. This leads to the following definition:

> Two events A and B are **independent** if the probability of the second event B is not affected by the occurrence or nonoccurrence of the first event A.

If the probability of the second event B does depend on the result of the first event A, then we say that A and B are not independent events or are *dependent events*. In the above card-picking experiment, we see that the two events are dependent.

EXAMPLE 4.21 Suppose a fair coin is flipped twice. What is the probability that the second flip is a head?

SOLUTION In this case, the result for the second flip does not depend on the outcome of the first flip. The results are independent, and the answer is simply

$$P(\text{second coin is a head}) = \tfrac{1}{2}$$

regardless of the outcome on the first flip. ∎

Most of the situations we encounter in both probability and statistics throughout this book involve independent events. In fact, for most elementary statistical purposes, we want to avoid dependent events. For instance, suppose we are conducting a political poll to see how strongly people support the President on some issue of national concern. We will select a certain number of people as a sample, say 500. We certainly want the answers they give to be independent of each other. For that reason, we would not knowingly select a husband and wife since their opinions could be reflective of each other. The responses of the second person of such a couple might be dependent on the responses of the first.

While most of the events we consider are independent, it is surprising how few people actually are aware of the concept and its implications. For example, approximately one-half of all births are boys and one-half are girls. Thus, when a woman gives birth, the probability that the child is male is approximately .5. (In fact, from a mathematical point of view, the sex of a child and the outcome of flipping a fair coin are essentially equivalent since identical mathematics apply to both.) Moreover, the sex of a succession of children to the same mother is a series of independent events. The sex of a second child is not affected by the sex of the first child. However, consider how often you have heard the following type of story: A family has five children, all girls, and they want a boy. Because the first five children are all girls, the parents are convinced that the odds of the next child being a boy are significantly in their favor, so they decide to have a sixth child. Of course, approximately 50 percent of such families are going to have a sixth girl. The events are independent, and so the probability that the sixth child is a boy is still .5.

Unfortunately, many people carry this false line of reasoning to unreasonable lengths. Possibly the most notable example of this is King Henry VIII who went through six wives hoping to find one who would produce a large number of male heirs to the throne of England. Unfortunately for his wives, only one of them produced a son, Edward, who succeeded Henry.

When two events are independent, there is a very simple way to find the probability that both events will occur successively. It is known as the *Multiplication Rule*.

Multiplication Rule

If A and B are independent events, then

$$P(\text{both } A \text{ and } B) = P(A) \times P(B)$$

That is, when A and B are independent events, the probability that both hold is just the product of the individual probabilities. We often think of independent events A and B as occurring one after the other.

EXAMPLE 4.22 Use the fact that coin flipping represents independent events to calculate the probabilities involved in flipping two fair coins.

SOLUTION To obtain 2 heads, we note that

$$P(2H) = P(\text{head on first and head on second})$$
$$= P(\text{head on first}) \times P(\text{head on second})$$
$$= \tfrac{1}{2} \times \tfrac{1}{2} = \tfrac{1}{4}$$

The identical argument applies to the probability of getting 0 heads, so that $P(0H) = \tfrac{1}{4}$ also. Now, we know that in any experiment, the sum of the probabilities of all the possible distinct nonoverlapping events is 1, so that

$$P(0H) + P(1H) + P(2H) = \tfrac{1}{4} + P(1H) + \tfrac{1}{4} = 1$$

Thus, since we have already accounted for two of the three possible events, the remaining event represents the complementary event and so

$$P(1H) = 1 - \tfrac{2}{4} = \tfrac{2}{4}$$

Alternatively, we could calculate $P(1H)$ using the fact that the event 1 head arises from either a head followed by a tail or a tail followed by a head. Further, these are mutually exclusive events. Thus, since the individual outcomes in each case are independent, we find that

SECTION 4.4 • Mutually Exclusive and Independent Events

$$P(1H) = P(HT \text{ or } TH)$$
$$= P(HT) + P(TH) \quad \text{(mutually exclusive events)}$$
$$= P(H) \times P(T) + P(T) \times P(H) \quad \text{(independent events)}$$
$$= \tfrac{1}{2} \times \tfrac{1}{2} + \tfrac{1}{2} \times \tfrac{1}{2}$$
$$= \tfrac{1}{4} + \tfrac{1}{4} = \tfrac{2}{4}$$

∎

EXAMPLE 4.23 Suppose a coin is weighted in such a way that it comes down heads 60% of the time. Calculate the probabilities involved in flipping it twice.

SOLUTION Despite the fact that the coin is no longer fair, the outcomes on successive flips still represent independent events. All that changes is that the probability of getting 1 head on each flip is now .6. Therefore, the probability of getting 2 heads can be obtained from

$$P(2H) = P(\text{first is a head}) \times P(\text{second is a head})$$
$$= .6 \times .6 = .36$$

Similarly, the probability of getting 0 heads or 2 tails is

$$P(0H) = P(\text{first is a tail}) \times P(\text{second is a tail})$$
$$= .4 \times .4 = .16$$

Since the probabilities associated with these two events total .52, we conclude that the probability of getting 1 head and 1 tail by using this coin is $1 - .52 = .48$. We summarize the results of this experiment in the following table, where x represents the number of heads:

x	$P(x)$
0	.16
1	.48
2	.36
	1.00

∎

EXAMPLE 4.24 NASA estimates that the probability that a certain component in a communications satellite fails is .005. There is an independent backup system, and the probability that it fails is .01. What is the probability that both systems fail?

SOLUTION We are told that

$$P(\text{system 1 failure}) = .005$$
$$P(\text{system 2 failure}) = .01$$

Since these two systems are independent of each other, we can treat them as independent events. Therefore,

$$P(\text{both systems fail}) = P(\text{system 1 failure}) \times P(\text{system 2 failure})$$
$$= .005 \times .01 = .00005$$

and so the chance of both the system and its backup failing is just 5 out of 100,000. ∎

We note that the type of reasoning used here with independent events applies in many different fields. As the newspaper clipping in Figure 4.16

FIGURE 4.16

> The courts should apply a similarly skeptical "show me" attitude toward testimony that the chances of a random match in DNA type are only one in billions. DNA is only one of the many genetic markers now used for forensic identification purposes. Tests for red blood cells, white blood cells, proteins and enzymes are already in widespread use.
>
> Typically, when an attorney wants the expert to testify to the statistical probability of a random match, the testimony goes along these lines: "I tested the sample and found that Markers 1 and 4 were present. Only 1 percent of the population has Marker 1, and a mere 10 percent of the population possesses Marker 4. Moreover, these markers are independent; the fact that you have Marker 1 doesn't affect the likelihood that you'd have 4. My conclusion is therefore that only one-tenth of 1 percent of the population has the set of markers found in the stain at the crime scene—and it just so happens that the defendant tests out as having those same two markers."
>
> Once again, there are certain questions to be answered: How does the expert know that only 1 percent of the population has Marker 1? What studies of the frequency of that marker have been conducted? How does the expert know that Markers 1 and 4 are independent? Has the markers' independence been experimentally verified?

illustrates, it has even become a major line of reasoning in many criminal trials to demonstrate that a defendant must have committed a crime just on the basis of the probabilities involved. For example, in the murder trial of Wayne Williams (who was accused of killing 28 young black boys in Atlanta in 1981), the prosecution used probabilistic arguments based on, among other things, the likelihood of Williams having a certain green rug and the victims' bodies being wrapped in a rug from the same production and dye lot to convict him.

When we introduced the notion of the complementary event A', we mentioned that it is often much easier to find the probability of an event A by first finding the probability of A'. We now consider one last example involving independent events which requires this approach.

EXAMPLE 4.25 Find the probability that, in a group of three people—Al, Betty, and Carol—two have the same birthday.

SOLUTION It is extremely difficult to calculate this probability directly. Instead, we concentrate on the complementary event that they all have different birthdays. Suppose we select any one of the three, say Al, and then consider a second person in the group, say Betty. In order for Betty to have a different birthday from Al's, she must have been born on any one of the 364 dates other than Al's birthday. Thus, the probability that Betty's birthday is different from Al's is $\frac{364}{365}$.

Next, if Carol's birthday is to be different from both Al's and Betty's, her birthday must fall on any one of the remaining 363 days. Therefore, the probability that Carol has a birthday which is different from both Al's and Betty's is $\frac{363}{365}$.

We now calculate the probability that all three have different birthdays using the multiplication rule. This is

P(all different)

$= P$(Betty's is different from Al's and Carol's is different from both)

$= P$(Betty's is different from Al's) $\times$ P(Carol's is different from both)

$= \frac{364}{365} \times \frac{363}{365}$

$= .9973 \times .9945$

$= .9918$

Therefore, since this is the probability of the complementary event, the probability that there is a match among the three birthdays is

$$P(\text{match}) = 1 - P(\text{all different})$$
$$= 1 - .9918$$
$$= .0082$$

Clearly, this is a very unlikely eventuality.

We can extend this line of reasoning to determine the probability of a match in birthdays occurring within groups of any number of people. For instance, if we have a group of four people, then the probability of a match is

$$P(\text{match}) = 1 - P(\text{all different})$$
$$= 1 - \left(\tfrac{364}{365}\right)\left(\tfrac{363}{365}\right)\left(\tfrac{362}{365}\right) = .0164$$

Surprisingly, when this line of reasoning is continued for larger groups of people, the results become very much at odds with what our intuition suggests. In fact, it turns out that whenever there is a group of 23 people, the probability that a match in birthdays occurs is approximately .5. That is, in approximately one-half of such groups, we expect a match. Further, by the time the group is as large as 48 people, the probability of a match is virtually 1.

Computer Corner

We have included Program 8: Birthday Problem Simulation to allow you to experiment with the birthday problem we just discussed. To use this program, you must supply the size of the group and the number of runs or groups to be generated. Suppose you ask for 20 groups of size 15. The program generates 20 different sets of 15 birthdays and tests to see whether there is a match in each group. If there is a match, an appropriate message is shown including the date on which the match occurs. If there is no match, this fact is also shown. When the simulations are completed, the program summarizes the results and compares them to the theoretical results predicted based on probability theory. We suggest that you use this program with a variety of group sizes to see that the results we have discussed actually are correct despite the fact that they are so contrary to our intuitive feelings.

Exercise Set 4.4

Mastering the Techniques

1. Suppose a card is randomly selected from a standard deck. Find the following probabilities.
   (a) P(red card or a spade)
   (b) P(red card or a king)
   (c) P(red card or a face card)
   (d) P(spade or a red ace)
2. Suppose two fair dice are rolled. Find the following probabilities.
   (a) $P(\text{sum} = 3 \text{ or sum} = 5)$
   (b) $P(\text{sum} = 3, 4 \text{ or } 5)$
   (c) P(sum is even)
   (d) P(sum is prime number); that is, 2, 3, 5, 7, 11
   (e) P(sum is even or sum is prime)
   (f) P(sum is prime or sum is 7 or 11)
   (g) P(sum is odd or sum is 3, 4 or 5)

3. An experiment consists of flipping a fair coin and rolling a fair die.
   (a) What is the sample space?
   Find the probability of getting
   (b) A head and a 6
   (c) A head or a 6
   (d) A head and an odd number
   (e) A head or an odd number
   (f) A value of 4 or higher on the die
   (g) A tail and a value of 4 or higher on the die.
4. An experiment consists of flipping a fair coin and picking a card from a standard deck. Find the probability of getting
   (a) A head and an ace
   (b) A head or an ace
   (c) A head and a face card
   (d) A head or a face card
   (e) A tail and a diamond
   (f) A tail and an odd number

Applying the Concepts

5. In a group of 200 students, 120 are taking Spanish, 60 are taking French, and 10 are taking both. If a student is selected randomly, what is the probability that he or she
   (a) is taking both languages,
   (b) is taking either French or Spanish,
   (c) is taking neither language?
6. Out of 400 customers in a fast-food restaurant, 250 ordered hamburgers, 170 ordered chicken, and 40 ordered both.
   (a) What is the probability that a randomly selected customer ordered either a hamburger or chicken?
   (b) What is the probability that a randomly selected customer ordered neither one?
7. A coin is loaded in such a way that the probability of its landing heads is $\frac{3}{4}$. If this coin is flipped twice, find $P(2H)$, $P(1H)$ and $P(0H)$.
8. If the coin in Exercise 7 is flipped 3 times, find $P(3H)$ and $P(0H)$.
9. A child has developed her skill at flipping baseball cards to the point where she can get the face up 90% of the time.
   (a) If she flips two cards, what is the probability that both land face up? That both land face down?
   (b) If she flips three such cards, what is the probability that they will all land face up? That they will all land face down?
10. A soda bottling company finds that 2% of its 1-liter bottles are underfilled. If two bottles are selected at random, what is the probability that both are underfilled?
11. A professor estimates that 85% of her students pass an introductory chemistry course. If two students are selected at random, find the probability that (a) both will fail and (b) both will pass.
12. A survey finds that 40% of school buses on the road in a certain community are unsafe and should be repaired. If the police stop and check two school buses in

this area, what is the probability that **(a)** both are unsafe and **(b)** both are safe?

13. A baseball player's lifetime batting average is .300. If he comes up to bat 4 times during a game, what is the probability that he **(a)** gets four hits and **(b)** gets no hits? (Assume that his average does not change as the result of several additional at-bats.)

14. A basketball player's career foul-shooting percentage is 70%. If he is awarded five foul shots during a game, what is the probability that he connects on all five shots? (Assume that the percentage does not change as a result of several additional foul shots.)

15. An airline claims that 90% of its planes land on time. If three flights are selected randomly, what is the probability that all three are on time?

16. Consider the newspaper clipping in Figure 4.17 on the relative effectiveness of a variety of common birth control devices. Use the theoretical effectiveness values (the largest if there is a range) to determine the probability of conception in the following situations.
    (a) The pill and a condom are used.
    (b) An IUD and a condom are used.
    (c) The pill and spermicides are used.
    (d) The pill or spermicides are used.

17. Calculate the probability that two people in a group of five have the same birthday.

18. The length of the year on Mars is approximately 686 days. Find the probability that, in a group of three Martians, two have the same birthday.

FIGURE 4.17

New Options in Birth Control

Some of the more common birth control methods, their effectiveness and drawbacks. The effectiveness rating is the number of pregnancies per 100 users per year.

Method	Effectiveness in theory	Effectiveness in practice	Drawbacks
Male sterilization	0.15	0.2-0.5	Swelling, pain common; reversible in only about 50 percent of patients
Female sterilization	0.05	0.2-1	Surgery required; reversibility limited
Intrauterine Device	1-3	1-5	Complications, infections can lead to infertility, death
The pill	0.5	1-8	Must be taken regularly; increased risk of cardiovascular disease
The Minipill	1	3-10	Less effective than the pill; some menstrual irregularity; high rate of ectopic pregnancy
Condom	1-2	3-15	May tear; sometimes unpopular
Diaphragm	2	4-25	Bladder infection; may become dislodged
Vaginal spermicides	3-5	10-25	Relatively unreliable
Periodic abstinence	2-5	10-30	Unreliable; requires careful record keeping
Vaginal sponge	11	15-30	May cause allergic reaction; should be removed within 24 hours

SOURCE: Population Crisis Committee, Washington, D.C.

19. The length of the year on Mercury is 88 days. Find the probability that, in a group of four Mercurians, two have the same birthday.
20. Select a major league baseball team. Consult a baseball yearbook or encyclopedia and check the birthdates of the approximately 25 players on the roster. Is there a match? Compare your results with other students' and see what percentage of teams contain birthday matches.

Computer Applications

21. Use Program 8: Birthday Problem Simulation to experiment with the probabilities involved in the birthday problem. In particular, run the program with groups of 10, 15, 20, 25 and 30 people. Use at least 30 groups or runs in each case, and compare the actual percentage of matches that occurred to the theoretical predictions.

4.5 Dependent Events and Conditional Probability

In most of the work we have done with probability to this point, the experiments in question usually involved independent events. As a result, determining the probability of such an event was reasonably straightforward. However, if the events are dependent, then the reasoning can become more complicated. This was evident in Section 4.4 when we considered the problem of picking two cards from a deck. The probability that the second card is an ace clearly depends on the outcome of the first pick, so that the answer is either $\frac{3}{51}$ or $\frac{4}{51}$, depending on what the first outcome was. Thus, we cannot answer such a question definitively unless we know the result of the first outcome.

We often face a situation where we want to find the probability of an event given some partial knowledge about the outcomes. For instance, suppose someone picks a card from a deck and you happen to notice that the card is red. If you are then asked to determine the probability that the card is a heart, the answer is $\frac{13}{26}$, not $\frac{13}{52}$, since you have some information that reduces the number of possible outcomes.

Consider the problem of finding the probability of an event B given that a "previous" event A has occurred or that you have some information on a related event A. This type of problem is known as *conditional probability* and is written as

$$P(B|A)$$

and read as the "probability of B given A". If events A and B are independent, then there is no difficulty in solving such a problem. Since the probability that the second event B occurs is not affected by the occurrence or nonoccurrence of A, the result is simply equal to $P(B)$, the probability that B occurs.

Suppose, though, that events A and B are not independent. The key to answering such a conditional probability question in relatively simple cases involves considering the sample space. We have previously used the idea that if the outcomes associated with the events in question are equally likely, then the probability that an event E occurs is given by

$$P(E) = \frac{\text{number of ways that } E \text{ can occur}}{\text{total number of possible outcomes}}$$

where the total number of possible outcomes is the size of the sample space. However, when we consider conditional probability, the occurrence or non-occurrence of the first event A alters the original sample space. For example, with the problem of picking two cards from a deck, the original sample space consists of the 52 cards in the full deck. When we consider the probability associated with picking the second card, we no longer have the full deck of 52 cards. Rather, we are now selecting a card from a reduced or modified deck, and so the corresponding sample space consists of just 51 cards. As a consequence, the associated probability that the second card is an ace is based on this reduced sample space and the number of ways that we obtain the ace is either 3 or 4 out of the 51.

EXAMPLE 4.26 Consider a family with three children. Find the probability that there are two boys given that at least one child is known to be a girl.

SOLUTION The original sample space for this situation is

{BBB, BBG, BGB, BGG, GBB, GBG, GGB, GGG}

However, the fact that we are told that at least one child is a girl means that we are actually working with a reduced sample space, namely,

{BBG, BGB, BGG, GBB, GBG, GGB, GGG}

since one of the original outcomes, BBB, is clearly impossible. Therefore, the conditional probability that such a family has two boys, given that at least one child is a girl, is

$$P(2 \text{ boys} \mid \text{at least one child is a girl}) = \tfrac{3}{7}$$

■

EXAMPLE 4.27 Repeat Example 4.26, given that the oldest child is a girl.

SOLUTION In this case, the given information leads to a reduced sample space consisting of

{GBB, GBG, GGB, GGG}

and so the associated probability is

$$P(2 \text{ boys} \mid \text{oldest child is a girl}) = \tfrac{1}{4}$$

■

EXAMPLE 4.28 Find the probability that, when you roll two fair dice, the sum is 7 given that one die landed with the 5 showing.

SOLUTION To find the reduced sample space, we consider two possible cases. The first die could land with the 5 showing to produce the possible outcomes (5, 1), (5, 2), (5, 3), (5, 4), (5, 5), and (5, 6); or the second die could land with the 5 showing to produce the possible outcomes (1, 5), (2, 5), (3, 5), (4, 5), (5, 5), and (6, 5). Since one of these outcomes, (5, 5), is duplicated, the reduced sample space consists of the following 11 possible outcomes:

$$\{(1, 5), (2, 5), (3, 5), (4, 5), (5, 5), (6, 5), (5, 1), (5, 2), (5, 3), (5, 4), (5, 6)\}$$

Among these 11 possible equally likely outcomes, those corresponding to a sum of 7 are (2, 5) and (5, 2). Therefore, the desired probability is

$$P(\text{sum is 7} \mid \text{at least one came up 5}) = \tfrac{2}{11}$$

∎

EXAMPLE 4.29 Find the probability that, when two fair dice are rolled, one comes up showing a 5 given that the total is 7.

SOLUTION The reduced sample space consists of all possible outcomes where the total of the two faces is 7. These are

$$\{(1, 6), (2, 5), (3, 4), (4, 3), (5, 2), (6, 1)\}$$

Among these six equally likely possible outcomes, those corresponding to one face showing a 5 are (2, 5) and (5, 2). Therefore, the probability that one of the dice shows a 5 is

$$P(5 \mid \text{sum is 7}) = \tfrac{2}{6} = \tfrac{1}{3}$$

∎

Most of the problems in conditional probability that we consider can be solved easily by enumerating the reduced sample space. However, it is also possible to approach these ideas in a more sophisticated manner by using the following formula.

The **conditional probability** of B given A is

$$P(B \mid A) = \frac{P(\text{both } A \text{ and } B)}{P(A)}$$

To see where this formula comes from, we consider a Venn diagram to display the two events A and B, as shown in Figure 4.18. If we are given that event A has occurred, then we must restrict our attention exclusively to what happens within the circle representing A. Thus, the probability of B given A reduces to that portion of the circle for event B that lies within the circle for event A. This fraction is just the probability that A and B both occur, divided by the probability that A occurs, and so gives the above formula.

FIGURE 4.18

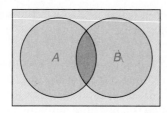

EXAMPLE 4.30 Repeat Example 4.29, using the formula for conditional probability.

SOLUTION Let

A: sum of the two faces is 7

B: one die comes up 5

The conditional probability is

$$P(B \mid A) = P(\text{one face is a 5} \mid \text{sum of the faces is 7})$$

$$= \frac{P(\text{one face is a 5 and the sum is 7})}{P(\text{sum is 7})}$$

Using the original sample space of all 36 equally likely possible outcomes, we see that

$$P(\text{sum is 7}) = \tfrac{6}{36}$$

and

$$P(\text{sum is 7 and one die comes up 5}) = \tfrac{2}{36}$$

Therefore,

$$P(\text{one face is 5} \mid \text{sum is 7}) = \frac{\tfrac{2}{36}}{\tfrac{6}{36}} = \tfrac{2}{6} = \tfrac{1}{3}$$

which is the same result as we obtained above. ∎

Exercise Set 4.5

Applying the Concepts

1. In a family with four children, find the probability that there are two boys given that the oldest child is a boy.
2. In a family with four children, find the probability that there are two boys given that the oldest child is a girl.
3. In a family with four children, find the probability that there are two boys given that at least one child is a boy.
4. In a family with four children, find the probability that there are three boys given that the first two children are girls.
5. When two fair dice are rolled, find the probability of getting a total of 6 given that one die came up 4.

6. When two dice are rolled, find the probability that a 4 comes up given that the sum of the two faces is 6.
7. When two dice are rolled, find the probability that the sum is 8 given that only even-number faces show.
8. When two dice are rolled, find the probability that the sum is 8 given that only odd-number faces show.
9. Find the probability that a card selected from a regular deck is a king given that it is a face card (jack, queen or king).
10. Find the probability that a card selected from a deck is a 10 given that it is an even number.

4.6 Random Patterns in Chaos (Optional)

Our usual conception of random events is that they are completely random and so display no patterns or predictability whatsoever. In this section, we consider two very different situations in which we give a visual representation for random events and see radically different results. The first is a phenomenon known as the *random walk* or *drunkard's walk* problem.

Consider a drunk who is so besotted that he doesn't know which way he is going when he leaves a bar at a particular corner in a city. Consequently, he simply staggers on to the next corner without regard to whether he is going north, south, east or west. Thus, his choice of direction is essentially purely random. At the next corner, he again chooses a direction purely by chance, so that he might even end up going back down the street he came from. The probability that he will choose that direction or any of the other three possibilities is $\frac{1}{4}$. Suppose he then continues this process indefinitely. That is, he staggers one block at a time and then randomly selects any of the four possible directions to continue to the next corner, and so forth.

You can experiment with possible outcomes in such a random walk or drunkard's walk process by using Program 9: Drunkard's (Random) Walk Simulation. You must enter the number of such steps and the length of each step (essentially, the length of the block). The program then carries out the associated random walk simulation using a random process based on a probability of $\frac{1}{4}$ for selecting each possible direction at each corner. Suppose we select 200 steps of length 8. The results of two separate runs of the program are shown in Figures 4.19 and 4.20. The drunkard's starting point is indicated by the larger circle in the center, and his ending point after the 200 steps is indicated by the smaller circle.

Notice that the two sets of results are dramatically different. The drunkard randomly takes very different courses and goes in very different directions as he meanders up and down the streets. Moreover, since the choice of direction at each corner is random, it is highly unlikely that he will cover any great distance from his starting point, and this is quite clear from the computer displays.

Incidentally, while we discussed the random walk process from the point of view of a drunkard, it is actually an extremely important concept. For in-

FIGURE 4.19

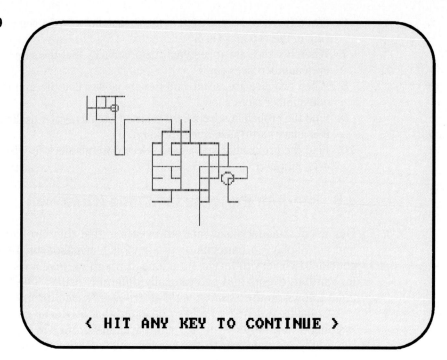

FIGURE 4.20

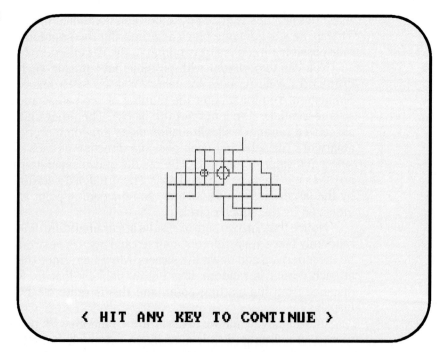

stance, scientists routinely use random walk methods to describe the random motion of individual gas molecules or the random motion of electrons in a wire. In fact, they have developed a formula for predicting the most likely distance that the drunkard will end up from his starting point. It is given by

$$\text{Most likely distance} = L \times \sqrt{N}$$

where L is the length of each step and N is the total number of steps. Unfortunately, the derivation of this result is beyond the scope of this book. However, this theoretical value is shown as part of the program. You will notice that it is usually quite different from the actual distance achieved in any single run. Nevertheless, if you repeat the random walk program many times with the same choices for L and N and average the resulting distances, you will find that the mean comes quite close to the predicted value.

The previous discussion on the drunkard's walk problem has probably convinced you that random events are totally unpredictable and display no patterns whatever. While this is true for individual events, it is not quite the case when we consider a large number of random events. Sometimes, there are very definite patterns that we should expect. For instance, if we flip two fair coins 10,000 times, we expect that the outcomes should consist of two heads approximately 2500 times, two tails approximately 2500 times, and one of each approximately 5000 times. In this case, the pattern is reasonably obvious.

Occasionally, however, the result of repeating a random process many times is a pattern which is impossible to predict and of such an unexpected nature that it is often difficult to believe. We illustrate such a case in the following discussion involving a process which can be thought of as an introduction to the *mathematical theory of chaos*.[1]

Suppose that A, B, and C are three points in the plane. We select another point P_0 as a starting point for a random process, as shown in Figure 4.21. We then randomly select one of the three vertices A, B, or C, and we plot the midpoint of the line segment connecting P_0 to that vertex to generate P_1. We then randomly select any one of the three vertices and again plot the midpoint between P_1 and it; this point is labeled P_2. This process is then repeated indefinitely.

We would naturally assume that the resulting collection of points would be a completely random array of dots. However, this is not the case. As shown in Figure 4.22, the result is an incredibly intricate and symmetric pattern.

In fact, if any portion of this pattern is magnified, we find that the identical shapes continually recur. Such a repeating pattern is said to *replicate* indefinitely. Moreover, the identical pattern will emerge no matter what initial point P_0 we select.

---

[1] This discussion is based on an investigation conducted by Kenneth S. Gordon, and the associated program is due to him. These results appear in his article "Finding Mathematical Patterns in Chaos", in the *International Journal of Mathematical Education in Science and Technology*, vol 24, 1993.

FIGURE 4.21

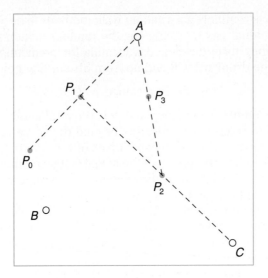

We have included a program on your disk, Program 10: Patterns in Chaos, to allow you to experiment with such random processes and see the visual effects that result. We urge you to try the program; you will be amazed by the patterns that appear.

Before leaving you to your explorations, we will suggest several changes that are possible in the original procedure. The first question we raise is, What

FIGURE 4.22

will be the effect if the points generated are not at the midpoint, but at some other fraction of the distance from the preceding point to a vertex? For instance, what if they are two-thirds of the way to the vertex or one-quarter of the way to the vertex or even 1.3 times the distance to the vertex? We call this quantity the *distance factor* and note that it is reasonable to use anything from 0 to 2, expressed as a decimal. We suggest that you try the program by using some systematic sequence of values for this distance factor, say $d = .8$, $d = .6, d = .4, d = .3$, etc. Can you explain what is happening based on your investigations? What happens if $d = 1$? Can you explain why you get what you do? You may want to try to anticipate the patterns that will appear as you make the various changes.

We next consider the possibility of using a distance factor d which is *greater than* 1. Intuitively, you might expect that the points generated would move farther and farther from the vertices and eventually go shooting off to infinity in all directions. However, this is not always the case. In fact, that happens only if the distance factor $d \geq 2$. We leave it to you to see what happens when d is between 1 and 2.

The third question we raise is, What will be the effect if there are more than three vertices? For instance, what happens if there are four vertices A, B, C and D? In the program, you are able to select a minimum of three and a maximum of six fixed points and then any desired value for the distance factor. For example, in Figure 4.23, we show the results of using four vertex points (arranged in a square) and a distance factor which is slightly less than 1.

FIGURE 4.23

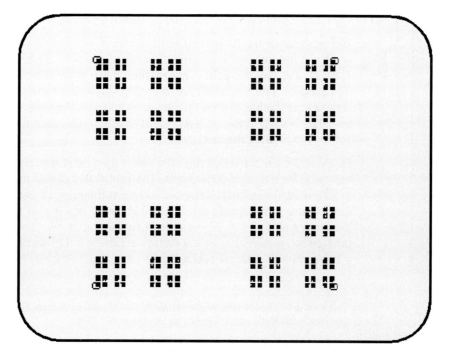

We note that other interesting geometric effects can be produced when the vertices are not symmetrically located (that is, not in an equilateral triangle or a square). Similarly, it is also possible to obtain other fascinating effects if different or weighted probabilities are used in determining which vertex to move toward. However, the program on the disk does not provide for such cases.

The rapidly developing field of chaos is one of the most exciting areas in mathematics today. It has found application in such diverse areas as medicine and meteorology as well as the movies. The interested reader is encouraged to look at the book *Chaos: Making a New Science* by James Gleick (Viking Press, 1987).

Student Projects

Conduct a study to determine the average or most likely waiting time for a car at a red light. You will need the following relatively simple mathematical analysis.[2]

Suppose the total fixed cycle for a traffic light is N seconds and that a particular segment of that cycle—green in one direction or a left-turn signal—lasts for m seconds. For example, a cycle may take a total of 120 seconds and the light is green for 30 of them. We assume that cars can arrive at the intersection at any given second during the cycle, and so we measure all times to the nearest second. Further, to keep things simple, we assume that cars stop and start instantly.

We concentrate on the particular case with $N = 120$ and $m = 30$. If a car arrives at the intersection during any one of the $m = 30$ seconds ($t = 0$, 1, 2, ..., 29) that the light is green, then there is a 0-second wait and this happens with probability $m/N = 30/120 = .25$. However, if the car arrives at any other instant t between 30 and 119, then there is a corresponding waiting time. (If it arrives at time $t = 120$, then it effectively arrives at $t = 0$ on the next cycle.)

Suppose the car arrives at time $t = 30$. It then must wait through the full red cycle for a total of 90 seconds. The probability of this happening is $1/120 = 1/N$. Similarly, if the car arrives at time $t = 31$ seconds, then it must wait for 89 seconds and the probability of this is also $\frac{1}{120}$. Further, if it arrives at time $t = 32$ seconds, then the wait is 88 seconds with probability $\frac{1}{120}$. Eventually, if it arrives at time $t = 119$ seconds, then it has a 1 second wait with probability $\frac{1}{120}$.

---

[2]This section is based on the article "Mathematical Models of Waiting Time" by the authors which appeared in the *Mathematics Teacher*, vol. 83, November 1990.

The mean or expected waiting time W for a randomly arriving car is given by

$$W = (\text{0-second wait}) \times (\text{probability of 0-second wait})$$
$$+ (\text{1-second wait}) \times (\text{probability of 1-second wait})$$
$$+ (\text{2-second wait}) \times (\text{probability of 2-second wait})$$
$$+ \cdots + (\text{90-second wait}) \times (\text{probability of 90-second wait})$$

$$= 0 \times \tfrac{30}{120} + 1 \times \tfrac{1}{120} + 2 \times \tfrac{1}{120} + 3 \times \tfrac{1}{120} + \cdots + 90 \times \tfrac{1}{120}$$

$$= \tfrac{1}{120}(1 + 2 + 3 + \cdots + 90)$$

since $\tfrac{1}{120}$ is a common factor. Further, the term in parentheses on the last line above is just the sum of the first 90 integers, and it is a mathematical fact that this sum is precisely

$$1 + 2 + 3 + \cdots + 90 = \frac{90 \times 91}{2}$$

$$= \frac{8190}{2} = 4095$$

Therefore, the average waiting time for a car is

$$W = \tfrac{1}{120} \times 4095 = 34.125 \text{ seconds}$$

at this particular intersection.

Note that this includes cars which arrive during the green portion of the cycle and go right through with no wait.

In general, suppose we consider a traffic cycle which lasts for N seconds (instead of 120 seconds) and which is green for m seconds (instead of 30 seconds). We can carry out the same derivation as above to find that the average or expected waiting time is given by

$$W = \frac{1}{N}[1 + 2 + 3 + \cdots + (N - m)]$$

Since the sum of the first k integers, for any k, is given by the formula

$$1 + 2 + 3 + \cdots + k = \tfrac{1}{2}k(k + 1)$$

we obtain, with $k = N - m$,

$$W = \frac{(N - m)(N - m + 1)}{2N}$$

For instance, if we apply this formula to the above example, we must use $N = 120$ and $m = 30$ to obtain

$$W = \frac{(120 - 30)(120 - 30 + 1)}{2(120)}$$

$$= \frac{(90)(91)}{240} = 34.125$$

as we found directly above.

For your project, you should identify an appropriate intersection and time the cycle to determine the values for N and m. You can then use the above formula to calculate the theoretical mean waiting time for all cars that come to your intersection.

You should randomly select a set of at least 30 cars approaching the intersection and measure the length of time each car waits at the intersection. If it arrives while the light is green, then the waiting time is 0. Be sure that your selection process is random (use Table I, the table of random digits, or Program 2: Random Number Generator) to provide a means of selecting cars randomly.

Once you have collected your experimental data, calculate the mean and the standard deviation for them. You should then compare the observed mean to the predicted value given by W when you use the formula above. How close is the actual result to the predicted one? Is the prediction an effective one?

Once you have collected these data, you should then:

1. Organize the set of data.
2. Construct an appropriate relative frequency distribution.
3. Construct the associated histogram, bar chart, and/or frequency polygon.

Finally, you should incorporate all these items into a formal statistical project report including:

A statement of the topic being studied.

The source of the data.

A discussion of how the data were collected and why you believe that it is a random sample.

The tables and diagrams displaying the data.

Your calculations of the mean and standard deviation for the data.

The calculation of the theoretical mean waiting time W based on the values for N and m at your intersection.

A discussion of the comparison between the actual and the predicted mean waiting times.

> A discussion of any surprises you may have noted in connection with collecting the data, organizing them, or the results of the calculations. In particular, identify any special circumstances (behavior of drivers, traffic conditions, weather conditions, road conditions, etc.) that may have affected the results.
>
> A discussion of how effective and accurate the predicted value for W is, based on the mathematical model compared to the actual outcomes.

CHAPTER 4 SUMMARY

Probability is used to measure how likely an event is to occur. It is always a number between 0 and 1.

In an experiment, the **sample space** consists of all possible outcomes. If the outcomes in an experiment are **equally likely**, then the probability of an event A is

$$P(A) = \frac{\text{number of ways } A \text{ can occur}}{\text{total number of possible outcomes}}$$

In an experiment where the events are distinct and include all possible outcomes, the sum of the probabilities must be 1.

The **Fundamental Counting Principle** is: If an event A can occur in m ways and an event B can occur in n ways, then A followed by B can occur in $m \times n$ ways.

A **permutation** is a specific arrangement of a group of objects in some order. The number of permutations possible when n objects are taken r at a time is

$$_nP_r = \frac{n!}{(n-r)!}$$

where

$$n! = n(n-1)(n-2) \cdots 3 \cdot 2 \cdot 1 \quad \text{and} \quad 0! = 1$$

A **combination** is any collection of objects without regard to order. The number of combinations possible when n objects are taken r at a time is

$$_nC_r = \frac{n!}{r!(n-r)!}$$

Two events A and B are **mutually exclusive** if they have no outcomes in common. The **Addition Rule** for probability states:

If A and B are mutually exclusive events, then

$$P(A \text{ or } B) = P(A) + P(B)$$

If A and B are not mutually exclusive events, then
$$P(A \text{ or } B) = P(A) + P(B) - P(\text{both } A \text{ and } B)$$

Two events A and B are **independent** if the probability of B is not affected by the outcome of A. The **Multiplication Rule** for probability states: If A and B are independent events, then
$$P(A \text{ and } B) = P(A) \times P(B)$$

The **complement** A' of an event A is the event that A does not occur. Its probability is
$$P(A') = 1 - P(A)$$

The **conditional probability** of B given A is
$$P(B \mid A) = \frac{P(\text{both } A \text{ and } B)}{P(A)}$$

Review Exercises

1. Suppose you select one month randomly. What is the probability that: **(a)** it has 30 days? **(b)** it starts with a J? **(c)** it ends in an *r*?
2. The clearance bin at an audio shop contains the following numbers of tapes: 34 rock, 22 country and western, 15 classical, 16 light listening, 8 jazz and 5 comedy. If you reach in and select one tape at random, what is the probability that you pick: **(a)** a classical tape? **(b)** a musical tape? **(c)** either classical or jazz?
3. Two fair dice are rolled. What is the probability of getting **(a)** a sum of either 7 or 11? **(b)** a sum greater than 9? **(c)** a sum of 5 or less? **(d)** a sum between 6 and 9 inclusive? **(e)** a sum that is neither 5 nor 8?
4. The student cafeteria at a school has 2 lines for hot entrees, 3 lines for fast food items, 1 line for desserts, 2 lines for salads and 4 lines for beverages. In how many different ways can a student line up to get: **(a)** one item in each category? **(b)** either a hot entree or a fast-food item, but not both, plus one item in each of the other categories?
5. A nursery school has 12 children who all live in a certain neighborhood. How many different carpools of 4 children each are possible?
6. In a certain community, 43% of adults are registered Democrats, 38% are registered Republicans and the rest are registered with other parties. If a telephone poll is conducted by calling people at random, what is the probability of getting **(a)** 2 Democrats in a row? **(b)** 2 registrants of the major parties in a row? **(c)** 2 registrants from other parties in a row? **(d)** 3 Republicans in a row?

5 Discrete Probability Distributions

With elections only a month away, politicians are feeling very uncertain about the mood of the electorate. They can no longer count on the breakdown of party affiliations of registered voters to translate into a similar distribution of votes for their respective candidates. Instead, voters are increasingly affected by many diverse issues which have splintered the traditional voting patterns in most local and regional elections.

In one locality, for example, the distribution of party affiliations among registered voters is: Republican: 43%; Democrat: 32%; Conservative: 5%; Liberal: 7%; Independent: 13%. In times past, most voters could be counted on to vote the party line unhesitatingly and so the final vote tallies would usually closely reflect the party distribution. Thus, the Republican party could likely expect that roughly 43% of the votes would be cast for their candidates and the Democrats could expect that about 32% of the votes would go to their candidates and so forth.

Today's voters are far more apt to be single-issue voters and make their decisions on the basis of economics (bread-and-butter issues), race, abortion (pro-choice or pro-life) or women's issues rather than on the basis of their party membership. Consequently, politicians can no longer anticipate that the distribution of votes will reflect the distribution of party registrations. Therefore, it is not only a matter of trying to convince a relatively small number of voters (the independents and undecideds) to vote for their party's candidates. They now also have to work to make sure that even the voters who are registered in their own party will vote for their candidates.

In this chapter, we will see how probability distributions (the breakdown of the likelihoods associated with different events) arise and how they are interpreted and applied in a wide variety of fields.

5.1 Random Variables and Probability Distributions

Whenever we conduct a statistical study, we are interested in determining information about some particular quantity. It might be the heights of individuals in some group, or the time measured for an Olympic downhill ski run, or the number of heads obtained in flipping a set of 10 coins, or the number of people in a sample poll of 500 prospective voters who prefer a certain candidate. In any such case, the quantity being studied is the variable in question.

In an algebraic context, a variable typically is a quantity which is unknown, but which presumably can be determined by solving one or more equations. The fact that an equation exists involving the variable suggests that there is some predictable relationship or pattern for that quantity.

A variable in a statistical experiment is not quite the same as a variable in the algebraic sense. When we collect data about a certain quantity, there is usually no precisely predictable pattern or relationship for them. Thus, when we collected the heights of individuals in a class, there was no way of anticipating the value for the height of any individual student. Similarly, if we collect data on the number of heads obtained when a coin is flipped repeatedly, we cannot predict the individual outcomes. In all such situations, the values associated with the statistical variable are random. For this reason, the variable that is studied in any statistical experiment is called a *random variable*.

> A **random variable** takes on different values, each with an associated probability.

Furthermore, it is necessary to distinguish between two types of random variables:

> 1. A **discrete random variable** is one which can assume any of a set of possible values which can be counted or listed.
> 2. A **continuous random variable** is one which can assume any of an infinite spectrum of different values across an interval and which cannot be counted or listed.

For example, if we flip five coins and the random variable x represents the number of heads that occur, then the only possible values it can assume are $x = 0, 1, 2, 3, 4$ or 5. This is a discrete random variable. Similarly, if we poll 500 people to determine how many prefer a particular candidate, then the random variable can take on only the values $x = 0, 1, 2, \ldots, 500$. Again, it is a discrete random variable. In general, we can think of a discrete random variable as one that arises from a situation in which some quantity is *counted*.

On the other hand, consider the timing of an Olympic skier. The time for a downhill race is typically measured in hundredths of a second. This is done only because that degree of accuracy is usually adequate to distinguish between the times for different racers. In the bobsled and luge competitions, the times are measured in thousandths of a second, since it is necessary to use the additional accuracy to distinguish one sled from another. If the occasion demanded it, it is possible to time events to a far greater level of accuracy. Physicists and computer scientists routinely time events to within millionths or even billionths of a second. Thus, theoretically, when an event involves time, or equivalently when time is the random variable in question, then time can assume any of an infinite array of uncountably many values across an interval. It is a continuous random variable.

Consider the problem of measuring individual heights of students. In the sample group we discussed in Chapters 2 and 3, five people were all recorded as being 66 inches tall. However, what does that mean precisely? Is it likely that any individual is exactly 66.0000000000... inches tall? If the measurement is carried out to a sufficiently high degree of accuracy, the answer is virtually certain to be *no*. Instead, when we talk about someone being 66 inches tall, we actually mean that the person's height is within a range of heights, probably 65.5 to 66.49 inches, and the height is then rounded to 66 inches. Thus, we again have a situation involving a continuous random variable. In general, a continuous random variable arises from a statistical experiment in which we *measure* some quantity.

For our purposes, we can distinguish between the two types of random variables as follows:

> A discrete random variable is one that arises when we **count** how many times a certain quantity occurs.
>
> A continuous random variable is one that arises when we **measure** a certain quantity.

In either case, it is important to realize that the possible values for any random variable must be numerical values. Thus, on a 20-question true-false test, we could not use either true or false as a value for a random variable. However, we can use a random variable to represent the *number* of correct answers $x = 0, 1, 2, \ldots, 20$.

Further, the two types of random variables, discrete and continuous, give rise to two different types of statistical populations. When we have a continuous random variable and consider all the data pertaining to it, we generate a *continuous population* or a *continuous distribution*. Thus, for instance, data on the heights of all students at your school would constitute a continuous distribution of heights. Data on the age of everyone in the country constitute a continuous population of ages. We will consider some particular continuous distributions in detail beginning in Chapter 6.

On the other hand, when we deal with a discrete random variable and consider all the possibilities associated with it, we generate a *discrete population* or a *discrete distribution* or a *probability distribution*. Suppose, for instance, the experiment involves flipping a set of three coins and counting the number of heads. There are four possible outcomes—0H, 1H, 2H or 3H—each with an associated probability $P(0H)$, $P(1H)$, $P(2H)$ and $P(3H)$, respectively, which we can display in a table as follows:

x	P(x heads)
0	$\frac{1}{8}$
1	$\frac{3}{8}$
2	$\frac{3}{8}$
3	$\frac{1}{8}$

Such a table of probability values is called a *probability distribution* since it gives the probabilities associated with all the possible values of the random variable x. We note that there are other ways to specify probability distributions. They can also be given by a formula or even a graphical display such as a histogram which provides the probability associated with each possible outcome of the experiment. We use the phrase *probability distribution* to represent all these possibilities.

> For any discrete probability distribution,
>
> 1. The probability associated with each possible value of the random variable x is between 0 and 1.
> 2. The sum of all the associated probabilities must be 1.

In fact, if a table of probabilities were produced in which the total of the individual probabilities was not equal to 1, then it would not be a probability distribution.

EXAMPLE 5.1 Consider the statistical experiment of rolling a fair die. The possible outcomes are 1, 2, 3, 4, 5 and 6, and so the random variable x giving the number on the top face is a discrete random variable. Since all these outcomes are equally likely, the probability of each outcome (or each value of the random variable) is $\frac{1}{6}$. Therefore, the associated probability distribution is as follows:

x	P(x)
1	$\frac{1}{6}$
2	$\frac{1}{6}$
3	$\frac{1}{6}$
4	$\frac{1}{6}$
5	$\frac{1}{6}$
6	$\frac{1}{6}$

EXAMPLE 5.2 Consider the statistical experiment involving the sex of children in a family with four children. Let the random variable x be the number of boys. It assumes only the possible values 0, 1, 2, 3 or 4 and so is discrete. Further, since

the probability of a child being a boy is .5, this is equivalent to the problem of flipping four coins and counting the number of heads. Therefore, the probability distribution is

x	P(x boys)
0	$\frac{1}{16}$
1	$\frac{4}{16}$
2	$\frac{6}{16}$
3	$\frac{4}{16}$
4	$\frac{1}{16}$

In reality, approximately 52% of all live births are boys. However, the infant mortality rate among boys is considerably higher than that among girls, so that by 6 months of age, about 51% of all surviving children are girls. This accounts, in part, for the greater proportion of females in our population. To simplify things here, we are assuming that the proportions of boys and girls among newborn babies are both 50%.

EXAMPLE 5.3 Consider the experiment of rolling a pair of fair dice. Let the item of interest be the sum of the two faces. We have previously seen that there are 36 equally likely outcomes, and the associated probability distribution is as follows:

x	P(x)
2	$\frac{1}{36}$
3	$\frac{2}{36}$
4	$\frac{3}{36}$
5	$\frac{4}{36}$
6	$\frac{5}{36}$
7	$\frac{6}{36}$
8	$\frac{5}{36}$
9	$\frac{4}{36}$
10	$\frac{3}{36}$
11	$\frac{2}{36}$
12	$\frac{1}{36}$

Instead of the tabular forms we have been using, it is also possible to display a probability distribution visually by using a variation on the histograms we considered in Chapter 2. We construct a series of boxes, each of whose heights represents the probability of the associated value of the random variable. Thus, in Figure 5.1, we show the histogram corresponding to the probability distribution for the experiment in Example 5.1 on rolling a fair die. In Figure 5.2, we show the histogram for the probability distribution of Example 5.2 involving the sex of four children. In Figure 5.3, we show the histogram for the probability distribution associated with the sum of the faces on two dice.

FIGURE 5.1

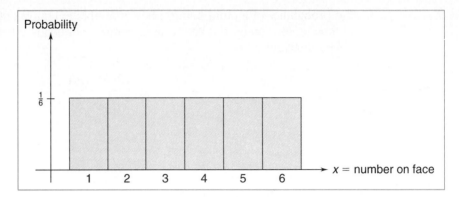

In later sections, we will discuss ways in which a probability distribution can be determined in terms of a formula that specifies the probability associated with each value of a random variable. Such a formula is an alternate representation for a probability distribution.

A probability distribution consists of the probabilities of all possible values of a random variable. It therefore can be thought of as a population consisting of those probability values. As a population, it has a mean μ and a standard deviation σ. They are given by the following formulas:

> The **mean** of a probability distribution is
> $$\mu = \Sigma x P(x)$$
> The **standard deviation** of a probability distribution is
> $$\sigma = \sqrt{\Sigma (x - \mu)^2 P(x)} \qquad (5.1)$$
> This is equivalent to
> $$\sigma = \sqrt{\Sigma x^2 P(x) - \mu^2} \qquad (5.2)$$

FIGURE 5.2

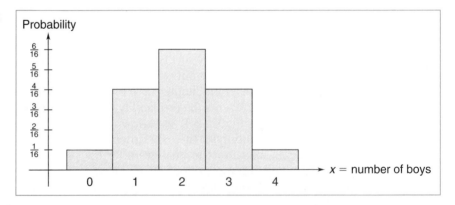

FIGURE 5.3

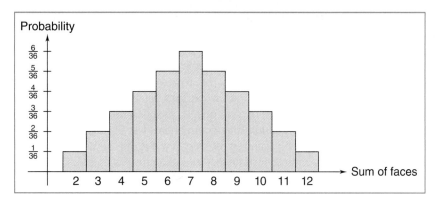

We illustrate how to use these formulas in the case of the probability distribution associated with rolling three dice. As with calculations for the mean and standard deviation in Chapter 3, it is usually easiest to carry out the work in tabular form. We use formula (5.2) for computing the standard deviation.

x	$P(x)$	$xP(x)$	x^2	$x^2P(x)$
0	$\frac{1}{8}$	0	0	0
1	$\frac{3}{8}$	$\frac{3}{8}$	1	$\frac{3}{8}$
2	$\frac{3}{8}$	$\frac{6}{8}$	4	$\frac{12}{8}$
3	$\frac{1}{8}$	$\frac{3}{8}$	9	$\frac{9}{8}$
		$\mu = \frac{12}{8} = 1.5$		$\frac{24}{8}$

Notice that the entries under $xP(x)$ represent the product of each value of x and its associated probability. The mean for the probability distribution is the sum of the entries in this column and so is $\mu = 1.5$.

The last column represents the product of each x^2 and the associated probability for x. Consequently, the variance is given by

$$\sigma^2 = \Sigma x^2 P(x) - \mu^2$$
$$= \frac{24}{8} - (1.5)^2$$
$$= 3 - 2.25 = .75$$

and so the standard deviation is

$$\sigma = \sqrt{.75} = .866$$

Exercise Set 5.1

Mastering the Techniques

Determine which of the following random variables are discrete and which are continuous. Give reasons in each case.

1. The number of movies a person has seen during the previous month
2. The weight lost on a particular diet plan
3. The gas mileage (in miles per gallon) for a set of cars
4. The number of people in a group who prefer Zippi Cola
5. The number of cars that pass a police radar trap during a day
6. The speeds of the cars as they pass a radar trap
7. The lengths of a set of 2 × 4's in a lumber yard
8. The amount of time commercials last during an hour of prime-time television
9. The temperatures recorded for a group of patients
10. The number of tickets sold at a movie theater each day for a month.

Determine which tables in Exercises 11 to 16 represent probability distributions and which do not.

11.

x	$P(x)$
0	.3
1	.4
2	.3

12.

x	$P(x)$
0	.5
1	.4
2	.1

13.

x	$P(x)$
1	.2
2	.4
3	.3
4	.2

14.

x	$P(x)$
1	.4
2	.35
3	.25
4	.10

15.

x	$P(x)$
0	$\frac{15}{100}$
1	$\frac{24}{100}$
2	$\frac{28}{100}$
3	$\frac{20}{100}$
4	$\frac{12}{100}$

16.

x	$P(x)$
0	$\frac{5}{82}$
1	$\frac{16}{82}$
2	$\frac{32}{82}$
3	$\frac{24}{82}$
4	$\frac{8}{82}$

In Exercises 17 to 20, determine the number p which makes it a probability distribution.

17.

x	$P(x)$
1	.2
2	.3
3	.4
4	p

18.

x	$P(x)$
1	.40
2	.25
3	.25
4	p

19. x	P(x)
0	.15
1	.24
2	p
3	.16
4	.08

20. x	P(x)
0	.05
1	.15
2	.31
3	p
4	.22

21. Use your answer to Exercise 17 to calculate the mean and standard deviation of the resulting probability distribution.
22. Use your answer to Exercise 19 to calculate the mean and standard deviation of the resulting probability distribution.

Applying the Concepts

Construct the probability distribution for each of the following statistical experiments as a table and as a histogram:

23. The distribution of the number of girls in families with five children.
24. A loaded coin is constructed in such a way that $P(H) = .4$ and $P(T) = .6$. The coin is flipped twice and the number of heads is observed.
25. A four-sided die is rolled and the number on the side facing "down" is recorded.
26. The four-sided die in Exercise 25 is rolled twice and the sum of the two "down" faces is recorded.
27. Of the residents of a certain community, 10% are immigrants who arrived in the last 10 years. Groups of two people are randomly selected and the number of immigrants is counted.
28. Of the managerial employees of a large organization, 40% are women. Groups of two such employees are selected randomly and the number of women is counted.
29. Of the teachers in a particular school district, 75% are women. Groups of two teachers are selected at random and the number of females is counted.
30. It is known that the breakdown of registered voters in a certain community is one-third Republican, one-third Democrat, and one-third Independent. Groups of two voters each are studied with respect to party registration, and the number of Republicans is observed.
31. Find the mean and standard deviation for the probability distribution in Exercise 23.
32. Find the mean and standard deviation for the probability distribution in Exercise 24.
33. Find the mean and standard deviation for the probability distribution in Exercise 27.
34. Show that formulas (5.1) and (5.2) for the standard deviation are algebraically equivalent. [*Hint:* Use the facts that $\mu = \Sigma x P(x)$ and $\Sigma P(x) = 1$.]

5.2 The Binomial Distribution

While there are many different discrete distributions that are useful, the most important is the *binomial distribution*. It is based on the notion of a binomial experiment or binomial process. We introduce these concepts and study some of their properties and applications in this section.

> A **binomial experiment** or **binomial process** is characterized by the following four conditions:
>
> 1. A certain event or activity is repeated a fixed number of times. These are called the **repetitions** or **trials**. The number of trials is denoted by **n**.
> 2. There are exactly two possible outcomes for each trial. We can think of these two outcomes as "success" and "failure."
> 3. The probability of success is the same for each trial. It is denoted by π.
> 4. The individual trials are independent of one another.

Thus, in a binomial process, there can be only two possible outcomes for each repeated independent trial, and the probability for those outcomes must be constant from trial to trial.

We have actually encountered a number of binomial processes previously. Probably the simplest involves coin flipping. Suppose we flip a set of five coins, or equivalently we flip a single coin 5 times in succession. The act of flipping the coin is repeated 5 times, so there are $n = 5$ repetitions or trials. On each of the five trials, there are two possible outcomes, either heads or tails. If getting a head is considered a success, then the probability of success is $\pi = \frac{1}{2}$ and is the same on each of the five trials. Finally, each successive trial or flip is independent of the previous outcomes. Therefore, this is a binomial process.

Now consider the process of rolling a fair die 5 times and recording the number on the face. There are five trials to this experiment. They are independent of one another. The probability of each outcome—1, 2, 3, 4, 5 and 6—remains the same from trial to trial (it is $\frac{1}{6}$). But this is not a binomial process because there are *six* possible outcomes on each trial, not the two required for a binomial process.

Sometimes it is possible to reinterpret a process to transform it to a binomial experiment. In the dice rolling example, suppose we are interested only in obtaining a 3 on the die. There are thus two possible outcomes, either getting a 3 or not getting a 3. The probability of success is $\pi = \frac{1}{6}$ (and the probability of failure is $\frac{5}{6}$), there are still $n = 5$ trials, and they are independent. The experiment is now binomial.

Alternatively, if success involves getting either a 1 or a 2 on the die, then we have a different binomial process with two possible outcomes: Success means getting a 1 or a 2, and failure means getting a 3, 4, 5, or 6. We still have $n = 5$ independent trials, but now $\pi = \frac{2}{6} = \frac{1}{3}$ is the fixed probability of success which is the same from trial to trial.

Furthermore, we note that if the probability of success in a binomial process is π, then the corresponding probability of failure is

$$\text{Probability of failure} = 1 - \pi$$

EXAMPLE 5.4 Suppose that 60% of the registered voters in a particular county are Republicans. A random sample of 200 voters in this county is selected, and the voters are asked whether they are Republicans. Is this a binomial process?

SOLUTION In this problem, the possible responses are either yes or no, so there are two possible outcomes. Since the question is asked of 200 randomly selected individuals, there are $n = 200$ repetitions of the basic action. Presumably, the responses obtained will be independent of one another; that's essential in choosing an appropriate random sample. Finally, since the overall population in the county is known to be 60% Republican, the probability that any individual voter is a Republican is .60. Therefore, $\pi = .60$, and this is constant for each of the 200 people surveyed. Consequently, this survey is a binomial process.

Incidentally, care must be taken when constructing such a poll if we want to ensure that it will be a binomial experiment. Thus, if the responses available were Republican, Democrat, and Independent or Republican, Democrat, Independent, and undecided, then it would no longer be binomial since there are more than two possible outcomes.

EXAMPLE 5.5 Efforts at proving the existence of telepathy and other ESP (extrasensory perception) phenomena are based on a statistical argument. The type of test administered involves a special deck of 25 cards made up of five cards for each of five symbols: a circle, a square, a triangle, a series of wavy lines, and a star, as shown in Figure 5.4. One person (the sender or transmitter) goes through the shuffled deck, card by card, attempting to transmit an image of the symbol to another person (the receiver or the subject) who tries to receive the image on each card and who makes a record of each symbol received.

Suppose the subject is simply guessing wildly or randomly. There are two possible outcomes on each trial—either the correct answer or an incorrect answer. On each card, his or her chance of guessing correctly is 1 out of 5, since the five symbols are equally likely to occur. That is, his or her probability of success is $\pi = \frac{1}{5} = .2$. Since this is repeated a total of 25 times (once for each of the 25 cards in the deck), there are $n = 25$ trials to this experiment. Finally, the guess made on each card is independent of that for any other card. As a consequence, we see that this type of ESP test is a binomial process.

FIGURE 5.4

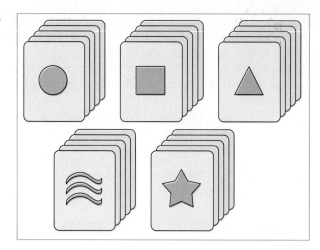

Now that we have studied the idea of a binomial process, we can turn to the related notion of a binomial distribution.

> A **binomial distribution** is the probability distribution associated with a binomial process.

It consists of either a table or a histogram-type display or a mathematical formula giving all the probabilities associated with the binomial process. In particular, since a binomial process involves n repetitions of a basic event, the possible outcomes consist of the possibility of 0 successes, 1 success, 2 successes, ..., n successes. Therefore, the corresponding binomial distribution must contain $n + 1$ items together with the associated $n + 1$ probabilities, if written in a table, or it must contain $n + 1$ boxes, if displayed as a histogram.

EXAMPLE 5.6 Consider the binomial process involving the sex of the children in a family with four children. If having a girl is interpreted as success, then the binomial distribution using $n = 4$ and $\pi = \frac{1}{2}$ is

x	P(x)
0	$\frac{1}{16}$
1	$\frac{4}{16}$
2	$\frac{6}{16}$
3	$\frac{4}{16}$
4	$\frac{1}{16}$

It is important to realize that a binomial distribution is based on two distinct quantities: the number n of trials and the probability π of success on each trial. Thus, for each different pair of values for n and π, there is a different binomial distribution. Note that we use the Greek letter π to represent the probability of success since this quantity is a parameter for the binomial population.

Furthermore, every distribution or population has a mean μ and a standard deviation σ. For a binomial distribution, the mean is equivalent to the number of successes we would expect out of n trials. This is also known as the *expected value*. For example, suppose we flip a fair coin 1000 times. Since the probability of success is $\frac{1}{2}$ on each flip, we expect approximately 500 heads to occur.

Another way of looking at this is the following: Suppose the binomial process of flipping the fair coin 1000 times is conducted over and over again. Thus, we might get 518 heads on the first series, 493 on the second, 511 on the third, and so forth. If we average the number of heads that occur on each series of 1000 flips, then in the long run we expect the overall number of heads on the 1000 flips to average 500. Thus, any unusual series of either heads or tails in the short run are likely to average out over the long run. This is the mean μ for this binomial distribution.

In a similar way, suppose we consider the binomial experiment of rolling a fair die 6000 times and we interpret success as getting a 3 on the die. In this case, $n = 6000$ and $\pi = \frac{1}{6}$. The number of 3s we expect to obtain is approximately 1000 out of the 6000 rolls in the long run, and this is the mean μ of this binomial distribution.

The above two situations illustrate the following general principle for any binomial distribution:

> Given any binomial distribution with n trials and probability of success π, the mean is
>
> $$\mu = n\pi$$
>
> and the standard deviation is
>
> $$\sigma = \sqrt{n\pi(1-\pi)}$$

Note that in the formula for σ we use both the probability of success π and the probability of failure $1 - \pi$.

EXAMPLE 5.7 In the ESP experiment we discussed in Example 5.5 where $n = 25$ and $\pi = \frac{1}{5} = .2$, the mean number of correct answers is

$$\mu = n\pi = 25(.2) = 5$$

and the standard deviation is

$$\begin{aligned}
\sigma &= \sqrt{n\pi(1-\pi)} \\
&= \sqrt{25(.2)(1-.2)} \\
&= \sqrt{25(.2)(.8)} \\
&= \sqrt{4} = 2
\end{aligned}$$

Having introduced the idea of the standard deviation for a binomial distribution, let's now consider just what it signifies. The same interpretation we used for the standard deviation in Chapter 3 applies here as well. Consider the results on the ESP experiment. If a person is guessing randomly on the test, we expect her or him to obtain a mean of $\mu = 5$ correct answers with a standard deviation of $\sigma = 2$ correct answers. Therefore, using the Empirical Rule, we expect that approximately 68% of all people who take this test will score within 1 standard deviation of the mean, that is, from 3 right answers to 7 right answers. Further, we expect that approximately 95% of all people who take the test will score within 2 standard deviations of the mean, or from 1 to 9 correct answers.

It is the consequence of the above observations, incidentally, that is used to "prove" the existence of ESP. If approximately 95% of all people who take the test are expected to score between 1 and 9 correct answers, then only about 5% should score either 0 right or 10 or more right. However, if an individual consistently scores 12 right or 15 right or 20 right, then that indicates something far too unlikely to have happened purely by chance, from a probabilistic standpoint. In turn, this suggests that some factor other than purely random guessing is present. This additional factor is believed by some to be ESP, by others to be some type of subconscious cues picked up by the subject from the sender, and by still others to be outright cheating, probably by the person conducting the tests.

Since it is easy enough to make up such a set of ESP test cards, we suggest that you do so and see what kinds of scores you can achieve. In the next section, we will see how to calculate the actual probabilities involved in achieving different scores, so you will be able to measure just how high or how unlikely your score is.

Computer Corner

We have included Program 11: The Binomial Distribution on your disk to let you develop an understanding of the shape of the binomial distribution and how it depends on the choices of n and π. To use it, you must enter your choices for n and π as well as the probability of failure $1 - \pi$. The program then displays the corresponding histogram for the binomial distribution. We suggest that you experiment with it by changing only one of the values, either n or π, to see the effects on the shape of the distribution. What happens, for fixed n, if π increases or decreases? What happens, for fixed probability of success π, if n increases?

In addition, this program includes an opportunity for developing your skills at using the formulas to find the mean and standard deviation of a binomial distribution. After each graph is complete, the program asks you if you want to perform the calculations for μ and σ. As you do the calculations by hand and enter your values, the program checks that you have the correct answers based on your choice of n and π. If you have made any errors, it attempts to identify what you did wrong and prompts you with the relevant formula for calculating the correct answer. We note that you will need to develop facility in calculating these values for applications of the binomial distribution in later sections.

In a separate direction, the formula for the mean $\mu = n\pi$ of a binomial distribution makes sense intuitively. However, the formula for the standard deviation $\sigma = \sqrt{n\pi(1 - \pi)}$ is less obvious. You can check the accuracy of this formula on an experimental basis by using Program 12: Binomial Simulation. It allows you to experiment with a variety of different binomial distributions to see the effects of different values for n and π on the distribution and its mean and standard deviation. To use the program, you must enter your choice for n, the number of trials, the probability of success π (as a decimal) and the probability of failure $1 - \pi$.

For example, suppose you choose $n = 4$ and $\pi = .5$. This is equivalent to the binomial experiment of flipping a set of four fair coins. The program simulates this experiment by repeating this process several hundred times. For each trial, the program counts the number of heads that result, between 0 and 4, and graphs the outcome, as shown in the typical display in Figure 5.5. The program then repeats this process randomly several hundred times until the display reaches the top of the screen. As the program proceeds, the shape of the outcomes more and more mirrors the histogram for the corre-

FIGURE 5.5

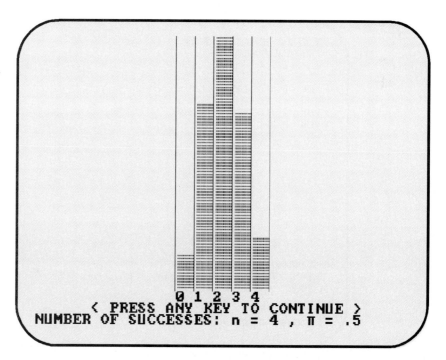

sponding binomial distribution. When the program is complete and you hear the beep, you can proceed to a text screen where the numerical results obtained are listed, as reproduced in Figure 5.6. From this, you can see the actual breakdown of the results and compare them to the theoretical predictions $P(0H) = \frac{1}{16} = .0625$, $P(1H) = \frac{4}{16} = .25$, $P(2H) = \frac{6}{16} = .375$, and so forth.

In addition, the program supplies the mean and standard deviation for all the runs conducted (242 in this particular case). The mean number of successes for these 242 simulated outcomes is 2.04 compared to the predicted mean $\mu = n\pi = (4)(\frac{1}{2}) = 2$, which is quite close. Similarly, the standard deviation for the number of successes in these 242 trials is 1 compared to the predicted value of

$$\sigma = \sqrt{n\pi(1 - \pi)} = \sqrt{4(\tfrac{1}{2})(\tfrac{1}{2})} = 1$$

We can repeat this procedure by changing the values for n and π. Suppose we use $n = 12$ and $\pi = .25$. A typical graphical result is shown in Figure 5.7, and the corresponding numerical output is shown in Figure 5.8. From these, we can again compare the mean for the 327 simulated outcomes of 3.02 successes per trial to the theoretical mean $\mu = (12)(.25) = 3$ and compare the standard deviation of the number of successes in the 327 random repetitions, 1.439, to that predicted by the formula

$$\sigma = \sqrt{12(.25)(.75)} = \sqrt{2.25} = 1.5$$

FIGURE 5.6

```
OUT OF  242  REPETITIONS OF THE
        BINOMIAL PROCESS
        WITH n =  4    AND π =  .5

% WITH   0  SUCCESSES =   4.95 %
% WITH   1  SUCCESSES =  26.44 %
% WITH   2  SUCCESSES =  35.95 %
% WITH   3  SUCCESSES =  25.2  %
% WITH   4  SUCCESSES =   7.43 %




MEAN NUMBER OF SUCCESSES =  2.04
STANDARD DEVIATION =  1.008
BY THEORY:  μ = 2    σ = 1


< SPACE BAR => MENU; OTHER => GRAPH >
```

FIGURE 5.7

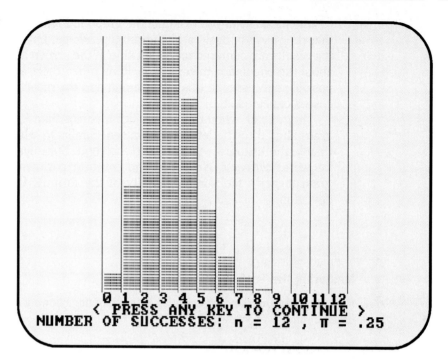

FIGURE 5.8

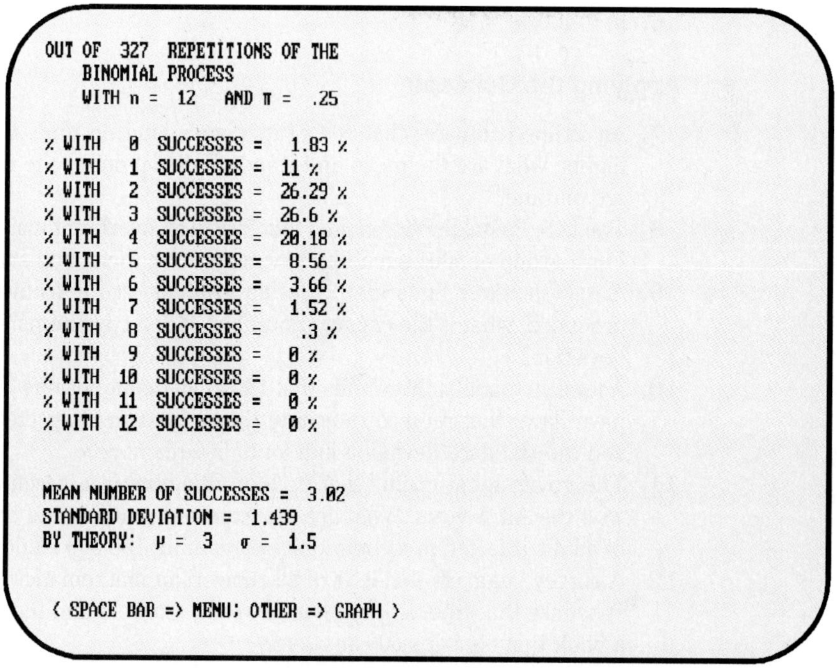

Incidentally, if you examine the graph in Figure 5.7, you will note that the result is again a histogram-type shape, but that the bulk of the results are centered to the left side of the screen and are centered about the population mean $\mu = 3$. Remember, there are 0 through 12 possible successes in this experiment, and the mean number of successes expected is 3.

In general, when the probability π is less than $\frac{1}{2}$, we get a histogram shifted to the left on the screen. Similarly, when π is greater than $\frac{1}{2}$, the histogram is shifted to the right. Statistically, this shift is known as *skewness*. We encourage you to experiment with this program to get a feel for the validity of the formula for the standard deviation.

Exercise Set 5.2

Mastering the Techniques

Find the mean and standard deviation for each of the following binomial distributions based on the given values for n and π.

1. $n = 100$, $\pi = .5$
2. $n = 100$, $\pi = .25$
3. $n = 180$, $\pi = \frac{1}{3}$
4. $n = 200$, $\pi = .4$
5. $n = 1000$, $\pi = .2$
6. $n = 400$, $\pi = .8$

Applying the Concepts

7. An airline estimates that 90% of its planes land on time. Out of a group of 400 flights, what are the mean and standard deviation for the number of flights that are on time?
8. The U.S. Postal Service claims that 85% of all overnight mail arrives the next day. For a group of 5000 such items, what are the mean and standard deviation?
9. A manufacturer finds that 3% of its products are defective. Out of 1000 items produced, what is the expected number of defective items? What is the standard deviation?
10. A jewelry manufacturer finds that 18% of all diamonds she gets from her supplier have flaws that require returning them. Find the expected number of returns and the standard deviation in a lot of 500 diamonds.
11. The government claims that 0.4% of all adults in a certain region are infected with the AIDS virus. What are the expected number and the standard deviation of adults infected in a community containing 100,000 adults?
12. A survey suggests that 0.5% of all answers in mathematics textbooks are wrong. What are the expected number of wrong answers and the standard deviation in a book that contains 600 answers?

Computer Applications

13. Consider a binomial distribution with $n = 10$ trials. Use Program 11: the Binomial Distribution to investigate the different binomial distributions (shape, mean and standard deviation) that result from a choice for probability of success of $\pi = .1, \pi = .3, \pi = .5, \pi = .7,$ and $\pi = .9$.
14. Consider a binomial distribution with probability of success $\pi = 0.6$. Use Program 11: the Binomial Distribution to investigate the different binomial distributions (shape, mean and standard deviation) that result from a choice for the number of trials of $n = 3, n = 6, n = 9, n = 12,$ and $n = 15$.
15. Consider the binomial distribution based on $n = 10$ and $\pi = .4$. Calculate the mean and standard deviation, and then compare these values to the results obtained by using Program 12: Binomial Simulation.
16. Repeat Exercise 15 five different times. Each time, record the mean and standard deviation for the results of each run; then average both sets of values to see how the averages compare to the predicted values for μ and σ.
17. Repeat Exercise 15 for the binomial distribution with $n = 15$ and $\pi = \frac{1}{3}$. (*Note:* You cannot enter $\frac{1}{3}$ as the value for π in the program—you must first convert it to a decimal equivalent, say .333, and then enter it.)

5.3 Binomial Probabilities

We have seen that a discrete probability distribution consists of all the possible outcomes in an experiment along with their associated probabilities. In practice, we are usually interested in finding the probability associated with one particular outcome. For instance, with a binomial distribution, we want to find the probability of getting x successes out of n trials, assuming that the probability of success is π. Such questions arise in many different fields.

Suppose that we have a binomial process based on n trials with probability π of success. A typical *binomial probability problem* takes this form:

> Out of n trials of a binomial process with probability π of success, what is the probability of obtaining precisely x successes?

The number of successes we seek, x, must be an integer between 0 and n. For example, if the binomial process involves flipping a set of 10 coins where $\pi = \frac{1}{2}$, then we might ask: What is the probability of obtaining 7 heads? In this case, $x = 7$. Alternatively, in the ESP experiment, a typical question might be: What is the probability of obtaining 12 right answers out of 25? In this case, $n = 25, \pi = \frac{1}{5} = .2,$ and $x = 12$. As another example, the binomial process might involve a study of 500 people to see if they have any adverse reaction to a new drug. If the company producing the drug believes that only 2% should have such reactions, what is the probability that 18 people out of this group of 500 will encounter problems? Here $n = 500, \pi = .02,$ and $x = 18$.

To answer such questions, we first require the following mathematical shorthand, known as *factorial notation* (which some of you may have seen in optional Section 4.3). The expression 3!, read "3 factorial", is defined as

$$3! = 3 \times 2 \times 1 \; (= 6)$$

Similarly,

$$4! = 4 \times 3 \times 2 \times 1 \; (= 24)$$
$$5! = 5 \times 4 \times 3 \times 2 \times 1 \; (= 120)$$

and so forth. In general, if n represents any positive integer, then

$$n! = n(n-1)(n-2) \cdots 3 \cdot 2 \cdot 1 \qquad n \geq 1$$

Finally, by convention, we define

$$0! = 1$$

Using this notation, we now present a mathematical formula that answers any binomial probability problem. Suppose that the binomial distribution is based on n trials and that π is the probability of success, so that $1 - \pi$ represents the probability of failure. Moreover, if we are interested in x successes out of the n trials, there are $n - x$ failures. The probability of obtaining x successes out of n trials is

$$P(x) = \frac{n!}{x!(n-x)!} \pi^x (1-\pi)^{n-x}$$

Although initially this formula may appear quite intimidating, it is not as bad as first appearances suggest. The formula starts with a set of factorials. The numerator is the factorial for n, the number of trials. The denominator has the factorials for both the number of successes x and the number of failures $n - x$. The entire term

$$\frac{n!}{x!(n-x)!}$$

is known as a *binomial coefficient* and is often written as either

$$_nC_x \quad \text{or} \quad \binom{n}{x}$$

The next term in the formula, π^x, is just the probability of success, π, raised to the power equal to the number of successes x. The last term $(1 - \pi)^{n-x}$ is the probability of failure, $1 - \pi$, raised to the power equal to the number of failures, $n - x$. We show how to use this formula in the following examples.

EXAMPLE 5.8 What is the probability of getting 3 heads when 4 fair coins are flipped?

SOLUTION This is a binomial probability problem with $n = 4$, $\pi = \frac{1}{2}$, and $x = 3$. Therefore,

$$1 - \pi = \frac{1}{2} \quad \text{and} \quad n - x = 4 - 3 = 1$$

so that the binomial probability formula yields

$$P(x = 3) = \frac{4!}{3!(1)!} \left(\frac{1}{2}\right)^3 \left(\frac{1}{2}\right)^1$$

$$= \left[\frac{4 \cdot 3 \cdot 2 \cdot 1}{(3 \cdot 2 \cdot 1)(1)}\right]\left(\frac{1}{8}\right)\left(\frac{1}{2}\right)$$

Instead of multiplying the binomial coefficient terms within the square brackets, it is always much easier to perform all possible cancellations first. In this case, we obtain

$$P(x = 3) = [4]\left(\tfrac{1}{16}\right) = \tfrac{4}{16} = \tfrac{1}{4}$$

We note that this is the same answer as we obtained for this problem in earlier sections. ∎

EXAMPLE 5.9 A student takes a 10-question multiple-choice test completely unprepared, so that he must guess at all answers. If there are 5 choices for each question, what is the probability that he will get 6 correct answers?

SOLUTION This is a binomial probability problem with $n = 10$, $\pi = \frac{1}{5} = .2$, and $x = 6$. Consequently, $1 - \pi = .8$ and $n - x = 4$. Therefore,

$$P(x = 6) = \frac{10!}{6!(4!)} (.2)^6(.8)^4$$

$$= \frac{10 \cdot 9 \cdot 8 \cdot 7 \cdot 6 \cdot 5 \cdot 4 \cdot 3 \cdot 2 \cdot 1}{(6 \cdot 5 \cdot 4 \cdot 3 \cdot 2 \cdot 1)(4 \cdot 3 \cdot 2 \cdot 1)} (.2)^6(.8)^4$$

$$= \left[\frac{10 \cdot \overset{3}{9} \cdot 8 \cdot 7}{4 \cdot 3 \cdot 2 \cdot 1}\right](.2)^6(.8)^4$$

$$= [210](.000064)(.4096)$$

$$= .0055 \approx .006$$

Clearly, his chances of passing the test are extremely poor. ∎

EXAMPLE 5.10 In the ESP experiment, what is the probability of getting 12 right answers based solely on random guessing?

SOLUTION This is a binomial probability problem with $n = 25$, $\pi = .2$, $1 - \pi = .8$, $x = 12$, and $n - x = 13$. Therefore, the binomial probability formula gives

$$P(x = 12) = \frac{25!}{12!\,13!} (.2)^{12}(.8)^{13}$$

$$= \frac{25 \cdot 24 \cdot \ldots \cdot 1}{(12 \cdot 11 \cdot \ldots \cdot 1)(13 \cdot 12 \cdot \ldots \cdot 1)} (.2)^{12}(.8)^{13}$$

$$= (5{,}200{,}300)(.000000004096)(.05497)$$

$$= .00117$$

Thus, 12 right guesses are quite unlikely to occur purely by chance. ∎

Note that many calculators express the value for the middle term $(.2)^{12}$ in scientific notation as

$$4.096 - 09 = 4.096 \times 10^{-9} = .000000004096$$

Incidentally, the calculations involved in determining binomial probabilities are usually not reasonable to do by hand. Minimally, they require the use of a handheld calculator, preferably one that has an exponential key of the form y^x. However, since such problems are very important in statistics, alternatives are available that provide us with the answers without our actually having to perform the calculations. One approach involves the use of computers, and we discuss this later. Another approach involves the use of special binomial tables which contain the calculated values for a wide variety of binomial probabilities. Such a table, Table II, is found at the end of this book. If you examine it, you will find that it contains many of the common values for the probability $\pi = .05, .1, .2, .3, .4, .5, .6, .7, .8, .9,$ and $.95$. In addition, there are entries for various values of $n = 1, 2, 3, \ldots, 15$ and 20. Finally, under each of these values for n are the possible values of x, the number of successes. Thus, under $n = 10$ trials, there are listings for $x = 0, 1, 2, \ldots, 10$ successes, as reproduced in Figure 5.9. These probabilities are given to three-decimal-place accuracy. Further, any probability which is zero to three decimal places is left blank and is interpreted as being 0.

Therefore, to find the probability of obtaining $x = 4$ successes out of $n = 10$ trials with probability of success $\pi = .1$, we locate the corresponding entry in Table II ($\pi = .1$, $n = 10$, $x = 4$) and so read .011 as the probability. If the probability π of success is .7, then the corresponding result is $P(x = 4) = .037$.

EXAMPLE 5.11 A student takes a 10-question multiple-choice test purely by guessing. If there are 5 choices for each question, what is the probability of his obtaining 6 or more correct answers?

SOLUTION To answer this, we must consider

$$P(x \geq 6) = P(x = 6) + P(x = 7) + P(x = 8) + P(x = 9) + P(x = 10)$$

FIGURE 5.9

n	x	.05	.1	.2	.3	.4	.5	.6	.7	.8	.9	.95
10	0	.599	.349	.107	.028	.006	.001					
	1	.315	.387	.268	.121	.040	.010	.002				
	2	.075	.194	.302	.233	.121	.044	.011	.001			
	3	.010	.057	.201	.267	.215	.117	.042	.009	.001		
	4	.001	.011	.088	.200	.251	.205	.111	.037	.006		
	5		.001	.026	.103	.201	.246	.201	.103	.026	.001	
	6			.006	.037	.111	.205	.251	.200	.088	.011	.001
	7			.001	.009	.042	.117	.215	.267	.201	.057	.010
	8				.001	.011	.044	.121	.233	.302	.194	.075
	9					.002	.010	.010	.121	.268	.387	.315
	10						.001	.006	.026	.107	.349	.599

since the individual results are mutually exclusive, and hence we can add the individual probabilities using the Addition Rule. We have already found that $P(x = 6) = .006$ by direct calculation as well as from the binomial table. We now consider $x = 7$ and find from the table that $P(x = 7) = .001$, while all subsequent probabilities are effectively 0. As a consequence, we find that

$$P(x \geq 6) = .006 + .001 + 0 = .007$$

and hence is not much better than his chance of getting just 6 right answers. ∎

Obviously, the use of such a table makes the problem of determining binomial probabilities much easier than actually calculating the values by hand. But there are definite limitations in using binomial tables. First, Table II only goes from $n = 1$ to $n = 15$ and includes $n = 20$, so that other values of n cannot be handled with this table. For example, the ESP experiment requires $n = 25$.

A further disadvantage to depending on such tables is that they list only relatively few values for the probability of success π. If π is .22 or .6178 or even such relatively common values as .25, .75, $\frac{1}{3}$, $\frac{2}{3}$, or $\frac{1}{6}$, then the table is useless. While more detailed tables are available that include a greater range of values for n and for π, they are usually found in books devoted exclusively to statistical tables. Needless to say, there are no tables available which are all-inclusive.

Of course, we can always resort to hand calculations whenever the tabular approach is not applicable. However, the work involved in calculating each of the individual probabilities is considerable, and as we will see in Chapter 6, there is a much easier method available for handling the more complicated problems of this form.

Computer Corner

The other alternative available to you for determining binomial probabilities is Program 13: Binomial Probability. This program allows you to work with binomial distributions where π is *any* probability value whatsoever (in decimal form) and n is as large as 100. You must enter the values for the number of trials n, for the probability of success π, and for the desired number of successes x. In addition, the program asks for the probability of failure, $1 - \pi$, and the number of failures, $n - x$. It then draws the histogram for the indicated binomial distribution, colors in the particular box corresponding to x successes, and calculates the associated probability. A typical display of this program corresponding to Example 5.10 with $n = 25$, $\pi = .2$, and $x = 12$ is shown in Figure 5.10. Note that the shaded box (which represents obtaining $x = 12$ right answers out of $n = 25$ guesses) is extremely small, as we would expect based on the associated probability of .0012.

Further, since the height of the box is equal to the probability of x successes and the width of the box is 1 unit, the area of the box is numerically equal to the desired probability. That is, there is a direct link between the probability of an event and the area of a box in the histogram. We will make extensive use of this type of relationship in Chapter 6.

After the graphical display is complete, you can transfer to a menu screen where you are presented with a variety of options including

FIGURE 5.10

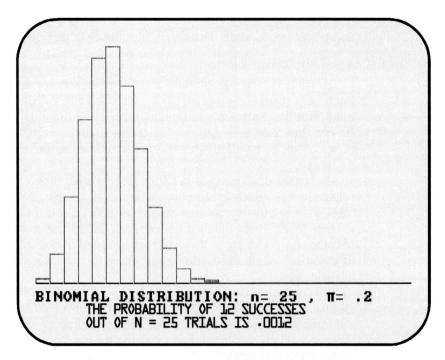

changing the number of successes x, changing the probability of success π, and changing the number of trials n.

Incidentally, the program will actually accept values of n larger than 100; in fact, you can use values up to 250. However, in such cases, there is numerical output, but no graphical display. Furthermore, the quantities involved in some binomial calculations exceed the capacity of the computer (just as they will exceed the capacity of a handheld calculator). When this occurs, the program provides an appropriate message, and you will have to change the values you selected for n and π.

Minitab Methods

Minitab can also be used to calculate binomial probabilities. In fact, rather than giving you the probability associated with one particular number of successes x out of n trials with probability of success π, Minitab produces a full table of probabilities (actually the entire binomial distribution based on n and π) for all possible values of $x = 0, 1, 2, \ldots, n$. In the sample output in Figure 5.11, we show the results of using Minitab for the binomial distribution based on $n = 10$ and $\pi = .35$. Notice, however, that to obtain this table, you must use the

FIGURE 5.11

```
MTB > PDF;
SUBC> BINOMIAL n = 10  p = .35.

   BINOMIAL WITH n = 10  p = 0.350000
        K         P( X = K)
        0          0.0135
        1          0.0725
        2          0.1757
        3          0.2522
        4          0.2377
        5          0.1536
        6          0.0689
        7          0.0212
        8          0.0043
        9          0.0005
       10          0.0000
```

> Minitab command
>
> > PDF;
>
> (for probability density function) and the associated subcommand
>
> > BINOMIAL n = 10   pi = .35.
>
> or more simply, just
>
> > BINOMIAL 10   .35.
>
> Either way, remember that the primary command (**PDF;**) must end with a semicolon to tell Minitab that you want to go on to a subcommand, and the subcommand must end with a period.

Exercise Set 5.3

Mastering the Techniques

Use the binomial probability formula to calculate the following probabilities:

1. $n = 6, \pi = .5, P(x = 1)$
2. $n = 6, \pi = .5, P(x = 2)$
3. $n = 8, \pi = .5, P(x = 2)$
4. $n = 8, \pi = .5, P(x = 3)$
5. $n = 6, \pi = .3, P(x = 1)$
6. $n = 6, \pi = .3, P(x = 2)$
7. $n = 7, \pi = .8, P(x = 5)$
8. $n = 7, \pi = .8, P(x = 6)$

9–16. Repeat Exercises 1 to 8, using Table II of binomial probabilities.

Applying the Concepts

17. A coin is loaded in such a way that it lands with a head showing 60% of the time. If it is flipped 5 times, find the probability of getting 4 heads.
18. A die is loaded so that the probability of getting a 6 is .25 and the probability of a 1, 2, 3, 4 or 5 is .15. If it is rolled 4 times, find the probability of getting three 6s.
19. A student answers 10 questions on a true-false test purely by guesswork. What is the probability that he will get 7 right?
20. A multiple-choice test contains 10 questions with four choices each. If a student guesses at each, what is the probability that she will get six right? At least six right?
21. A hamburger chain runs a contest in which the chance of winning a free burger is .1. If a person collects 10 entry cards, what is the probability that he will win three burgers? Three or more burgers?

22. A reading skills test indicates that 30% of the children in a certain school require remediation. If a group of eight children is selected at random, what is the probability that four need remediation? That no more than two need remediation?
23. The U.S. Postal Service estimates the 85% of all overnight mail items are delivered the next day. If a person sends out five such items, what is the probability that they will all arrive the following day? That at least three will arrive the following day?
24. A TV rating service estimates that 30% of the population watched a particular situation comedy one evening. If a random group of nine people are selected, what is the probability that five of them saw the show? That at most two saw the show?
25. Repeat Exercise 24 if the group in question comprises 12 people. What if the group contains 15 people?
26. If the probability that a newborn child will be a boy is .52, construct the binomial distribution table based on $n = 3$ births. How do these probabilities compare with the values we previously found based on $\pi = .5$?

Computer Applications

Use Program 13: Binomial Probability to answer the following questions:

27. A true-false test has 100 questions. If a student guesses wildly at all of them, what is the probability that he will score 70 right? 65 right? 60 right?
28. A lottery is designed so that 5% of all players win some type of prize. If a person buys 60 tickets, what is the probability that she will win 7 prizes? 5 prizes? 2 prizes? 0 prizes?
29. 65% of the adults in a community favor the construction of a new garbage disposal plant. If 80 randomly selected people are surveyed, what is the probability that 50 will favor the plant? That 60 of them will? That 40 of them will?
30. Approximately 42% of the population have type A blood. If 30 people are selected at random, what is the probability that 10 are type A, that 12 are, that 15 are?
31. A study finds that 67% of all households have VCRs. If a randomly selected group of 50 families is surveyed, what is the probability that 40 have VCRs, that 35 do, that 30 do?
32. The Internal Revenue Service estimates that 6% of all tax returns are audited. If 72 taxpayers are surveyed, what is the probability that none, that 5, and that between 10 and 12 have been audited?

5.4 The Hypergeometric Distribution (Optional)

Since the binomial distribution is probably the single most important discrete probability distribution, we devoted the last several sections to its properties and applications. However, there are many other discrete distributions which are useful. We consider one of them, the *hypergeometric distribution*, in this section.

Recall that a binomial experiment is characterized by the following four requirements:

1. A certain basic action is repeated n times.
2. There are two possible outcomes for each trial.
3. The probability of success π remains constant from trial to trial.
4. The n trials are independent.

For example, the experiment of picking 5 cards from a standard 52-card deck and determining the probability that exactly 3 are hearts is a binomial experiment with $n = 5$, $\pi = \frac{13}{52} = \frac{1}{4}$ and $x = 3$ *provided* that each card picked is replaced in the deck before the next card is drawn.

However, if the individual cards are not replaced, then each successive pick is taken from a smaller deck and so the probability of choosing a heart changes from one pick to the next. Furthermore, the probabilities involved in each pick depend on the previous picks as well. Thus, the probability that the first card is a heart is definitely $\frac{1}{4}$. However, the probability that the second card is a heart is either $\frac{13}{51}$ or $\frac{12}{51}$, depending on what happened on the first pick. The probability that the third card is a heart is $\frac{13}{50}$, $\frac{12}{50}$ or $\frac{11}{50}$, depending on the two previous cards picked. The individual events are not independent and the probability of success is not the same from trial to trial. Obviously, then, this problem is not a binomial process and we cannot use the results of the previous sections to solve it.

We can solve this problem by using some of our previous notions of probability from Sections 4.2 and 4.3, especially the Fundamental Counting Principle and the ideas involving combinations. First, since we are picking 5 cards from a 52-card deck, we know that the total number of possible "hands" is given by

$$_{52}C_5 = \frac{52!}{5!(52 - 5)!}$$

since order is irrelevant. Next we have to determine the number of these hands that contain 3 hearts and 2 nonhearts. Consider those hands containing just 5 cards. The total number of ways that 3 hearts can be selected from the 13 hearts in the deck is given by

$$_{13}C_3 = \frac{13!}{3!(13 - 3)!}$$

Moreover, the total number of ways that the remaining 2 nonheart cards can be selected out of the 39 nonheart cards in the deck is given by

$$_{39}C_2 = \frac{39!}{2!(39 - 2)!}$$

Therefore, the total number of ways that we can select a 5-card hand containing 3 hearts and 2 nonhearts is given by

$$_{13}C_3 \times {}_{39}C_2$$

by the Fundamental Counting Principle. Thus, the probability that such a hand occurs is

$$\frac{_{13}C_3 \times {_{39}C_2}}{_{52}C_5}$$

Alternatively, suppose we want to find the probability of picking a 10-card hand from a standard 52-card deck in such a way that the hand contains 7 hearts and therefore 3 nonhearts. Using the same reasoning as before, we see that the probability is

$$\frac{_{13}C_7 \times {_{39}C_3}}{_{52}C_{10}}$$

We can extend the above procedure to solve a more general type of problem. The numbers shown in parentheses refer to the values of the quantities in the first problem above on 5-card hands with 3 hearts. Suppose a sample of size n ($n = 5$ in the first problem) is drawn without replacement from a larger group of N items ($N = 52$ in the above case). Suppose further that of the N items, k are considered successes ($k = 13$ hearts) and $N - k$ are considered failures ($N - k = 52 - 13 = 39$ nonhearts). Then the probability of getting precisely x successes out of the n selections is

$$P(x) = \frac{_{k}C_x \times {_{N-k}C_{n-x}}}{_{N}C_n}$$

This formula is known as the *hypergeometric probability formula*. Just as the binomial probability formula defines the binomial distribution based on n and π (because the formula supplies all possible probabilities for $x = 0, 1, \ldots, n$), this hypergeometric probability formula defines a distribution known as the *hypergeometric distribution* by providing a way to determine all possible probabilities for $x = 0, 1, 2, \ldots, n$.

We illustrate the use of this formula in the following examples.

EXAMPLE 5.12 A supermarket display consists of 300 containers of yogurt. It is known that 25 of them are outdated. If a customer selects 12 containers at random, what is the probability that exactly 2 are outdated?

SOLUTION Since the containers are being selected without replacement, we must use the hypergeometric distribution. In this case, we have

A total of $N = 300$ containers

A sample size of $n = 12$

A total of $k = 25$ successes out of 300

And $N - k = 300 - 25 = 275$ failures out of 300

Since we are interested in the probability that $x = 2$ of the 12 containers selected are outdated, we find from the hypergeometric probability formula that

$$P(x = 2) = \frac{{}_kC_x \times {}_{N-k}C_{n-x}}{{}_NC_n}$$

$$= \frac{{}_{25}C_2 \times {}_{275}C_{10}}{{}_{300}C_{12}}$$

$$= \frac{[25!/(2! \times 23!)][275!/(10! \times 265!)]}{300!/(12! \times 288!)}$$

After a considerable amount of simplification, this probability is equal to .195. ■

EXAMPLE 5.13 A lottery is set up in such a way that 50 prizes are awarded for every 500 tickets sold. If a person buys 20 tickets, what is the probability that she wins precisely 3 prizes?

SOLUTION Since this problem does not involve replacement, it is a hypergeometric probability problem with $N = 500$, sample size $n = 20$, $k = 50$, and therefore $N - k = 450$. As a result, the probability of having precisely 3 winning tickets is

$$P(x = 3) = \frac{{}_kC_x \times {}_{N-k}C_{n-x}}{{}_NC_n}$$

$$= \frac{{}_{50}C_3 \times {}_{450}C_{17}}{{}_{500}C_{20}}$$

$$= \frac{[50!/(3! \times 47!)][450!/(17! \times 433!)]}{500!/(20! \times 480!)}$$

This eventually simplifies to .194. ■

Since the hypergeometric distribution is itself a population, it has a mean μ and a standard deviation σ that are given by

$$\mu = \frac{nk}{N}$$

$$\sigma = \frac{1}{N}\sqrt{\frac{nk(N-k)(N-n)}{N-1}}$$

where N = population size
n = sample size
k = number of successes in the entire population

We illustrate these formulas using the values from Example 5.13 on the lottery where $N = 500$, $n = 20$, and $k = 50$. Therefore,

$$\mu = \frac{(20)(50)}{500} = 2$$

is the average number of prizes that a person would expect to win out of every 20 tickets purchased. Further,

$$\sigma = \frac{1}{500}\sqrt{\frac{(20)(50)(500-50)(500-20)}{500-1}}$$

$$= \frac{1}{500}\sqrt{\frac{(20)(50)(450)(480)}{499}}$$

$$= 1.316$$

Computer Corner

Since the calculations involved in solving hypergeometric probability problems are so complicated, we have included Program 14: The Hypergeometric Distribution. This program functions much as Program 13: Binomial Probability. You must supply the desired values for the population size N, sample size n, number of successes within the population k, and desired number of successes in the sample x. The program then draws the histogram representing the hypergeometric distribution, colors in the particular box corresponding to x successes, and calculates the value for the probability. A typical display is shown in Figure 5.12 based on $N = 30$, $k = 18$, $n = 12$ and $x = 7$. The resulting probability of 7 successes out of 12 trials is then .291. Notice, incidentally, that the graphical display only ranges from $x = 0$ to $x = 12$ successes, since this is the maximum number of successes possible in a sample of size $n = 12$.

After the display is complete, you can transfer to a menu screen that presents you with a variety of options including changing the desired number of successes x, changing the sample size n, changing the number of successes in the population k, and changing the size of the population N.

Exercise Set 5.4

Mastering the Techniques

For each of the following sets of values for N, n and k, calculate the probability of obtaining precisely x successes in a hypergeometric distribution.

1. $N = 12$, $n = 5$, $k = 8$, $x = 3$
2. $N = 12$, $n = 5$, $k = 8$, $x = 1$
3. $N = 10$, $n = 6$, $k = 5$, $x = 3$

FIGURE 5.12

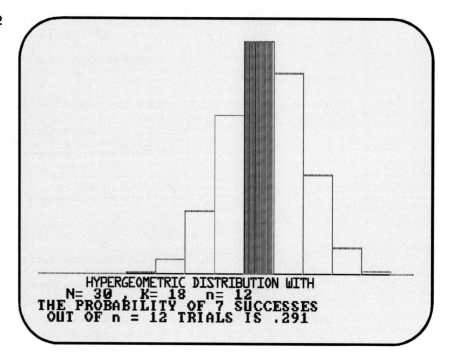

4. $N = 10, n = 5, k = 7, x = 3$
5. $N = 10, n = 7, k = 5, x = 3$
6. $N = 10, n = 3, k = 8, x = 0$

7. Calculate the mean and standard deviation for the hypergeometric distribution in Exercise 1.
8. Calculate the mean and standard deviation for the hypergeometric distribution in Exercise 3.

Applying the Concepts

9. The mathematics faculty at a college consists of 20 people, 7 of whom are women. If a committee of five people is set up at random, what is the probability that precisely two of the members will be women?
10. In a group of 16 unrelated people who go to a mountain resort, 12 are skiers. If a group of four of them are randomly selected and seated together for breakfast, what is the probability that three of them are skiers?
11. Of the 30 people working out in a health club one afternoon, 25 are nonsmokers. If six are randomly selected, what is the probability that none are smokers?
12. In a class of 24 students, 16 prefer Democrats to Republicans. If a random group of six students is selected, what is the probability that three prefer Democrats?

Computer Applications

13. In a group of 50 sports fans in New York, 35 prefer the Mets to the Yankees. If a random sample of 20 is selected from among the group of 50, find the probability that precisely 18 prefer the Mets.
14. A used car lot has 72 cars, 48 of which were made in the United States. Of the 14 cars sold one weekend, what is the probability that 10 were made in the United States?
15. A video rental store receives a delivery consisting of 46 new titles, 15 of which were never shown in theaters. If a clerk takes 12 of the tapes at random to make a window display of new releases, what is the probability that more than 6 of the tapes will be nontheater releases?
16. Of the 58 patients scheduled to be seen in a clinic for senior citizens, 17 are there to receive flu shots. If a group of 20 of these patients are randomly selected to be seen by one particular doctor, what is the probability that she will administer fewer than five flu shots?

5.5 The Poisson Distribution (Optional)

We now consider another discrete probability distribution known as the *Poisson distribution*. Before introducing it, we first consider several probability problems that cannot be solved by using any of the methods developed previously.

Problem 1 An orthopedic surgeon's records show that she is called to the hospital to set an average of three broken limbs per day. What is the probability that she will set five broken limbs during a particular day?

Problem 2 The loan officer in a bank finds that he interviews an average of eight loan applicants per day. Find the probability that he interviews only three applicants during a certain day.

Problem 3 A major league baseball team typically postpones an average of five games per season due to rain. What is the probability that the team will have more than nine rain-outs during a particular season?

All of these three problems have certain characteristics in common. First, each involves finding the probability of a certain number of successes *during a given interval of time*. (Compare this to the typical binomial probability problem of finding the probability of a certain number of successes *out of n trials*.)

Second, in each case, we are looking at situations where there are relatively few successes during the indicated time period. Thus, the probability of success in any sufficiently small time interval will be quite small.

Third, the individual successes are independent of one another. Thus, the fact that one person breaks a leg should not influence whether any other person breaks an arm. The fact that one baseball game is rained out should not affect the probability that any other game on a different day is also rained out. (We obviously ignore double-headers, multiple fractures, multiple injuries in an accident, and comparable complications.)

Finally, while it may not be an obvious condition based on the above examples, we also assume that the successes will occur uniformly over the entire time period under consideration. Thus, we assume that a rain-out will occur at any point in a baseball season with equal likelihood or that a loan applicant will walk into the bank at any time during the day with equal likelihood.

Under the above conditions, the probabilities involved follow a *Poisson probability distribution*. As we have seen, any probability distribution is completely known once a formula or other method for calculating all probabilities involved is available. For the Poisson distribution, the appropriate probabilities are given as follows:

> The **Poisson probability formula** giving the probability that x successes occur during a given time interval is
>
> $$P(x \text{ successes}) = \frac{e^{-\mu}\mu^x}{x!} \qquad x = 0, 1, 2, 3, \ldots$$
>
> where μ is the average number of successes occurring in the given time interval and the symbol e represents the number
>
> $$e \approx 2.71828$$

This number e is a mathematical constant (analogous to $\pi \approx 3.14159$) which occurs in many applications in mathematics. (Those of you who have studied calculus will recognize the number e as the base of the natural logarithm system.)

Many calculators are designed to work with e directly. You may have a key on your calculator marked e^x or EXP or EX. In that case, all you need do to calculate the value for e^{-3}, for example, is to key in -3 and then press the e^x key to get .0497871. Other calculators do not show e^x explicitly but provide for it in terms of a related mathematical function using a key marked ln X. If you have such a calculator, you also have a key marked either 2nd or INV or SHIFT. To find e^{-3}, simply key in -3 and then press INV followed by ln X to get the same numerical value.

If your calculator does not have either of these features, then you must key in the value of $e = 2.71828$ each time you need it and work with it as with any other number. Thus, to calculate e^{-3}, it is necessary to raise the number 2.71828 to the -3 power. Since the Poisson probability formula involves negative exponents for e, you should be aware that when you raise e to a negative power, the result will always be a small decimal. If it is not, check your arithmetic.

SECTION 5.5 • The Poisson Distribution (Optional)

To illustrate the use of the Poisson probability formula, we apply it to solve each of the three problems posed at the beginning of this section.

EXAMPLE 5.14 An orthopedic surgeon's records show that she is called to the hospital to set an average of three broken limbs per day. What is the probability that she will set five broken limbs during a particular day?

SOLUTION We are told that the average number of broken limbs per day (one time period) is $\mu = 3$ and that we want the probability that $x = 5$. Therefore, from the Poisson probability formula,

$$P(x = 5) = \frac{e^{-\mu}\mu^x}{x!}$$

$$= \frac{e^{-3} \times 3^5}{5!}$$

$$= \frac{(.0497871)(243)}{120}$$

$$= .1008188 \approx .1008$$

EXAMPLE 5.15 The loan officer in a bank finds that he interviews an average of eight loan applicants per day. Find the probability that he interviews only three applicants during a certain day.

SOLUTION In this case, we are given $\mu = 8$ and want to find the probability that $x = 3$. Thus,

$$P(x = 3) = \frac{e^{-8} \times 8^3}{3!}$$

$$\frac{(.0003355)(512)}{6}$$

$$= .0286261 \approx .0286$$

EXAMPLE 5.16 A major league baseball team typically postpones an average of five games per season due to rain. What is the probability that the team will have more than nine rain-outs during a particular season?

SOLUTION In this case, we are told that $\mu = 7$ and we want the probability of more than 9 rain-outs. Thus, we must find the probability that $x = 10$, $x = 11, \ldots$. Let's start with $x = 10$, so that

$$P(x = 10) = \frac{e^{-5} \times 5^{10}}{10!}$$

$$= \frac{(.006738)(9,765,625)}{3,628,800}$$

$$= .01813$$

Similarly,

$$P(x = 11) = \frac{e^{-5} \times 5^{11}}{11!}$$

$$= \frac{(.006738)(48{,}828{,}125)}{39{,}916{,}800}$$

$$= .00824$$

$$P(x = 12) = \frac{e^{-5} \times 5^{12}}{12!}$$

$$= \frac{(.006738)(2.4414 \times 10^8)}{4.79 \times 10^8}$$

$$= .00343$$

$$P(x = 13) = \frac{e^{-5} \times 5^{13}}{13!}$$

$$= \frac{(.006738)(1.2207 \times 10^9)}{6.227 \times 10^9}$$

$$= .00132$$

and

$$P(x = 14) = \frac{e^{-5} \times 5^{14}}{14!}$$

$$= \frac{(.006738)(6.1035 \times 10^9)}{8.7178 \times 10^{10}}$$

$$= .00047$$

If all we want is an answer correct to three decimal places, it should be apparent that all successive probabilities will be much smaller and so will have no effect on the total. Thus, the answer to this question is given by the sum of the above values, namely,

$$P(x > 9) = .01813 + .00824 + .00343 + .00132 + .00047$$
$$= .03159 \approx .032$$

■

Since the Poisson distribution is a population, it has a mean and a standard deviation. They are given by the following:

> A Poisson distribution with mean μ has standard deviation
> $$\sigma = \sqrt{\mu}$$

For example, if the mean μ of a Poisson distribution is 12, then its standard deviation is

$$\sigma = \sqrt{\mu} = \sqrt{12} = 3.464$$

Further, the standard deviation for a Poisson distribution is equal to $\sqrt{\mu}$ so that the variance is equal to μ. It is interesting to note that the Poisson distribution is the only discrete distribution which has the property that its mean is equal to its variance.

Computer Corner

As with any discrete probability distribution, the Poisson distribution can also be displayed as a histogram showing the probabilities for different values of $x = 0, 1, 2, \ldots$ based on a given value for the mean μ. For instance, in Figure 5.13, we show a computer-generated histogram for the Poisson distribution based on $\mu = 8$.

Many of you will see no difference between a histogram such as this one and any of the histograms we produced previously for the binomial distribution. To explore the differences between these two distributions, we have included Program 15: The Poisson Distribution. Recall that the mean μ for a binomial distribution is $\mu = n\pi$. You must supply the values for n and π, and the program draws the corresponding binomial distribution with mean $\mu = n\pi$ and standard deviation $\sigma = \sqrt{n\pi(1 - \pi)}$ as a histogram. It then uses the same

FIGURE 5.13

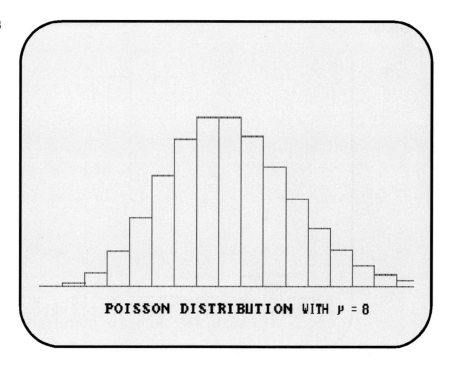

value for μ to superimpose the histogram for the associated Poisson distribution. For example, if you select $n = 16$ and $\pi = .5$, then $\mu = 8$ and so we obtain the same Poisson distribution shown in Figure 5.13 superimposed on the corresponding binomial distribution. This is shown in Figure 5.14 where the lighter lines represent the Poisson distribution and the darker lines represent the binomial distribution. (In the actual computer display, the different distributions are shown in different colors.) From the figure or from actually running the program, you will observe that, while the shapes of the two histograms are similar, there is a significant difference between the two.

You may notice that the two histograms reach their maximum heights in roughly the same location. This corresponds to the mean for each distribution.

Suppose we explore the relationship between the two distributions by taking $n = 20$ and $\pi = .2$. The two histograms are displayed in Figure 5.15, where we see that the two distributions are seemingly much more similar than those shown in Figure 5.14. Further, if we use $n = 50$ and $\pi = .1$, as shown in Figure 5.16, then we see that the two distributions are extremely close to one another. In fact, whenever π is very small and n is relatively large, the Poisson distribution is usually very close to the corresponding binomial distribution. Until quite recently when calculators and computers became widely available, statisticians and scientists used the values predicted by the Pois-

FIGURE 5.14

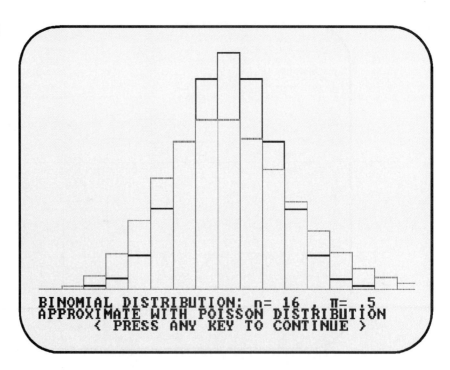

SECTION 5.5 • The Poisson Distribution (Optional)   **219**

FIGURE 5.15

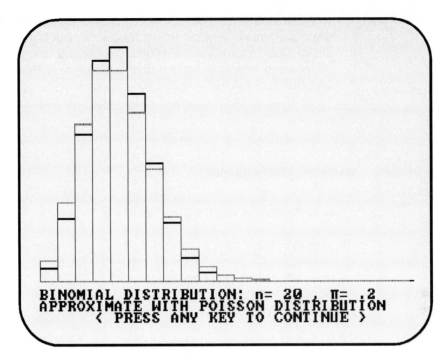

FIGURE 5.16

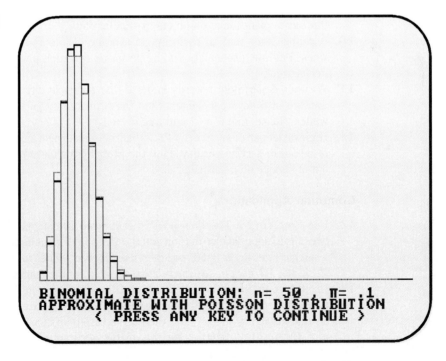

son distribution as approximations to binomial probabilities (at least when π is small), since it is much easier to calculate Poisson probabilities than binomial probabilities. We suggest that you explore this relationship between the two distributions by running the program with different values for n and π.

Exercise Set 5.5

Mastering the Techniques

In each of the following Exercises, calculate the probability of x successes based on a Poisson distribution with the given value for μ.

1. $\mu = 3, x = 2$
2. $\mu = 3, x = 4$
3. $\mu = 5, x = 3$
4. $\mu = 5, x = 4$
5. $\mu = 5, x = 5$
6. $\mu = 5, x = 6$
7. Calculate the standard deviation for the Poisson distribution in Exercise 1.
8. Calculate the standard deviation for the Poisson distribution in Exercise 3.

Applying the Concepts

9. A small rural airport has an average of five takeoffs per day. Find the probability that seven planes take off on one particular day.
10. A dentist finds that he performs an average of four root canal operations per week. Find the probability that he performs two such procedures during a particular week.
11. Weather department records indicate that a certain city experiences snow an average of 6 days per winter. Find the probability that there is no snow one winter. (Recall that $0! = 1$ and that $a^0 = 1$.)
12. Highway department records show that they can expect an average of four trucks to be out of service per day. Find the probability that more than six trucks are out of service on a particular day.

Computer Applications

13. Use Program 15: The Poisson Distribution to investigate the relationship between the binomial distribution and the Poisson distribution. In particular, select a sample size, say $n = 50$, and then use different values for π, say $\pi = .5, .4, .3, .2, .1, .05, .02$, etc., to see the relative accuracy of the approximation.
14. Use the computer to investigate the relationship between the binomial distribution and the Poisson distribution by varying the sample size n. In particular, select a value for π, say $\pi = .1$, and then use a variety of increasing values for n, say $n = 10, 20, 30, \ldots, 100$, to see the relative accuracy of the approximation.

5.6 The Trinomial Distribution (Optional)

In most of our work with probability distributions to this point, we have studied situations that involve two outcomes, either success or failure. The most important of these distributions are those involving binomial experiments and binomial probability. In several instances, we were able to reinterpret some situations to produce a binomial experiment. For example, in rolling a fair die, there are six possible outcomes. However, if we are interested in obtaining a 5, say, then we interpret a 5 as representing success and any other outcome as representing failure.

Unfortunately, it is not always possible to do this. For example, suppose a poll is conducted in which respondents are asked their opinion on a certain issue. The possible choices might be *agree*, *disagree*, and *no opinion*. Suppose further that we want success to stand for agreement with the position. To make this into a binomial experiment, it would be necessary to combine all disagree and no opinion responses and consider all of them as failure. Clearly, this is not reasonable. In an analogous situation, we might be studying an election campaign with three major candidates Wong, Fontanez, and McGee. While in theory we could combine Wong and Fontanez as opposed to McGee, it might not be reasonable to do this either.

Consequently, it often becomes necessary to handle situations having three (or more) different outcomes. We consider such an eventuality in this section. In parallel to the underlying conditions for a binomial experiment, we introduce the notion of a *trinomial experiment*.

A **trinomial experiment** is one in which

1. A certain basic action is performed with n trials or repetitions.
2. Each repetition involves three possible outcomes A, B and C.
3. Each of the three outcomes has a fixed probability

$$p = P(A) \quad q = P(B) \quad \text{and} \quad r = P(C)$$

which remains the same from trial to trial.
4. The individual trials are independent.

For example, consider a poll on the issue of imposing import tariffs. Suppose that in the entire population, 50% favor such a move, 20% oppose it, and 30% have no opinion. Suppose that 500 adults are polled. The process of conducting the poll is a trinomial experiment since the basic act of asking people their positions on the issue is conducted 500 times; there are three possible outcomes on each trial—agree, disagree, and no opinion; there are fixed probabilities for each response of .50, .20, and .30, respectively, which remain the same from trial to trial; finally, if the sample is chosen appropriately, each trial is independent of the others.

As with the binomial distribution, a typical question involving a trinomial experiment is to determine the probability of particular sets of outcomes. For instance, in the above poll of 500 people, we might want to determine the probability that 240 people agree with the tariff, that 90 disagree, and that 170 have no opinion.

The set of all possible probabilities connected with a trinomial experiment forms a new probability distribution known as the *trinomial distribution*. Before we consider how such probabilities can be calculated, let's recall some comparable ideas relating to binomial probability. In the binomial case, there are two outcomes: success and failure. A typical problem asks for the probability of x successes (and hence $n - x$ failures) out of n trials. If the probability of success is π (and hence the probability of failure is $1 - \pi$), then

$$P(x \text{ successes out of } n) = \frac{n!}{x!(n-x)!} \pi^x (1 - \pi)^{n-x} \quad (5.3)$$

Let's change the notation used here slightly. If there are to be x successes out of the n trials, then we denote the number of failures $n - x$ by the letter y. Obviously, $x + y$ must total n. Further, let the probability of success be denoted by p instead of π, and let the probability of failure be q instead of $1 - \pi$ so that $p + q = 1$. With these changes, the binomial probability formula becomes

$$P(x \text{ successes out of } n) = \frac{n!}{x!\, y!} p^x q^y \quad (5.4)$$

where

$$x + y = n \quad \text{and} \quad p + q = 1$$

We now turn to the development of a comparable formula to yield trinomial probabilities. Since there are three possible outcomes A, B and C in any trinomial trial, there must be a fixed probability for each outcome, namely,

$$p = P(A) \quad q = P(B) \quad \text{and} \quad r = P(C)$$

as indicated above. Moreover, it is necessary that

$$p + q + r = 1$$

Furthermore, in a set of n trials, we are interested in the probability of some combination of outcomes involving x A's, y B's and z C's, where it is necessary that

$$x + y + z = n$$

For instance, in the discussion of the poll above, out of the $n = 500$ trials, we wanted to consider the case where $x = 240$ who agree, $y = 90$ who disagree, and $z = 170$ with no opinion. Clearly,

$$x + y + z = 240 + 90 + 170 = 500 = n$$

If we now look at the alternate binomial probability formula (5.4) given above, the corresponding trinomial probability formula should be fairly evident. In particular, as you may expect, the probability of getting x A's, y B's and z C's is

$$P(x, y, z) = \frac{n!}{x!\, y!\, z!}\, p^x q^y r^z \qquad (5.5)$$

where

$$x + y + z = n \quad \text{and} \quad p + q + r = 1$$

The term

$$\frac{n!}{x!\, y!\, z!}$$

is called a *trinomial coefficient*. As with the binomial coefficient, it is always a positive integer.

EXAMPLE 5.17 Suppose it is known that 50% of the population approves of an import tariff, 20% is opposed to it, and 30% has no opinion in the matter. Set up the expression for the probability that, in a poll of 500 people, precisely 240 support the tariff, 90 are opposed to it, and 170 have no opinion.

SOLUTION Using the trinomial probability formula (5.5) with

$$p = .5 \quad q = .2 \quad \text{and} \quad r = .3$$

and with

$$x = 240 \quad y = 90 \quad \text{and} \quad z = 170$$

we find that

$$P(x = 240, y = 90, z = 170) = \frac{500!}{240!\, 90!\, 170!} (.5)^{240} (.2)^{90} (.3)^{170}$$

Due to the extreme complexity of these numbers, we do not attempt to work out the result numerically, but simply leave it as is. ∎

EXAMPLE 5.18 Suppose the same poll is conducted with 10 people. Find the probability that seven respondents agree with the tariff, one is opposed to it, and two have no opinion.

SOLUTION We now have $n = 10$, $x = 7$, $y = 1$, and $z = 2$, so that the trinomial probability formula (5.5) gives

$$P(x = 7, y = 1, z = 2) = \frac{10!}{7!\,1!\,2!}(.5)^7(.2)^1(.3)^2$$

$$= \frac{10 \times 9 \times 8 \times 7 \times \cancel{6} \times \ldots \times \cancel{2} \times \cancel{1}}{(7 \times \cancel{6} \times \ldots \times \cancel{2} \times \cancel{1})(1)(2 \times 1)}(.0078125)(.2)(.09)$$

$$= (360)(.0001406)$$

$$= .050625 \approx .0506 \quad \blacksquare$$

EXAMPLE 5.19 In a three-way election race, one candidate's campaign manager claims that her candidate, Wong, has the support of 40% of the voters, that Fontanez has 33%, and that McGee has 27%. If a random sample of 11 voters is polled, find the probability that precisely 6 of them prefer Wong, 3 of them prefer Fontanez, and 2 of them prefer McGee.

SOLUTION In this case, we are given

$$n = 11 \qquad p = .40 \qquad q = .33 \qquad r = .27$$
$$x = 6 \qquad y = 3 \qquad z = 2$$

so that the trinomial probability formula (5.5) yields

$$P(x = 6, y = 3, z = 2) = \frac{11!}{6!\,3!\,2!}(.40)^6(.33)^3(.27)^2$$

$$= \frac{11 \times 10 \times 9 \times 8 \times 7 \times \cancel{6} \times \ldots \times \cancel{2} \times \cancel{1}}{(\cancel{6} \times \ldots \times \cancel{2} \times \cancel{1})(3 \times 2 \times 1)(2 \times 1)}(.0041)(.03594)(.0729)$$

$$= (4620)(.0000107)$$

$$= .049576 \approx .0496 \quad \blacksquare$$

The above ideas for trinomial probability and the trinomial distribution can be extended to cases where there are four or more outcomes on each trial. The resulting distributions are known as *multinomial distributions* and the corresponding formula for multinomial probability should be fairly easy to predict based on what we have done so far. However, we do not go into this here at all.

Computer Corner

We have included Program 16: Trinomial Distribution Simulation to let you experiment with the trinomial distribution. To use this program, you must supply the number of trials n (limited to a maximum of $n = 10$), the number of such samples (say 50 or 100), and the probabilities p and q (which therefore determine r since $p + q + r = 1$). The program performs a random simulation of the corresponding trinomial process and keeps track of the number of outcomes x, y and z that result on each trial. It then displays the individual results in a three-dimensional analog of a histogram. A sample run based on $n = 6$ with $p = .5$ and $q = .3$ (and hence $r = .2$) is shown in Figure 5.17. To interpret the result, notice that the number of A's (equal to x) is plotted to the right while the number of B's (equal to y) is plotted along the diagonal axis. Since $p = .5$ is the dominant probability, there is a greater tendency for A's to occur than for B's or C's, and hence the stacks toward the right are higher than those toward the left.

We suggest that you experiment with this program by first selecting $n = 4$ or 5 or 6, say, and by using $p = q = .5$. In that case, r must be 0, and hence the result reduces to a binomial distribution. Notice the shape of the resulting graph, as typified by the result in Figure 5.18. The stacks representing the results all lie along a diagonal and the shape of the stacks should be very suggestive of a binomial distribution. Suppose you then introduce a nonzero value for r, say by selecting $p = q = .45$. What happens to the display? Suppose you select $p = .8$ and $q = .1$. How does this skew the resulting shape? What if you take $p = .1$ and $q = .8$?

Incidentally, after each graphical display is complete, you can transfer to an accompanying text screen where the percentage of each (nonzero) outcome is listed. While these results can give you a feel for the probabilities involved, we believe you will get a much better appreciation for them from the visual displays.

FIGURE 5.17

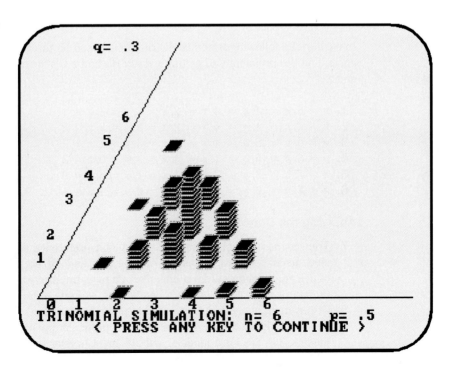

FIGURE 5.18

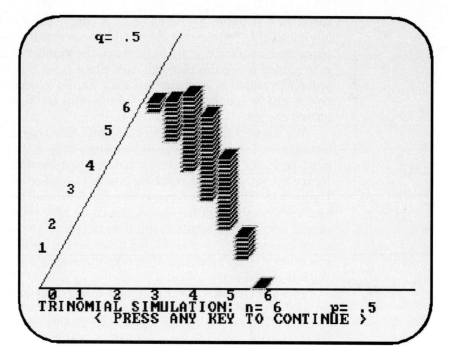

Exercise Set 5.6

Mastering the Techniques

In each of the following trinomial distributions based on the given values for n, p, q, and r, find the probability of getting x A's, y B's and z C's for the given values of x, y and z.

1. $n = 6, p = .4, q = .4, r = .2, x = 3, y = 2, z = 1$
2. $n = 6, p = .4, q = .4, r = .2, x = 2, y = 2, z = 2$
3. $n = 6, p = .5, q = .1, r = .4, x = 3, y = 1, z = 2$
4. $n = 6, p = .5, q = .1, r = .4, x = 3, y = 0, z = 3$
5. $n = 7, p = .2, q = .7, r = .1, x = 1, y = 6, z = 0$
6. $n = 7, p = .2, q = .7, r = .1, x = 2, y = 4, z = 1$

Applying the Concepts

7. The results of a school survey on fast-food preference shows that 50% of the students prefer pizza, 25% prefer hamburgers, and 25% prefer chicken. If a group of five students go out for a bite after a movie, find the probability that three of them order pizza, one orders a hamburger, and one orders chicken.
8. Government agencies are required to purchase cars made in the United States. Records indicate that 50% of the cars at a particular agency are General Motors models, 30% are Ford models, and 20% are Chrysler models. If a certain office

orders seven cars from a motor pool on a particular day, find the probability that they receive four General Motors cars, two Ford cars, and one Chrysler car.

9. Industry records indicate that 60% of all computers purchased by schools are IBM and compatibles, 25% are Apple computers, and 15% are all other types. If a group of eight teachers from different schools meet at a national conference, what is the probability that four use IBM computers and that the remaining four use Apple computers?

10. A record store finds that it has the following breakdown of its music sales: 50% of purchases are tapes, 10% are records, and 40% are compact disks (CDs). In a group of eight randomly selected sales, find the probability that 3 tapes, 2 records, and 3 CDs are sold.

11. Conjecture a formula for multinomial probability with four possible outcomes A, B, C and D with respective probabilities of p, q, r and s. Let the sample size be n, and write a formula for the probability of getting x A's, y B's, z C's and w D's. What relations exist between p, q, r, s, n, x, y, z and w?

12. Apply the formula you conjectured in Exercise 11 to determine the probability of getting 5 A's, 3 B's, 1 C and 1 D if $P(A) = .4$, $P(B) = .3$, $P(C) = .2$ and $P(D) = .1$.

Computer Applications

13. Use Program 16: Trinomial Distribution Simulation to conduct your investigations of the trinomial distribution along the lines suggested in the text above.

Student Projects

For this project, you should conduct a series of ESP experiments and analyze the results statistically.

To do this, first you have to construct a deck of ESP test cards. This can be done most simply by using a set of 3 × 5 file cards or by cutting out 25 cards from several sheets of cardboard. Then draw each of the five shapes—circle, square, triangle, star, and wavy lines—onto five cards.

You should conduct the ESP test a minimum of 20 times using the same or different individuals for the test. If any individual subject scores exceptionally high or exceptionally low, you might want to repeat the test several more times with that person. Conduct the test in as careful a setting as possible. Try to avoid any distractions or any physical arrangement which might allow the subject to see the cards. In each instance, record the number of correct responses obtained by the subject.

Once you have collected these data, you should

1. Organize the set of data.
2. Construct an appropriate relative frequency distribution.

3. Construct the associated histogram, bar chart and/or frequency polygon.
4. Calculate (either by hand or by using Program 13: Binomial Probability) the probability of obtaining each number of successes that occurred.
5. Calculate the mean and standard deviation for the number of successes you obtained on all the tests.
6. Compare the values for the mean and standard deviation you calculated for your sample data to the theoretical values $\mu = 5$ and $\sigma = 2$ and discuss the comparison.

Finally, you should incorporate all these items into a formal statistical project report. The report should include

A statement of the topic being studied.

The source of the data.

A discussion of how the data were collected and why you believe that it is a random sample.

The tables and diagrams displaying the data.

All the pertinent calculations.

A discussion of any surprises you may have noted in connection with running the experiments or organizing and analyzing the data. In particular, identify any outcomes that are extremely unlikely.

Your conclusions regarding the existence of ESP based on your experimental results.

CHAPTER 5 SUMMARY

In any statistical study, the quantity being investigated is a **random variable**. A **discrete random variable** arises when a quantity is counted. A **continuous random variable** arises when a quantity is measured. In turn, they correspond to discrete populations and continuous populations, respectively.

A **probability distribution** gives the probabilities associated with all possible values of a discrete random variable. The sum of all the probabilities in a probability distribution must be 1. The mean for a probability distribution is

$$\mu = \Sigma x P(x)$$

and the standard deviation is

$$\sigma = \sqrt{\Sigma (x - \mu)^2 P(x)} = \sqrt{\Sigma x^2 P(x) - \mu^2}$$

A **binomial distribution** is based on a **binomial experiment** having n trials, two possible outcomes (success and failure) on each trial, and a fixed **probability of success** π on each of the independent trials. The mean of a binomial distribution is

$$\mu = n\pi$$

The standard deviation of a binomial distribution is

$$\sigma = \sqrt{n\pi(1-\pi)}$$

In the binomial distribution with n trials and probability of success π, the probability of obtaining exactly x successes is

$$P(x) = \frac{n!}{x!(n-x)!} \pi^x (1-\pi)^{n-x}$$

Review Exercises

1. Which of the following tables represent a probability distribution?

   a.
x	P(x)
0	.25
1	.35
2	.20
3	.30

   b.
x	P(x)
0	.20
1	.30
2	.30
3	.20

2. Determine the value of the probability p that makes the following a probability distribution:

x	P(x)
0	.15
1	p
2	.42
3	.22
4	.08

3. Find the mean and standard deviation of the probability distribution you completed in Review Exercise 2.

4. Suppose that 35% of all students at a college take statistics. If groups of 3 students are studied, construct the probability distribution for the number who take statistics.

5. A study finds that 16% of all teenagers in a certain area have driven before they are legally of age. What is the probability that in a randomly selected group of 8 teenagers in this area, none have driven while they were underage?

6. An FBI study reports that the crime rate in Omaha is 70 cases per 1000 residents, or 7%. Assume that this figure applies equally to all residents. If groups of 250

Omaha residents are considered, what are the mean and standard deviation of the associated binomial distribution?

*7. A child has a box containing 650 baseball cards, all mixed up. 54 of these cards are labeled as All-Stars. If she randomly picks 20 cards from the box, what is the probability that she gets 4 All-Stars?

*8. The employees in a company use an average of 8 sick days each year. Find the probability that a particular employee will be out 12 days during the year.

---

*Indicates a question based on an optional topic.

The Normal Distribution

Officials of the Motorola Corporation are calling on other leading companies in the United States to follow Motorola's lead in implementing a stringent policy of product quality control. Under its "Sigma Six" program, Motorola has drastically increased the quality of its electronics components such as computer chips by reducing the number of defective items produced. As a result, Motorola is one of the few U.S. companies that is able to compete effectively with electronics manufacturers in Japan and elsewhere in the Far East who have lower operating costs.

The Motorola executives hope that their success using statistical quality control procedures will be repeated at other U.S. firms. They stress the need for other companies to improve quality control or face the loss of business.

Motorola's executives believe that the only way the United States can compete effectively with manufacturers in other nations is by increasing the reliability of the products we produce. To do this requires the use of statistical methods to control the variability of the items produced. Motorola's "Sigma Six" program is intended to reduce the occurrence of defective chips and other electronic products to an incredible level of at most one flaw out of every million items produced.

The ideas involved are primarily statistical in nature. The term "Sigma Six" refers to six standard deviations. The frequency of defective items produced follows a bell-shaped distribution pattern known as the normal distribution having a mean μ and a standard deviation σ. Motorola's efforts are designed to minimize the variability in their production process to keep σ very small and to improve the level of the process so that the only defective items produced are more than six standard deviations away from the mean. Statistically, this is equivalent to producing no more than one defective item out of every million components manufactured.

In this chapter, we will see how many different data sets are distributed according to this bell-shaped curve and learn how to assess probabilities associated with a normal distribution and so make statistical decisions based on it.

6.1 Introduction to the Normal Distribution

When we introduced the idea of a random variable in Chapter 5, we saw that there are two types:

> 1. **Discrete random variables** typically arise from *counting* some type of outcome. They can take on only one of a countable set of possible values.
> 2. **Continuous random variables** typically arise from *measuring* some type of quantity. They can take on any of an infinite spectrum of possible values across an interval.

In turn, we saw that a discrete random variable corresponds to a discrete probability distribution, the most important of which is the binomial distribution. In a similar way, a continuous random variable corresponds to a *continuous distribution*. We examine the most important of these, the normal distribution, in this chapter.

First, let's put several additional ideas into perspective. One of the primary applications of the binomial distribution involves finding the probabilities in a binomial process. Thus, we developed techniques to determine the probability of getting x successes out of n trials. We visualize this by means of a histogram. Each box represents one of the possible outcomes $x = 0, 1, 2, \ldots, n$ successes.

> Since each box has width equal to 1 and height equal to the probability of that number of successes, the *area of each box* is precisely the same, numerically, as the *probability of success*.

Suppose we flip three coins. The corresponding histogram is shown in Figure 6.1. The second box corresponds to the outcome $x = 1$ head out of the three coins. The probability of this happening is $P(x = 1) = \frac{3}{8}$. Therefore, the height of the second box is $\frac{3}{8}$, and the area of this box is

$$\text{Base} \times \text{height} = 1 \times \tfrac{3}{8} = \tfrac{3}{8}$$

Thus, the area of the box and the probability of the outcome are numerically the same. In fact, the sum of the areas of all the boxes in the histogram is

$$\tfrac{1}{8} + \tfrac{3}{8} + \tfrac{3}{8} + \tfrac{1}{8} = \tfrac{8}{8} = 1$$

Further, if we want to find the probability of obtaining either 1 or 2 heads out of the three coins, then we can either add the individual probabilities $\tfrac{3}{8} + \tfrac{3}{8} = \tfrac{6}{8}$ or add the individual areas of the corresponding boxes $\tfrac{3}{8} + \tfrac{3}{8} = \tfrac{6}{8}$. These ideas have direct counterparts when we study continuous distributions.

FIGURE 6.1

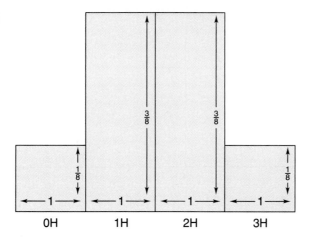

Suppose we now want to determine the probability of a particular outcome in a continuous distribution. The distribution might involve heights of individuals, say, and we wish to determine the probability that a randomly selected individual has height $x = 66$ inches. We interpret height as a quantity that can be measured to *any* conceivable degree of accuracy, so that infinitely many different heights are possible. Thus, the likelihood of a person being exactly any single height, say 66.0000000... inches or 71.5000000... inches, is zero. Since this argument applies to any specific height, we must conclude that the probability of being any precise height is zero. The same reasoning holds true for any other continuous random variable x.

> The probability that a continuous random variable assumes any single value will always be zero.

Instead of considering individual values for continuous random variables, we must consider intervals of values. With heights, we interpret 66 inches as representing a range of values, say, from 65.5 to 66.49 inches. We then consider the probability that the height of an individual falls within such an interval of values

$$P(65.5 < x < 66.49)$$

In general, for continuous random variables, we consider the probability that a measurement x falls between any two given values x_L and x_R

$$P(x_L < x < x_R)$$

By far, the most important continuous distribution is the *normal distribution* or *Gaussian distribution* having mean μ and standard deviation σ. The normal distribution is represented by the well-known *bell-shaped curve* shown in Figure 6.2. The essential properties of the normal distribution are as follows:

FIGURE 6.2

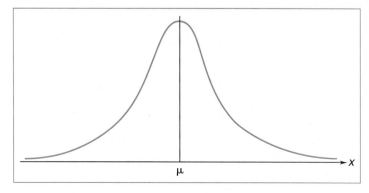

FIGURE 6.3

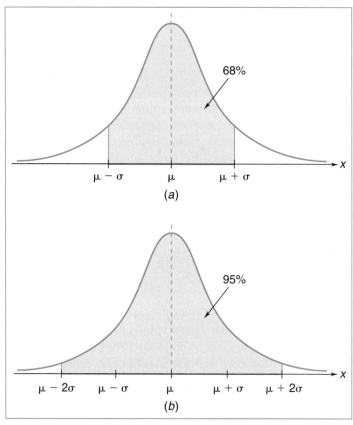

Properties of the Normal Distribution

1. It is centered at the mean μ.
2. It is symmetric about the mean (the left and right halves are mirror images of each other).
3. Approximately 68% of a normal population lies within 1 standard deviation (sd) of the mean. See Figure 6.3a.

SECTION 6.1 • Introduction to the Normal Distribution   235

4. Approximately 95% of a normal population lies within 2 standard deviations of the mean. See Figure 6.3b.
5. The total area under the normal distribution curve is 1.
6. The probability that a randomly selected member x of a normal population lies between two values x_L and x_R

$$P(x_L < x < x_R)$$

is precisely equal to the area under the normal curve between x_L and x_R.

FIGURE 6.4

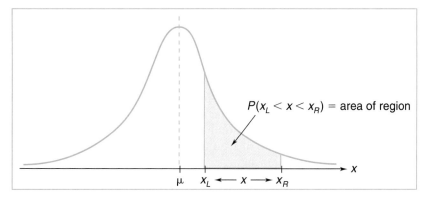

In many ways, the last property will be the most useful to us, although admittedly it makes the least sense initially. It states that there is a direct relationship between the probability that x falls between two values

$$P(x_L < x < x_R)$$

and the corresponding shaded area under the normal distribution curve, as shown in Figure 6.4. For instance, we might want to calculate the probability

FIGURE 6.5

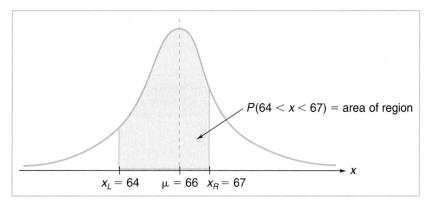

that a person's height is between 64 and 67 inches,

$$P(64 < x < 67)$$

This can be found by determining the corresponding shaded area under the normal distribution curve, as shown in Figure 6.5.

This approach is the precise analog to the corresponding ideas we discussed regarding the binomial distribution at the beginning of the section.

> There is a direct relationship between the probability of an event and the associated area. For the binomial distribution, the probability is equal to the area of one or several boxes in the corresponding histogram. For the normal distribution, the probability is equal to the area of a region under the normal distribution curve.[1]

We now illustrate the power of the normal distribution, using the information provided by the above six properties. The scores on one standard IQ test are normally distributed with mean $\mu = 100$ and standard deviation $\sigma = 20$. In Figure 6.6 we display the information guaranteed by these properties regarding this particular normal distribution.

Consider a randomly selected individual drawn from this population who has an IQ score of x on this test. From Figure 6.6, we can immediately conclude the following:

1. The probability that a randomly selected individual has an IQ between 80 and 120, or $P(80 < x < 120)$, is approximately .68. (The interval extends 1 standard deviation on each side of the mean $\mu = 100$.)
2. The probability that a randomly selected individual has an IQ between 100 and 120, or $P(100 < x < 120)$, is approximately .34. (Because of the symmetry of the normal distribution, one-half of the 68%, or 34%, of the population lies on each side of the mean $\mu = 100$).
3. The probability that a randomly selected individual has an IQ between 80 and 100, or $P(80 < x < 100)$, is approximately .34.
4. The probability that a randomly selected individual has an IQ between 60 and 140, or $P(60 < x < 140)$, is approximately .95. (The interval extends 2 standard deviations on each side of the mean.)

---

[1]Those of you who have studied calculus will recognize the problem of finding the area under a curve between two given values as one of the fundamental ideas in calculus. The answer is given by the definite integral of the function. In this case, the equation for the normal distribution curve is

$$f(x) = \frac{1}{\sigma\sqrt{2\pi}} \exp\left[-\frac{1}{2}\left(\frac{x-\mu}{\sigma}\right)^2\right]$$

While this function cannot be integrated in closed form, there are many effective ways to approximate the value of a definite integral to any desired degree of accuracy, and these are employed routinely to obtain the values we will use.

FIGURE 6.6

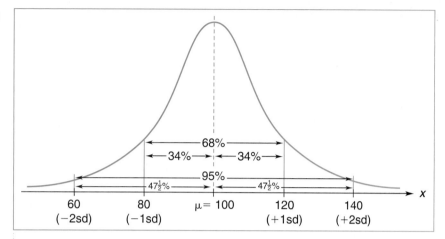

5. The probability that a randomly selected individual has an IQ between 100 and 140, or $P(100 < x < 140)$, is approximately .475. (Because of the symmetry of the normal distribution, one-half of the 95%, or 47.5%, of the population lies on each side of the mean.)
6. The probability that a randomly selected individual has an IQ between 60 and 100, or $P(60 < x < 100)$, is approximately .475.
7. The probability that a randomly selected individual has an IQ above 100, or $P(x > 100)$, is .5. (Since the mean is at the center of the normal distribution, one-half of the total population lies on each side of it.)
8. The probability that a randomly selected individual has an IQ below 100, or $P(x < 100)$, is .5.

We strongly suggest that you not continue until you are absolutely certain you see where each of these results comes from. We will be using these results in the following examples based on the same situation of a normal distribution of IQ scores having mean $\mu = 100$ and standard deviation $\sigma = 20$.

EXAMPLE 6.1 What is the probability that a randomly selected individual has an IQ over 140?

SOLUTION We begin with the following observation. The total area under the normal curve is 1 and the curve is symmetric about the mean, so that the area under the normal curve on each side of the mean is .5. Further, an IQ over 140 represents a score more than 2 standard deviations *above* the mean μ. From Figure 6.6 and item 5 above, we know that

$$P(100 < x < 140) \approx .475$$

Therefore,

$$P(x > 140) \approx .500 - .475 = .025$$

as shown in Figure 6.7.

FIGURE 6.7

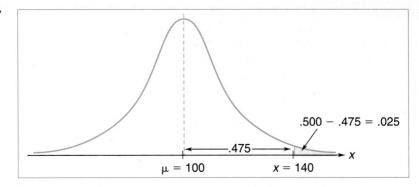

Alternatively, since 50% of the normal population lies above the mean $\mu = 100$ and approximately $47\frac{1}{2}\%$ of the population lies between 100 and 140, only about

$$50\% - 47\frac{1}{2}\% = 2\frac{1}{2}\%$$

of the population can fall above 140.

EXAMPLE 6.2 What is the probability that a randomly selected individual has an IQ between 60 and 120?

SOLUTION A score of 60 represents 2 standard deviations *below* the mean while a score of 120 represents 1 standard deviation *above* the mean. We know that the probability that x falls between 60 and 100, or $P(60 < x < 100)$, is approximately .475. The probability that x is between 100 and 120, or $P(100 < x < 120)$, is approximately .34. Consequently, from an examination of Figure 6.8, we see that we must *add* the two probabilities (equivalently, the two areas) and so obtain

$$P(60 < x < 120) \approx .475 + .34 = .815$$

FIGURE 6.8

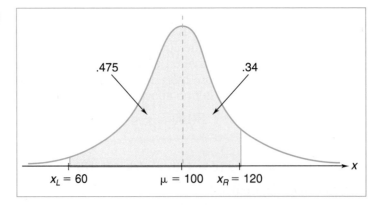

FIGURE 6.9

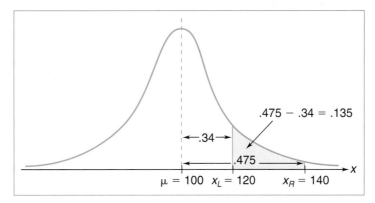

EXAMPLE 6.3 What is the probability that a randomly selected individual has an IQ score between 120 and 140?

SOLUTION As shown in Figure 6.9, we seek the area of the portion of the graph between 120 and 140. We know that approximately 34% of the population lies between 100 and 120 [$P(100 < x < 120) \approx .34$]. Approximately 47.5% of the population lies between 100 and 140 [$P(100 < x < 140) \approx .475$]. From Figure 6.9, we see that the desired area is simply the area from 100 to 140 *minus* the area from 100 to 120, so that

$$P(120 < x < 140) \approx .475 - .34 = .135$$

■

The above examples are typical of all problems involving normal distribution probabilities. In fact, there are really only five distinct cases:

1. The value is between the mean μ and some value on one side only. (See Figures 6.10a and b.)
2. The value x is beyond some value on one side only. (See Figure 6.10c and d.) In this case, you must *subtract* the known probability from .5.
3. The value x is between two values on either side of the mean μ. (See Figure 6.10e.) In this case, you must *add* the respective probabilities.
4. The value x is between two values on the same side of the mean μ. (See Figures 6.10f and g.) In this case, you must *subtract* the respective probabilities.
5. The value x is above some value to the left of the mean μ, or x is below some value to the right of the mean μ. (See Figure 6.10h and i.) In this case, you must *add* the known probability to .5.

We now consider an example illustrating all these possibilities.

EXAMPLE 6.4 Suppose that in a certain population group serum cholesterol levels are normally distributed with mean 160 and standard deviation 25. Find the approximate probability that the cholesterol level for a randomly selected individual from this group is

(a) Between 160 and 185,
(b) Between 110 and 160,
(c) Above 210,
(d) Between 135 and 210,
(e) Between 110 and 135,
(f) Above 135.

FIGURE 6.10

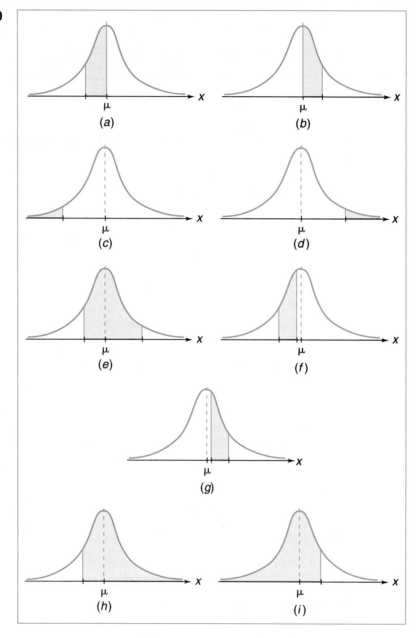

SOLUTION For parts *a* and *b*, since the mean $\mu = 160$ and the standard deviation $\sigma = 25$, we know that approximately 68% of the population has cholesterol levels between 135 and 185 and that about 95% of the population has levels between 110 and 210. Consequently, from Figure 6.11*a* and *b*, it is clear that

$$P(160 < x < 185) \approx .34$$

and that

$$P(110 < x < 160) \approx .475$$

For part *c*, we see from Figure 6.11*c* that we must subtract the area between the mean $\mu = 160$ and the right-hand value $x = 210$, or .475, from the total area on one side of the mean, .5. Thus,

$$P(x > 210) \approx .500 - .475 = .025$$

(Notice that when you subtract probabilities or areas, as we did here, the result must come out *positive* since all probabilities and all areas are positive.)

For part *d*, we consider Figure 6.11*d* and observe that 135 is 1 standard deviation below the mean $\mu = 160$ while 210 is 2 standard deviations above it. Therefore, we must add the two areas to obtain

$$P(135 < x < 210) \approx .34 + .475 = .815$$

For part *e*, we examine Figure 6.11*e* and observe that 110 is 2 standard deviations below the mean and 135 is 1 standard deviation below it. Consequently, we must subtract the two areas to obtain

$$P(110 < x < 135) \approx .475 - .34 = .135$$

For part *f*, we note that 135 is 1 standard deviation below the mean. As shown in Figure 6.11*f*, we must include the 50% of the population above the mean as well as the 34% of the population between 135 and the mean $\mu = 160$. Therefore,

$$P(x > 135) \approx .34 + .50 = .84$$ ■

Let's now consider one more problem involving normal distribution probabilities which is also based on the IQ scores with mean $\mu = 100$ and standard deviation $\sigma = 20$ used earlier in this section: What is the probability that a randomly selected individual has an IQ between 100 and 130?

We cannot answer this question with the information at our disposal. All the previous questions involved values that were either 1 or 2 standard deviations away from the mean μ. However, a score of 130 represents one and a half standard deviations above the mean $\mu = 100$, and we do not know the value for the area under the normal curve corresponding to this. But if this value were available, then the problem would be easy to solve. Consequently, we need to develop a more detailed approach to such problems that allows us to handle cases involving other than simply 1 or 2 standard deviations. We do this in the next section.

FIGURE 6.11

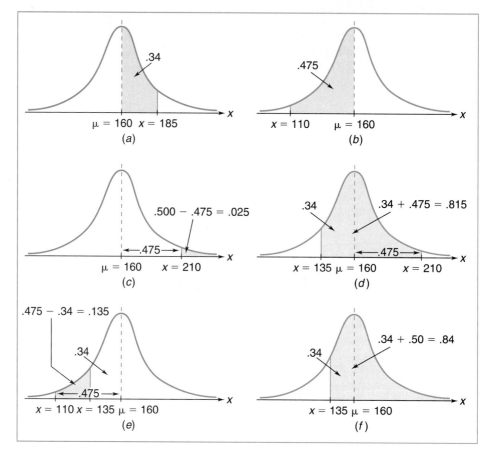

Finally, it is important to realize that when we speak of a population, we do not usually mean a population of people. In particular, the population in question consists of the set of numbers or measurements for the quantity under consideration. Thus, when we discuss IQ scores, the population consists of the IQ scores themselves, not the people whose scores are measured. Similarly, the population of cholesterol levels consists of the cholesterol values themselves, not the people who have these levels.

Exercise Set 6.1

Mastering the Techniques

Given a normally distributed population with mean $\mu = 10$ and standard deviation $\sigma = 2$, find the following probabilities by first drawing an appropriate diagram.

1. $P(x > 10)$
2. $P(10 < x < 12)$
3. $P(10 < x < 14)$
4. $P(8 < x < 10)$

5. $P(8 < x < 14)$
6. $P(6 < x < 12)$
7. $P(x > 12)$
8. $P(x > 14)$
9. $P(x < 8)$
10. $P(x < 6)$
11. $P(12 < x < 14)$
12. $P(6 < x < 8)$
13. $P(x > 6)$
14. $P(x < 12)$

Applying the Concepts

The mathematics SAT scores for the high school seniors in a particular city are normally distributed with mean 450 and standard deviation 100. Illustrate Exercises 15 to 20 with appropriate diagrams and answer the questions.

15. What is the probability that a randomly selected student has an SAT score between 450 and 550?
16. What is the probability that a randomly selected student has an SAT score between 450 and 650?
17. What is the probability that a randomly selected student has an SAT score between 350 and 650?
18. What is the probability that a randomly selected student has an SAT score above 550?
19. What is the probability that a randomly selected student has an SAT score between 250 and 350?
20. What is the probability that a randomly selected student has an SAT score below 250?

A bank finds that its savings account balances are approximately normally distributed with mean $600 and standard deviation $150. Illustrate Exercises 21 to 28 with appropriate diagrams and answer the questions.

21. What is the probability that a randomly selected account has a balance between $450 and $600?
22. What is the probability that a randomly selected account has a balance between $600 and $900?
23. What is the probability that a randomly selected account has a balance above $750?
24. What is the probability that a randomly selected account has a balance between $450 and $750?
25. What is the probability that a randomly selected account has a balance between $300 and $750?
26. What percentage of all accounts have a balance between $450 and $900?
27. What percentage of all accounts have a balance below $300?
28. What percentage of all accounts have a balance above $450?

A study shows that the cost of new hardcover textbooks is approximately normally distributed with mean $42 and standard deviation $6.50. Illustrate Exercises 29 to 34 with an appropriate diagram, and answer the questions.

244 CHAPTER 6 • The Normal Distribution

29. What is the probability that a randomly selected text costs between $35.50 and $48.50?
30. What is the probability that a randomly selected text costs between $29 and $35.50?
31. What is the probability that a randomly selected text costs more than $55?
32. What is the probability that a randomly selected text costs between $29 and $35.50?
33. What percentage of all textbooks cost more than $55?
34. What percentage of all textbooks cost between $35.50 and $55?

6.2 Applications of the Normal Distribution

In the last section, we saw that a wide variety of useful questions can be answered regarding any set of normally distributed data as long as we focus our attention on intervals which extend either 1 or 2 standard deviations from the mean μ. To do this, we used the rough estimates for the area under the normal curve—the fact that approximately 68% of the population lies within 1 standard deviation of the mean and approximately 95% lies within 2 standard deviations of the mean.

In this section, we extend these ideas considerably to develop totally parallel ways to handle situations involving any number of standard deviations (not just 1 and 2). At the same time, we refine the accuracy of our work so that our answers are exact rather than approximate. To do this, we use the fact that the number of standard deviations that a given measurement x lies away from the mean μ is precisely the associated z value that we introduced in Section 3.4. Recall that

$$z = \frac{x - \mu}{\sigma}$$

In the discussion at the very end of Section 6.1, we were unable to find the probability $P(100 < x < 130)$ based on a normal distribution with $\mu = 100$ and $\sigma = 20$ because the value 130 lies 1.5 standard deviations above the mean. That is, the z value associated with $x = 130$ is

$$z = \frac{x - \mu}{\sigma} = \frac{130 - 100}{20} = \frac{30}{20} = 1.5$$

However, if the corresponding value for the area under the normal curve were available, then the problem could be easily solved.

Since problems of this type are important in so many different disciplines in modern life, we must be able to find, easily, the areas of many different regions under the normal distribution curve. However, every possible combination of μ and σ produces a different normal distribution. It is clearly impossible to have values available for every conceivable case since there are an infinite number of them. Instead, we use the z value to transform *any*

FIGURE 6.12

z	.00	.01	.02	.03	.04	.05	.06	.07	.08	.09	z
.0	.0000	.0040	.0080	.0120	.0160	.0199	.0239	.0279	.0319	.0359	.0
.1	.0398	.0438	.0478	.0517	.0557	.0596	.0636	.0675	.0714	.0753	.1
.2	.0793	.0832	.0871	.0910	.0948	.0987	.1026	.1064	.1103	.1141	.2
.3	.1179	.1217	.1255	.1293	.1331	.1368	.1406	.1443	.1480	.1517	.3
.4	.1554	.1591	.1628	.1664	.1700	.1736	.1772	.1808	.1844	.1879	.4
.5	.1915	.1950	.1985	.2019	.2054	.2088	.2123	.2157	.2190	.2224	.5
.6	.2257	.2291	.2324	.2357	.2389	.2422	.2454	.2486	.2517	.2549	.6
.7	.2580	.2611	.2642	.2673	.2704	.2734	.2764	.2794	.2823	.2852	.7
.8	.2881	.2910	.2939	.2967	.2995	.3023	.3051	.3078	.3106	.3133	.8
.9	.3159	.3186	.3212	.3238	.3264	.3289	.3315	.3340	.3365	.3389	.9
1.0	.3413	.3438	.3461	.3485	.3508	.3531	.3554	.3577	.3599	.3621	1.0
1.1	.3643	.3665	.3686	.3708	.3729	.3749	.3770	.3790	.3810	.3830	1.1
1.2	.3849	.3869	.3888	.3907	.3925	.3944	.3962	.3980	.3997	.4015	1.2
1.3	.4032	.4049	.4066	.4082	.4099	.4115	.4131	.4147	.4162	.4177	1.3
1.4	.4192	.4207	.4222	.4236	.4251	.4265	.4279	.4292	.4306	.4319	1.4
1.5	.4332	.4345	.4357	.4370	.4382	.4394	.4406	.4418	.4429	.4441	1.5

normal distribution with mean μ and standard deviation σ into a single normal distribution having mean 0 and standard deviation 1; it is known as the *standard normal distribution*. Statisticians have computed the areas under this standard normal curve corresponding to all values of z calculated to two decimal places. These values are widely available in tables such as Table III in the appendix and reproduced for your convenience on the inside front cover. A portion of this table is reproduced here in Figure 6.12.

In this normal distribution table, observe that the left-hand and right-hand columns contain values for $z = 0, .1, .2, \ldots, 3.0$. The entries in the body of the table represent the area under the corresponding portion of the normal curve from the center at $z = 0$ (which is equivalent to $x = \mu$) to the desired number of standard deviations z. This is the shaded region both at the top of the normal distribution table and in Figure 6.13.

FIGURE 6.13

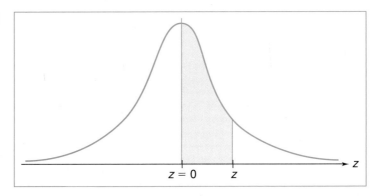

If we move down the left column of the table to $z = 1.0$, we see that the associated value for the area is .3413. Notice that $z = 1.0$ corresponds to 1 standard deviation above the mean μ, and we have previously used a value of *approximately* 34% = .34 as the associated area. Similarly, from the table, we see that the area corresponding to $z = 2.0$ is .4772 compared to the approximate value of .475 we have used until now. These values from the table are correct to four decimal places and are considered accurate for all standard statistical applications. Consequently, from this point on, we will work exclusively with the values from the normal distribution table, so that we no longer have to use the words *approximately equal to*.

Furthermore, from the table, we see that the value corresponding to $z = 1.5$ is .4332. We can now easily answer the question of finding the probability that a randomly selected individual has an IQ between 100 and 130. This is

$$P(100 < x < 130)$$

which is equivalent to

$$P(0 < z < 1.5)$$

which is .4332. See Figure 6.14.

If you examine the normal distribution table, you will certainly notice the row of entries across the top, .00, .01, .02, ..., .09. These entries represent the second decimal place in a z value. Thus, to find the area from the mean μ ($z = 0$) at the center of the normal distribution to an x value which lies 1.25 standard deviations above the mean ($z = 1.25$), we move down the left column of the table until we reach $z = 1.2$ and then move across that row to the column headed by .05—the appropriate entry for the area is .3944. Similarly, the area corresponding to $z = .83$ is .2967. The area corresponding to $z = 1.07$ is .3577, and so forth.

CAUTION Be certain that you see where all these entries come from and what they represent before you read on.

FIGURE 6.14

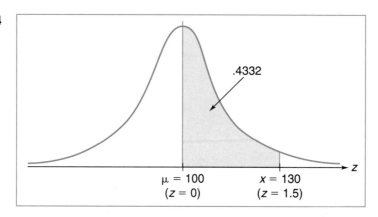

With the normal distribution table at our disposal, we can handle with ease and with complete accuracy all the types of problems we discussed in Section 6.1. We are no longer restricted to cases involving precisely 1 or 2 standard deviations. First we illustrate this capability in terms of probability problems involving z values only. After that, we will see how problems involving any desired normal distribution can be treated by first calculating the appropriate z value which is equivalent to the associated number of standard deviations from the mean.

As in the examples in the last section, you must always include a diagram with each situation.

EXAMPLE 6.5 Find the following probabilities and illustrate each case with a diagram:

(a) $P(0 < z < 1.25)$,
(b) $P(z > 1.43)$,
(c) $P(-.87 < z < 0)$,
(d) $P(z < -.57)$,
(e) $P(-2.17 < z < 1.08)$,
(f) $P(.64 < z < 2.11)$.

SOLUTION For part a, we simply consult the normal distribution table for the area corresponding to $z = 1.25$ and so conclude that

$$P(0 < z < 1.25) = .3944$$

as shown in Figure 6.15a.

For part b, we see that the table entry corresponding to $z = 1.43$ is .4236, and this represents the area under the curve from the center at $z = 0$ to the desired point at $z = 1.43$, as shown in Figure 6.15b. Consequently, since the total area on each side of the center in a normal distribution is .5, we see that the area beyond $z = 1.43$ is given by $.5 - .4236 = .0764$. Since the associated probability is equal to the area under the curve,

$$P(z > 1.43) = .0764$$

For part c, we first notice that there are no negative values for z listed in the table. However, we do not need such values tabulated explicitly since the normal distribution is symmetric about the mean: whatever happens on the left side is precisely mirrored on the right side. Therefore, the area under the curve between $z = -.87$ and the mean at $z = 0$ is exactly the same as that between $z = 0$ and $z = .87$, as shown in Figure 6.15c. Consequently,

$$P(-.87 < z < 0) = P(0 < z < .87) = .3078$$

For part d, we first observe that the area between the mean at $z = 0$ and $z = -.57$ is equal to the area between $z = 0$ and $z = .57$ and so is .2157, as we see from Figure 6.15d. Therefore, the area beyond $z = -.57$ is

$$P(z < -.57) = P(z > .57) = .5 - .2157 = .2843$$

FIGURE 6.15

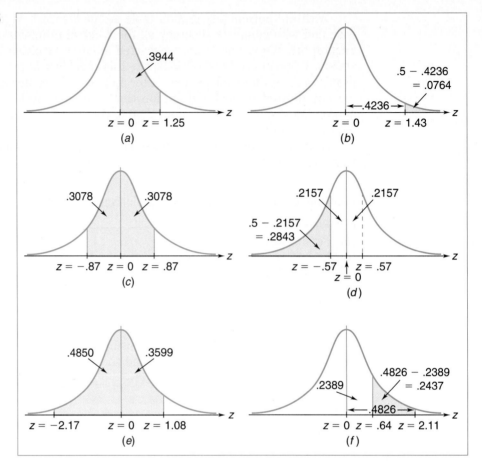

For part *e*, we first examine Figure 6.15*e*. We see that we must consider two regions under the normal curve, one between $z = -2.17$ and the mean $z = 0$ and the other between the mean $z = 0$ and $z = 1.08$. For the left side, we find that the area is .4850; for the right side it is .3599. Consequently,

$$P(-2.17 < z < 1.08) = .4850 + .3599 = .8449$$

For part *f*, we first examine Figure 6.15*f*. We have two regions on the same side of the mean. Corresponding to $z = .64$, we find that the associated area is .2389, while corresponding to $z = 2.11$, the associated area is .4826. From the diagram, it is clear that the final answer is the difference between these two quantities, so that

$$P(.64 < z < 2.11) = .4826 - .2389 = .2437$$

While Example 6.5 illustrates how we can utilize the normal distribution table directly to calculate probabilities involving values for *z*, this is usually only an intermediate stage. It is far more typical to encounter a question about probability relating to some measurement x which is normally distributed with mean μ and standard deviation σ. We then convert the x value to the equivalent z value and read off the appropriate entry from the table. Depending on the particular probability desired, we perform one of the types of analyses described in the list on p. 239 Section 6.1.

For instance, suppose we return once more to the situation involving IQ scores which are normally distributed with mean $\mu = 100$ and standard deviation $\sigma = 20$. Suppose we want to find the probability that an individual's IQ score is between 100 and 145

$$P(100 < x < 145)$$

We first find that the associated z value corresponding to $x = 145$ is

$$z = \frac{145 - 100}{20} = \frac{45}{20} = 2.25$$

and so, as shown in Figure 6.16,

$$P(100 < x < 145) = P(0 < z < 2.25) = .4878$$

EXAMPLE 6.6 The hourly total sales at each register in a supermarket are normally distributed with mean $\mu = \$650$ and standard deviation $\sigma = \$60$. Find the probability that during a randomly selected hour, one of the registers totals (*a*) between $650 and $750, (*b*) over $800, and (*c*) between $600 and $650.

SOLUTION For part *a*, we first note that the lower value $x = 650$ is the mean, so that the corresponding z value is $z = 0$. Further, corresponding to $x = 750$, we find that

$$z = \frac{750 - 650}{60} = \frac{100}{60} = 1.67$$

FIGURE 6.16

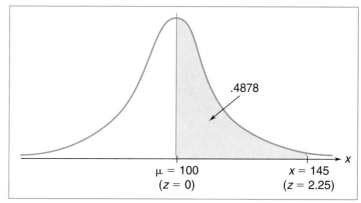

FIGURE 6.17

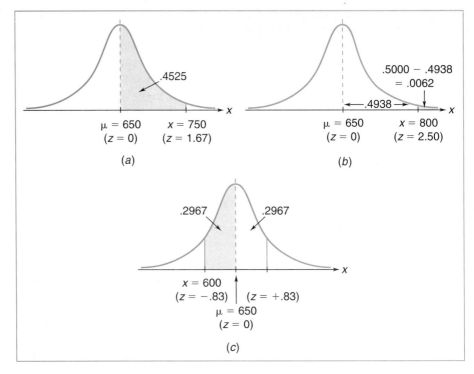

and so, as shown in Figure 6.17a,

$$P(650 < x < 750) = P(0 < z < 1.67) = .4525$$

For part b, we first find the z value corresponding to x = 800 as

$$z = \frac{800 - 650}{60} = \frac{150}{60} = 2.50$$

so that, as shown in Figure 6.17b,

$$P(x > 800) = P(z > 2.50) = .5 - .4938 = .0062$$

For part c, the z value corresponding to x = 600 is

$$z = \frac{600 - 650}{60} = \frac{-50}{60} = -.833333 \approx -.83$$

Since the area under the normal curve between $z = -.83$ and the mean $z = 0$ is exactly the same as that between $z = 0$ and $z = +.83$, as shown in Figure 6.17c, it follows that

$$P(600 < x < 650) = P(-.83 < z < 0) = .2967$$

FIGURE 6.18

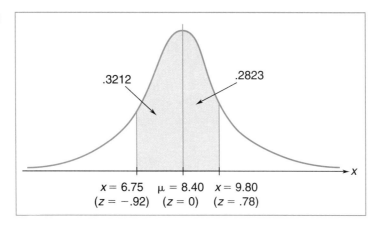

EXAMPLE 6.7 The hourly wages paid to the employees of a large corporation are normally distributed with mean and standard deviation of $8.40 and $1.80, respectively. What percentage of the workers earn between $6.75 and $9.80 per hour?

SOLUTION We must find

$$P(6.75 < x < 9.80)$$

as illustrated in Figure 6.18. The z value corresponding to $x = 9.80$ is

$$z = \frac{9.80 - 8.40}{1.80} = \frac{1.40}{1.80} = .78$$

and so the area to the right of the mean is .2823. Similarly, the z value corresponding to $x = 6.75$ is

$$z = \frac{6.75 - 8.40}{1.80} = \frac{-1.65}{1.80} = -.92$$

and the area to the left of the mean is .3212. We add the two areas to get

$$P(6.75 < x < 9.80) = P(-.92 < z < .78) = .3212 + .2823 = .6035$$

so that slightly over 60% of the workers earn between $6.75 and $9.80 an hour. ∎

EXAMPLE 6.8 A study finds that the time spent on commercials per hour on a certain radio station is approximately normally distributed with mean $\mu = 12.8$ minutes and standard deviation $\sigma = 2.4$ minutes. During a randomly selected hour, what is the probability that between 14 and 16 minutes were devoted to commercials?

SOLUTION We must determine $P(14 < x < 16)$.

FIGURE 6.19

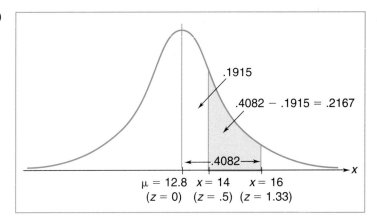

Corresponding to $x = 16$, we find that

$$z = \frac{16 - 12.8}{2.4} = \frac{3.2}{2.4} = 1.33$$

and so the area under the normal curve from the center to $x = 16$ is .4082. Similarly, corresponding to $x = 14$, we find that

$$z = \frac{14 - 12.8}{2.4} = .50$$

and so the area under the curve from the center to $x = 14$ is .1915. From Figure 6.19, we see that it is necessary to take the difference between these two values, hence

$$P(14 < x < 16) = P(.50 < z < 1.33)$$
$$= .4082 - .1915 = .2167 \qquad \blacksquare$$

In each of these examples, we were given values for the measurements x and then had to find the associated probabilities. Occasionally, we are given the probability and asked to determine the value of x corresponding to it. To do this, we must reverse the above procedure and use the formula

$$x = \mu + z\sigma$$

We illustrate this approach in the following examples.

EXAMPLE 6.9 Suppose that speeds on a highway are normally distributed with mean $\mu = 52$ miles per hour and standard deviation $\sigma = 6$ miles per hour. If the police will stop and ticket anyone going in the fastest 1%, what is the fastest speed someone can drive without being stopped?

FIGURE 6.20

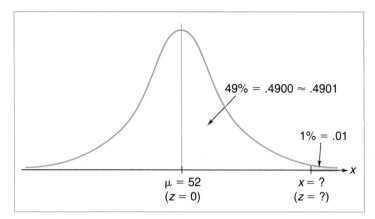

SOLUTION To solve this problem, we must identify the point where the fastest 1% of the population begins. Thus, we have a normal distribution centered at $\mu = 52$, and the portion of the population in question is shown in Figure 6.20. The problem is to find the value of x indicated.

If 1% of the population is to the right of this point, then 49% of it lies between the mean μ and the point. So we need to determine the z value corresponding to 49% of the population. To do this, we find the entry in the normal distribution table closest to .4900, and this is .4901; this corresponds to $z = 2.33$. As a consequence, the above formula yields

$$x = 52 + 2.33(6) = 52 + 13.98 = 65.98$$

Therefore, we conclude that anyone driving over 66 miles per hour will be stopped for speeding. ∎

EXAMPLE 6.10 State Education Department guidelines require that students who score in the bottom 15% on a standardized test be given special remediation services. If the test scores on this examination are normally distributed with mean $\mu = 136$ and standard deviation $\sigma = 27$, find the cutoff value below which a student should be identified for extra academic assistance.

SOLUTION To identify the cutoff for the bottom 15%, as shown in Figure 6.21, we immediately see that 35% of the population lies between this point and the mean μ. From the table, we find that the entry closest to .3500 is .3508, which corresponds to a z value of 1.04. However, since we are concerned with the *bottom* portion of this population which is *below* the mean, we must use $z = -1.04$. Consequently, the desired cutoff is at

$$\begin{aligned} x &= \mu + z\sigma \\ &= 136 + (-1.04)(27) \\ &= 136 - 28.08 = 107.92 \end{aligned}$$

Any student who scores below 108 should be given the additional help. ∎

FIGURE 6.21

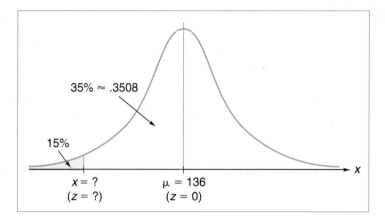

Computer Corner

The calculations in the above examples can be handled very simply by using a computer. Moreover, the computer has the additional advantage of being able to guide you through the steps necessary to solve such problems. When you select Program 17: Normal Distribution Probability, you will see these capabilities. The program first asks you for the mean and standard deviation of the normal population with which you are working. It then asks for the left-hand value of x and the right-hand value of x that you desire. For instance, you might enter 100 and 20 for the mean and standard deviation, respectively, and 88 and 111 for the two values of x. Incidentally, if you are interested in values of x beyond 123, say, then you should enter 123 as the left-hand value and any value of x more than 3 standard deviations to the right of the mean, say 200.

Finally, the program asks if you want to do some of the calculations by hand. If you answer NO, then it proceeds directly to a graphics display. Otherwise, it asks you to calculate and type in the value for z corresponding to each value of x you supplied. The program then proceeds to the graphics display. It then asks you for the appropriate probability or equivalently the corresponding area under the normal curve that you must calculate by hand using the normal distribution table. The program calculates this probability and compares your answer to the correct one. If you made an error in logic, you will be given two more chances to come up with the correct answer. After your third wrong try, the computer will show you the correct answer. If you are having trouble doing normal distribution probability problems by hand, you should use this feature as a tutorial to identify the types of mistakes you are making and to improve your skills.

The graphics display for this program starts with the graph of the normal distribution curve. Vertical lines are drawn corresponding to the values of x you supply, and information regarding the corresponding *z* values is displayed. The indicated region under the curve is colored in, and finally the correct value for the probability is calculated and displayed. After the graphics page is completed, you can transfer to an accompanying text screen for additional detail by pressing any key. From the text screen, you can go to a menu screen by pressing the space bar. You will then have the choice of changing the *x* values, changing the parameters μ and σ, or other possibilities.

You may also want to experiment with Program 18: Normal Distribution Simulation. This program has the same input as Program 17—the mean, the standard deviation, and the values of *x* desired. However, it is designed to simulate the normal distribution by randomly generating a series of normally distributed points, plotting each under the graph of the normal distribution curve, and keeping track of how many points (and hence what percentage of points) fall inside the indicated region. See Figure 6.22 for typical output. Before the simulation process begins, you must indicate the number of random points you want generated. We suggest starting with several hundred. Obviously, the accuracy of the simulation increases as the number of random points increases, but that also takes more time. The maximum number of points you should select is around 2000 or 2500, or else the points graphed will rise too high on the screen for an effective display.

FIGURE 6.22*a*

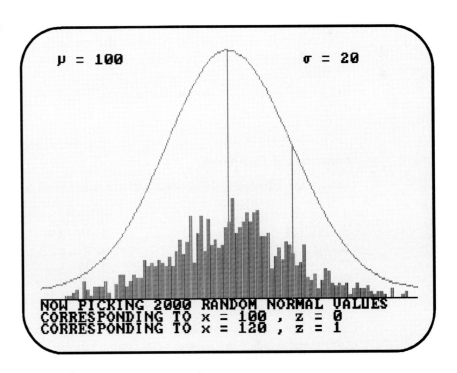

FIGURE 6.22b

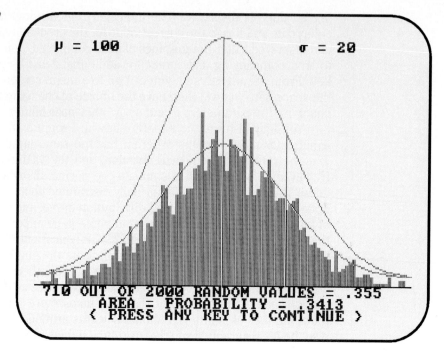

You may want to try this program using some of the better known sets of values. For instance, you may set up an interval of length 1 standard deviation on both sides of the mean to see how close the simulation comes to reproducing the theoretical probability value of .6826.

Exercise Set 6.2

Mastering the Techniques

Calculate the following probabilities involving z values. Draw a diagram in each case.

1. $P(0 < z < .53)$
2. $P(0 < z < 2.06)$
3. $P(-1.28 < z < 0)$
4. $P(-2.56 < z < 0)$
5. $P(z > 1.63)$
6. $P(z > 2.24)$
7. $P(-.82 < z < .41)$
8. $P(-1.31 < z < 2.73)$
9. $P(.64 < z < 2.71)$
10. $P(1.48 < z < 2.15)$
11. $P(-.96 < z < -.08)$
12. $P(-2.21 < z < -1.39)$

Calculate the following probabilities based on a normal distribution having mean $\mu = 100$ and standard deviation $\sigma = 15$. Draw a diagram in each case.

13. $P(100 < x < 105)$
14. $P(100 < x < 128)$
15. $P(74 < x < 100)$
16. $P(86 < x < 100)$
17. $P(96 < x < 108)$
18. $P(82 < x < 111)$
19. $P(x > 135)$
20. $P(x < 80)$
21. $P(x > 90)$
22. $P(x < 125)$
23. $P(112 < x < 128)$
24. $P(60 < x < 80)$

Applying the Concepts

25. A bank finds that the balances in its savings accounts are normally distributed with mean $\mu = \$800$ and standard deviation $\sigma = \$150$. Find the probability that a randomly selected account has a balance under $400.
26. A county government finds that the daily amounts that it receives in payment for parking tickets are normally distributed with mean and standard deviation $1400 and $320, respectively. What is the probability that the amount collected on a randomly chosen day is between $1000 and $1800? What is the probability that the amount collected is more than $2000?
27. A video rental store finds that the number of its daily rentals is normally distributed with mean $\mu = 187$ tapes and standard deviation $\sigma = 34$ tapes. What is the probability that the store rents more than 200 tapes on a given day?
28. A brand of tires lasts an average of 32,000 miles with a standard deviation of 5400 miles. If the lifetime of such tires is normally distributed, what is the probability that a randomly selected tire lasts more than 40,000 miles?
29. The bottling operation for a certain brand of soda is set up in such a way that the mean contents for a supposed 2-liter bottle is actually 2.08 liters with a standard deviation of .06 liter. Suppose that the contents of such bottles are normally distributed. What is the probability that a randomly selected bottle contains less than 2 liters? What is the probability that a bottle contains more than 2.1 liters?
30. A survey indicates that new car prices have mean $15,225 with a standard deviation of $3600. If such prices are approximately normally distributed, what percentage of all new cars will sell for between $13,000 and $17,000? What percentage will sell for more than $20,000?
31. A member of a college track team finds that his mean time for running a 100-yard dash is 10.8 seconds with a standard deviation of .3 second. If his times are normally distributed, what is the probability that he will run such a race in under 10.2 seconds?
32. A waiter finds that his average tip from a party of two diners is $5.40 with a standard deviation of $2.10. If the tips are normally distributed, what is the probability that he receives between $6.00 and $7.00 from a randomly selected couple?
33. A professor administers a midterm examination in which the grades are normally distributed with mean 42 and standard deviation 18. If she plans to curve the grades in such a way that the top 20% of students receive A's, what is the lowest test grade that will earn an A?
34. The same professor wants the bottom 12% to receive F's. What is the lowest passing grade on her examination?

Computer Applications

35. Use Program 18: Normal Distribution Simulation with $\mu = 100$, $\sigma = 20$, $x_L = 100$, $x_R = 120$, and 1000 points. Run the program six different times with these values, and record the percentage of random points that fall into the given interval each time. Calculate the mean and standard deviation for these results, and compare the individual values and their mean to the table entry.
36. Repeat the procedure in Exercise 35 with other sets of values of your choice.

6.3 The Normal Approximation to the Binomial Distribution

We turn our attention once more to the binomial distribution that we first treated in Section 5.2. Recall that this distribution is based on n independent trials of a particular experiment having two possible outcomes, success or failure. The probability of success π is fixed from one trial to the next, and the probability of failure is $1 - \pi$. We visualize a binomial distribution as a histogram having $n + 1$ boxes representing 0 successes, 1 success, 2 successes, ..., n successes. Furthermore, a binomial distribution has a mean and a standard deviation, respectively, given by

$$\mu = n\pi$$
$$\sigma = \sqrt{n\pi(1 - \pi)}$$

A binomial probability problem typically asks for the probability of obtaining x successes out of n trials in such an experiment. This can be found by using the binomial probability formula from Section 5.3 or by referring to Table II of binomial probabilities. Each of these approaches has significant drawbacks, especially if n is large (the table may not go far enough or the formula becomes unrealistically complicated) or if π is not a relatively simple decimal (the table contains only a few selected values for π). These difficulties are compounded if we want a range of successes, say, the probability of getting between 200 and 215 successes out of 300 trials when the probability of success $\pi = .637$.

To find a simple method for solving such problems, let's first consider the shape of a variety of binomial distributions. In Figure 6.23, we display a series of computer-generated histograms for a group of binomial distributions all having probability $\pi = .5$, but with $n = 5$, $n = 10$, $n = 15$, $n = 20$, $n = 25$, $n = 30$, $n = 40$, $n = 50$, $n = 75$, and $n = 100$ trials. As the number of trials increases, the individual boxes that make up the histogram become less distinct and the overriding shape of the distribution becomes more evident. In particular, the shape of the binomial distribution looks more and more like a normal distribution in general outline as the number of trials increases.

As a consequence, we will compare the normal and binomial distributions using Program 19: Normal Approximation to Binomial. The program draws

FIGURE 6.23a

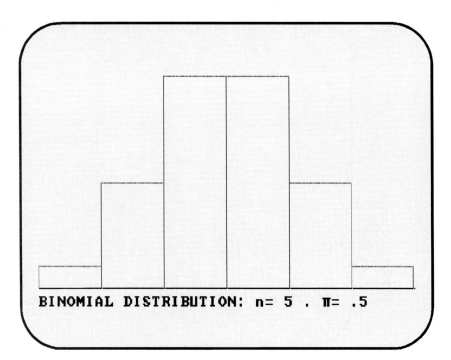

FIGURE 6.23b

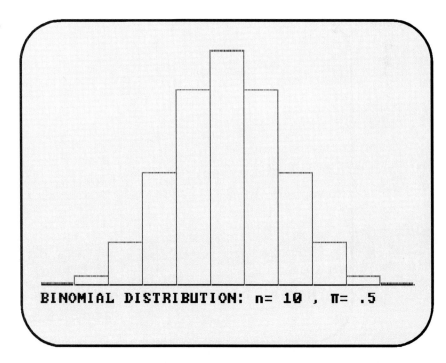

FIGURE 6.23c

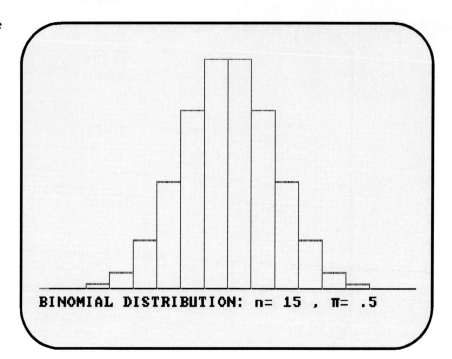

FIGURE 6.23d

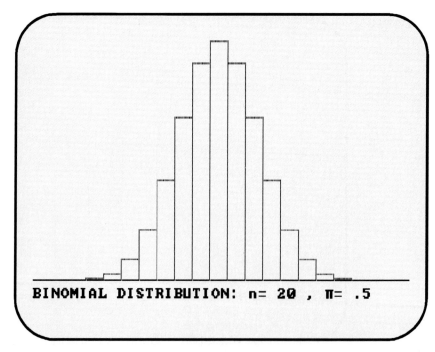

FIGURE 6.23e

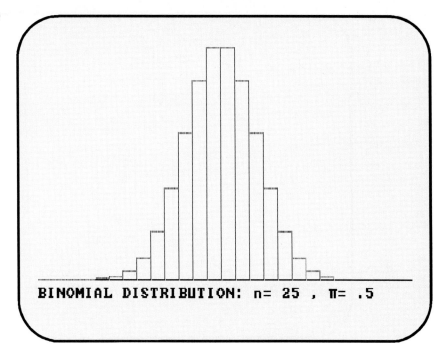

FIGURE 6.23f

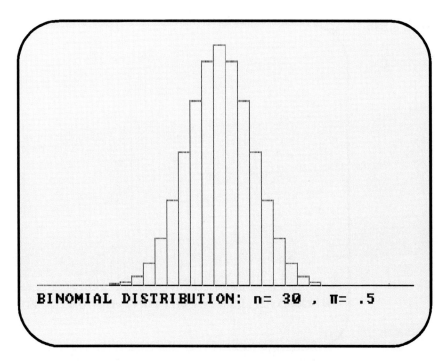

FIGURE 6.23g

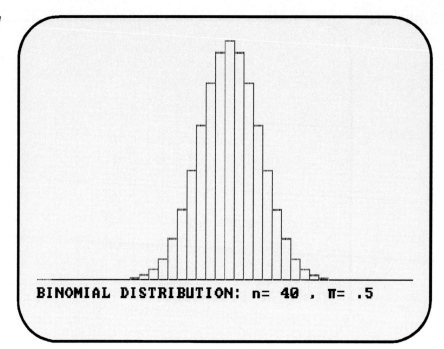

FIGURE 6.23h

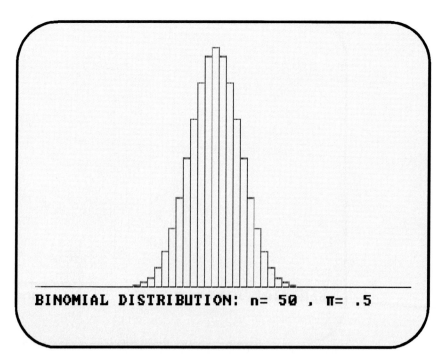

FIGURE 6.23i

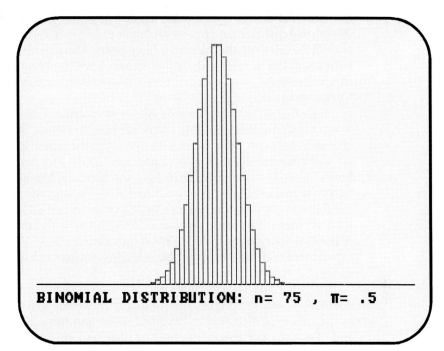

FIGURE 6.23j

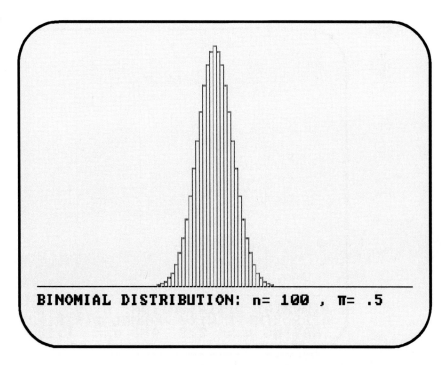

the binomial distribution for any choice of n and π and then superimposes the normal distribution curve with mean $\mu = n\pi$ and standard deviation $\sigma = \sqrt{n\pi(1-\pi)}$ over the binomial histogram. Thus, both the binomial distribution and the normal distribution shown have the identical values for these two parameters. (We describe the actual use of this program in the Computer Corner at the end of this section.)

Suppose we use the program with $\pi = .5$ and $n = 5$. The graphics output is reproduced in Figure 6.24, where we see that there is a considerable discrepancy between the two distributions. On the other hand, in Figure 6.25, we show the results with $\pi = .5$ and $n = 20$. In this case, the normal distribution curve is a better match or fit to the binomial histogram. In Figure 6.26, we show the results with $\pi = .5$ and $n = 50$, and we see that the normal distribution curve is a very close match to the binomial histogram. In Figure 6.27, we show the results with $\pi = .283$ and $n = 75$, and again the outlines of the two distributions seem almost indistinguishable. We encourage you to experiment with this program by selecting various combinations of n and π to see the results.

In summary, we conclude that when n is fairly small, there is a major distinction between the two distributions. However, as n increases, the shape of the binomial distribution becomes more and more normal in appearance and hence the corresponding normal distribution is a better and better fit to the binomial histogram. In general, the match between the two distributions is close enough that the normal distribution can be used in place of the binomial distribution provided that both $n\pi \geq 5$ and $n(1-\pi) \geq 5$.

FIGURE 6.24

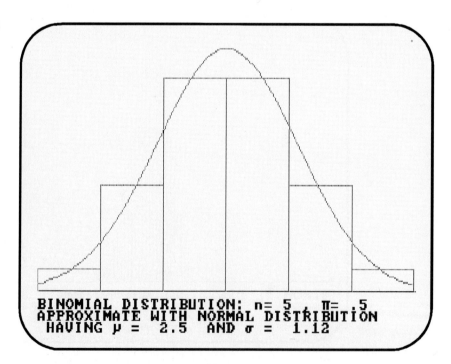

SECTION 6.3 • The Normal Approximation to the Binomial Distribution **265**

FIGURE 6.25

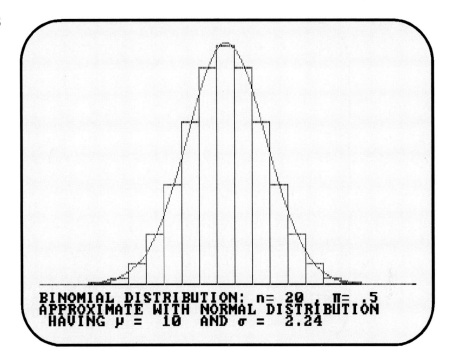

FIGURE 6.26

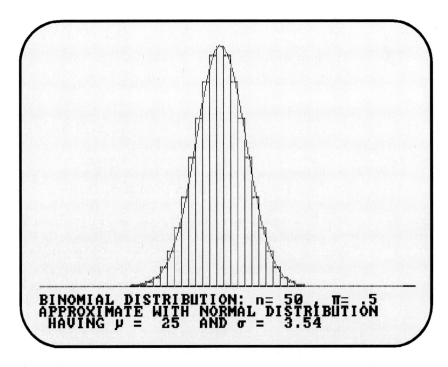

FIGURE 6.27

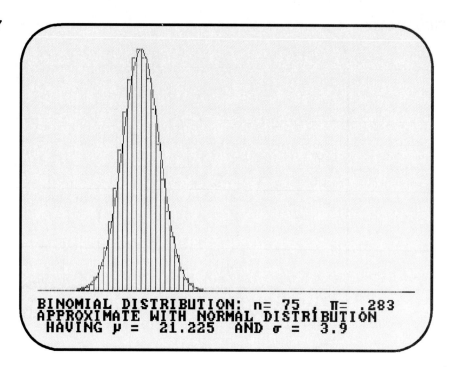

A binomial distribution based on n successes with probability π of success can be approximated by a normal distribution with mean

$$\mu = n\pi$$

and standard deviation

$$\sigma = \sqrt{n\pi(1-\pi)}$$

provided that

$$n\pi \geq 5 \quad \text{and} \quad n(1-\pi) \geq 5$$

In this regard, notice that in Figure 6.24 where $n = 5$ and $\pi = .5$, $n\pi = 5(.5) = 2.5 < 5$ and the match is not good. On the other hand, in Figure 6.25 where $n = 20$ and $\pi = .5$, $n\pi = 20(.5) = 10 \geq 5$ and $n(1-\pi) = 20(.5) = 10 \geq 5$, the match is reasonably close, and we can use the normal distribution to approximate the binomial distribution. Similarly, in Figure 6.26 where $n = 50$ and $\pi = .5$, $n\pi = 25$ and $n(1-\pi) = 25$, and the match is also good and we can approximate the binomial distribution with the normal distribution.

Nevertheless, there are some essential differences between the two distributions, no matter what the values for n and π. By its very definition, the binomial distribution is a *discrete* probability distribution so that the associated random variable can take on any of a finite number of possible values: the number of successes $x = 0, 1, 2, \ldots, n$. On the other hand, the normal

FIGURE 6.28

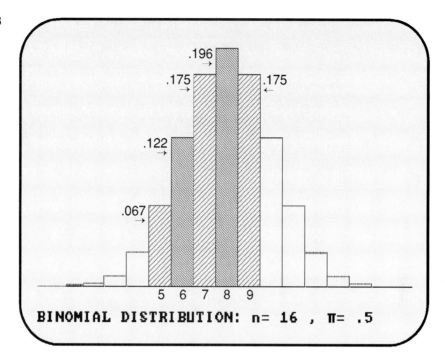

distribution is a *continuous* distribution, so that its associated random variable can assume any of an infinite spectrum of possible values. This underlying difference in the nature of the two distributions is something we must take into account.

For the moment, however, let's not concern ourselves with this complication, but continue in a rather naive manner. We pointed out in Section 5.3 and again in Section 6.1 that the probability of x successes in a binomial experiment is precisely equal to the area of the corresponding box in the binomial histogram. Further, if we are interested in the probability of any of a series of successes, then we must add the areas of the corresponding boxes. Consider the binomial distribution with $n = 16$ and $\pi = .5$. Suppose we want the probability of getting between 5 and 9 successes. (We interpret the word "between" to include both the lower and upper values in this and all related problems.) This probability is the sum of the areas of the boxes corresponding to $x = 5, 6, 7, 8$ and 9 in the histogram, as shown in Figure 6.28. However, since the associated normal distribution is a close fit to the binomial histogram, we might conclude that the area under the corresponding portion of the normal curve will be very close to the area of these boxes.

Therefore, we might be tempted to pursue the following (somewhat erroneous) line of reasoning to determine

$$P(5 \leq x \leq 9)$$

The mean and standard deviation for the binomial distribution are, respectively,

$$\mu = n\pi = 16(.5) = 8$$
$$\sigma = \sqrt{n\pi(1-\pi)} = \sqrt{16(.5)(.5)} = \sqrt{4} = 2$$

Since the desired probability is the sum of the areas of the boxes in the histogram for $x = $ 5, 6, 7, 8 and 9 and the normal distribution is very close to the histogram, we might expect to get a very accurate approximation to this probability by finding the area under the corresponding portion of the normal distribution curve. We want to consider the associated normal probability problem since, as we have seen in the previous two sections, it is much easier to solve. Thus, we consider

$$P(5 \le x \le 9)$$

Using $\mu = 8$ and $\sigma = 2$, we find that the z value corresponding to $x = 5$ is

$$z = \frac{5-8}{2} = \frac{-3}{2} = -1.5$$

and the z value corresponding to $x = 9$ is

$$z = \frac{9-8}{2} = .5$$

Therefore, as we see from Figure 6.29, the associated probability is

$$P(5 \le x \le 9) = .4332 + .1915 = .6247$$

Since the numbers used here ($n = 16$ and $\pi = .5$) are relatively simple, we can calculate the correct answer for this problem by using Table II, and so we obtain

$$.067 + .122 + .175 + .196 + .175 = .735$$

Obviously, this is considerably different from our approach using the normal distribution, although the approximation should be good since $n\pi = 16(.5) = 8 \ge 5$ and $n(1-\pi) = 16(.5) = 8 \ge 5$.

FIGURE 6.29

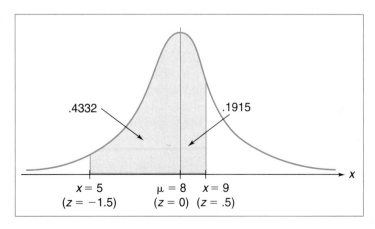

While the above method for using the normal distribution to approximate the binomial distribution seems particularly simple and logical, it must be modified slightly to produce accurate results. The problem is that we did not take into account the basic difference between the *discrete* binomial distribution and the *continuous* normal distribution. Consider Figure 6.30 which shows the output from Program 19 for $n = 10$ and $\pi = .5$. Suppose we are interested in the probability of getting 5 successes. This is the area of the center box in the histogram. For the binomial distribution, where x takes on only the values 0, 1, 2, ..., 10, $x = 5$ represents the *entire* middle box. On the other hand, for the normal distribution, $x = 5$ represents only one of an infinite array of possible values. Therefore, it actually represents just one vertical line located at the center of the box corresponding to $x = 5$ in the binomial histogram. Thus, if we want to find the area under the normal curve that corresponds to $x = 5$ successes in the binomial distribution, it is necessary to find the area under the normal curve corresponding to the full range of measurements covered by the box. To do this correctly, we must think of the box as extending from $4\frac{1}{2}$ to $5\frac{1}{2}$.

This type of adjustment is known as a *continuity correction* and must be made whenever we approximate the discrete binomial distribution with a continuous normal distribution. The integer values that the discrete random variable takes on must be spread across the entire continuous scale. In particular, each integer value of the discrete random variable x in the binomial distribution must be spread or stretched one-half a unit in each direction. Thus, $x = 2$ must be stretched from 1.5 to 2.5, $x = 3$ must be stretched from 2.5 to

FIGURE 6.30

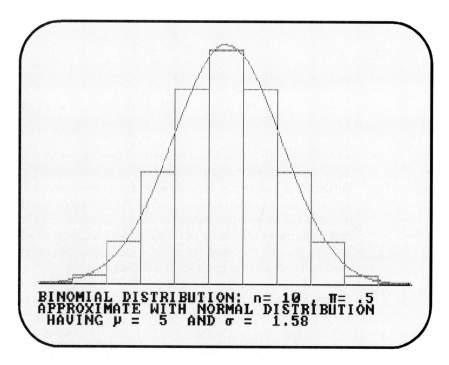

3.5, and so forth. The totality of the intervals so formed then covers the full spectrum of values of the continuous random variable.

> A *continuity correction* must be used to approximate a discrete binomial distribution with a continuous normal distribution. Each possible integer value x for the number of successes must be "stretched" to extend from $x - \frac{1}{2}$ to $x + \frac{1}{2}$.

With this idea of a continuity correction in mind, let's reconsider the problem we tried to solve previously, namely,

$$P(5 \leq x \leq 9)$$

where $\mu = 8$ and $\sigma = 2$. This time, however, we reinterpret what we mean by 5 through 9 successes in terms of a continuity correction. Thus, the lower value 5 must be thought of as the range of values from 4.5 to 5.5, and so the lower limit must begin at $x = 4.5$. Similarly, the upper value 9 must be thought of as extending from 8.5 to 9.5, and so the upper limit will be $x = 9.5$. Therefore, the problem we seek to solve is actually

$$P(4.5 < x < 9.5)$$

The z value corresponding to $x = 4.5$ is

$$z = \frac{4.5 - 8}{2} = \frac{-3.5}{2} = -1.75$$

and the z value corresponding to $x = 9.5$ is

$$z = \frac{9.5 - 8}{2} = \frac{1.5}{2} = .75$$

Therefore, as shown in Figure 6.31, the desired probability is given by

$$P(4.5 < x < 9.5) = P(-1.75 < z < .75)$$
$$= .4599 + .2734 = .7333$$

We note that this result is quite different from the erroneous answer of .6247 we obtained earlier without using the continuity correction. More importantly, it is actually very close to the correct answer of .735 we obtained by using the binomial tables directly. We therefore see that using the normal distribution as an approximation to the binomial distribution is an effective technique to obtain fairly accurate results as long as $n\pi$ and $n(1 - \pi)$ are both at least 5. Keep in mind, of course, the necessity of introducing the continuity correction in all such situations.

EXAMPLE 6.11 On an ESP test with $n = 25$ and $\pi = .2$, what is the probability of getting (*a*) more than 10 right answers and (*b*) 10 or more right answers?

FIGURE 6.31

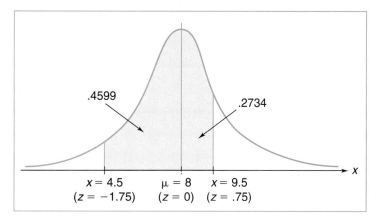

SOLUTION Because of the complexity of the arithmetic, we want to use the normal approximation to the binomial distribution. To see if it is applicable, we first check whether $n\pi$ and $n(1 - \pi)$ are both at least 5. Since

$$n\pi = 25(.2) = 5 \quad \text{and} \quad n(1 - \pi) = 25(.8) = 20 \geq 5$$

we can proceed. The mean and standard deviation of this binomial distribution are, respectively,

$$\mu = n\pi = 25(.2) = 5$$
$$\sigma = \sqrt{n\pi(1 - \pi)} = \sqrt{25(.2)(.8)} = 2$$

For part *a*, we observe that *more than* 10 *right* means 11 or 12 or ... 25 right answers (but 10 is *not* included). By the continuity correction, 11 right answers are spread so that they start at 10.5. Thus, we must find

$$P(x > 10.5)$$

The z value corresponding to $x = 10.5$ is

$$z = \frac{10.5 - 5}{2} = \frac{5.5}{2} = 2.75$$

and so the desired probability is

$$P(x > 10.5) = P(z > 2.75) = .5 - .4970 = .0030$$

as shown in Figure 6.32a.

For part *b*, we want to find the probability of 10 or more successes. In this case, we include $x = 10$ successes which start at 9.5 using the continuity correction, so that we consider

$$P(x > 9.5)$$

FIGURE 6.32

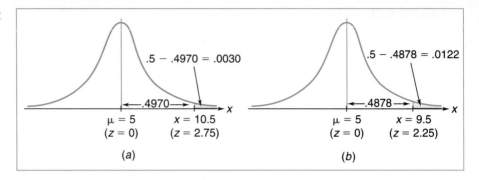

The corresponding z value is

$$z = \frac{9.5 - 5}{2} = \frac{4.5}{2} = 2.25$$

and so the desired probability is

$$P(x > 9.5) = P(z > 2.25) = .5 - .4878 = .0122$$

as shown in Figure 6.32b. ∎

EXAMPLE 6.12 Suppose that 40% of all students at a certain college take statistics at some point in their careers. If a group of 200 students are randomly selected, what is the probability that between 75 and 100 of them take statistics?

SOLUTION This is a binomial probability problem with $n = 200$ and $\pi = .4$, and we seek the probability of $x = 75, 76, \ldots, 100$ successes. Clearly, this problem requires using the normal approximation to the binomial distribution, so we first check to make sure it is applicable. In this case,

$$n\pi = 200(.4) = 80 \geq 5 \quad \text{and} \quad n(1 - \pi) = 200(.6) = 120 \geq 5$$

so we can proceed. The mean and standard deviation for this distribution are, respectively,

$$\mu = 200(.4) = 80$$
$$\sigma = \sqrt{200(.4)(.6)} = \sqrt{48} = 6.93$$

We now introduce the continuity correction. Since both 75 and 100 are included, the interval of values desired extends from 74.5 to 100.5, so that we seek to determine

$$P(74.5 < x < 100.5)$$

The z value corresponding to $x = 74.5$ is

$$z = \frac{74.5 - 80}{6.93} = \frac{-5.5}{6.93} = -.79$$

FIGURE 6.33

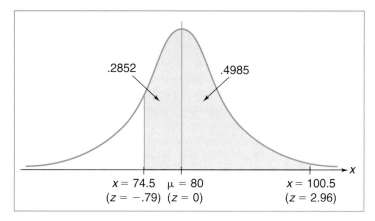

and the z value corresponding to $x = 100.5$ is

$$z = \frac{100.5 - 80}{6.93} = \frac{20.5}{6.93} = 2.96$$

Consequently, as shown in Figure 6.33, the desired probability is

$$P(74.5 < x < 100.5) = P(-.79 < z < 2.96)$$
$$= .2852 + .4985 = .7837 \quad\blacksquare$$

EXAMPLE 6.13 A survey finds that 63.7% of the people in a certain region favor nuclear power. If 300 people from that area are chosen randomly, find the probability that between 200 and 215 of them are in favor.

SOLUTION This is a binomial distribution with $n = 300$ and $\pi = .637$, and we must first determine whether we can approximate it with a normal distribution. Since

$$n\pi = 300(.637) = 191.1 \geq 5 \quad \text{and} \quad n(1 - \pi) = 300(.363) = 108.9 \geq 5$$

we can proceed. The mean and standard deviation, respectively, are

$$\mu = 300(.637) = 191.1$$

and

$$\sigma = \sqrt{300(.637)(.363)} = \sqrt{69.37} = 8.329$$

We approximate this binomial distribution with the associated normal distribution using continuity corrections at both endpoints. Since $x = 200$ is included, the lower limit becomes 199.5; and since $x = 215$ is included, the upper limit becomes 215.5. The problem to be solved is then

$$P(199.5 < x < 215.5)$$

FIGURE 6.34

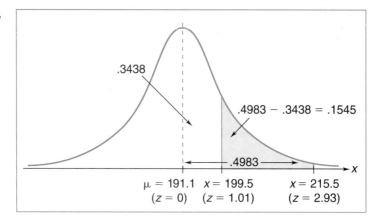

The z value corresponding to $x = 199.5$ is

$$z = \frac{199.5 - 191.1}{8.329} = \frac{8.4}{8.329} = 1.01$$

and the z value corresponding to $x = 215.5$ is

$$z = \frac{215.5 - 191.1}{8.329} = \frac{24.1}{8.329} = 2.93$$

Therefore, as seen from Figure 6.34,

$$P(199.5 < x < 215.5) = P(1.01 < z < 2.93)$$
$$= .4983 - .3438 = .1545 \quad \blacksquare$$

Computer Corner

We now discuss the details for using Program 19: Normal Approximation to Binomial. As we indicated in the text, this program draws the binomial histogram corresponding to any values you desire for n (the maximum is 100) and π and then superimposes over it the corresponding normal distribution curve. However, rather than producing only a graphical display, the program is designed to be a tutorial to lead you through the steps and decisions involved in the process.

When you run the program, it first asks you for the values of n and π. It then asks you if you want to do the work by hand. If so, you will be asked to calculate the mean μ and the standard deviation σ. If you make any errors in these calculations, the program displays the appropriate formula and prompts you to try again. After your third incorrect attempt, the program supplies the correct answers. It then transfers to the graphics display with the normal distribution curve superimposed over the binomial histogram.

Exercise Set 6.3

Mastering the Techniques

For each of the following combinations of n and π, determine whether the normal approximation to the binomial distribution can be used.

1. $n = 20$, $\pi = .5$
2. $n = 20$, $\pi = .2$
3. $n = 12$, $\pi = .4$
4. $n = 12$, $\pi = .75$
5. $n = 10$, $\pi = .5$
6. $n = 10$, $\pi = .7$
7. $n = 40$, $\pi = .1$
8. $n = 40$, $\pi = .8$
9. Consider the binomial distribution with $n = 36$ and $\pi = .5$. Use the normal approximation to determine the probability of getting
   (a) Between 15 and 20 successes
   (b) More than 22 successes
   (c) At most 12 successes
   (d) At least 18 successes
   (e) Fewer than 18 successes
   (f) No more than 15 successes
   (g) Exactly 20 successes (*Hint:* Interpret 20 as extending from 19.5 to 20.5.)
10. Consider the binomial distribution with $n = 100$ and $\pi = .5$. Use the normal approximation to the binomial distribution to find the probability of getting
    (a) Between 48 and 53 successes
    (b) Between 55 and 60 successes
    (c) 53 or more successes
    (d) At least 44 successes
    (e) At most 50 successes
    (f) No fewer than 54 successes
    (g) Exactly 50 successes

Applying the Concepts

11. A true-false test has 36 questions. If a student guesses at each question, what is the probability that he gets at least 22 correct?
12. Suppose a multiple-choice test has 72 questions, each with three choices. If a student guesses at each question, what is the probability that she gets at least 36 correct?
13. A school's enrollment is precisely 50% female. If a group of 64 students is randomly selected, what is the probability that there are between 25 and 35 women in the group?
14. A large corporation claims that 25% of its managerial employees are members of minority groups. If a group of 48 such employees is randomly selected, what is the probability that it includes between 5 and 10 minority individuals?
15. According to a recent survey, 65% of all households in the United States have VCRs. If a group of 80 households is randomly selected, find the probability that fewer than 50 of them have VCRs.

16. The survey reported in Exercise 15 also found that 18% of all households in this country have home computers. If a group of 80 households is randomly selected, find the probability that more than 50 of them have computers.
17. A conservation group claims that 20% of all maple trees in a certain forest area are suffering from the effects of acid rain. If a group of 225 trees in this forest is randomly tested, what is the probability that at least 50 of them are so afflicted?
18. Suppose a political poll indicates that 60% of adults support the President's stand on a certain issue. If 200 people are randomly selected, what is the probability that fewer than 100 support the President?
19. In a certain region, 40% of all teenage drivers are involved in traffic accidents before their twentieth birthday. If a group of 80 teenage drivers is studied, what is the probability that between 25 and 40 of them will be involved in an accident?
20. A consumer testing service estimates that 2.5% of all cans of tuna in a certain production batch are infected with a certain type of microorganism. If 500 cans are tested at random, what is the probability that more than 15 are infected? that no more than 10 are infected?

Computer Applications

21. Consider a variety of binomial distributions with $\pi = .4$. Use Program 19: Normal Approximation to Binomial to explore the shapes of these distributions corresponding to $n = 5, 10, 15, 20, 25, 30, 50,$ and 100. What can you conclude about the location of each of the distributions? At what point do you conclude that the normal distribution makes an effective approximation to the binomial histogram? Does this visual conclusion agree with the criteria given in the text?
22. Repeat Exercise 21 using $\pi = .2$.
23. Repeat Exercise 21 using $\pi = .1$.

6.4 When Is a Set of Data Normal?

Throughout this chapter, we have solved a wide variety of problems using methods that apply to a normal population. In each case, the question specifically stated that the quantity under investigation was normally distributed. However, when you work with your own data, how do you know that the underlying population is indeed normal? If it is not, then the methods we have developed do not apply; if you use them anyway, the results are very likely erroneous. Moreover, many of the methods to be developed in later chapters likewise require that the population in question be normal and so we will again be faced with the question of determining whether that is the case.

The basic criteria for a population with mean μ and standard deviation σ to be normal are

1. The population is centered at the mean μ.
2. The population is symmetric about the mean.

3. Approximately 68% of the population lies within 1 standard deviation of the mean.
4. Approximately 95% of the population lies within 2 standard deviations of the mean.

In practice, it is difficult to prove that a given set of data is normally distributed using these criteria. However, it is much simpler to use these criteria to show that a set of data is *not* normally distributed.

We begin by constructing the histogram representing the set of data and calculating the mean and standard deviation. First and foremost, if the population is normal, then the shape of the histogram should follow a normal distribution pattern. If it does not, then the population may not be normal. Next, if the histogram is not symmetric about the mean, then the data are *skewed*, either to the left or the right. In either case, we should suspect that the population may not be normally distributed. See Figure 6.35, where we show two histograms, one skewed to the left and the other to the right. Based on them, we suspect that neither one represents a normal population.

The two criteria based on the empirical rule are somewhat harder to apply since it is difficult to assess where 68% and 95% of the data lie. Instead, we consider what they tell us about a population. The fact that about 68% of any normal population lies within 1 standard deviation of the mean and that about 95% lies within 2 standard deviations of the mean indicates that the bulk of the population is near the mean and that the rest of the population drops off extremely rapidly in both directions. Picture a graph for the normal distribution curve. Thus, if the two tails of the histogram do not "die" rapidly, then we should suspect that the population is not normally distributed. We illustrate this in Figure 6.36, where the histogram is roughly symmetric but the tails do not drop off rapidly.

Finally, in any set of "real" data, there is always a chance of encountering one or more *outliers*, isolated values that lie far from the center of the data. The presence of one or two such outliers is usually not enough to cause us to decide that a set of data is not normal; they could arise from an error in measurement or from chance variation. For instance, suppose we use a computer to simulate flipping a set of 10 coins 200 times. We would expect the results to be centered about a mean of 5 heads per set. While it is extremely

FIGURE 6.35

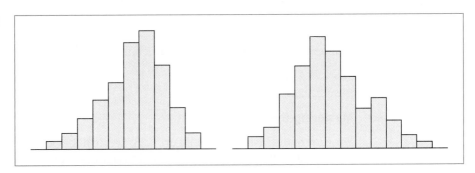

FIGURE 6.36

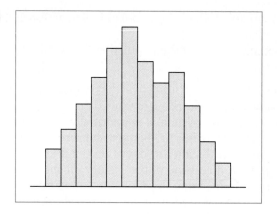

unlikely (1 out of 1024), we might get one particular outcome where 10 heads come up; this would be an outlier. However, if this happened three times out of the 200 simulations, then we should suspect that something is wrong. The presence of too many outliers would be cause to suspect that a population is not normal. See Figure 6.37.

The ideas presented here represent only a very rough eyeball test to see if a set of data appears to be normal. At best, we can suspect that the data are clearly not normal in appearance. In turn, this might suggest that the population may not be normal. Thus, the first step that should be taken with any set of data is: Look at it to see its shape and possible pattern. In later chapters, we will develop more sophisticated methods for testing whether a set of data is normal. However, such procedures should only be used *after* you have done a visual test of the data.

FIGURE 6.37

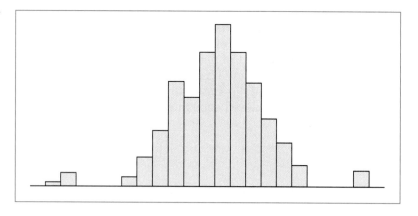

Exercise Set 6.4

Mastering the Concepts

For each of the histograms in Figure 6.38, decide if the population could be normally distributed or if it is likely not normal. In each case, indicate your reasons for your decision.

FIGURE 6.38

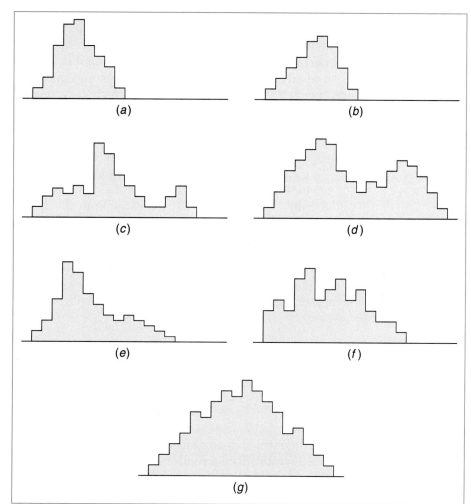

CHAPTER 6 SUMMARY

The most important continuous distribution is the **normal distribution** with mean μ and standard deviation σ. The normal distribution curve is symmetric about its center μ, and the total area under the curve is 1.

For any continuous distribution, the probability that a random variable x assumes any single value is 0. You therefore must consider the probability that x lies in some range of values, say, x_L to x_R. For a normally distributed random variable x, the probability $P(x_L < x < x_R)$ is equal to the area under the normal curve from x_L to x_R.

Given a measurement x from a normal distribution with mean μ and standard deviation σ, you can transform it to the corresponding measurement in the *standard normal distribution* having mean 0 and standard deviation 1 by using the associated z value,

$$z = \frac{x - \mu}{\sigma}$$

The corresponding probability values are found in Table III.

To find the probabilities associated with a normal distribution, you should first sketch the graph of the normal distribution curve and indicate the region in question. When you are looking at the picture, common sense should help you to decide if you can just use the table entry, add two normal probability values, take the difference between two such values, use the difference between such a value and .5, or use the sum of such a value and .5.

To locate the tail of a normal distribution, use the desired percentage in the tail to determine the percentage of the population from the center to the cutoff point, find the associated z value from the table, and then use the formula

$$x = \mu + z\sigma$$

to determine the desired value of x.

A binomial distribution based on n trials with probability of success π can be approximated by a normal distribution provided that

$$n\pi \geq 5 \quad \text{and} \quad n(1 - \pi) \geq 5$$

Both distributions have mean

$$\mu = n\pi$$

and standard deviation

$$\sigma = \sqrt{n\pi(1 - \pi)}$$

To use the normal approximation to the binomial distribution, you must introduce a **continuity correction** to stretch each value of the discrete random variable x to represent the interval $x - .5$ to $x + .5$.

Review Exercises

1. The daily caloric intake of college students is normally distributed with mean 1450 calories and standard deviation 300 calories. What is the probability that a randomly selected student will consume more than 2000 calories per day?
2. A city claims that an average of 400 couples are married each week by its marriage agents. If the weekly number of marriages are normally distributed with standard deviation 120, find the probability that between 350 and 500 couples marry during a particular week.
3. A study finds that the mean number of hours of sleep that white collar workers get is normally distributed with mean 6.8 hours and standard deviation 0.6 hours. What is the probability that a randomly selected individual from this group sleeps between 6 and 6.5 hours?
4. The waiters in a restaurant receive an average tip of $16 per table with a standard deviation of $3.50. Suppose that the tips are normally distributed. A waiter feels

that he has been stiffed if the tip is less than $10. What is the probability of this happening?

5. According to a report, the typical worker in Japan puts in an average of 46.5 hours per week on the job with a standard deviation of 5.4 hours. If the number of hours worked per week is normally distributed, find the minimum number of hours put in by the most dedicated 10% of the Japanese workforce.

6. A study claims that 10% of all adults in a certain community have been exposed to the HIV virus for AIDS. In a group of 200 people in this community, what is the probability that more than 25 have been exposed to HIV?

7. 17% of the children in sixth grade in a certain school district are reading below grade level. If a group of 75 sixth-graders from this district are tested, what is the probability that between 10 and 15 are reading below grade level?

Sampling Distributions

With election day only a week away, a new poll shows that incumbent Mayor Sam Smith is trailing challenger Glenda Johnson by 8 percentage points, 45% to 37%. According to the people polled, the main reasons that Smith is in trouble are soaring taxes, unemployment and his weak stand on sexual harassment issues. The poll found that Johnson's support is particularly strong among Democrats, women and working class individuals. Smith's support comes from Republicans, men, business and industry, and professionals.

On the basis of this poll, Johnson's camp is predicting her victory at the election booths next week.

The poll involved a series of telephone interviews with a randomly selected group of 300 city residents. The margin of error in the poll is 6 percentage points.

Reports such as these are standard fare before any election. Have you ever wondered how they are done, how they are analyzed and how accurate they are?

The key to any poll lies in selecting an appropriate random sample. What is an appropriate random sample? Did the people conducting the poll simply open the phone book and call the first 300 names listed or did they set up an automatic dialing machine to make the calls for them? If so, that is certainly not a random sample, nor is it an appropriate method. For instance, the very use of a telephone survey means that people without telephones are automatically excluded from the poll. Might they have different viewpoints from those with telephones?

Also, were the calls made at random times during the day or evening, or were they all made during the convenient (for the pollsters) morning hours. If so, then they likely would miss people who are working. Did they take care to get an appropriate demographic mix that represents the breakdown in the city's population of race, sex, religion, ethnic background, income, education, and so forth? If they didn't, then the results of the poll may not correctly reflect the true voting preferences of the people.

Moreover, the poll reported on the preferences of only 300 people. Is that adequate to reflect the voting pattern of a million people in the city? In fact, did the people polled all declare themselves as definite voters? If many of them only answered questions but will not actually cast a ballot, how legitimate can the sample be?

Further, the poll indicated the preference of only 82% of the people interviewed. What about the remaining 18%? Do they plan to vote or will they sit out the election? If they plan to vote, then the percentages given can change dramatically between now and election day. Isn't it a little premature for one candidate's camp to be declaring victory? In fact, might that not affect how, or if, people vote? Finally, what is the significance of the 6 percentage point margin of error?

Is Johnson's lead necessarily as impressive as it seems?

In this chapter, we will study the idea of drawing random samples and extracting essential information from them. In later chapters, we will see how to use such samples to make the kind of predictions and decisions that arise from election polls.

7.1 Sampling and the Distribution of Sample Means

The primary focus of modern statistics is on using data from a sample to make predictions or decisions about the underlying population from which the sample is drawn. This is known as *inferential statistics*. So far we have studied descriptive statistics and probability since they are the foundations for statistical inference. In this chapter, we develop the fundamental ideas about statistical sampling which are at the heart of inferential statistics. In the following chapters, we show how these ideas are applied in many different fields.

Suppose that the heights of all the students enrolled at your college make up the population of interest. This population of heights has mean μ and standard deviation σ. Typically, we know little about this underlying population and probably do not know its exact shape or the actual values for μ and σ.

Suppose we want to determine the value of the population mean μ. We could certainly find it by collecting the heights of all students at the college and calculating the mean. However, this can be an extremely time-consuming, complicated, and expensive procedure, particularly if your school has a very large enrollment. Instead, we settle for an accurate prediction or estimate of μ based on an appropriate random sample.

To illustrate, we consider the heights of the students in your statistics class as the sample. Realize that this sample is chosen for convenience for the purposes of this discussion; it is not a random sample. In practice, we would select a truly random sample and not just use an established group. Further, note that this sample is simply one of many possible samples that can be drawn from the entire population consisting of all student heights at your school.

The information we will need about this sample is its size n, its mean $\bar{x}$, and its standard deviation s. For this hypothetical class, suppose $n = 36$, $\bar{x} = 66$ inches, and $s = 3$ inches. If we had to make a prediction of the population mean μ based on this sample with sample mean $\bar{x} = 66$, then our best estimate for μ is 66 inches also. Such a single-value prediction for μ is known as a *point estimate*.

However, it is very unlikely that the population mean μ is precisely the same as the sample mean $\bar{x}$ for any particular random sample. Therefore, while we might accept this as an estimate, we certainly wouldn't put too much faith in its accuracy.

To see why a single sample can be very misleading, suppose your class contained the men's basketball team. In that case, the sample mean $\bar{x}$ would certainly be considerably larger than μ. Alternatively, if your class contained the women's gymnastics team (whose members are all under 5 feet tall), then the mean height $\bar{x}$ would be considerably smaller than μ. Therefore, a more accurate statement is that the population mean μ is probably *close to* the sample mean $\bar{x} = 66$. Even this is correct only if the sample group is fairly typical of the total population.

Consequently, we must be able to determine how typical your class (or sample) is for representing the heights of all members of the student body. One way to do this is to take a second "random" sample. Suppose you go to the next classroom, collect the heights of all the students, and find $\bar{x} = 65$ for this second "random" sample. You still don't know if your class is particularly tall or if the other class is somewhat short. Therefore, the wisest course of action might be to average the two sample means, 66 and 65, and use the value of 65.5 as the best prediction for μ. This will tend to reduce the effects of extremes in either group.

Proceeding in this fashion, we might take a third "random" sample from a third class, where $\bar{x} = 67$, say, and then average the three sample means. This would further reduce the effects on the prediction of any abnormally tall or short groups. In fact, we could theoretically continue this process indefinitely by taking more and more samples from the original population and then averaging all the sample means $\bar{x}$ from each of the samples. Intuitively, the resulting mean of all these sample averages should be far more accurate an estimate of μ than what we obtain from any single random sample. (Of course, this process does not make a lot of sense in practice because it actually involves more time and work than determining the value of μ by using all members of the original population directly.)

Let's consider the above ideas from a somewhat more sophisticated perspective. If we collect the sample means $\bar{x}$ from a large number of samples, then we are collecting a set of numbers: the means of all these samples. Suppose we continue this process indefinitely, so that we have the mean of *every* possible random sample of, say, 36 students drawn from the college's enrollment. The totality of all these $\bar{x}$'s can then be considered as forming a new population derived from the original population. That is, we start with an underlying population composed of the heights, x, of individual students at the college. This population has mean μ and standard deviation σ. We then construct a new population composed of the mean heights, $\bar{x}$, of every possible sample of size 36.

This new population is known as the *distribution of sample means* or the *sampling distribution of the mean*. Because it is a population in its own right, it has its own population parameters: a mean and a standard deviation. To distinguish these values from the corresponding parameters μ and σ for the original population, we denote the mean of the distribution of sample means by $\mu_{\bar{x}}$ and the standard deviation of this population by $\sigma_{\bar{x}}$.

The mean $\mu_{\bar{x}}$ of this distribution of sample means is the average of the means $\bar{x}$ of all possible samples of a given size. The process of averaging the various $\bar{x}$'s should produce a value close to μ for the original population. Thus, we might expect $\mu_{\bar{x}}$ to be very close to μ. It is considerably harder to predict how the value of $\sigma_{\bar{x}}$ for the distribution of sample means relates to the value of σ for the original population.

We now investigate the properties of the distribution of sample means. Specifically, three things are of interest to us:

1. What type of distribution is it? (What is its shape?)
2. What is its mean $\mu_{\bar{x}}$, and how does it relate to the mean μ of the original population?
3. What is its standard deviation $\sigma_{\bar{x}}$, and how does it relate to the standard deviation σ of the original population?

To discover the answers to these questions, we use Program 20: Central Limit Theorem Simulation. This program allows us to see the effects of repeated sampling from different underlying populations. First, we must select one of four different built-in, fixed populations:

1. Normal population
2. Skewed population (majority of the data are skewed to one side)
3. Uniform population (the data are spread evenly across the interval)
4. U-shaped population (majority of the data are clustered at the two ends)

Let's first select the normal population. We must supply the desired sample size (suppose we pick $n = 4$) and the number of such samples (suppose we pick 400). The program first draws a histogram for the underlying normal population and displays its parameters $\mu = 68.07$ and $\sigma = 2.76$. It then selects random samples of size 4 (groups of four drawn from the population), finds the mean of these four numbers, and plots the mean as a short horizontal line. Thus, the program is actually producing and displaying a large portion of the distribution of sample means. In Figure 7.1, we show a typical graph of what you might obtain when you use the program.

From this figure, several observations can be made. First, the overall shape of the distribution of sample means is approximately normal. Second, the center for these sample means is very close to the center for the original population. Third, the spread or variation in the sample means is considerably smaller than the spread in the original population. The sample means are clustered much more tightly together than the original population heights,

FIGURE 7.1

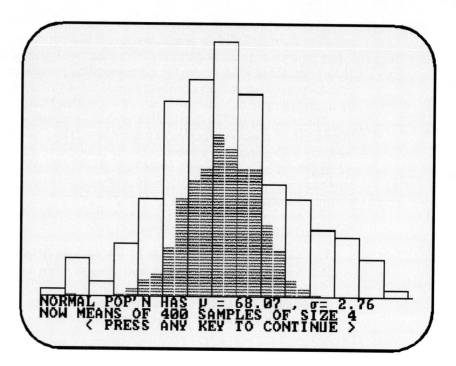

and so we expect their standard deviation to be *smaller* than σ for the original population. In fact, we can estimate the value of $\sigma_{\bar{x}}$ based on the screen display. The sample means are spread across approximately one-half of the screen compared to the original population which is spread across the entire screen. Therefore, we might estimate that for this situation, $\sigma_{\bar{x}}$ is approximately one-half of σ.

When the graph is completed, you can transfer to a text screen where you are given information about the actual mean and standard deviation of all 400 sample means. For the particular run shown in Figure 7.1, the numerical results show that the means of the 400 samples of size 4 had a mean of 68.02, which is very close to the population mean μ = 68.07. Moreover, the standard deviation for these 400 sample means is 1.44 compared to σ = 2.76 for the population. This is approximately the one-half that we estimated visually.

Suppose that we rerun the program using a different sample size, say, $n = 9$ instead of 4. A typical result is shown in Figure 7.2. We observe that the shape of the sampling distribution is again approximately normal and that it is still centered very close to the mean of the original population. However, the sample means are now clustered more closely together than in the previous case. In fact, the overwhelming majority of the sample means shown are within the middle 5 boxes of the histogram out of the 15 displayed across the screen. We therefore estimate that, in this case, $\sigma_{\bar{x}}$ is approximately one-third the value of σ. The accompanying numerical calculations bear out these

FIGURE 7.2

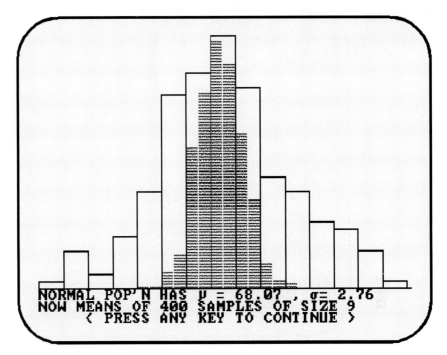

observations. The mean of the 400 sample means is 68.14 compared to the population value of 68.07, and the standard deviation is .91, which is roughly one-third of the value for $\sigma = 2.76$.

The fact that the standard deviation for the sample means $\sigma_{\bar{x}}$ is smaller when the sample size n is larger does make sense. If we take a sample of size $n = 4$ from a population, then it is easy for a single large or small entry to have a major effect on $\bar{x}$. Thus, if three numbers were selected from near the middle of the original population and one from the far end, then the corresponding sample mean $\bar{x}$ would be skewed considerably toward that end. Since this is reasonably likely to happen, there will be a wide spread in the location of all the $\bar{x}$'s. In fact, if you look back at Figure 7.1, you will notice that one particular sample mean fell far to the left and can only have arisen from a sample consisting of all very short heights. Such an isolated value is known as an *outlier* and is quite untypical of the distribution.

On the other hand, if the sample size is $n = 9$ instead of $n = 4$, then the effect of any single large or small number in the sample will be considerably less. To obtain a sample mean skewed far to the left or the right, the preponderance of members of the sample would have to be far to that side, and the probability of this happening is relatively small. If the sample size n is still larger, then the probability of this occurring will be smaller still.

Let's test this out by considering a still larger sample size in the program, say $n = 16$. A typical result is shown in Figure 7.3. As before, we note that

FIGURE 7.3

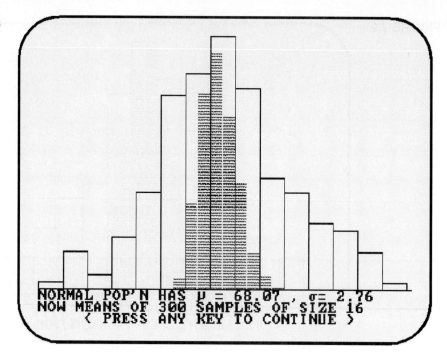

the shape of the distribution of sample means is seemingly normal and that it is centered very close to the original population mean μ. Further, we see that the spread is less than in the previous cases and is visually approximately one-quarter of σ. The accompanying numerical result for the average of the 300 sample means is 68.08 compared to $\mu = 68.07$. Further, the standard deviation of the sample means is .72, which is approximately one-quarter of $\sigma = 2.76$.

From these observations, we can predict what is likely to happen for larger values of the sample size n as long as we work with a normal population. Therefore, let us consider the case where the underlying population is not normal in shape. Suppose we select the U-shaped population whose mean $\mu = 68.61$ and whose standard deviation $\sigma = 4.80$. As before, we begin with 400 random samples of size of $n = 4$. A typical result is shown in Figure 7.4, from which we can make some interesting observations. First, the shape of the original population is definitely not normal, and most members of that population are located far from the center. Nevertheless, the sample means are centered near the middle of the graph, even though few members of the original population are there. Thus, we see how averaging tends to dilute the effects of extremes. Second, the shape of the distribution of sample means is clearly not normal in this case. If anything, it looks more like some surrealistic version of a city skyline. Third, the spread of the sample means is still considerably smaller than the spread in the original population. In fact, we can

FIGURE 7.4

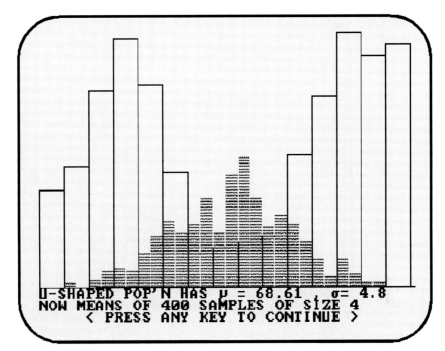

estimate visually that the spread is approximately one-half of the original spread. These conclusions are supported by the numerical calculations. The original population has mean $\mu = 68.61$ while the 400 sample means have a mean of 68.38, which is quite close. Further, the standard deviation of the original population is $\sigma = 4.80$ while the standard deviation of the 400 sample means, 2.39, is one-half of the population standard deviation.

In Figures 7.5, 7.6, and 7.7, we show typical results with samples of size $n = 9$, $n = 16$ and $n = 25$, respectively, drawn from this U-shaped population. In each case, the center of the distribution of sample means is close to the value of μ of the underlying population. Further, as the sample size n increases, the standard deviation in each case decreases and, in fact, is approximately one-third, one-fourth, and finally one-fifth of the value of σ. In addition, we observe that as the sample size increases from $n = 9$ to $n = 16$ to $n = 25$, the shape of the resulting distribution of sample means seems to smooth out and become ever more normal in appearance.

We encourage you to repeat this series of experiments using the other two populations available in the program, the skewed population and the uniform population. You will see that a similar set of conclusions can be drawn. Thus, in each case, the mean of the sample means is very close to the mean μ of the original population. The standard deviation $\sigma_{\bar{x}}$ is always a fraction of σ, and the fraction clearly depends on the sample size n. The larger the value of n, the smaller the fraction. The shape of the distribution is clearly not

FIGURE 7.5

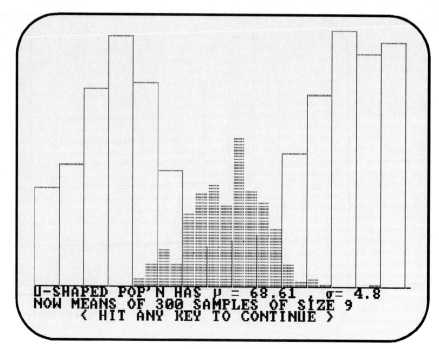

FIGURE 7.6

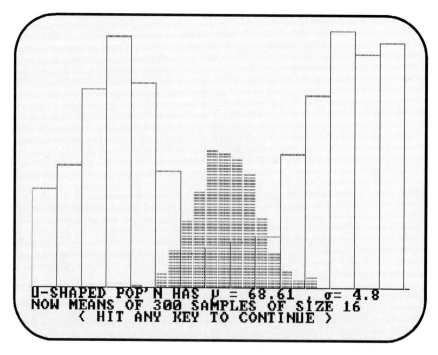

FIGURE 7.7

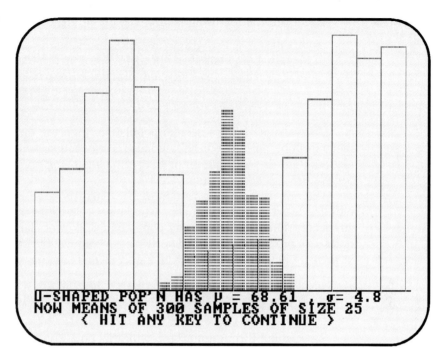

normal when the sample size n is small and the underlying population is not normal, but it becomes more and more normal in appearance as the sample size increases.

The above observations are essentially the basis of the single most powerful tool in inferential statistics, the Central Limit Theorem.

Central Limit Theorem

Consider the distribution of sample means having mean $\mu_{\bar{x}}$ and standard deviation $\sigma_{\bar{x}}$ based on random samples of size n from an underlying population having mean μ and standard deviation σ.

1. If the original population is **not** normally distributed, then the distribution of sample means **is** approximately normally distributed provided that the sample size n is large enough.

2. $$\mu_{\bar{x}} = \mu$$

3. $$\sigma_{\bar{x}} = \frac{\sigma}{\sqrt{n}}$$

Moreover, the following is also true of the distribution of sample means:

> If the original population **is** normally distributed, then the distribution of sample means is also normally distributed regardless of the sample size.

If we sample from a normal population, then the distribution of sample means will be normally distributed also. If we sample from a nonnormal population, then the distribution of sample means is *approximately* normal *if the sample size n is large enough.*

The only remaining question is, How large is large enough? When samples are drawn from a symmetric population, relatively small sample sizes are adequate. When the underlying population is skewed, larger samples are required. In practice, for most underlying populations, the sampling distribution is virtually indistinguishable from a normal distribution whenever $n > 30$.

> We use $n > 30$ as our criterion for the sample size being "large enough".

We assume throughout the rest of this book that all populations considered are such that $n > 30$ is sufficient for normality. In practice, you should examine a histogram for the data to make sure it is not very skewed. If it is, you are well advised to use a sample size considerably larger than 30.

The Central Limit Theorem and the above facts allow us to apply all the methods we previously developed for the normal distribution in a broader range of circumstances, as illustrated in the following examples.

EXAMPLE 7.1 Random samples of size $n = 64$ are drawn from a population having mean $\mu = 100$ and standard deviation $\sigma = 24$. Find the parameters for the resulting distribution of sample means. What can be concluded about this distribution?

SOLUTION For the distribution of sample means,
$$\mu_{\bar{x}} = \mu = 100$$
and
$$\sigma_{\bar{x}} = \frac{\sigma}{\sqrt{n}} = \frac{24}{\sqrt{64}} = \frac{24}{8} = 3$$

Further, we note that there is no explicit mention of the shape of the original population. However, since the sample size $n = 64 > 30$, the Central Limit Theorem guarantees that the distribution of sample means is approximately normal. ∎

EXAMPLE 7.2 Suppose that the batting averages of all players who ever played in the major leagues are normally distributed with mean 270 and standard deviation 30. If groups of 25 players are randomly selected, what are the mean and standard deviation of the corresponding sampling distribution of the mean? What can be said about the shape of this sampling distribution?

SOLUTION We immediately see that
$$\mu_{\bar{x}} = \mu = 270$$
and
$$\sigma_{\bar{x}} = \frac{\sigma}{\sqrt{n}} = \frac{30}{\sqrt{25}} = \frac{30}{5} = 6$$

Furthermore, since the original batting averages are normally distributed, the distribution of sample means is also normal despite the small sample size $n = 25$. ∎

EXAMPLE 7.3 A bank finds that its savings account balances have a distribution with mean $360 and a standard deviation of $120. If random groups of 16 accounts are checked as part of an audit, what are the mean and standard deviation of the distribution of sample means? What can be concluded about this distribution?

SOLUTION We have
$$\mu_{\bar{x}} = \mu = 360$$
and
$$\sigma_{\bar{x}} = \frac{\sigma}{\sqrt{n}} = \frac{120}{\sqrt{16}} = \frac{120}{4} = 30$$

However, since we are studying samples of size $n = 16 < 30$ and no indication is given that the original population is normally distributed, we cannot conclude anything about the shape of the distribution of sample means. ∎

In all the preceding discussion, we have implicitly assumed that the underlying population is so large that it is effectively infinite. However, in practice, we are sometimes sampling from a finite population of size N, and an adjustment has to be made when $\sigma_{\bar{x}}$ is calculated in certain cases. In particular, for finite populations of size N,

$$\sigma_{\bar{x}} = \frac{\sigma}{\sqrt{n}} \sqrt{\frac{N-n}{N-1}}$$

where n is the sample size. The quantity
$$\sqrt{\frac{N-n}{N-1}}$$
is called the *finite population correction factor*. If n is much smaller than N, this quantity is close to 1 and therefore can be ignored in calculating $\sigma_{\bar{x}}$. In

practice, it is necessary to use this factor whenever n is more than 5 % of N. Thus, if $N = 1000$, we have to make the adjustment only if n is greater than 50.

EXAMPLE 7.4 Suppose that the batting averages of the 400 current major league players are normally distributed with mean 270 and standard deviation 30. If groups of 25 players are randomly selected, what are the mean and standard deviation of the corresponding sampling distribution of the mean? What can be said about the shape of this sampling distribution?

SOLUTION The mean of the distribution of sample means is

$$\mu_{\bar{x}} = \mu = 270$$

However, the underlying population is a finite population with $N = 400$. Since the sample size $n = 25$ is more than 5% of 400, we must introduce the finite population correction factor. Thus,

$$\sigma_{\bar{x}} = \frac{\sigma}{\sqrt{n}} \sqrt{\frac{N - n}{N - 1}}$$

$$= \frac{30}{\sqrt{25}} \sqrt{\frac{400 - 25}{400 - 1}}$$

$$= \frac{30}{5} \sqrt{\frac{375}{399}}$$

$$= 5.817$$

Since the batting averages are normally distributed, the distribution of sample means is also normal with mean $\mu_{\bar{x}} = 270$ and standard deviation $\sigma_{\bar{x}} = 5.817$. (Notice that in Example 7.2 where the underlying population consisted of all baseball players, not just 400 current players, we did not need the correction factor and so $\sigma_{\bar{x}}$ was 6.) ∎

Finally, we consider one further question about the distribution of sample means: How large is it? Or equivalently, how many different samples are possible? Suppose your school has an enrollment of 3600 and we consider samples of size $n = 36$ students per sample. Many of you will undoubtedly conclude that there are precisely $3600/36 = 100$ possible samples. This is totally wrong. To see why, suppose we are interested in samples of size 2 rather than size 36. Suppose further that you insist that *you* personally must be part of the sample. Then you could be paired with any of 3599 different students, and so you alone would be part of 3599 different samples. (Of course, that would no longer be a random sample.) Next, suppose your best friend has to be part of every sample. He or she was already counted in one of the 3599 samples you "starred" in, but now can be paired with any one of the remaining 3598 students other than you. Thus, if we look at just the two of you, there are 3599

+ 3598 = 7197 different samples. We can continue this process, picking one student at a time as the "star," and so end up with a total of

$$3599 + 3598 + 3597 + 3596 + \cdots + 1 = 6{,}478{,}200$$

different samples of size 2. In fact, if we performed the same type of analysis with samples of size 36, then the total number of such samples would be far larger. Thus, we see that the distribution of sample means is a far larger population than the original one on which it is based. You will be asked to find an expression for the size of this distribution in the Exercise Set.

Fortunately, we will never have to be concerned with all members of this population. In practice, we use a single random sample drawn from the underlying population and so have to consider just one specific value $\bar{x}$ from the distribution of sample means.

Exercise Set 7.1

Mastering the Techniques

1. A population has mean $\mu = 80$ and standard deviation $\sigma = 12$. If random samples of size $n = 4$ are drawn from it, what are the mean $\mu_{\bar{x}}$ and the standard deviation $\sigma_{\bar{x}}$ of the corresponding distribution of sample means? What, if anything, can you conclude about the shape of this distribution?
2. Repeat Exercise 1 if $n = 9$.
3. Repeat Exercise 1 if $n = 16$.
4. Repeat Exercise 1 if $n = 36$.
5. Repeat Exercise 1 if $n = 50$.
6. Repeat Exercise 1 if $n = 72$.
7. A finite population of 600 measurements has mean 400 and standard deviation 30. If random samples of size $n = 16$ are drawn from it, give the mean and standard deviation of the corresponding distribution of sample means. What, if anything, can you conclude about the shape of the distribution of sample means?
8. Repeat Exercise 7 if $n = 25$.
9. Repeat Exercise 7 if $n = 100$.
10. Repeat Exercise 7 if $n = 144$.

Applying the Concepts

11. The mean GPA of the students at a college is normally distributed with mean 2.75 and standard deviation .60. Suppose groups of 25 students are studied. What are the mean and standard deviation for the distribution of sample means? What, if anything, can you conclude about the shape of this distribution?
12. The heights of the male students at a school are normally distributed with mean 69 inches and standard deviation 3 inches. If groups of 16 students are randomly selected, what are the parameters of the distribution of sample means? What, if anything, can you conclude about the shape of this distribution?
13. A recent study has found that the mean number of hours per week that people in the United States watch television is 41 with a standard deviation of 10 hours. If groups of 100 people are randomly selected, what are the parameters of the

distribution of sample means? What, if anything, can you conclude about the shape of this distribution?

14. Suppose that groups of 25 people are randomly selected from the population in Exercise 13. What are the parameters of the distribution of sample means? What, if anything, can you conclude about the shape of this distribution?

15. A certain brand of light bulb has a mean lifetime of 1200 hours with a standard deviation of 80 hours. If the bulbs are packaged 100 to a carton, what are the mean and standard deviation for the mean lifetime $\bar{x}$ of the 100 bulbs per carton?

16. Suppose that several dozen random samples of size $n = 25$ are drawn from a population whose mean μ and standard deviation σ are unknown. If the mean of all the $\bar{x}$'s found is 200 and the standard deviation of these $\bar{x}$'s is 15, what is your estimate of the true mean $\mu_{\bar{x}}$ and the true standard deviation $\sigma_{\bar{x}}$ of the distribution of sample means? What is your estimate of the mean μ and standard deviation σ of the original population?

17. Samples of size $n = 2$ are drawn from a population of size $N = 3600$, as in the text. Show that the number of possible samples is given by $_{3600}C_2$.

18. Samples of size n are drawn from a population of size N. Use combinations to express the total number of possible samples.

Computer Applications

19. Repeat the series of investigations we performed in the text using the skewed population (choice 2 in Program 20). In particular, use sample sizes of $n = 4, 9, 16, 25$ and 36, and observe the values for the mean and standard deviation and the shape of the distribution of sample means in each case.

20. Repeat Exercise 19 using the uniform distribution (choice 3).

21. Select 10 different random samples of size $n = 4$ from a set of stock market quotations in a recent newspaper. (Use Program 2: Random Number Generator to determine which entries to select for your samples.) Calculate the mean $\bar{x}$ for each of these 10 samples. Then calculate the mean and standard deviation for this set of 10 $\bar{x}$'s to approximate $\mu_{\bar{x}}$ and $\sigma_{\bar{x}}$ for the distribution of sample means. Finally, estimate the most likely values for μ and σ of the original population.

22. Repeat Exercise 21 using random samples from a current set of sports data, such as batting averages in baseball or number of yards gained rushing in football.

7.2 Applications of the Central Limit Theorem

In the last section, we introduced the distribution of sample means. It consists of the means $\bar{x}$ of all possible samples of size n drawn from an underlying population having mean μ and standard deviation σ. The mean $\mu_{\bar{x}}$ and the standard deviation $\sigma_{\bar{x}}$ of this sampling distribution of the mean are

$$\mu_{\bar{x}} = \mu$$

$$\sigma_{\bar{x}} = \frac{\sigma}{\sqrt{n}}$$

In addition, if the original population is normally distributed, then the distribution of sample means is also a normal distribution. Moreover, the Central Limit Theorem assures us that as long as the sample size $n > 30$, the distribution of sample means is approximately normal in shape no matter what the shape of the original population. Consequently, in either case, the techniques that we developed for normal distributions with mean μ and standard deviation σ can now be applied to the sampling distribution of the mean which has mean $\mu_{\bar{x}}$ and standard deviation $\sigma_{\bar{x}}$.

Consider the following question:

Problem The lifetime of a particular brand of light bulb is normally distributed with mean $\mu = 1200$ hours and standard deviation $\sigma = 80$ hours. The bulbs are packaged 100 to a carton.

(a) What is the probability that an individual bulb will last less than 1180 hours?
(b) What is the probability that the mean $\bar{x}$ of the 100 bulbs in a carton is less than 1180 hours?

Before we solve this problem, one important observation is in order. Part (a) is precisely the type of normal distribution probability problem we faced in Chapter 6 and so represents nothing new. The difference between parts (a) and (b) lies in the fact that the second question involves not the lifetime x for an individual bulb, but rather the mean lifetime $\bar{x}$ for a set of 100 bulbs.

To solve problems of this type, you must distinguish between probabilities for the value of a single normally distributed measurement x and probabilities for the mean $\bar{x}$ of a sample. In the first case, finding probabilities for a single measurement x requires using the original normal distribution with parameters μ and σ. The corresponding z values are calculated as

$$z = \frac{x - \mu}{\sigma}$$

In the second case, finding probabilities for the mean $\bar{x}$ of a sample requires using the distribution of sample means with parameters $\mu_{\bar{x}} = \mu$ and $\sigma_{\bar{x}} = \sigma/\sqrt{n}$. The corresponding z values are calculated as

$$z = \frac{\bar{x} - \mu_{\bar{x}}}{\sigma_{\bar{x}}} = \frac{\bar{x} - \mu_{\bar{x}}}{\frac{\sigma}{\sqrt{n}}}$$

To answer part (a), $P(x < 1180)$, we apply the normal distribution techniques of Chapter 6 to the normal population with $\mu = 1200$ and $\sigma = 80$. We first find

$$z = \frac{x - \mu}{\sigma} = \frac{1180 - 1200}{80} = -.25$$

FIGURE 7.8

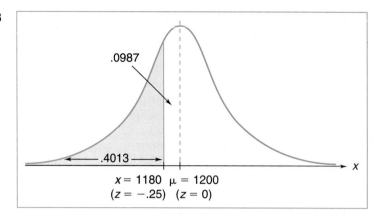

so that

$$P(x < 1180) = P(z < -.25) = .4013$$

as illustrated in Figure 7.8.

We now turn to part (b). What is the probability that the *mean* $\bar{x}$ of 100 of these bulbs is less than 1180? Symbolically, this is

$$P(\bar{x} < 1180)$$

To answer this question, we use the fact that the distribution of sample means is normally distributed since the original population is normal. Thus, the sampling distribution for the mean has parameters

$$\mu_{\bar{x}} = \mu = 1200 \quad \text{and} \quad \sigma_{\bar{x}} = \frac{\sigma}{\sqrt{n}} = \frac{80}{\sqrt{100}} = 8$$

We must find the area of the region to the left of $\bar{x} = 1180$ (as shown in Figure 7.9). The z value corresponding to $\bar{x} = 1180$ is

$$z = \frac{\bar{x} - \mu_{\bar{x}}}{\sigma_{\bar{x}}} = \frac{1180 - 1200}{8} = \frac{-20}{8} = -2.5$$

FIGURE 7.9

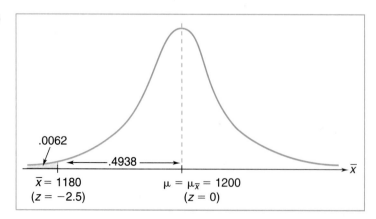

Therefore, the problem reduces to

$$P(\bar{x} < 1180) = P(z < -2.5) = .0062$$

Thus, we see that the probability that the *average* life of all 100 bulbs in a box is under 1180 hours is quite small. It is much smaller than the probability that a *single* bulb lasts less than 1180 hours. The reason for the difference in the probabilities lies in the difference between the values for σ and $\sigma_{\bar{x}}$ in the two normal populations. Since the sampling distribution has a much smaller standard deviation, 8 compared to 80, we expect the sample means $\bar{x}$ to be clustered much more tightly about the population mean 1200. Consequently, it is much less likely for the mean of a sample $\bar{x}$ to be below 1180 than it is for the value of a single measurement x to be below 1180.

EXAMPLE 7.5 A dentist finds that the mean elapsed time between regular checkups for her patients is 8.4 months with a standard deviation of 2.4 months. If a group of 36 of her patients is randomly selected, find the probability that the average time elapsed between office visits is between 8 and 9 months.

SOLUTION In this case, we are given

$$n = 36 \quad \mu = 8.4 \quad \text{and} \quad \sigma = 2.4$$

and are asked for the probability that the mean $\bar{x}$ of this sample is between 8 and 9:

$$P(8 < \bar{x} < 9)$$

While no specific mention is made about the shape of the original distribution, the Central Limit Theorem assures us that the distribution of sample means is approximately normally distributed since the sample size $n = 36 > 30$. Moreover, the mean and standard deviation of this sampling distribution are given by

$$\mu_{\bar{x}} = \mu = 8.4 \text{ months}$$

and

$$\sigma_{\bar{x}} = \frac{\sigma}{\sqrt{n}} = \frac{2.4}{\sqrt{36}} = .4 \text{ month}$$

See Figure 7.10. As a result, corresponding to $\bar{x} = 8$,

$$z = \frac{8 - 8.4}{.4} = \frac{-.4}{.4} = -1$$

and corresponding to $\bar{x} = 9$,

$$z = \frac{9 - 8.4}{.4} = \frac{.6}{.4} = 1.5$$

FIGURE 7.10

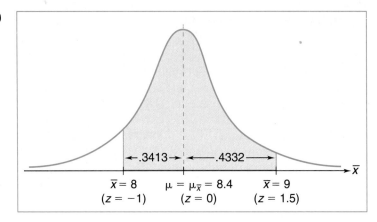

The desired probability is given by

$$P(8 < \bar{x} < 9) = P(-1 < z < 1.5)$$
$$= .3413 + .4332 = .7745$$

EXAMPLE 7.6 The grade-point averages for the students at a certain college are normally distributed with mean $\mu = 2.75$ and standard deviation $\sigma = .60$. For a randomly chosen class of 25 students, find the probability that the mean GPA is between 2.80 and 3.00.

SOLUTION Since the original population is normally distributed, the sampling distribution of the mean is also normally distributed with

$$\mu_{\bar{x}} = \mu = 2.75$$

and

$$\sigma_{\bar{x}} = \frac{\sigma}{\sqrt{n}} = \frac{.60}{\sqrt{25}} = .12$$

FIGURE 7.11

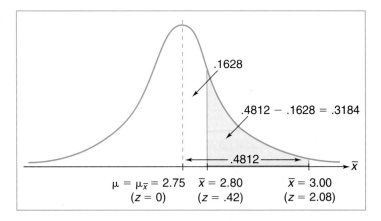

Consequently, as seen from Figure 7.11, the z values corresponding to $\bar{x} = 2.80$ and $\bar{x} = 3.00$ are, respectively,

$$z = \frac{2.80 - 2.75}{.12} = \frac{.05}{.12} = .42$$

and

$$z = \frac{3.00 - 2.75}{.12} = \frac{.25}{.12} = 2.08$$

Therefore,

$$P(2.80 < \bar{x} < 3.00) = P(.42 < z < 2.08)$$
$$= .4812 - .1628 = .3184 \quad \blacksquare$$

EXAMPLE 7.7 An airplane seats 144 passengers. It is designed for a maximum load of 215 pounds per person (individual weight plus luggage). Airline records indicate that the mean weight per person is 210 pounds with a standard deviation of 18 pounds. What percentage of full flights are overloaded?

SOLUTION A flight is overloaded if the mean weight per passenger exceeds 215 pounds. Thus, we have to find

$$P(\bar{x} > 215)$$

where the group of 144 people on any particular flight consists of a sample of size $n = 144$ drawn from the population of all airline passengers. This population has mean $\mu = 210$ and standard deviation $\sigma = 18$. The distribution of sample means, based on samples of size $n = 144$, has mean $\mu_{\bar{x}} = 210$ and standard deviation

$$\sigma_{\bar{x}} = \frac{18}{\sqrt{144}} = 1.5$$

Corresponding to $\bar{x} = 215$, we find that

$$z = \frac{215 - 210}{1.5} = \frac{5}{1.5} = 3.33$$

Therefore, as shown in Figure 7.12,

$$P(\bar{x} > 215) = P(z > 3.33)$$
$$= .5000 - .4996 = .0004$$

Therefore, .04% of full flights are overloaded. $\quad \blacksquare$

Incidentally, it is important to note that probability problems such as these can only be solved provided we are assured that the distribution of sample means is essentially normally distributed. That is, the standard deviation σ of the underlying population must be known and either the underlying population must be normal or the sample size n must be greater than 30.

FIGURE 7.12

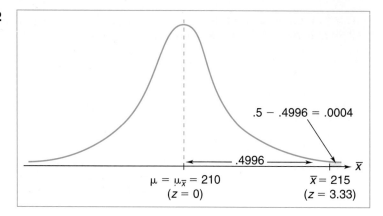

Exercise Set 7.2

Mastering the Techniques

1. Samples of size 36 are drawn from a population with mean 40 and standard deviation 12. Find
   (a) the mean and standard deviation of the distribution of sample means,
   (b) $P(40 \leq \bar{x} \leq 42)$
   (c) $P(40 \leq \bar{x} \leq 44)$
   (d) $P(38 \leq \bar{x} \leq 44)$
   (e) $P(37 \leq \bar{x} \leq 41)$
   (f) $P(41 \leq \bar{x} \leq 43)$
   (g) $P(\bar{x} \geq 43)$
   (h) $P(\bar{x} \leq 39)$
   (i) $P(\bar{x} \leq 41.5)$
2. Repeat Exercise 1 using a sample size of $n = 60$.
3. Samples of size 16 are drawn from a normally distributed population with mean 350 and standard deviation 60. Find
   (a) the mean and standard deviation of the distribution of sample means.
   (b) $P(350 \leq \bar{x} \leq 380)$
   (c) $P(350 \leq \bar{x} \leq 375)$
   (d) $P(338 \leq \bar{x} \leq 359)$
   (e) $P(336 \leq \bar{x} \leq 386)$
   (f) $P(368 \leq \bar{x} \leq 377)$
   (g) $P(\bar{x} \geq 370)$
   (h) $P(\bar{x} \leq 380)$
   (i) $P(\bar{x} \leq 337)$
4. Repeat Exercise 3 using a sample size of $n = 30$.

Applying the Concepts

5. A light bulb manufacturer claims that his bulbs last an average of 1000 hours with a standard deviation of 120. If 100 such bulbs are tested, find the probability that the mean lifetime of this sample is
   (a) Between 986 and 1000 hours
   (b) Between 992 and 1012 hours
   (c) More than 1020 hours
   (d) Under 990 hours
6. A movie theater estimates that the average patron spends $2.25 at the concession stand with a standard deviation of $1.20. If the theater seats 400 people, find the

probability that, for a randomly selected full-house performance, the mean expenditure for the group is (a) above $2.40; (b) between $2.10 and $2.20.

7. Management of a sports stadium estimates that the average fan spends $8.50 on food and other items with a standard deviation of $3.00. If the stadium typically attracts 10,000 fans to a game, find the probability that the mean amount spent by the crowd of 10,000 exceeds $8.55.

8. The weights of extra-large eggs are normally distributed with mean 2.5 ounces and standard deviation of .6 ounces. If a dozen such eggs are randomly selected, find the probability that the mean weight of the eggs is
   (a) Greater than 2.52 ounces
   (b) Less than 2.47 ounces
   (c) Between 2.49 and 2.51 ounces

9. As part of a reforestation project, a sample of 100 two-year-old saplings is tested. It is known that for such trees the mean diameter after 2 years is 1.6 cm with a standard deviation of .4 cm. Find the probability that, in this sample, the mean is
   (a) Greater than 1.5 cm
   (b) Greater than 1.7 cm
   (c) Under 1.55 cm

10. A restaurant owner has determined that the average dinner bill for her patrons is $38 with a standard deviation of $16. If the restaurant seats 64 people, find the probability that the mean bill for one complete seating (a) is under $40; (b) is between $40 and $43. (c) Find the probability that the total receipts for that seating are more than $2400.

11. The IRS claims that the average tax refund is $450 with a standard deviation of $240. If a random sample of 36 refunds are checked, find the probability that the mean is (a) under $400, (b) over $500, (c) between $430 and $490.

12. Suppose that the mean weight of cars (including passengers) on the road is 3500 pounds with a standard deviation of 1200 pounds. A narrow bridge over a river is designed for a maximum load of 144,000 pounds. If the length of the bridge is such that no more than 36 cars can fit on it at any given time, what is the probability that the bridge is overloaded? (*Hint:* The maximum capacity of the bridge should be averaged over the 36 cars.)

13. Explain why you cannot answer the following question by using the methods of this section: A carton containing 24 boxes of a certain cereal is randomly selected. Each box is supposed to contain 12 ounces with a standard deviation of .3 ounce. Find the probability that the mean weight of the boxes in this carton is below 11.9 ounces.

14. A tire manufacturer claims that its tires will last an average of 30,000 miles with a standard deviation of 3000 miles. A consumer organization tests a random sample of 36 tires and finds that they last an average of 28,000 miles.
   (a) What is the probability that the mean for such a sample is 28,000 miles or less?
   (b) Based on the probability in part (a) for a set of actual data, what is your assessment of the validity of the claim?

7.3 The *t*-Distribution for Small Samples

In the last two sections, we saw how the normal distribution arises and can be applied in sampling situations provided that the population standard deviation σ is known and that

1. We have a large enough sample ($n > 30$)   *or*
2. We start with an underlying population which is normally distributed

There is still the essential question of what happens when σ is not known. In this case we would like to use the sample standard deviation s (which can be calculated directly from the data) in place of the population standard deviation σ. There are two separate possibilities:

1. The underlying population is approximately normal   *or*
2. The underlying population is clearly not normal or is simply not known

In either event, we cannot assume that the distribution of sample means is normal. In the first case, the sampling distribution for $\bar{x}$ involves a new distribution which we will introduce in this section. In the second case, some of the nonparametric statistical methods to be developed in Chapter 13 may apply.

Suppose we start with a population which is approximately normal having mean μ and unknown standard deviation σ. The corresponding sampling distribution of the mean $\bar{x}$ involves what is known as the *t-distribution*, or *Student's t-distribution*. (The statistician who developed the theory, W. S. Gosset, used the pen name *Student* because his employers would not allow him to publish the results of his "top-secret" statistical work in a brewery.)

> We use the *t*-distribution whenever σ is unknown and the underlying population is approximately normal. It is based on **t values** which are calculated as
> $$ t = \frac{\bar{x} - \mu_{\bar{x}}}{s/\sqrt{n}} $$

Notice how this formula for t values parallels the one for calculating z values

$$ z = \frac{\bar{x} - \mu_{\bar{x}}}{\sigma/\sqrt{n}} $$

when the distribution of sample means is normal and σ is known. Further, as with z values, t values measure the number of standard deviations away from the mean $\mu_{\bar{x}}$ corresponding to a particular value of $\bar{x}$. While it is the values of the quantity

$$\frac{\bar{x} - \mu_{\bar{x}}}{(s/\sqrt{n})}$$

which follow a *t*-distribution and not the values of $\bar{x}$ themselves, we will not concern ourselves with this distinction. That is, we speak of the $\bar{x}$'s as following a *t*-distribution for the sake of simplicity, even though it is technically

$$\frac{\bar{x} - \mu_{\bar{x}}}{(s/\sqrt{n})}$$

which does so.

In actuality, for any given sample size n, there is a different *t*-distribution. That is, there is one *t*-distribution for samples of size $n = 2$, another for samples of size $n = 3$, and so on indefinitely. However, those based on samples of size $n > 30$ can be approximated by a normal distribution for all standard statistical purposes. These *t*-distributions, one for each value of the sample size n, clearly depend on n and on a related quantity called the *degrees of freedom*, which is defined as 1 less than the sample size. That is,

> Degrees of freedom (df) = $n - 1$

Thus, if we have samples of size $n = 18$ drawn from a roughly normal population, the corresponding distribution of sample means is a *t*-distribution having $n - 1 = 17$ degrees of freedom.

In all subsequent developments in this book involving small samples ($n \leq 30$) when σ is not known, unless specifically stated otherwise, we assume that the underlying population is approximately normal so that the corresponding distribution of sample means follows a *t*-distribution having $n - 1$ degrees of freedom.

You can get a feel for the shapes of a variety of *t*-distributions by using Program 21: *t*-Distributions. The program first draws a standard normal distribution curve for comparison (see Figure 7.13 for the sample output) and then graphs a sequence of *t*-distribution curves. The first curve corresponds to the *t*-distribution with 1 degree of freedom (sample size $n = 2$). Notice that the curve is much flatter than the normal curve; it doesn't rise as much at the center (at the mean μ), and it dies off much more slowly in both tails. However, as with the normal distribution, the *t*-distribution is centered at the mean μ, is symmetric about the mean, and the total area under the curve is also precisely 1 square unit.

The next graph drawn corresponds to the *t*-distribution with 5 degrees of freedom ($n = 6$). It is higher at the center and dies off somewhat more rapidly than the *t*-distribution with 1 degree of freedom. However, it is still quite different from the normal distribution curve. The subsequent graphs drawn

FIGURE 7.13

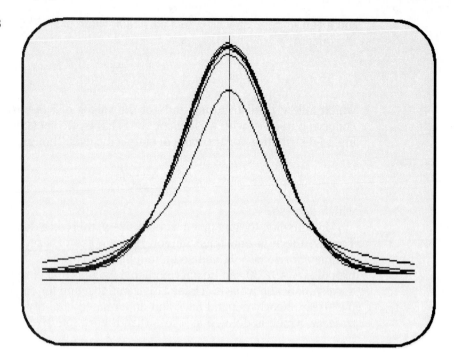

by the program are the curves for the *t*-distributions with 10 degrees of freedom ($n = 11$), 15 degrees of freedom ($n = 16$), 20 degrees of freedom ($n = 21$), and finally 25 degrees of freedom ($n = 26$). Each curve is successively closer to the normal distribution curve. If the demonstration were to proceed past 29 degrees of freedom ($n = 30$), you would not be able to distinguish visually between the *t*-distribution and the normal distribution. Thus, whenever $n > 30$, we are justified in using the normal distribution as an approximation to the *t*-distribution.

Incidentally, the initial display has a horizontal range extending from 4 standard deviations below the mean to 4 standard deviations above the mean. You can "zoom in" on any portion of the graph by changing this interval as one of the available options after the first display is completed.

EXAMPLE 7.8 Samples of size $n = 16$ are drawn from an approximately normal population with mean $\mu = 200$. Find the *t* values corresponding to sample means $\bar{x} = 224$ and $\bar{x} = 188$ if the sample standard deviation $s = 80$ in each case.

SOLUTION Since the sample size $n = 16 < 31$ and the original population is approximately normal, we have a *t*-distribution with $n - 1 = 15$ degrees of freedom. Thus,

$$\mu_{\bar{x}} = \mu = 200$$

and

$$\frac{s}{\sqrt{n}} = \frac{80}{\sqrt{16}} = \frac{80}{4} = 20$$

Therefore, corresponding to a sample mean of $\bar{x} = 224$,

$$t = \frac{224 - 200}{20} = 1.20$$

and corresponding to $\bar{x} = 188$,

$$t = \frac{188 - 200}{20} = \frac{-12}{20} = -.60$$

Although we could solve probability problems such as those in Section 7.2 involving means of small samples drawn from nearly normal populations with unknown standard deviation, this is not a particularly important application of the *t*-distribution and we will not consider it. Instead, we will return to the *t*-distribution in Chapter 8 to handle more significant situations involving sampling.

Exercise Set 7.3

Mastering the Techniques

1. Samples of size $n = 9$ are drawn from a roughly normal population with mean $\mu = 25$. Find the *t* values corresponding to sample means $\bar{x} = 48$, $\bar{x} = 54$, $\bar{x} = 40$, and $\bar{x} = 42.3$ if $s = 12$ in each case.
2. Samples of size $n = 25$ are drawn from a roughly normal population having mean $\mu = 2400$. Find the *t* values corresponding to sample means $\bar{x} = 2700$, $\bar{x} = 3300$, $\bar{x} = 2940$, $\bar{x} = 1993$, and $\bar{x} = 2222$ if $s = 800$ in each case.
3. Samples of size 18 are drawn from a roughly normal population having mean $\mu = 6.8$. Find the *t* values corresponding to sample means $\bar{x} = 7$, $\bar{x} = 7.1$, $\bar{x} = 6.5$, and $\bar{x} = 6.73$ if $s = 1.32$ in each case.
4. Samples of size 5 are drawn from a roughly normal population having mean $\mu = 9.4$. Find the *t* values corresponding to sample means $\bar{x} = 9.5$, $\bar{x} = 9.47$, and $\bar{x} = 9.32$ if $s = 2.35$ in each case.
5. Repeat Exercise 1, using samples of size $n = 4$. Compare the relative size of the *t* values in each case.
6. Repeat Exercise 2, using samples of size $n = 6$. Compare the relative size of the *t* values in each case.

Applying the Concepts

Each of the histograms in Figure 7.14 represents the display for a set of sample data. Indicate which ones suggest that the underlying population *may be* approximately normal and which do not. Give reasons in each case.

FIGURE 7.14

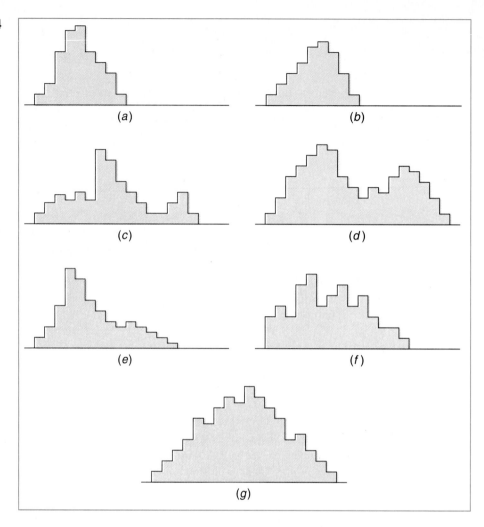

7.4 Overview of Probability Situations

Up to this point, we have considered a wide variety of situations involving probability. As such, it is not surprising if some of you feel that all these problems are beginning to look the same and you are having trouble distinguishing one from another. For that reason, we include this section to summarize and hopefully to provide some perspectives and suggestions on identifying some of the most important classes of probability problems covered.

There are primarily four interrelated types of probability problems that students tend to confuse:

1. A binomial probability problem with a relatively small number of trials
2. A binomial probability problem with a large number of trials

3. A probability problem involving a single measurement x drawn from a normal population
4. A probability problem involving the mean $\bar{x}$ of a sample drawn from a population

There are other types of probability problems that you may encounter. However, these four classes are more typical of situations that occur in statistics than in probability, and hence we key on them in this section. Typical problems of each type are as follows:

1. What is the probability of obtaining 4 successes out of 7 when the probability of success is .6?
2. What is the probability of obtaining between 41 and 48 successes out of 70 trials if the probability of success is .6?
3. A set of measurements is normally distributed with mean $\mu = 40$ and standard deviation $\sigma = 8$. What is the probability that a single measurement falls between 37 and 49?
4. If a random sample of size $n = 36$ is drawn from a population with mean $\mu = 40$ and standard deviation $\sigma = 8$, what is the probability that the mean of the sample is between 37 and 41?

Three of these four types of problems can be solved by using the normal distribution. The standard problem of this kind is question 3. You are given the mean and standard deviation for a normal distribution and are asked to find the probability that x falls between two given values. To solve this, find the corresponding z values for $x = 37$ and $x = 49$, consult the normal distribution table, and, in this case, add the values for the areas under the normal distribution curve. See Section 6.2.

Question 4 reduces to the same type of situation once you make use of the Central Limit Theorem and properties of the distribution of sample means. Remember, when σ is known, we can use a normal distribution to describe the distribution of sample means if either the underlying population is normally distributed or the sample size is greater than 30. Thus, we have to determine the mean and standard deviation for this sampling distribution. We know that

$$\mu_{\bar{x}} = \mu = 40$$

and

$$\sigma_{\bar{x}} = \frac{\sigma}{\sqrt{n}} = \frac{8}{\sqrt{36}} = 1.33$$

By using these parameters, the solution to the problem now involves precisely the same steps as the solution to question 3. See Section 7.2.

Question 2 is based on a binomial process with $n = 70$ trials and probability of success $\pi = .6$. Since

$$n\pi = 42 > 5 \quad \text{and} \quad n(1 - \pi) = 28 > 5$$

the binomial distribution can be approximated by a normal distribution having mean

$$\mu = n\pi = 70(.6) = 42$$

and standard deviation

$$\sigma = \sqrt{n\pi(1-\pi)} = \sqrt{70(.6)(.4)} = \sqrt{16.8} = 4.099$$

We must remember to introduce the continuity correction since this problem involves approximating the discrete binomial distribution with a continuous normal distribution. Since we seek the probability of obtaining between 41 and 48 successes, we "stretch" each of these values so that we consider x between 40.5 and 48.5 and then apply the normal distribution procedure to obtain the answer. See Section 6.3.

The only question that cannot be solved by using the normal distribution is question 1, which is also a binomial probability problem with $n = 7$ and $\pi = .6$. Since

$$n\pi = 4.2 < 5 \quad \text{and} \quad n(1-\pi) = 2.8 < 5$$

the normal approximation to the binomial distribution does not apply. However, this problem can be solved by

1. Using the binomial probability formula
2. Using the binomial distribution table, Table II (provided that the required entries for n and π appear in the table)
3. Using Program 13: Binomial Probability

See Section 5.3. If we solve this problem by using the binomial probability formula, we obtain

$$P(4 \text{ successes out of } 7) = \frac{n!}{x!(n-x)!} \pi^x (1-\pi)^{n-x}$$

$$= \frac{7!}{4!(7-4)!} (.6)^4 (.4)^{7-4} = .2903$$

To distinguish between these four types of probability problems, ask yourself:

Is it a binomial probability problem (typically containing a phrase such as *so many successes out of so many*)?

If it is binomial, is n large [check that both $n\pi \geq 5$ and $n(1-\pi) \geq 5$ for the normal approximation] or small (binomial formula)?

Is it a normal probability problem (typically containing values for the mean and standard deviation)?

If it is normal, does it involve the sample mean $\bar{x}$ (requiring the distribution of sample means with $\mu_{\bar{x}}$ and $\sigma_{\bar{x}}$)?

When you face a collection of mixed probability problems, such as the ones below or the ones that might appear on a test, the best strategy is to read through the entire list of problems before you tackle any of them. As you read through the problems, keep the preceding ideas in mind and jot down your initial identification of each problem. Also recall that there are other types of probability problems which do not fall into any of these four categories. Therefore, you should be alert to the possibility that a particular problem is of a different nature altogether.

After you have made these initial identifications, go back to solve each problem in turn. You will be pleasantly surprised at how much easier they will appear if you properly identify each one first.

Exercise Set 7.4

Applying the Concepts

1. A die is specially constructed so that two of its faces are colored red while the remaining four faces are colored green. If such a die is rolled 4 times, find the probability that a green face turns up on precisely 3 rolls.
2. The die in Exercise 1 is rolled 72 times. Find the probability of obtaining between 40 and 50 green faces.
3. A municipal bus company finds that the distance that its buses travel in a month is normally distributed with mean $\mu = 6000$ miles and standard deviation 800 miles. What is the probability that a randomly selected bus travels between 6400 and 7200 miles in a particular month?
4. The northside depot for the buses in Exercise 3 services 64 buses. What is the probability that the mean use of the buses housed at this depot is less than 5800 miles in a particular month?
5. An amusement park owner finds that 30% of all customers ride the Crazy Coaster. If a group of 8 customers are selected at random, what is the probability that 4 take the ride?
6. Airline records indicate that the mean weight of luggage checked by passengers is 42 pounds with a standard deviation of 15 pounds. If a plane carries 144 passengers, find the probability that the mean weight of their luggage exceeds 45 pounds.
7. The number of tickets sold for a college's basketball games is normally distributed with mean 2400 and standard deviation 500. Find the probability that at a randomly selected game, between 2000 and 2500 tickets are sold.
8. The number of gallons of gasoline pumped at a gas station each week is normally distributed with a mean of 21,000 gallons and a standard deviation of 6000 gallons. If a random sample of 25 weeks is selected, find the probability that an average of 18,000 to 20,000 gallons is pumped.

9. Health department records in a certain city show that 25% of all children entering public school have not been immunized against diphtheria. If a group of 200 such children are studied, what is the probability that more than 60 have not received the vaccine?
10. A certain university provides full tuition scholarships to 8% of its entering first-year students. If 10 students are selected at random, find the probability that 2 have such scholarships.
11. Government records indicate that 60% of all college graduates in a certain state have not repaid their student loans within 5 years of graduation. If a group of 40 such graduates are randomly selected, find the probability that 25 have not repaid their loans.
12. The number of hours per week that students at a college work at jobs is normally distributed with a mean of 26 hours and a standard deviation of 2 hours. What percentage of the students work more than 30 hours per week?
13. A computer store displays four IBM computer models, two Apple models, one Compaq model and one ATT model. If a customer comes in and sits down in front of one of the display models at random, what is the probability that he or she has picked an Apple?
14. A study shows that children watch television an average of 48 hours per week with a standard deviation of 14 hours per week. If a group of 80 children are randomly selected, find the probability that they watch an average of 44 to 46 hours per week.
15. In a supermarket, 24% of the customers purchase something from the appetizer department. In a random group of 7 customers, find the probability that 4 buy something at the appetizer department.
16. For the supermarket in Exercise 15, the total sales (in dollars) at its appetizer department are normally distributed with mean $8.40 and standard deviation $2.75. Find the probability that a randomly selected customer on line at the counter purchases more than $12 in appetizers.
17. Find the probability that the average sale for a randomly selected group of 10 customers at the appetizer department in Exercises 15 and 16 is below $8.00.
18. At the appetizer counter in the previous problems, it is known that 80% of the customers purchase meat items, 35% purchase cheese items, and 23% purchase both. If a customer is randomly selected, what is the probability that person purchases either meat or cheese?
19. A TV network executive estimates that the average number of hours of sports telecasts that the typical U.S. adult male watches each year is normally distributed with a mean of 440 hours and a standard deviation of 82 hours. Based on these figures, what percentage of the adult male population watches more than 400 hours of sports each year?
20. A political poll conducted in a certain state prior to a Presidential primary shows that 21% of the voters favor a particular candidate. If a random group of 80 voters are surveyed, find the probability that fewer than 12 favor that candidate.
21. A computer disk manufacturer estimates that .5% of all its disks are defective. What is the probability that more than 8 disks are defective in an order of 1000 disks?

7.5 Investigating Other Sampling Distributions (Optional)

In this chapter, we have introduced the distribution of sample means consisting of the means $\bar{x}$ of all possible samples of size n drawn from some underlying population. The properties of this sampling distribution are summarized in the Central Limit Theorem. These ideas form the basis of inferential statistics. Throughout the next several chapters, we will develop some extremely powerful statistical methods that enable us to make predictions and decisions based on sample data.

Before proceeding, we investigate some related ideas. In Chapter 3, when we considered numerical measures, we introduced a variety of ways to describe data. We indicated that the most useful measure of central tendency is the mean. The other numerical measures of central tendency (the median, the mode and the midrange) are also useful in a variety of contexts. We therefore investigate sampling these quantities from finite populations.

Suppose we have a situation in which the median is the indicated measure for a set of data. We want to draw a random sample from an underlying population and calculate the median. In parallel to what we did earlier with the distribution of sample means, we consider the comparable sampling distribution consisting of the sample medians m of all possible samples of size n drawn from the underlying population. Thus, we introduce the *distribution of sample medians* or the *sampling distribution of the median*. This population will have its own mean, its own standard deviation, and its own population median. To work with these quantities, we use the following notation. The median of any sample is denoted by m, while the median of the underlying population is M. The sampling distribution will have mean μ_m and standard deviation σ_m. Furthermore, as we saw with the distribution of sample means, we must relate these parameters of the sampling distribution to the corresponding parameters μ, σ, and M of the original underlying population.

Although a detailed study of this distribution is beyond the scope of this course, we can experiment with the sampling distribution of the median using Program 22: Distribution of Sample Medians. The program is similar to Program 20: Central Limit Theorem Simulation. It has the same four underlying populations built in: a normal population, a skewed population, a uniform population and a U-shaped population. You can select any sample size up to $n = 50$ and any number of such samples.

For example, we might start with the normal population and consider samples of size $n = 4$. The program generates repeated random samples, calculates the *median* of each, and plots the median as a horizontal line segment. A typical set of graphical results based on 300 such samples is shown in Figure 7.15. We observe that the shape of this sampling distribution appears roughly normal, the average of the sample medians seems to be close to both the mean and the median of the original population, and the spread of the sampling distribution of the median is considerably smaller than that for the original population. In fact, the standard deviation for this sampling distribu-

FIGURE 7.15

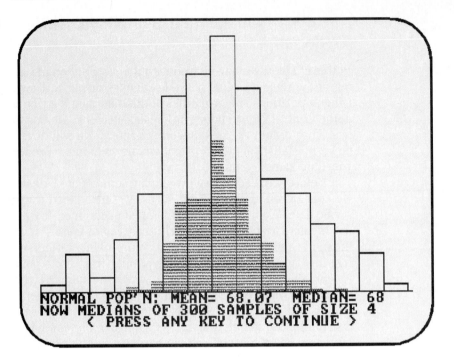

tion appears to be approximately one-half of the standard deviation of the underlying population. The accompanying numerical data show that the mean for the 300 sample medians is 68.09 compared to the population median $M = 68$ and the population mean $\mu = 68.07$. (Note that for symmetrical populations, such as the normal, the mean μ and the median M are usually very close.) Similarly, the standard deviation of these 300 medians is 1.39 compared to the population value of $\sigma = 2.76$. This is certainly close to the one-half that we concluded by visual inspection. (Note that the program does not calculate the median of all the sample medians because it would take far too much time.)

If we repeat this experiment with a larger sample size, say $n = 9$, then we obtain a result such as that shown in Figure 7.16. The overall shape of the distribution of sample medians is again roughly normal in appearance. The mean of all the sample medians is close to the population median M. Finally, the spread is considerably smaller than that for the underlying population and, in fact, appears to be roughly one-third of the spread. The corresponding numerical values for these 300 sample medians show a mean of 67.89 and a standard deviation of 1.12.

If we run the program once more, using 250 samples of size $n = 16$, then we obtain a graph such as that in Figure 7.17. The mean and standard deviation for the 231 sample medians are, respectively, 67.93 and .69. Thus, we again see that the mean of the sample medians is close to the population

FIGURE 7.16

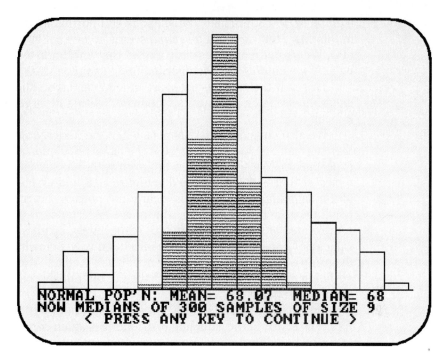

FIGURE 7.17

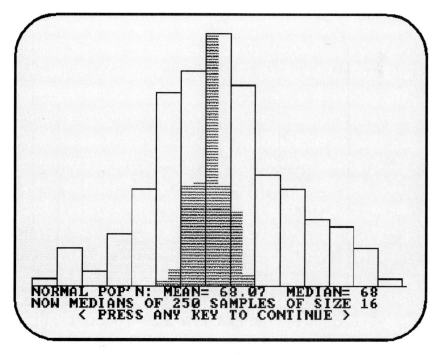

median $M = 68$. Further, the shape is still normal in appearance. However, in this case, the standard deviation seems somewhat larger than one-quarter of σ. We should not necessarily expect that a formula for σ_m will turn out to be $\sigma/\sqrt{n}$ as it does for the distribution of sample means where $\sigma_{\bar{x}} = \sigma/\sqrt{n}$.

To see if these conjectures are indeed valid, you should continue these experiments using the normal population with still larger values for n, say, $n = 25$ or $n = 36$.

We also consider what happens when the underlying population is other than normal. Suppose we select the U-shaped population and repeat the above experiments. We first run the program with 300 samples of size $n = 4$. A typical set of results is displayed in Figure 7.18, and we observe that the shape of the sampling distribution is clearly not normal. Furthermore, the corresponding numerical values for the mean and standard deviation of the 300 sample medians are 68.96 and 3.41, respectively, compared to the population values of $M = 69$ and $\sigma = 4.80$. Consequently, we conclude that the mean of the sample medians is close to the population median. Moreover, the standard deviation of the medians is smaller than σ, but not approximately the one-half we might expect.

Similarly, the results of 250 simulations with $n = 10$ are shown in Figure 7.19. The shape of the sampling distribution is again certainly not normal. The mean and standard deviation of the sample medians are 68.72 and 3.04, respectively. The mean of the medians is still close to M and the standard de-

FIGURE 7.18

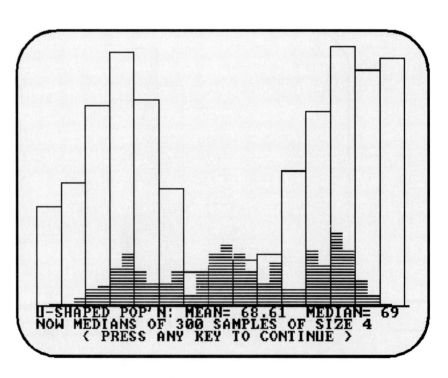

FIGURE 7.19

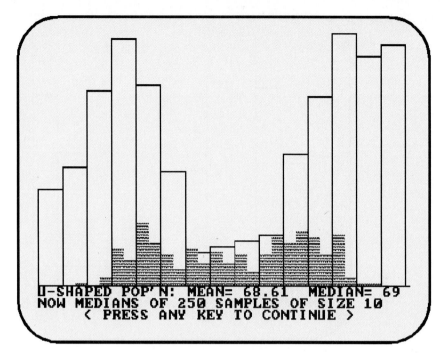

viation is somewhat smaller, but there appear to be no predictable patterns to them.

Finally, a typical set of results with $n = 20$ is shown in Figure 7.20, with accompanying numerical values of 68.68 for the mean of the medians and 2.6 for the standard deviation.

We suggest that you continue these experiments to see the effects of using a still larger sample size with the U-shaped distribution. Similarly, repeat the entire series of experiments with the other two populations. Based on these cases, what conclusions can *you* come to regarding the shape of this distribution, especially as the sample size *n* increases? What can you conclude about the values for the parameters μ_m and σ_m for the distribution of sample medians?

Just as we studied the distribution of sample medians, we can consider the *distribution of sample modes* or the *sampling distribution of the mode*. We denote the sample modes by *m*. Moreover, this sampling population has a mean μ_m, a standard deviation σ_m, and a mode M (or possibly several modes if it is bimodal or even multimodal). We again use the computer as an exploratory tool to see what we can conjecture about this distribution using Program 23: Distribution of Sample Modes. To do so, we repeat the steps we used above to explore the distribution of sample medians.

We select the normal population and have the program generate 300 random samples of size $n = 4$. A typical set of results is shown in Figure 7.21.

FIGURE 7.20

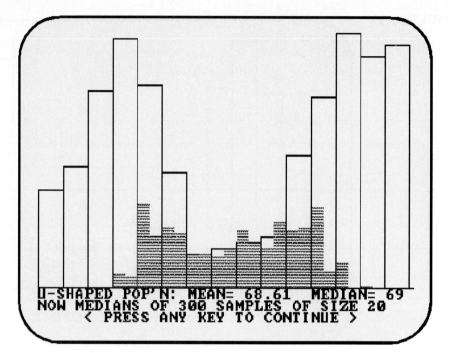

FIGURE 7.21

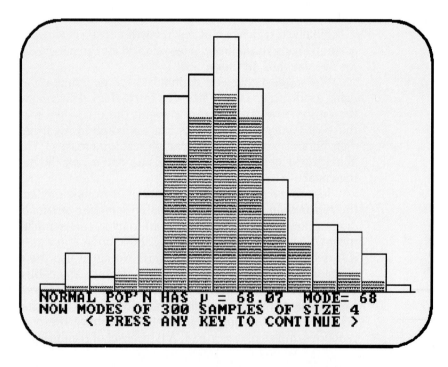

One rather striking observation that we can make immediately is that the shape of this sampling distribution seems to follow the shape of the underlying normal population very closely. The accompanying numerical data seem to bear this out. The mean for the 300 sample modes is 68.03 compared to the population mode $\mathcal{M} = 68$. The standard deviation for the sample modes is 2.03 compared to the population standard deviation $\sigma = 2.76$. The value for the sample modes is smaller.

If we repeat this series of experiments with a larger sample size, say $n = 9$, then the results are fairly comparable. A typical display is shown in Figure 7.22, where we again see that the shape of the sampling distribution is remarkably similar to that of the original population. The numerical values for these 300 sample modes show a mean value of 67.85, a standard deviation of 2.13, and a modal value of 68. Other series of sample runs based on 250 samples with sample sizes of $n = 16$ and $n = 25$ are shown in Figures 7.23 and 7.24, respectively. The corresponding numerical values are summarized as follows:

$n = 16$ mean $= 67.91$ sd $= 1.71$ mode $= 68$

$n = 25$ mean $= 67.84$ sd $= 1.39$ mode $= 68$

We might therefore conjecture that the shape of the sampling distribution for the mode is almost the same as the original normal population, but with a smaller spread which appears to decrease slightly as the sample size n in-

FIGURE 7.22

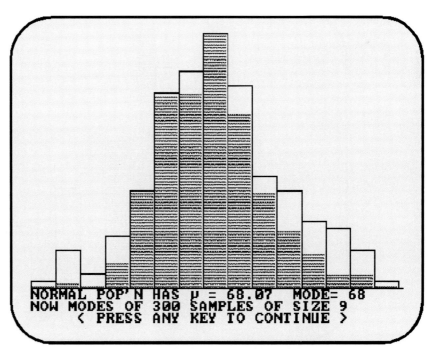

FIGURE 7.23

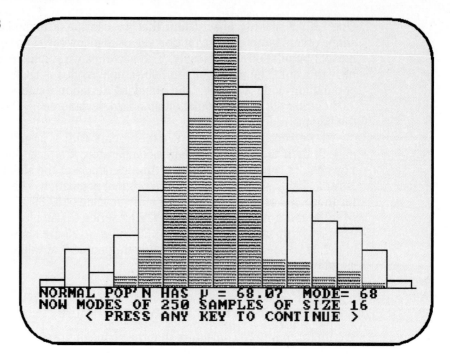

FIGURE 7.24

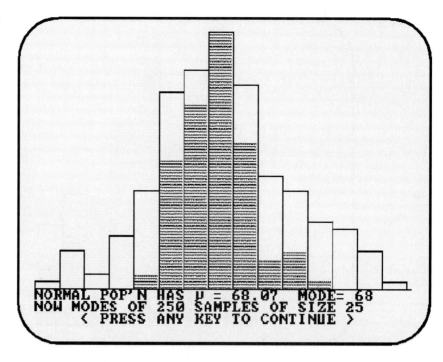

creases. Moreover, the mean of the modal values is extremely close to the population mode $M = 68$.

In Figures 7.25, 7.26, and 7.27, we illustrate typical results when using the program with the U-shaped population with samples of size $n = 4, 9$, and 16. The corresponding numerical results are

$n = 4$:   mean = 69.22   sd = 4.84   mode = 73

$n = 9$:   mean = 69.15   sd = 4.92   mode = 64

$n = 16$:   mean = 69.42   sd = 4.83   mode = 73

Compare these to the population values of $M = 73$, $\sigma = 4.80$ and $\mu = 68.61$. Notice that the mode of the 300 samples of size 9 was 64 which is quite different from the population mode of 73. If you examine the shape of the underlying U-shaped population, you will see that it is approximately bimodal. That is, while there is only one mode $M = 73$, the value $x = 64$ has almost the same frequency. In this particular run of the program, this value happened to predominate.

Therefore, the mode of the sample modes seems to come out close to the highest values, but may not approximate the population mode M if the underlying population is roughly multimodal. Further, the mean of the sample modes seems to be closer to the population mean μ.

We suggest that you continue this series of experiments with the distribution of sample modes by first increasing the sample size n and then utilizing

FIGURE 7.25

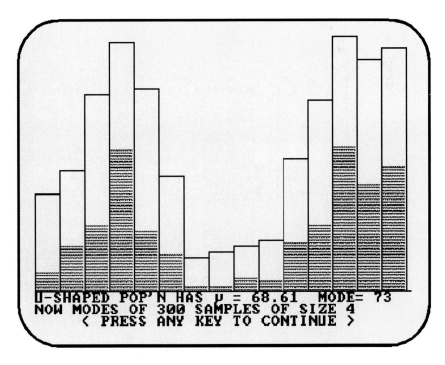

FIGURE 7.26

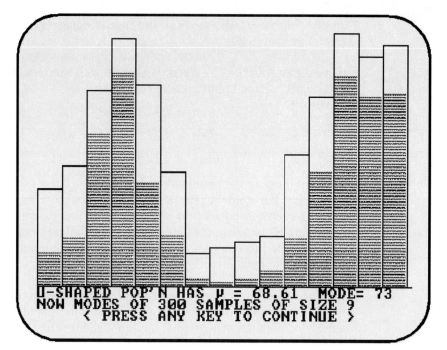

FIGURE 7.27

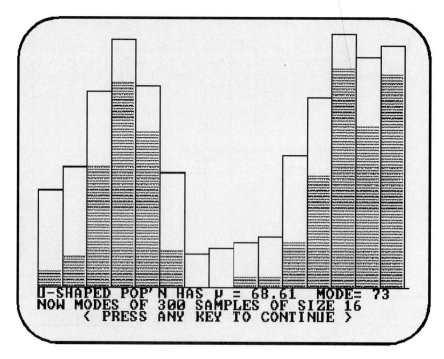

the other two underlying populations available. Are the above conclusions still appropriate when you do this?

In a totally similar way, we can also consider the *distribution of sample midranges* or the *sampling distribution of the midrange*. Rather than investigate it here, we leave it for you to experiment with this sampling distribution as an independent exercise at the end of the section using Program 24: Distribution of Sample Midranges.

Before leaving this subject, we should point out that many other sampling situations arise throughout statistics. For example, we might want to study the distribution of the sample standard deviations s to get information on the spreads within samples. We might want to consider the possibility of sampling from a binomial population to study the effects on the sample proportion. We might want to consider comparing statistics obtained from simultaneous sampling from two distinct populations. All these cases (as well as many others) occur later in this book, and we will experiment with each one of them using computer simulations.

Exercise Set 7.5

Computer Applications

Repeat the exploration process we used above to study and come to some tentative conclusions regarding the sampling distribution of the midrange by using Program 24: Distribution of Sample Midranges. As you proceed with this investigation, remember the primary questions to answer:

1. What can you conclude about the shape of this distribution, particularly as the sample size n increases?
2. What can you conclude about the mean of this distribution?
3. What can you conclude about the standard deviation of this distribution?
4. What can you conclude about the midrange of this distribution?

Have your entire statistics class select a topic such as one of the following for a joint student project on sampling:

1. Heights of female students at your college
2. Weights of male students at your college
3. Shoe sizes of women in your community
4. Stock prices (probably the closing prices) for companies listed in the current stock market reports in your local newspaper
5. Current sports results (batting averages in baseball, field goal percentage in basketball, total yards gained by players in football, number of goals scored by players in hockey, etc.)

Each of you should select a random sample from the chosen population of size $n = 36$ and collect the pertinent data from that sample. If appropriate, use either the table of random digits, Table I, or Program 2: Random Number Generator to select the sample randomly.

Once you have these data, calculate the sample mean for your sample. Your class should then pool all your sample results together so that each one of you contributes a single sample mean $\bar{x}$ from the distribution of sample means from the selected population.

Now that each of you has the full set of sample means collected by the entire class, do the following:

(a) Organize the set of data.
(b) Construct an appropriate relative frequency distribution.
(c) Construct the associated histogram, bar chart, and/or frequency polygon.
(d) Calculate the mean and the standard deviation for the set of all sample means.
(e) Use these results to estimate the most likely values for the parameters $\mu_{\bar{x}}$ and $\sigma_{\bar{x}}$ of the distribution of sample means.
(f) Estimate the most likely values for the parameters μ and σ of the original population from which you sampled.

Finally, you should incorporate all these items into a formal statistical project report. The report should include

- A statement of the topic being studied
- The source of the data
- A discussion of how the data were collected and why you believe that they represent a set of random samples
- Two sets of tables and diagrams displaying your set of sample data and the combined sets of all the sample means
- From the display on your set of sample data, a discussion of whether the display suggests that the underlying population being studied appears to be approximately normal
- Your predictions about the values for the parameters of both the original population and the sampling distribution
- A discussion of any surprises you may have noted in connection with collecting or organizing the data

CHAPTER 7 SUMMARY

The **distribution of sample means** or the **sampling distribution of the mean** consists of the means $\bar{x}$ of every possible sample of some size n drawn from a population having mean μ and standard deviation σ.

If the underlying population is normal, then the distribution of sample means is also normal.

The Central Limit Theorem states:

1. If the underlying population is not normal, then the distribution of sample means is approximately normal provided the sample size n is large enough. (Large enough is usually $n > 30$.)
2. $\mu_{\bar{x}} = \mu$
3. $\sigma_{\bar{x}} = \dfrac{\sigma}{\sqrt{n}}$

If samples of size n are drawn from a finite population of size N, then it is necessary to use the **finite population correction factor** to calculate $\sigma_{\bar{x}}$:

$$\sigma_{\bar{x}} = \dfrac{\sigma}{\sqrt{n}} \sqrt{\dfrac{N-n}{N-1}}$$

You can answer questions regarding probabilities involving the sample mean $\bar{x}$ when μ and σ are given by using the fact that the distribution of sample means is approximately normal whenever $n > 30$ or if the original population is normal. Simply use $\sigma_{\bar{x}}$ as the standard deviation instead of σ.

If the original population is approximately normal and σ is not known, then the statistic

$$t = \dfrac{\bar{x} - \mu_{\bar{x}}}{s/\sqrt{n}}$$

follows a *t*-distribution with $n - 1$ **degrees of freedom**. As n increases, the *t*-distribution approaches a normal distribution.

Review Exercises

1. The typical American adult consumes an average of 25 pounds of bananas each year with a standard deviation of 6 pounds. If a random group of 100 adults is surveyed, what are the mean and standard deviation of the sampling distribution? What can you conclude about the shape of the sampling distribution? What can you conclude about the shape of original distribution?
2. Given the information in Exercise 1 plus the fact that the consumption of bananas is normally distributed, find that probability that
   (a) A single individual consumes between 23 and 26 pounds of bananas in a given year
   (b) The average consumption for a random group of 100 adults is between 23 and 26 pounds in a given year
3. The average response times to emergency calls to a city's 911 line is 5.4 minutes with a standard deviation of 2.8 minutes. If the response times are normally distributed, find the probability that
   (a) An individual caller waits more than 8 minutes for an emergency services vehicle to arrive
   (b) The average wait on a random sample of 36 calls is greater than 8 minutes
4. Suppose the ESP test that we considered in Chapter 5 is administered to groups of 36 people.

(a) What are the mean and standard deviation for the binomial population associated with the ESP test?
(b) What are the mean and standard deviation for the distribution of sample means for the average number of correct guesses for groups of 36 people?
(c) Based on what you know about the binomial distribution for this experiment, what can you conclude about the shape of the sampling distribution?
(d) What is the probability that the average number of correct guesses by a group of 36 people is more than 7?

8 Estimation

In the NEWS

Today's college graduate leaves school with much more than the traditional sheepskin diploma. New graduates with a bachelor's degree receive an average starting salary of $27,037, ranging from the lowest average starting salary of $20,030 for retail workers to the highest average of $38,394 for chemical engineers.

Most of the new grads also leave college owing a large debt to the bank and the government. 63% of all graduates have to pay back a variety of educational loans which average over $18,000 per person. Also, while the bachelor's degree is typically thought of as a 4-year program, the average student now takes 5.4 years to complete it.

In addition, graduates holding a master's degree can expect to receive an average starting salary of $33,660 while those with a Ph.D. can expect to receive an average of $38,100. While this may not seem to be a big difference compared with the money offered to graduates with just a bachelor's degree, it can turn into an extremely large difference over the course of an entire career in some fields. For instance, an engineer with a master's degree can expect to make, on the average, over a million dollars more during his or her entire career than an engineer with only a bachelor's degree. An engineer with a Ph.D. can expect to make an average of over a million dollars more than one with only a master's degree.

The margin of error in this study is 3 percentage points.

Studies such as this one are conducted routinely to determine our achievements, our opinions and our behavior patterns in all aspects of our lives. Important decisions are made based on the results of such surveys, which then affect our lives. For instance, studies such as the one above are used by Congress in formulating policies about student aid programs. They are also used by large companies in determining what salaries to offer to graduates. They are used to counsel incoming students to make academic and career decisions based on current salaries.

But how accurate are such studies? Do they involve a sample that is truly representative of the entire population being studied? Do they apply "across the board" or are there other factors such as geographic location, the school a person graduates from, gender, race or age that are significant? What about the variation in the individual data? Does the overall average, in this case $27,037, tell us the whole story or should we also wonder about the spread? Certainly, the report above indicates a huge difference (almost double) in annual starting salary between people who go into retailing and those in chemical engineering. Within each of these subgroups, how large is the spread? Finally, what is the meaning of the 3 percent "margin of error" quoted in the above news item?

In this chapter, we will discuss the statistical methods used to make the kinds of estimates of averages and percentages reported on in this type of study.

8.1 Confidence Intervals for Means

We have previously developed the necessary foundation for applications of statistical inference. There are two primary branches. The first uses the data from a single sample to estimate or predict the mean or other parameter for an underlying population. The second uses the data from a sample to come to some decision about a population by testing a claim or hypothesis regarding a parameter of that population. This chapter is devoted to estimation. Hypothesis testing is introduced in Chapter 9.

We consider, once more, the problem of predicting or estimating the mean height of the student body at your school based on the data we collected about your hypothetical class. We used a value for the sample mean $\bar{x}$ of 66 inches with a standard deviation s of 3 inches for a group of $n = 36$ students. The population of all students at your school has unknown mean height μ and standard deviation σ.

> Our objective is to predict or **estimate** the value for the unknown mean μ based on the data from this one sample.

We previously indicated that the best *point estimate* for μ is 66 inches also. However, it is unreasonable to expect that the mean for a large population will precisely equal the mean of a randomly chosen sample drawn from it. It is more reasonable to conclude that the population mean μ is close to the sample mean $\bar{x}$. The problem we face is in deciding just what we mean by the phrase *close to*.

If a point estimate does not suffice, the alternative is to use an interval and conclude that the population mean μ lies within that interval. If we want to estimate the value of μ based on our sample mean $\bar{x} = 66$, then it is clear that μ may be larger than $\bar{x}$ or smaller than $\bar{x}$. We certainly cannot anticipate which direction to go. As a result, any prediction we make should take the form of an interval or range of values for the height centered at the sample mean $\bar{x}$. For instance, we might believe that an appropriate estimate for μ is the interval

	from 65.5 to 66.5
or	from 65 to 67
or	from 64 to 68
or	from 60 to 72

In each case, the interval is centered at $\bar{x} = 66$ inches. The first of these possibilities, 65.5 to 66.5, is certainly a more accurate prediction than just the single value of the point estimate 66 inches. However, it probably seems too "narrow" in the sense that it is not likely that the true mean μ will be that close to $\bar{x} = 66$. That is, if this interval were the prediction, you likely would not have a high degree of confidence in its accuracy.

The second suggested interval, from 65 to 67, is perhaps better in the sense that it is far more likely that this interval will contain the population mean μ. Consequently, you would likely have a greater degree of confidence in this prediction.

> As the length of the estimating interval increases, our degree of confidence in its actually containing the population mean μ also increases.

However, this line of reasoning should not be continued indefinitely. At the other extreme, the interval from 60 to 72 is virtually certain to contain μ. It is such a wide interval that there would be little doubt that μ is in it. For that matter, we could have reached this conclusion without having gone to the trouble of collecting a sample and performing the statistical calculations. In other words, this interval is so large that it really makes little use of the sample data. Furthermore, if it were used as the basis for a prediction about the value of μ, it is such a large interval that the prediction would be essentially worthless. After all, what information can be gained by stating that the mean height of all students at a college is between 60 and 72 inches?

> If an interval is too large, then the corresponding prediction is worthless.

What we require is a compromise between the two extremes. We have to give up total certainty regarding the accuracy of the prediction in order to achieve a small enough interval to give the estimate a semblance of usefulness. But we cannot demand too narrow an interval, for then it would be too unlikely for the interval to actually contain the unknown mean μ. Consequently, it is necessary to settle on some intermediate interval. In the process, we must accept a reasonable chance of being wrong in making this estimate. The question is, How large a chance of error are we willing to accept? Are we willing to accept a 90% chance of being right? A 95% chance? A 99% chance? Such a percentage is known as the *confidence level* of the estimate, and the corresponding interval we construct is called a *confidence interval*.

Suppose we select a confidence level of 95%. That is, we want to construct a confidence interval centered at the sample mean $\bar{x}$ which contains the true population mean μ with 95% certainty. This is equivalent to saying that the probability is .95 that the 95% confidence interval actually contains the unknown mean μ. Alternatively, if this procedure were conducted repeatedly by constructing many different estimates or confidence intervals based on different random samples, then we would be correct 95% of the time.

The figure of 95% undoubtedly reminds you of one of the primary characteristics of the normal distribution—namely, that about 95% of the population falls within 2 standard deviations of the mean. This is no coincidence, as you will see.

The sample mean $\bar{x}$ for the particular sample we selected is just one of the many sample means possible. It is one value from the distribution of sample means based on samples of size $n = 36$ drawn from the underlying population having unknown population mean μ and standard deviation σ. According to the Central Limit Theorem, the distribution of sample means has mean

$$\mu_{\bar{x}} = \mu$$

and standard deviation

$$\sigma_{\bar{x}} = \frac{\sigma}{\sqrt{n}}$$

Moreover, since $n = 36 > 30$, the distribution of sample means is essentially a normal distribution whatever the shape of the underlying population. As a result, approximately 95% of all sample means $\bar{x}$ fall within 2 standard deviations of the unknown mean $\mu_{\bar{x}} = \mu$. That is, with about 95% certainty, the sample mean $\bar{x}$ from our random sample will lie within 2 standard deviations of μ. Of course, having calculated $\bar{x}$, we don't know for certain if it does in fact lie within 2 standard deviations of μ, since we don't know the value of μ. Moreover, we don't even know if $\bar{x}$ is above or below μ. See Figure 8.1.

FIGURE 8.1

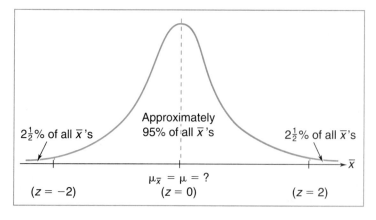

Suppose that $\bar{x}$ is actually within 2 standard deviations of μ. If we construct an interval centered at $\bar{x}$ and extend it 2 standard deviations, or $2\sigma_{\bar{x}}$, in each direction, then it will certainly contain μ. This occurs for approximately 95% of all the possible $\bar{x}$ values. See Figure 8.2a.

On the other hand, suppose that $\bar{x}$ is more than 2 standard deviations away from μ. If we likewise construct an interval centered at $\bar{x}$ which extends 2 standard deviations in each direction, then it will not reach far enough to contain μ. This will occur for approximately 5% of all possible $\bar{x}$ values. See Figure 8.2b. This argument provides the basis for constructing confidence intervals with any desired level of confidence.

Suppose we are not satisfied with having an interval which contains μ with only 95% certainty, but rather we insist on having a confidence level of 98%. To achieve it, we must extend the interval that we construct more than 2 standard deviations in each direction. See Figure 8.3. In particular, we have to find the value of z which corresponds to 98% of the normal population of all sample means. Since we want to extend such an interval in both directions,

FIGURE 8.2

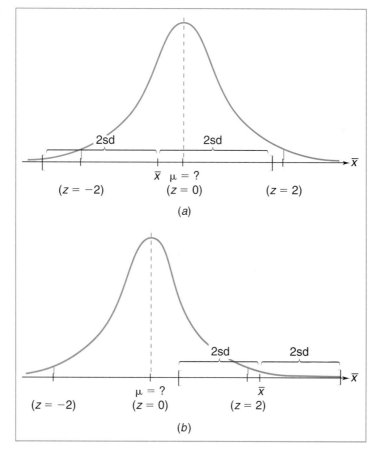

FIGURE 8.3

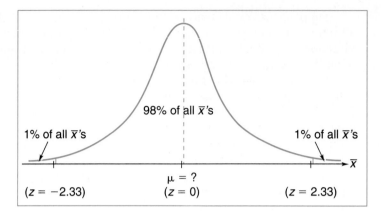

it is necessary to consider half of 98%, or 49%. From the normal distribution table, Table III, we find that the appropriate value for z is 2.33.

The above analysis is typical of how we find the z value necessary to construct a confidence interval with any desired level of confidence. To construct a 90% confidence interval, we want 45% of the population on either side of the mean μ. We see that there are two entries in the normal distribution table, Table III, .4495 and .4505 (corresponding to $z = 1.64$ and $z = 1.65$), which are equally close to $.45 = .4500$. Consequently, we can use either $z = 1.64$ or $z = 1.65$ (or, if we want to be particularly accurate, we could take $z = 1.645$). See Figure 8.4.

In a similar way, if we want a 95% confidence interval, then we want 47.5% of the population on either side of the mean μ. From Table III, the value of z corresponding to an area of .4750 is $z = 1.96$. We note that this is the precise value that we should use for a 95% confidence interval rather than the approximate value of $z = 2$ standard deviations that we considered in the preliminary discussion of confidence intervals. See Figure 8.5.

FIGURE 8.4

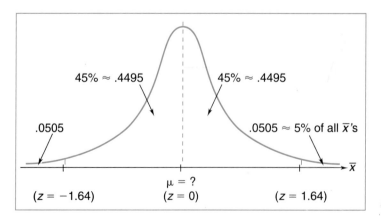

FIGURE 8.5

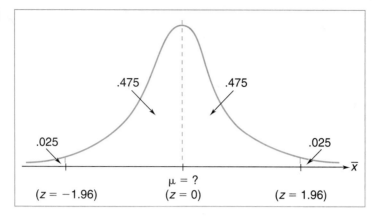

The remaining common confidence level is 99%. This requires having 49.5% of the population on either side of the mean μ. The corresponding value of z is $z = 2.58$, as seen in Figure 8.6.

> The most common confidence levels and the corresponding values of z are:
>
> | | |
> |---|---|
> | 90% | $z = 1.64$ or $z = 1.65$ |
> | 95% | $z = 1.96$ |
> | 98% | $z = 2.33$ |
> | 99% | $z = 2.58$ |

As you work through a variety of problems involving confidence intervals, you will become very familiar with these values.

FIGURE 8.6

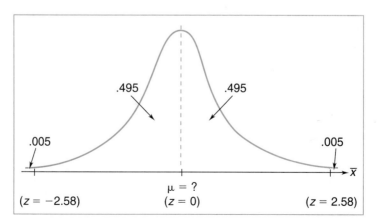

The confidence interval for the mean μ is an interval centered at the sample mean $\bar{x}$ and extending $z \cdot \sigma_{\bar{x}}$ in both directions. The confidence interval is given by

$$\bar{x} \pm z\sigma_{\bar{x}} = [\bar{x} - z\sigma_{\bar{x}}, \bar{x} + z\sigma_{\bar{x}}]$$

We illustrate how these ideas are applied in the following examples.

EXAMPLE 8.1 Construct a 95% confidence interval for the mean height μ of students based on the sample of size $n = 36$ with sample mean $\bar{x} = 66$ inches. Assume the standard deviation σ for adult heights is 3 inches.

SOLUTION We are given

$$n = 36$$
$$\bar{x} = 66$$
$$\sigma = 3$$

Since the sample size $n = 36 > 30$, the sampling distribution of the mean is approximately normal with mean $\mu = \mu_{\bar{x}}$ and standard deviation

$$\sigma_{\bar{x}} = \frac{\sigma}{\sqrt{n}} = \frac{3}{\sqrt{36}} = \frac{3}{6} = .5$$

Since we are constructing a 95% confidence interval for the mean μ, we want to encompass 47.5% of the population on either side of the mean, as shown in Figure 8.7. This corresponds to a z value of $z = 1.96$. The desired 95% confidence interval is centered at $\bar{x}$ and extends 1.96 standard deviations in each direction. Since $\sigma_{\bar{x}} = .5$, we have

FIGURE 8.7

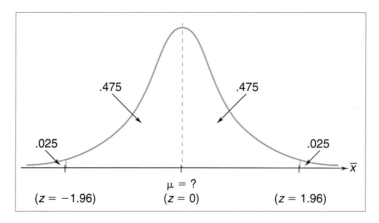

$$\bar{x} \pm z\sigma_{\bar{x}} = 66 \pm 1.96(.5)$$
$$= 66 \pm .98$$

or [65.02, 66.98]

as a 95% confidence interval for the mean. That is, with 95% confidence or certainty, we conclude that the true population mean μ is between 65.02 and 66.98 inches. Equivalently, the probability that this interval contains μ is .95. ■

Notice that the confidence interval is expressed mathematically as extending from the smaller value to the larger value.

EXAMPLE 8.2 Construct a 99% confidence interval for the mean height based on the same data on student heights.

SOLUTION As in Example 8.1, we have

$$n = 36$$
$$\bar{x} = 66$$
$$\sigma = 3$$

so that again

$$\sigma_{\bar{x}} = .5$$

To construct a 99% confidence interval, we must include 49.5% of the population of sample means on either side of the mean μ, as shown in Figure 8.8. We therefore use $z = 2.58$, so that the corresponding 99% confidence interval is

$$\bar{x} \pm z\sigma_{\bar{x}} = 66 \pm 2.58(.5)$$
$$= 66 \pm 1.29$$

or [64.71, 67.29] ■

FIGURE 8.8

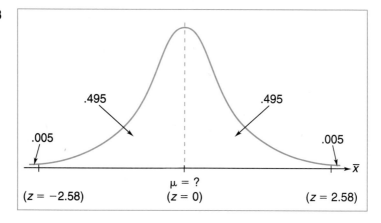

If we compare this result to the one in Example 8.1, we see that we have produced a larger interval. In retrospect, this is what we should expect.

> To achieve a greater degree of confidence in the estimate based on samples of a particular size, it is necessary to take a larger interval of values centered at the sample mean $\bar{x}$.

EXAMPLE 8.3 Construct a 99% confidence interval for the mean height of all students, assuming that the sample class contains $n = 100$ students instead of 36 students.

SOLUTION The sample data are now

$$n = 100$$
$$\bar{x} = 66$$
$$\sigma = 3$$

Consequently, the standard deviation for the distribution of sample means is

$$\sigma_{\bar{x}} = \frac{\sigma}{\sqrt{n}} = \frac{3}{\sqrt{100}} = \frac{3}{10} = .3$$

Since we want a 99% confidence interval for the mean μ, we use $z = 2.58$. See Figure 8.9. Therefore, the interval is now

$$\bar{x} \pm z\sigma_{\bar{x}} = 66 \pm 2.58(.3)$$
$$= 66 \pm .774$$

or
$$[65.226, 66.774]$$

or, rounded to two decimal places,
$$[65.23, 66.77]$$

∎

FIGURE 8.9

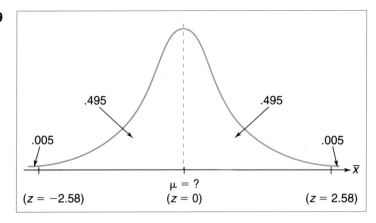

When we compare this result to the one in Example 8.2, we see that this confidence interval is considerably smaller than the previous one, [64.71, 67.29]. The only thing that has changed is the sample size n. From a purely mathematical point of view, n enters into the process through the formula $\sigma_{\bar{x}} = \sigma/\sqrt{n}$, and as n increases in size, the value for $\sigma_{\bar{x}}$ decreases. When $\sigma_{\bar{x}}$ decreases, the length of the confidence interval also decreases.

> For a fixed level of confidence, when the sample size increases, the length of the confidence interval decreases. When the sample size decreases, the length of the confidence interval increases.

We can interpret this from a different point of view. When we construct a confidence interval based on a relatively small sample, there is a degree of uncertainty in the prediction—the sample might be atypical in some sense or skewed high or low by several extreme values. However, if the sample size is relatively large, then $\bar{x}$ is far less likely to be affected by extreme values, and so we expect the prediction based on it to be more accurate. Consequently, it is not necessary to have as large an interval to achieve a desired degree of accuracy as with a smaller sample. In general, bigger samples yield better results.

There is one other point which bears mentioning. The above examples were all based on an assumption that the population standard deviation $\sigma = 3$. However, it is unrealistic to assume that we know the standard deviation σ for a population when we don't know the mean μ and, in fact, are trying to estimate it. In most practical situations, σ is also not known. Therefore, the best information we usually have available is the value of the sample standard deviation s. We can use s instead of σ in the formula for $\sigma_{\bar{x}}$ as long as $n > 30$ and the underlying population is roughly normal. Under these conditions, we will use $s/\sqrt{n}$ instead of $\sigma/\sqrt{n}$, so that

$$\sigma_{\bar{x}} = \frac{\sigma}{\sqrt{n}} \approx \frac{s}{\sqrt{n}}$$

EXAMPLE 8.4 The gas mileage for a sample of 64 randomly selected cars has a mean of 25 miles per gallon with a standard deviation of 6 miles per gallon. Find a 90% confidence interval for the mean gas mileage of all cars.

SOLUTION We have

$$n = 64$$
$$\bar{x} = 25$$
$$s = 6$$

We assume the underlying population is roughly normal and use

$$\frac{s}{\sqrt{n}} = \frac{6}{\sqrt{64}} = \frac{6}{8} = .75$$

instead of $\sigma_{\bar{x}}$. Consequently,

$$\frac{\bar{x} - \mu_{\bar{x}}}{(s/\sqrt{n})}$$

follows a *t*-distribution with $n - 1 = 63$ degrees of freedom. However, since $n = 64 > 30$, it is well approximated by a normal distribution. Since we seek a 90% confidence interval for the mean μ, we use $z = 1.64$ (see Figure 8.10) to obtain

$$\bar{x} \pm z\left(\frac{s}{\sqrt{n}}\right) = 25 \pm 1.64(.75)$$

$$= 25 \pm 1.23 \quad \text{or} \quad [23.77, 26.23]$$

Thus, with 90% certainty, the average gas mileage is between 23.77 and 26.23 miles per gallon. ∎

EXAMPLE 8.5 A lumberyard checks a random sample from a large delivery of plywood boards. The mean thickness of 49 boards is .98 inch with a standard deviation of .056 inch. Find a 95% confidence interval for the thickness of all such boards.

SOLUTION We have

$$n = 49$$
$$\bar{x} = .98$$
$$s = .056$$

FIGURE 8.10

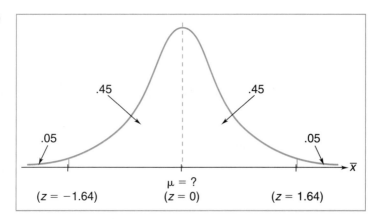

We use

$$\frac{s}{\sqrt{n}} = \frac{.056}{\sqrt{49}} = \frac{.056}{7} = .008$$

and approximate the resulting *t*-distribution with a normal distribution since $n = 49 > 30$, as shown in Figure 8.11. We use $z = 1.96$ so that the confidence interval is

$$\bar{x} \pm z\left(\frac{s}{\sqrt{n}}\right) = .98 \pm 1.96(.008)$$
$$= .98 \pm .0157$$

or
[.9643, .9957]

Upon rounding, this 95% confidence interval is

[.964, .996]

Thus, we estimate that the true mean thickness of all such boards is between .964 inch and .996 inch with 95% certainty. ∎

EXAMPLE 8.6 A university is conducting a study to determine the average amount spent by students each month on long-distance telephone calls. The result will be included in a list of anticipated costs for new students. The data collected in a random study, rounded to the nearest dollar, are the following:

14, 18, 22, 30, 36, 28, 42, 79, 36, 52, 15, 47,
95, 16, 27, 111, 37, 63, 127, 23, 31, 70, 27, 11,
30, 147, 72, 37, 25, 7, 33, 29, 35, 41, 337, 48,
15, 29, 73, 26, 15, 26, 31, 57, 40, 18, 85, 28,
32, 22, 37, 60, 41, 35, 26, 20, 58, 33, 23

FIGURE 8.11

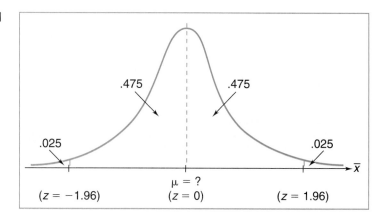

Analyze the data and construct a 95% confidence interval for the mean monthly long-distance call charges of all students at the university.

SOLUTION We begin by analyzing the data using Program 1: Statistical Analysis of Data. When we enter the $n = 60$ data values, we obtain the histogram shown in Figure 8.12a based on 10 classes: under 35, 35–69, 70–104, ... The associated mean and standard deviation are $\bar{x} = 45.97$ and $s = 47.11$, respectively. However, note that the entry $x = 337$ is an outlier; it may be the result of recording a faulty value or it might represent the actual phone bill of a student who makes an inordinate number of long-distance calls. As such, it is probably unrepresentative of the data and so distorts the results.

When we remove this entry from the list, we obtain the modified histogram shown in Figure 8.12b based on 10 classes: under 15, 15–29, 30–44, ... The associated values for the mean and standard deviation for this sample of size $n = 59$ are then $\bar{x} = 41.05$ and $s = 27.99$. Observe how much these values, particularly the standard deviation, changed after our having removed the single outlier. Further, note that the histogram has a more representative shape without the outlier. Moreover, since the histogram is skewed to the right, the sample data is not normally distributed. Consequently, we would not conclude that the underlying population is normal.

FIGURE 8.12a

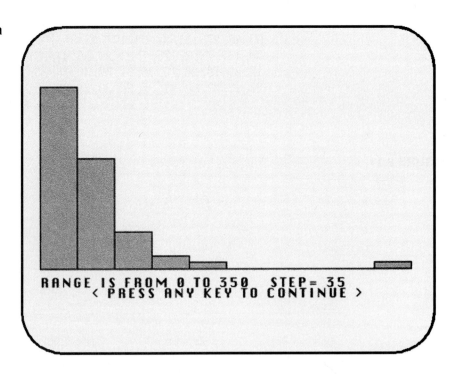

FIGURE 8.12b

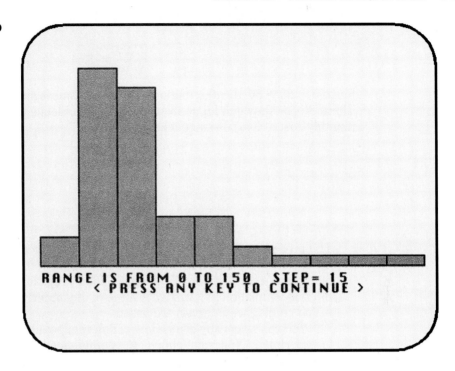

FIGURE 8.12c

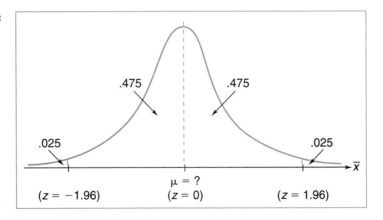

We now construct the 95% confidence interval based on the sample data using:

$$n = 59$$
$$\bar{x} = 41.05$$
$$s = 27.99$$

so that

$$\frac{s}{\sqrt{n}} = \frac{27.99}{\sqrt{59}} = 3.64$$

Since $n = 59 > 30$, the distribution of sample means is approximately normal. Therefore, as shown in Figure 8.12c, the 95% confidence interval for the mean monthly cost of long distance telephone calls is

$$41.05 \pm 1.96(3.64)$$

or

$$[\$33.92, \$48.18]$$

∎

Computer Corner

The procedure involved in constructing a confidence interval for the mean μ is quite straightforward. However, some of the ideas involved may seem a bit abstract. We have included Program 26: Confidence Interval Simulation to investigate some of the underlying concepts on which confidence intervals are based.

You can select any of a variety of different populations with which to experiment. You must choose the confidence level you want. The program first draws a vertical line corresponding to the mean of the population you picked. It then randomly selects almost 100 different samples of size $n = 31$ from this population, calculates the mean $\bar{x}$ and standard deviation s for each sample, and constructs the corresponding confidence interval. These confidence intervals are then graphed as horizontal lines on the screen. The location of the sample mean $\bar{x}$ at which each confidence interval is centered is indicated with a vertical mark.

From the typical display shown in Figure 8.13a, you can observe how most of the confidence intervals actually contain the mean μ in the sense that they cross the center line representing the mean. The relatively few confidence intervals that do not contain the mean are drawn in a different color and may be accompanied by a beep. If, for instance, you opt for a 90% confidence interval, you will see that approximately 90% of the confidence intervals constructed do in fact contain the mean μ. Similarly, if you select a 99% confidence level, then approximately 99% of all the confidence intervals constructed contain the mean. Details on the actual outcomes are displayed on an accompanying text screen. See Figure 8.13b.

Moreover, notice that the various confidence intervals displayed have different lengths. This is due to the fact that each sample has a different sample standard deviation s which determines the length of the interval. In addition, notice that even in cases where the confidence interval does not contain the mean μ, it usually misses by little.

SECTION 8.1 • Confidence Intervals for Means **343**

FIGURE 8.13a

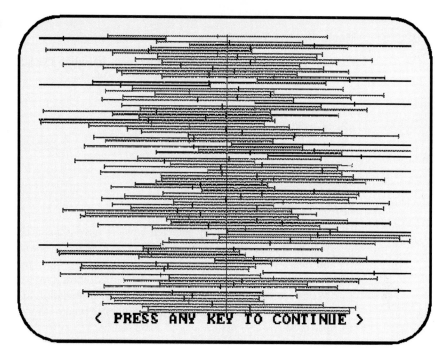

FIGURE 8.13b

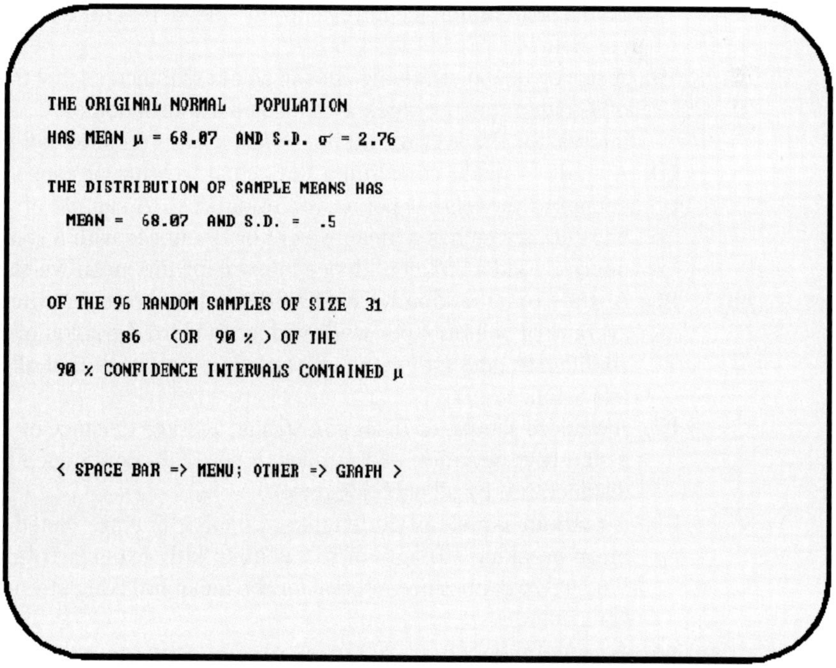

```
THE ORIGINAL NORMAL   POPULATION
HAS MEAN µ = 68.07  AND S.D. σ = 2.76

THE DISTRIBUTION OF SAMPLE MEANS HAS
  MEAN = 68.07  AND S.D. = .5


OF THE 96 RANDOM SAMPLES OF SIZE  31
        86    (OR  90 % ) OF THE
 90 % CONFIDENCE INTERVALS CONTAINED µ



 < SPACE BAR => MENU; OTHER => GRAPH >
```

Exercise Set 8.1

Mastering the Techniques

Find the 90%, 95% and 99% confidence intervals based on the following sets of sample data:

1. $n = 100, \bar{x} = 250, s = 80$
2. $n = 36, \bar{x} = 85, s = 24$
3. $n = 64, \bar{x} = 250, s = 80$
4. $n = 64, \bar{x} = 85, s = 24$

Applying the Concepts

5. A sample of 36 randomly selected bowlers have a mean bowling average of 125 pins with a standard deviation of 30 pins. Find a 90% confidence interval for the bowling average of all such bowlers.
6. A random sample of 100 patrons at a public library who check out books take out an average of 4.6 books with a standard deviation of 2 books. Find a 95% confidence interval for the average number of books taken out by all such library patrons.
7. In a random sample of 64 customers at a fast-food restaurant, the mean waiting time for being served is 3 minutes with a standard deviation of 1.5 minutes. Construct a 98% confidence interval for the average waiting time in this restaurant.
8. A traffic survey finds that, during 36 randomly selected hours, the mean number of cars that drive along a certain street is 110 with a standard deviation of 32. Find a 90% confidence interval for the mean number of cars that drive this street per hour.
9. A survey of 150 randomly chosen smokers indicates that they smoke an average of 97 cigarettes per week with a standard deviation of 36. Find a 99% confidence interval for the average number of cigarettes smoked per week.
10. A study is made concerning the actual weights of packs of carrots which are supposed to weigh 1 pound (or 16 ounces). A sample of 40 randomly selected bags of carrots has a mean weight of 19 ounces with a standard deviation of 1.3 ounces. Find a 95% confidence interval for the mean weight of all such bags.
11. A study of 50 randomly chosen 6-year-olds shows that they watch television an average of 38 hours per week with a standard deviation of 6.4 hours. Find a 99% confidence interval for the average time per week that all such children watch television.
12. A random sample of 75 joggers run an average distance of 2.4 miles per day with a standard deviation of 1.1 miles. Find a 90% confidence interval for the mean distance run by all such joggers.
13. A random sample of 40 lengths of fishing line are tested and found to have a mean breaking strength of 18.3 pounds with a standard deviation of 2.7 pounds. Find a 94% confidence interval for the mean breaking strength of all such fishing line.
14. A random sample of 200 cars crossing a bridge into a city during rush hour contain an average of 1.8 people per car with a standard deviation of .84. Find

a 97% confidence interval for the average number of people per car on the bridge during rush hour.

15. A large company wants to estimate the average number of pages typed per day by the secretaries in a typing pool. A random sample of 50 secretaries are chosen, and their average production is 32 pages with a standard deviation of 6 pages. Find a 90% confidence interval for the mean production of all secretaries in this office.

Computer Applications

16. Use Program 2: Random Number Generator to select a random sample of the number of home runs hit by 36 baseball players from the data set in Appendix I. Use Program 1: Statistical Analysis of Data to calculate $\bar{x}$ and s for these data and use the results to construct a 90% confidence interval for the average number of home runs of all players listed. Does this confidence interval contain the true population mean μ that you found in Exercise 21 of Section 3.1? Compare your results with those obtained by other people in your class and see whether approximately 90% of you did generate confidence intervals which contained the mean.

17. Repeat Exercise 16 using a sample of size 36 drawn from the data set of temperatures to generate a 95% confidence interval for the average temperatures of all cities listed in Appendix II. Compare the estimate to the actual result found in Exercise 22 of Section 3.1.

8.2 Confidence Intervals for Means Based on Small Samples

We have seen that we can use the normal distribution to construct confidence intervals for a population mean μ when σ is not known provided that

1. the underlying population is roughly normally distributed

and

2. the sample size n is larger than 30.

However, many cases arise where we have only a "small sample" of size $n \leq 30$. Consequently, we cannot construct a confidence interval to estimate the population mean μ as we did in Section 8.1. Of course, if we could simply increase the sample size, the problem would be solved, but this is not always possible either practically or economically. Therefore, we must modify the procedures from Section 8.1 to handle small-sample cases.

When we discussed sampling in Section 7.3, we saw that if the underlying distribution is approximately normal, then the values of

$$\frac{\bar{x} - \mu}{(s/\sqrt{n})}$$

follow a t-distribution with $n - 1$ degrees of freedom. Furthermore, if the sample size $n > 30$, we can approximate this t-distribution with a normal distribution. However, if $n \leq 30$, we must use the t-distribution.

Recall that when working with a *t*-distribution, we use *t* values

$$t = \frac{\bar{x} - \mu}{s/\sqrt{n}}$$

to measure the number of standard deviations that a sample mean $\bar{x}$ lies away from the population mean μ. As you will see, our use of *t* values precisely parallels the procedures with *z* values we developed in Section 8.1 in the large sample case.

Since there are many different *t*-distributions, one for each sample size *n*, you might expect a separate table of *t*-distribution values for each value of *n* (each comparable to the normal distribution table, Table III). However, this is unnecessary. The types of problems we encounter involving the *t*-distribution are all comparable to confidence interval problems. Think back to what we did in Section 8.1 with large-sample confidence intervals. We actually used only a very few entries from the normal distribution table—specifically, those corresponding to a 90% confidence interval ($z = 1.64$), a 95% confidence interval ($z = 1.96$), a 98% confidence interval ($z = 2.33$), and a 99% confidence interval ($z = 2.58$).

The same type of analysis applies for *t*-distributions, so that only the corresponding entries are needed from each of the different *t*-distributions. Thus, in Table IV for the *t*-distributions (reproduced on the inside front cover) we have listed just the specific values needed. A portion of this table is shown in Figure 8.14. For each degree of freedom, this table provides the value of *t*

FIGURE 8.14

df	$t_{.05}$	$t_{.025}$	$t_{.01}$	$t_{.005}$	df
1	6.314	12.706	31.821	63.657	1
2	2.920	4.303	6.965	9.925	2
3	2.353	3.182	4.541	5.841	3
4	2.132	2.776	3.747	4.604	4
5	2.015	2.571	3.365	4.032	5
6	1.943	2.447	3.143	3.707	6
7	1.895	2.365	2.998	3.499	7
8	1.860	2.306	2.896	3.355	8
9	1.833	2.262	2.821	3.250	9
10	1.812	2.228	2.764	3.169	10
11	1.796	2.201	2.718	3.106	11
12	1.782	2.179	2.681	3.055	12
13	1.771	2.160	2.650	3.012	13
14	1.761	2.145	2.624	2.977	14
15	1.753	2.131	2.602	2.947	15

FIGURE 8.15

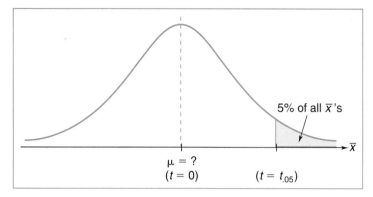

corresponding to each of the percentages or levels of confidence we commonly use. For reasons that will become apparent in the next chapter, the t values shown correspond to the area under the t-distribution curve beyond a given point; that is, in the tail to the right. See Figure 8.15.

Both the left- and right-hand columns of the t-distribution table list the number of degrees of freedom which distinguishes one t-distribution from another. The entries across the top of the table represent the areas *beyond* the appropriate cutoff point. They are listed as $t_{.05}$, $t_{.025}$, $t_{.01}$, and $t_{.005}$ and indicate, respectively, that 5%, $2\frac{1}{2}$%, 1%, and .5% of the t-distribution is in the tail.

Suppose we wish to construct a 90% confidence interval for the mean μ based on a small sample from an approximately normal population. This leaves 10% of the population in both tails and therefore 5% of the population in each tail. See Figure 8.16. If the sample size is $n = 13$, we look at the row corresponding to $n - 1 = 12$ degrees of freedom and at the column corresponding to 5% (.05) in a single tail. The appropriate value for t is $t = t_{.05} = 1.782$.

To construct a 98% confidence interval based on a sample of size $n = 6$, we look for $n - 1 = 5$ degrees of freedom and the value corresponding to 1% of the population in the tail, and so we find $t = t_{.01} = 3.365$. See Figure 8.17.

FIGURE 8.16

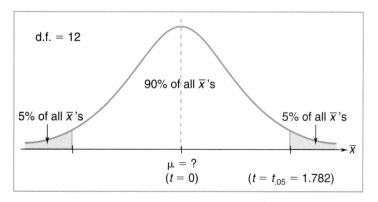

FIGURE 8.17

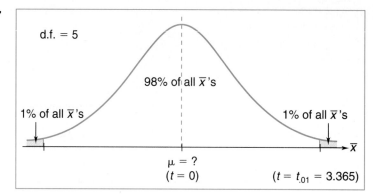

We now illustrate how confidence interval problems based on small samples from approximately normal populations can be solved by using the *t*-distribution. As you will see, the procedure is virtually identical to what we did with large-sample problems in Section 8.1. In all the following examples and problems, we assume that the underlying population is approximately normal so that we can use a *t*-distribution.

EXAMPLE 8.7 A new anti-rejection drug is tried on a group of 16 patients undergoing a certain transplant operation. The mean time until rejection is 23 months with a standard deviation of 12 months. Find a 90% confidence interval for the mean time until rejection for all such patients.

SOLUTION The sample statistics are

$$n = 16$$
$$\bar{x} = 23$$
$$s = 12$$

We calculate

$$\frac{s}{\sqrt{n}} = \frac{12}{\sqrt{16}} = \frac{12}{4} = 3$$

Since the population is approximately normal and the sample size $n = 16 < 31$, we use a *t*-distribution with $n - 1 = 15$ degrees of freedom.

To construct a 90% confidence interval, we need 5% of the population in each tail. Therefore, from Table IV, using 15 degrees of freedom and .05 as the area in one tail, we find that $t = t_{.05} = 1.753$. See Figure 8.18. As a result, the 90% confidence interval for μ is

$$\bar{x} \pm t\left(\frac{s}{\sqrt{n}}\right) = 23 \pm 1.753(3)$$
$$= 23 \pm 5.259$$

FIGURE 8.18

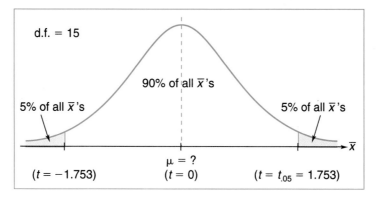

or

$$[17.741, 28.259]$$

or more realistically,

$$[17.7, 28.3]$$

That is, with 90% certainty, we conclude that the true mean time until rejection is between 17.7 and 28.3 months. ∎

In an experimental procedure such as in this example, the researcher would certainly want to keep the sample size as small as possible until the new procedure was known to be effective. Thus, there is a clear need for working with small samples. Further, in most such experimental situations, little is known about the underlying population. If it is reasonable to assume the population is approximately normal, then this procedure applies. Otherwise, methods such as the nonparametric techniques we introduce in Chapter 13 may apply.

EXAMPLE 8.8 A small computer software business puts an advertisement into nine different computer magazines and keeps track of the total orders generated by each advertisement. If the mean sales are $1200 with a standard deviation of $450, find a 95% confidence interval for the mean sales expected from all such advertisements.

SOLUTION The sample statistics are

$$n = 9$$
$$\bar{x} = 1200$$
$$s = 450$$

First we find that

$$\frac{s}{\sqrt{n}} = \frac{450}{\sqrt{9}} = \frac{450}{3} = 150$$

FIGURE 8.19

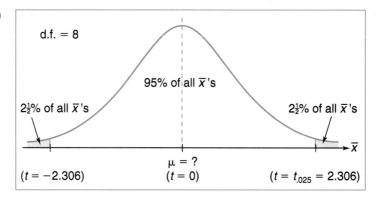

Since $n = 9$ and we assume that the total orders generated by an advertisement are approximately normally distributed, we use a *t*-distribution with $n - 1 = 8$ degrees of freedom. For a 95% confidence interval, we want 5% of the population in the two tails, or 2.5% in each tail. Using the *t*-distribution table with 8 degrees of freedom, we find that $t = t_{.025} = 2.306$. See Figure 8.19. Therefore, the desired 95% confidence interval for μ is

$$\bar{x} \pm t\left(\frac{s}{\sqrt{n}}\right) = 1200 \pm 2.306(150)$$

$$= 1200 \pm 345.90$$

or

[$854.10, $1545.90]

Consequently, we conclude that the average amount of sales generated by an advertisement would be between $854.10 and $1545.90, with 95% confidence. ∎

EXAMPLE 8.9 At one point in the voyage of the infamous Long Island "garbage barge", a study of the levels of toxic wastes was conducted. Find a 95% confidence interval for the mean amount of lead present, in parts per million (ppm), based on the data given in the article shown in Figure 8.20.

SOLUTION We are not given the calculated values for $\bar{x}$ and s, but rather the actual raw data. Before we can construct a confidence interval for the mean amount of lead present, we must compute the values for the sample mean and the sample standard deviation. We calculate:

$$\bar{x} = 37.28 \quad \text{and} \quad s = 3.82$$

Further, we find that

$$\frac{s}{\sqrt{n}} = \frac{3.82}{\sqrt{6}} = \frac{3.82}{2.45} = 1.56$$

FIGURE 8.20

What's in the Garbage
Amount of cadmium and lead, in parts per million, that was found in samples of garbage from the barge, and the federal government's maximum allowable levels for those materials

| | \------ Sample \------ | | | | | | Maximum |
	1	2	3	4	5	6	allowable
Cadmium	20.88	5.93	19.33	22.42	0	13.72	1.00
Lead	35.60	32.40	38.00	44.00	36.40	37.28	5.00

SOURCE: Volumetric Techniques Ltd.

High Levels of Toxics Found in Barge Ash

Since $n = 6$ and we assume that the underlying population is approximately normal, we use the t-distribution with $n - 1 = 5$ degrees of freedom. For a 95% confidence interval for μ, we want $2\frac{1}{2}\%$ of the population in each tail. See Figure 8.21. Using the t-distribution with 5 degrees of freedom, we find that $t = t_{.025} = 2.571$, so that the desired confidence interval for μ is

FIGURE 8.21

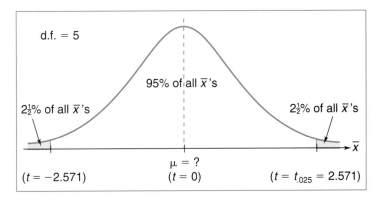

$$\bar{x} \pm t\left(\frac{s}{\sqrt{n}}\right) = 37.28 \pm 2.571(1.56)$$

$$= 37.28 \pm 4.01$$

or [33.27 ppm, 41.29 ppm]

Based on this example, we conclude with 95% certainty that the average level of lead is between 33.27 and 41.29 parts per million. Since the maximum allowable limit imposed by the federal government is 5.00 parts per million, as stated in the article, the level of lead in this garbage clearly exceeded the limit. ■

Another application of confidence intervals for means arises when we study two different populations and want to estimate how close their respective means are to each other. We consider this situation in Section 9.4 after we develop the necessary theory.

Computer Corner

Program 25: Confidence Intervals: Means/Proportions is designed to assist you in learning how to solve confidence interval problems. The program leads you through all the necessary steps and decisions involved in solving the various types of problems that you will encounter in this chapter. While the end result is to produce the actual confidence interval and to depict the result graphically, the intermediate steps in the procedure may be more important to ensure that you are doing the work correctly.

The program first asks if the confidence interval is for means or proportions. At this stage, you should respond with 1 for means. We discuss confidence intervals for proportions in Section 8.3. The program asks for the confidence level you desire. For a normal distribution, you can use any value from 90, 91, 92, ..., 99; for a small-sample t-distribution, you must select 90, 95, 98, or 99. The program then asks for your sample size n. It next asks whether you should use a normal distribution or a t-distribution. Then you are asked for the sample mean and the standard deviation s. The program then asks *you* to find the corresponding value for either z or t from the appropriate table. You will be given three chances to enter the correct z or t value for your selected confidence interval; if you cannot, the program will give you the correct answer. Finally, the program asks you to type in the confidence interval *you* calculate based on all the previous information. At each stage, it attempts to identify any errors you may make and notifies you if your answers are incorrect.

After the analysis is complete, the program draws the corresponding distribution, either normal or t. It displays the locations of all the different confidence levels as tick marks along the horizontal axis. Notice how the lower confidence levels (near 90%) are spaced closer together than the ones near 99%. This illustrates how fast the area under the curve drops off. It should also give you a feel for the price we pay to gain a higher degree of confidence in the prediction.

The program marks off (with two vertical lines) the cutoff points for the confidence level you specified and then paints in the appropriate region under the curve. Finally, it shows the corresponding confidence interval.

We suggest that you work through this program several times with different sets of data to develop familiarity with the steps needed to solve any confidence interval problem for the mean.

Minitab Methods

Minitab also provides a simple procedure to construct confidence intervals for the population mean μ. Suppose we want a 95% confidence interval for the mean based on a set of actual sample data values that we enter in column C1. The Minitab command

TINTERVAL 95 C1

is all that is needed, as shown in the sample Minitab session in Figure 8.22. Notice that Minitab displays the sample mean $\bar{x}$, the sample standard deviation s (STDEV), the standard deviation for the sampling distribution of the mean (which is sometimes called the standard error of the estimate, or SE MEAN), and the desired confidence interval.

FIGURE 8.22

```
MTB > PRINT C1
WEIGHT
    136    137    157    144    190    164    147    150    136    163    148
    174    211    169    148    184    163    144    130    181    156    147
    170    148    182    159    140    137    122    158

MTB > TINTERVAL 95% confidence interval for C1

                N      MEAN     STDEV   SE MEAN      95.0 PERCENT C.I.
WEIGHT         30    156.50     19.84      3.62    ( 149.09,  163.91)
```

Exercise Set 8.2

Mastering the Techniques

Construct 90%, 95% and 99% confidence intervals based on the following sets of sample data drawn from roughly normal populations.

1. $n = 9, \bar{x} = 300, s = 120$
2. $n = 25, \bar{x} = 37, s = 10$
3. $n = 18, \bar{x} = 300, s = 120$
4. $n = 12, \bar{x} = 37, s = 10$

Applying the Concepts

5. A stamp collector finds that the average price asked for a certain stamp in nine advertisements in a stamp magazine is $17 with a standard deviation of $3. Construct a 90% confidence interval for the mean price for this stamp from all stamp dealers.
6. A major league pitcher has a mean speed of 90 miles per hour with a standard deviation of 4 miles per hour on a random sample of 16 fastballs thrown during a game. What is a 95% confidence interval for the average speed of his fastballs?
7. A sample of 12 classes at a college have a mean enrollment of 28 students with a standard deviation of 5 students. Construct a 99% confidence interval for the average enrollment of all classes.
8. A sample of 20 private colleges around the country show a mean cost for a year's tuition of $9200 with a standard deviation of $2800. Find a 95% confidence interval for the average tuition costs at private colleges.
9. A sample of 18 corporate presidents show a mean annual salary of $275,000 with a standard deviation of $62,000. Find a 90% confidence interval for the average salary of all such individuals.
10. A sample of 20 taxis from a large taxi fleet in a certain city show the mean total fares collected per shift is $420 with a standard deviation of $163. Find a 98% confidence interval for the mean fare collected per shift by each such taxi.
11. An individual planning to buy a compact disc (CD) player surveys the sale prices for a particular model in various newspaper advertisements and catalogs. She finds the following set of prices: $138, $149, $129, $135, $145, $125, $139, $142. Construct a 90% confidence interval for the mean selling price of this CD player.
12. A psychology major is conducting a series of experiments involving the time, in seconds, it takes for a rat to navigate a maze. He obtains the following measurements: 42, 55, 48, 36, 43, 39, 31, 36. Construct a 98% confidence interval for the mean time needed for the rat to go through the maze.

Computer Applications

13. Use Program 2: Random Number Generator to select a random sample of the number of home runs hit by 16 baseball players from the data set in Appendix

I. Use Program 1: Statistical Analysis of Data to calculate $\bar{x}$ and s for these data, and use the results to construct a 90% confidence interval for the mean number of home runs of all players listed. Does this confidence interval contain the true population mean μ that you found in Exercise 21 of Section 3.1? Compare your results with those obtained by others in your class and see whether approximately 90% of you did generate confidence intervals which contained the mean.

14. Repeat Exercise 13 using a sample of size 12 drawn from the data set of temperatures to generate a 95% confidence interval for the average temperature of all cities listed in Appendix II. Compare the confidence interval estimate to the actual mean found in Exercise 22 of Section 3.1.

8.3 Confidence Intervals for Proportions

In the last two sections, we developed the techniques needed to estimate the value for a population mean μ using a random sample to construct a confidence interval for the mean. We now modify these ideas to consider a similar situation involving proportions or percentages within a population.

Consider a binomial population whose population proportion π is not known. For instance, we might want to know the proportion π of all adults in the United States who favor a particular policy of the President or the proportion π of all households in the United States that watch the Academy Awards telecast this year. To estimate this parameter π, we select a random sample from the underlying binomial population. The statistics of interest from this sample consist of the sample size n, the number of successes x, and the sample proportion $p = x/n$. From these values we wish to estimate the true population proportion π by constructing an appropriate confidence interval for π. The method is very similar to that used to construct a confidence interval for the population mean μ in Section 8.1.

Suppose that a poll is conducted among U.S. adults to determine whether they support a proposed constitutional amendment. Suppose further that, in this sample, 40% of the individuals polled support the amendment, so that the sample proportion $p = .40$. A confidence interval for the unknown population proportion is a range of values centered at .40. It might be from .38 to .42, or from .35 to .45, or from .30 to .50. Thus, as with confidence intervals for the mean, there are different confidence intervals for the population proportion π depending on the confidence level we select.

The principles for constructing such a confidence interval for the population proportion π are based on the following considerations. The particular sample of size n that we select is just one of many different possible samples of size n that can be drawn from the underlying binomial population. Each of these possible samples has an associated sample proportion p. The totality of all possible sample proportions based on samples of size n represents a new sampling population, the distribution of sample proportions.

> The **distribution of sample proportions** is approximately normal with mean
> $$\mu_p = \pi$$
> and standard deviation
> $$\sigma_p = \sqrt{\frac{\pi(1-\pi)}{n}}$$
> provided that both
> $$n\pi > 5 \quad \text{and} \quad n(1-\pi) > 5 \tag{8.1}$$

Since we don't know the value of π (in fact, we are trying to estimate it), we cannot calculate σ_p directly. However, we can estimate the value of σ_p by using the sample proportion p in place of π. Thus, we use

$$\sigma_p \approx s_p = \sqrt{\frac{p(1-p)}{n}}$$

as an approximation to σ_p throughout this section. To use this approximation, we require that both

$$np > 10 \quad \text{and} \quad n(1-p) > 10$$

rather than the conditions in formula (8.1).

> The confidence interval for the population proportion π is
> $$p \pm zs_p \quad \text{or} \quad \text{from } p - zs_p \text{ to } p + zs_p$$

With these minor modifications, the techniques we developed previously can now be applied to the present situation, as illustrated in the following examples.

EXAMPLE 8.10 A political poll is conducted in a certain locality just before election day. Of the 100 randomly selected people surveyed, 60 say they will vote for candidate Santana. Find a 90% confidence interval for the proportion of all voters in this community who will vote for Santana.

SOLUTION The sample statistics are

$$n = 100$$
$$x = 60$$
$$p = \frac{60}{100} = .60$$

Moreover, since $1 - p = 1 - .60 = .40$, we find that

$$np = 100(.6) = 60 > 10 \quad \text{and} \quad n(1 - p) = 100(.4) = 40 > 10$$

Therefore, the distribution of sample proportions is approximately normal with standard deviation σ_p, which we approximate by

$$s_p = \sqrt{\frac{(.60)(.40)}{100}}$$
$$= \sqrt{.0024} = .049$$

Since we seek a 90% confidence interval for the population proportion π, we must include 45% of the normal population on each side of the mean. This corresponds to $z = 1.64$, as shown in Figure 8.23. Therefore, the desired 90% confidence interval is

$$p \pm zs_p = .60 \pm 1.64(.049)$$
$$= .60 \pm .080$$

or

$$[.52, .68]$$

or, in terms of percentages,

$$[52\%, 68\%]$$

Thus, we conclude with 90% certainty that Santana will get a minimum of 52% of the vote. We therefore declare that Santana will win the election with 90% certainty. ∎

FIGURE 8.23

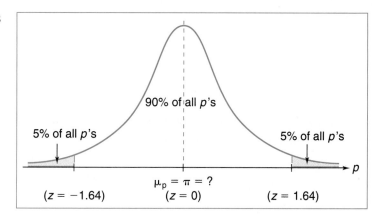

FIGURE 8.24

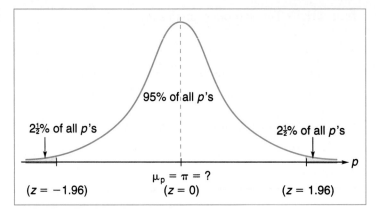

This is basically the method used by the media in predicting the winning candidates on election night. If the resulting confidence interval were [.48, .60], say, then a two-candidate race would be declared "too close to call." In practice, of course, special care must be taken that the random sample selected is truly representative of the entire population being studied.

EXAMPLE 8.11 In a random sample of 200 people who are having new eyeglasses made, 162 select plastic lenses rather than glass lenses. Find a 95% confidence interval for the percentage of all new eyeglasses made with plastic lenses.

SOLUTION The sample statistics are

$$n = 200$$

$$x = 162$$

$$p = \frac{162}{200} = .81$$

Since $1 - p = .19$, we find that $np = 162 > 10$ and $n(1 - p) = 38 > 10$, so we can use a normal distribution. We find that

$$s_p = \sqrt{\frac{(.81)(.19)}{200}}$$

$$= \sqrt{.0007695} = .028$$

Since we want a 95% confidence interval for the population proportion π, we use $z = 1.96$, as shown in Figure 8.24. Therefore, a 95% confidence interval is

$$p \pm zs_p = .81 \pm 1.96(.028)$$

$$= .81 \pm .055$$

so that the interval is [.755, .865]

or in percentages [75.5%, 86.5%]

That is, with 95% certainty, we conclude that between 75.5% and 86.5% of the people choose plastic lenses. ∎

EXAMPLE 8.12 Of a randomly chosen group of 300 air flights, 74% arrive on time.

(a) Find a 99% confidence interval for the proportion of all flights that arrive on time.

(b) Find a 95% confidence interval for the proportion of all flights that arrive late.

SOLUTION (a) The sample statistics are

$$n = 300 \qquad p = .74$$

Notice that no explicit value is given for x. We could calculate it if we needed it ($x = 74\%$ of $300 = 222$), but we do not require it to solve the problem. Further, since $1 - p = .26$, it follows that both $np = 222$ and $n(1 - p) = 78$ are greater than 10, so we can use a normal distribution. We find that

$$s_p = \sqrt{\frac{(.74)(.26)}{300}}$$

$$= \sqrt{.0006423} = .025$$

Since we want a 99% confidence interval for the proportion of all on-time flights, this corresponds to $z = 2.58$, as shown in Figure 8.25. Therefore, the confidence interval for the population proportion of all flights that arrive on time is

FIGURE 8.25

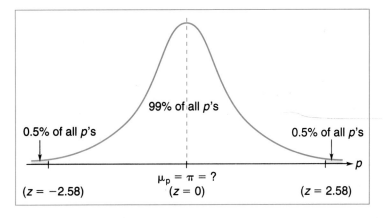

$$p \pm zs_p = .74 \pm 2.58(.025)$$
$$= .74 \pm .0645$$

or   [.6755, .8045]

or in percentages,   [67.55%, 80.45%]

That is, the true proportion of all flights that are on time is between 67.55% and 80.45%.

(b) To find a 95% confidence interval for the proportion of all flights that are late, we observe that if 74% of the sample flights arrive on time, then 26% are late. Therefore, we have the sample statistics

$$n = 300 \quad p = .26$$

so that $1 - p = .74$ and, as before, both np and $n(1 - p)$ are greater than 10. Further,

$$s_p = .025$$

as before. For the 95% confidence interval, we use $z = 1.96$, as shown in Figure 8.26. Therefore, the confidence interval for the proportion of late flights is

$$p \pm zs_p = .26 \pm 1.96(.025)$$
$$= .26 \pm .049$$

or   [.211, .309]

or in percentages,   [21.1%, 30.9%]

Therefore, with 95% certainty, we conclude that the proportion of late flights is between 21.1% and 30.9%.

FIGURE 8.26

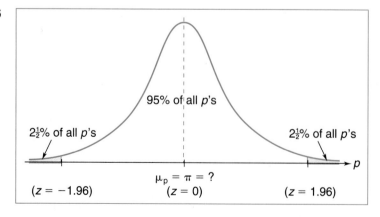

SECTION 8.3 • Confidence Intervals for Proportions

EXAMPLE 8.13 Out of 300 randomly selected people, 6 test positive for exposure to a certain disease. Find a 98% confidence interval for the proportion of all people in the population who have been exposed to this disease.

SOLUTION The sample statistics are

$$n = 300$$
$$x = 6$$
$$p = \frac{6}{300} = .02$$

and so $1 - p = .98$. Since $np = 300(.02) = 6$, which is not greater than 10, we cannot assume that the distribution of sample proportions is approximately normal, and so the methods of this section do not apply to construct a confidence interval. ∎

Throughout all the development in this section and previous sections, we have referred to the results of a sample. It is therefore interesting to consider the details of how a professional sample or survey is conducted. We discussed some of these ideas in Chapter 2 when we introduced the notion of a random sample. We considered a particular political survey conducted for the *New York Times* (see Figure 8.27). Read through this article carefully to see how the ideas and methods that we have been using are essentially the ones that are applied in practice. Notice that the result of such a survey is basically an estimate of the population proportion π for some issue. Often the results quoted in such an article represent a point estimate rather than a confidence interval estimate for the population proportion. However, if the important statistical details are included, as they are in this article, the possible error is usually given as well. This simply represents the size of the term $\pm zs_p$ that we use in constructing a confidence interval.

Another application of confidence intervals for proportions arises when we study two different populations and want to estimate how close their proportions are to each other. We consider this situation in Section 9.6 after we develop the necessary theory.

Computer Corner

In the Computer Corner of Section 8.2, we described some of the aspects of Program 25: Confidence Intervals: Means/Proportions. This program leads you through the steps needed to construct confidence intervals for proportions as well as for means. The program first asks if the confidence interval is for means or proportions. If you respond with 2 for proportions, it asks for the sample statistics—the sample size n and the sample proportion p. It then asks for the value of $1 - p$ and the desired level of confidence. This can be any value from 90, 91, 92, ..., 99. The program then asks *you* to find the corresponding z value from the normal distribution table. It also asks you to calculate and enter the values for the confidence interval. You will

FIGURE 8.27

How the Poll Was Conducted

The latest New York Times Poll is based on telephone interviews conducted Jan. 30, and 31 with 1,187 adults around the United States, excluding Alaska and Hawaii.

The sample of telephone exchanges called was selected by a computer from a complete list of exchanges in the country. The exchanges were chosen to insure that each region of the country was represented in proportion to its population. For each exchange, the telephone numbers were formed by random digits, thus permitting access to both listed and unlisted residential numbers. The numbers were then screened to limit calls to residences.

The results have been weighted to take account of household size and number of residential telephones and to adjust for variations in the sample relating to region, race, sex, age and education.

In theory, in 19 cases out of 20 the results based on such samples will differ by no more than three percentage points in either direction from what would have been obtained by interviewing all adult Americans. The potential error for smaller subgroups is larger. For example, for Republican primary or caucus voters, it is plus or minus five percentage points.

In addition to sampling error, the practical difficulties of conducting any survey of public opinion may introduce other sources of error into the poll.

be given three opportunities to provide the correct calculations; if you cannot, the program will provide the correct answer.

After this analysis is complete, the program draws the graph of the approximating normal distribution representing the distribution of sample proportions, marks all the possible confidence levels with tick marks, draws vertical lines corresponding to the confidence level you requested, and paints the appropriate portion of the region under the curve.

We have also included Program 27: Distribution of Sample Proportions to allow you to experiment with the distribution of sample proportions. This program is a simulation of the effects of selecting repeated random samples from a binomial population with parameters

π and n. It calculates and displays the proportion of successes in each series of trials.

First you supply the values for π and n for the underlying binomial population. Then the program asks for the number of repetitions you desire. It performs the random simulation the indicated number of times and displays the proportion of successes in each repetition. It also draws a vertical line to indicate the location of the population proportion π.

For example, suppose you select the binomial process with $\pi = .4$ and $n = 10$ and ask for 200 repetitions. The program randomly performs a binomial experiment 10 times with probability $\pi = .4$ of success and calculates the proportion of successes that occur. It repeats this process 200 times.

Notice that of the 10 trials, the only possible number of successes is 0, 1, 2, ..., 10, and hence the only possible proportions that can occur are $p = 0, .1, .2, ..., 1.0$. Consequently, for this set of values, the program can produce only these 11 possible results, and so the corresponding display consists of just 11 columns, one for each possible outcome. A typical display is shown in Figure 8.28.

By way of comparison, we also show the results from using $n = 43$ and $\pi = .4$ with 300 repetitions in Figure 8.29. In this case, there are many more possible proportions of success that can occur, hence many more columns of results appear in the simulation display. Further, as the number of trials n increases, observe how the resulting distribution of sample proportions becomes more normal in shape. Notice in the first case (with $n = 10$ and $\pi = .4$) that $n\pi = 4$, which is smaller than 5 and so violates condition (8.1) for approximate normality. In the second case (with $n = 43$ and $\pi = .4$), $n\pi$ and $n(1 - \pi)$ are both greater than 5, and so the approximation is good.

Moreover, an accompanying text screen displays some numerical results on the simulation. In particular, the program calculates the mean and standard deviation of the different sample proportions p and prints the results. In addition, it prints the theoretical values for the mean $\mu_p = \pi$ and standard deviation

$$\sigma_p = \sqrt{\frac{\pi(1 - \pi)}{n}}$$

for this sampling distribution for comparison. Notice that the simulated results are usually extremely close to the predicted values for these parameters.

We encourage you to experiment with this program by selecting a variety of different pairs of values for n and π to see the effects on the sampling distribution. In this way, you should develop a better understanding of the properties of the distribution of sample proportions.

FIGURE 8.28

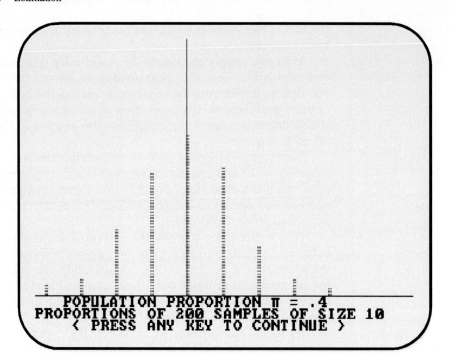

FIGURE 8.29

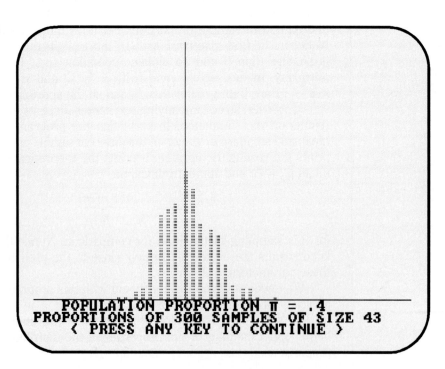

Exercise Set 8.3

Mastering the Techniques

Construct 90%, 95% and 99% confidence intervals for the population proportion π based on the following sets of sample data.

1. $n = 100, x = 25$
2. $n = 100, x = 65$
3. $n = 250, x = 50$
4. $n = 250, x = 100$
5. $n = 1000, x = 225$
6. $n = 1000, x = 480$

Applying the Concepts

7. In a taste test of two brands of soda administered to 100 randomly selected people, 55 preferred brand A to brand B. Construct a 90% confidence interval for the proportion of all people who prefer A to B.
8. A random survey of 60 people waiting for service in a fast-food restaurant found that 15 people waited more than 5 minutes to be served. Find a 95% confidence interval for the proportion of all customers who would have to wait more than 5 minutes.
9. A police department conducts a test on the brakes of 40 randomly selected trucks on a highway. They find that 14 of them are in need of repairs.
   (a) Find a 98% confidence interval for the proportion of all trucks driving with faulty brakes.
   (b) Find a 98% confidence interval for the proportion of all trucks driving with good brakes.
10. The Internal Revenue Service randomly selects 400 tax returns and reports that 120 contain errors of some type.
    (a) Find a 95% confidence interval for the proportion of all tax returns that have errors.
    (b) Find a 98% confidence interval for the proportion of all tax returns that do not have errors.
11. Of 200 randomly selected students surveyed at a university, 37% report that they do not live in dorms. Find a 90% confidence interval for the proportion of all students who live off-campus.
12. Of a group of 250 randomly selected individuals, 36% report that they own CD players. Find a 95% confidence interval for the proportion of all people who own CD players.
13. A survey conducted by an electric utility finds that 373 of the 2074 respondents own home computers. Construct a 96% confidence interval for the proportion of all households in this area that own computers.
14. A pre-election poll of 355 randomly selected people finds that 135 prefer a certain candidate. Construct a 97% confidence interval for the proportion of all voters who prefer her.

Computer Applications

15. Use Program 2: Random Number Generator to select a random sample of 40 baseball batting averages from the list in Appendix I. Count how many players in the sample are batting over .280. Use these sample data to construct a 90% confidence interval for the proportion of all players hitting over .280. Does this confidence interval contain the true population proportion π that you can find by counting? Compare your result with those obtained by others in your class, and see whether approximately 90% of you did generate confidence intervals which contain the population proportion π.

16. Repeat Exercise 15 using a sample of size 25 drawn from the data set of temperatures in different cities in Appendix II to generate a 95% confidence interval for the proportion of all cities whose high temperature is above 70°F on the day in question.

8.4 Determining the Sample Size n

Whenever we construct a confidence interval, there is a relationship between the sample size n and the length of the corresponding interval. The larger the sample size, the narrower the confidence interval for a fixed level of confidence. A narrow confidence interval is desirable since it represents a sharper estimate for either μ or π. On the other hand, the larger the confidence interval, the less sharp or precise the estimate.

Up to this point, we have been given a set of sample data and constructed the corresponding confidence interval based on it. Suppose you are personally conducting a survey to collect the data necessary to estimate the mean μ or the proportion π of some population of interest. In such a circumstance, you are free to take as large a sample as you need, assuming that factors such as cost or time are not critical. Since a larger sample results in a sharper estimate for either μ or π, it is desirable to take a relatively large sample. At the other extreme, taking a sample can be difficult and time-consuming, so you also want to use the smallest sample that is adequate to give an accurate estimate. The question then becomes, How large should the sample be?

Actually, there is a more important decision that we may have to make. We often want to achieve a certain predetermined level of accuracy in making an estimate. Since the size of the sample determines the accuracy of the estimate (the length of the confidence interval), this involves knowing how large a sample we must use. For instance, in a political poll, we may want the prediction to be accurate within 3 percentage points. How large must the sample be to achieve this?

First we consider the case for estimating the population mean μ. If you were estimating the mean height μ of all students at your school, you might want the result to be accurate to within 0.5 inch. Is it possible to determine a sample size n which will guarantee this degree of accuracy? As we will see, the answer is yes.

When we construct a confidence interval for the mean μ based on a large sample of size $n > 30$, the interval ranges from

$$\bar{x} - z\sigma_{\bar{x}} \quad \text{to} \quad \bar{x} + z\sigma_{\bar{x}}$$

Therefore, the maximum possible error in the estimate is just the distance from the center at $\bar{x}$ to either end of the interval, and this is precisely $z\sigma_{\bar{x}}$. Once we choose a given level of confidence, say 90%, this determines the value of z that we will use. In this case, $z = 1.64$. For convenience, we denote this specific value of z by z_0. Consequently, the maximum error in the estimate is equal to

$$z_0 \sigma_{\bar{x}} = z_0 \left(\frac{\sigma}{\sqrt{n}} \right)$$

Suppose that the maximum error we are willing to accept is E. (We suggested above that we might want $E = 0.5$ inch when estimating the mean height of students.) We therefore set this maximum error equal to the above expression

$$E = \frac{z_0 \sigma}{\sqrt{n}}$$

and then solve this equation algebraically for n. We first multiply through on both sides of the equation by $\sqrt{n}$ to obtain

$$E\sqrt{n} = z_0 \sigma$$

We then divide both sides by E, so that

$$\sqrt{n} = \frac{z_0 \sigma}{E}$$

and we finally square both sides to obtain

$$n = \left(\frac{z_0 \sigma}{E} \right)^2 \tag{8.2}$$

If the population standard deviation σ is known, this formula allows us to determine the minimum value for n which will achieve the desired degree of accuracy. Any sample of this size or larger will guarantee at least that level of accuracy.

EXAMPLE 8.14 Suppose that the standard deviation for adult heights is $\sigma = 3$ inches. We want to construct a 90% confidence interval for the mean μ and have it accurate to within 0.5 inch. Find the minimum sample size n needed.

SOLUTION In this case, we have

$$\sigma = 3$$
$$E = 0.5$$
$$z_0 = 1.64$$

Consequently, formula (8.2) yields

$$\begin{aligned} n &= \left(\frac{z_0 \sigma}{E}\right)^2 \\ &= \left[\frac{(1.64)(3)}{.5}\right]^2 \\ &= (9.84)^2 \\ &= 96.826 \end{aligned}$$

Obviously, we cannot have a fractional sample size, and so we round this result *up* to 97. Therefore, any sample of size $n \geq 97$ will give us the desired degree of accuracy with 90% certainty. ∎

The major problem with the above formula is that we usually do not know the value of σ for the population in question. One way to deal with this is to take a preliminary random sample of some arbitrary size $m > 30$ from the population and calculate $\bar{x}$ and s for it. We then use the value for s from this initial sample as an estimate for σ in Equation (8.2) and so approximate n as

$$n = \left(\frac{z_0 s}{E}\right)^2 \tag{8.3}$$

so long as n turns out to be larger than 30. We then take a second random sample of size n to construct our confidence interval with the desired level of accuracy. We illustrate this procedure in the next example.

EXAMPLE 8.15 A 95% confidence interval for the mean speed of cars on a highway is to be constructed which must be accurate to within 2 miles per hour. A preliminary sample yields a value of $s = 9$ miles per hour. What is the smallest sample size n that will provide the desired accuracy with 95% certainty?

SOLUTION We use

$$s = 9$$
$$E = 2$$
$$z_0 = 1.96 \quad \text{(for a 95\% confidence interval)}$$

Therefore, from equation (8.3),

$$n = \left[\frac{(1.96)(9)}{2}\right]^2$$
$$= (8.82)^2$$
$$= 77.79$$

Hence, any sample of size 78 or larger will give the desired accuracy with approximately 95% certainty. ∎

Incidentally, if the value for n determined by equation (8.2) or (8.3) turns out to be less than 31, then the procedure becomes invalid. The above derivation is based on using the z value for the desired level of confidence, which in turn presupposes that the sample size is at least 31. Consequently, if the formula produces a value of $n \leq 30$, the best strategy is to use a sample size of 31 or larger to achieve the desired level of accuracy.

We can also determine the sample size n needed to achieve a given level of accuracy in constructing a confidence interval for the population proportion π. The appropriate value for n is then given by

$$n = \frac{z_0^2 \pi(1-\pi)}{E^2} \tag{8.4}$$

where z_0 is the z value corresponding to the desired confidence level, E is the maximum error which we will accept, and π is the population proportion. (You will be asked to derive this result in Exercise Set 8.4.)

Since the confidence interval is used to estimate π, we will not know its value. There are two ways we can proceed. The first approach is to collect a preliminary sample, calculate the sample proportion p based on it, and use this as an initial estimate for π. In this case, equation (8.4) becomes

$$n = \frac{z_0^2 p(1-p)}{E^2} \tag{8.5}$$

provided that np and $n(1-p)$ are both greater than 10.

The second approach makes use of a fact which can be established using calculus: The maximum value that the product $\pi(1-\pi)$ can attain is .25, and this occurs when $\pi = .5$. For example, if $\pi = .3$, then $1 - \pi = .7$ and their product is $(.3)(.7) = .21 < .25$. Using this fact, we rewrite equation (8.4) as

$$n = \frac{z_0^2 \pi(1-\pi)}{E^2} \leq \frac{z_0^2(.25)}{E^2} = \frac{z_0^2(.5)^2}{E^2}$$

so that

$$n = \left(\frac{\frac{1}{2}z_0}{E}\right)^2 \quad (8.6)$$

EXAMPLE 8.16 The 90% confidence interval for the proportion of bluegrass seeds in bags of mixed grass seeds is to be constructed and must be accurate to within .02. A preliminary sample provides an initial estimate of $p = .64$. What is the smallest sample size n that will provide the desired accuracy with 90% certainty?

SOLUTION We are given

$$p = .64$$
$$E = .02$$
$$z_0 = 1.64 \quad \text{(for a 90\% confidence interval)}$$

Therefore, equation (8.5) yields

$$n = \frac{(1.64)^2(.64)(.36)}{(.02)^2}$$
$$= 1549.2$$

As a result, we would have to sample a minimum of 1550 seeds from such a mixture to achieve the desired accuracy.

If we use the alternate approach with Equation (8.6), then we obtain

$$n = \left[\frac{\frac{1}{2}(1.64)}{.02}\right]^2$$
$$= 1681$$

This result is larger than the one obtained by using equation (8.5) because equation (8.6) is the "worst-case scenario" using .25 as the greatest value that $\pi(1-\pi)$ could possibly attain. ∎

We note that when π is relatively close to .5, the difference between equations (8.5) and (8.6) is not major. However, if π is close to either 0 or 1, then the approximate results for n given by equation (8.6) will be considerably different from that given by equation (8.5). Therefore, if you suspect that the unknown proportion is either very small or very large, it makes sense to collect a preliminary sample from which to calculate an initial estimate p of π rather than run the risk of having to use a considerably larger sample than is necessary.

Let's see what happens if we construct a 90% confidence interval for the proportion of bluegrass seeds based on a sample of size $n = 1000$ (instead of the predicted minimum value of $n = 1550$). Since we are given $p = .64$, we find

$$s_p = \sqrt{\frac{(.64)(.36)}{1000}}$$

$$= \sqrt{.0002304}$$

$$= .0152$$

Further, for a 90% confidence interval, $z = 1.64$, so that the corresponding interval is

$$p \pm z s_p = .64 \pm 1.64(.0152)$$

$$= .64 \pm .0249$$

The accuracy is not yet within the desired $E = .02$. Clearly, to achieve this level of accuracy, we must use a still larger sample size; in particular, we would need a sample of size 1550 or greater.

Exercise Set 8.4

Mastering the Techniques

Determine the minimum value of n which produces the given level of accuracy E for a confidence interval for the population mean μ based on the given data:

1. 90% confidence interval, $\sigma = 2$, $E = .5$
2. 95% confidence interval, $\sigma = 6$, $E = 1.5$
3. 98% confidence interval, $s = 10$, $E = 2.5$
4. 99% confidence interval, $s = 24$, $E = 8$

Determine the minimum value of n which produces the given level of accuracy E for a confidence interval for the population proportion π based on the given data:

5. 90% confidence interval, $p = .25$, $E = .05$
6. 95% confidence interval, $p = .40$, $E = .06$
7. 98% confidence interval, $p = .65$, $E = .025$
8. 99% confidence interval, $p = .54$, $E = .01$

Applying the Concepts

9. A 90% confidence interval for the mean sixth-grade reading score in a school district is to be constructed and must be accurate within .2. A preliminary sample provides a value of $s = 1.2$. What is the smallest sample size n that provides the desired accuracy with 90% certainty?
10. A 95% confidence interval for the mean time that workers in a fast-food restaurant spend serving a customer is to be constructed and must be accurate to within 0.5 minute. A preliminary sample provides a value of $s = 2$ minutes. What is the smallest sample size n that provides the desired accuracy with 95% certainty?

11. A 95% confidence interval for the proportion of all holders of a particular credit card who are behind on their payments is to be constructed which must be accurate to within .04. A preliminary sample provides a value of $p = .72$. What is the smallest sample size n that provides the desired accuracy with 95% certainty?

12. A 98% confidence interval for the proportion of all cars which do not pass a state's emission control test is to be constructed which must be accurate to within .02. A preliminary sample provides a value of $p = .42$. What is the smallest sample size n that provides the desired accuracy with 98% certainty?

13. Derive Equation (8.4) for the sample size n needed to achieve a level of accuracy E when constructing a confidence interval for the proportion π. (*Hint:* Set E equal to $z_0 \sigma_p = z_0 \sqrt{\pi(1-\pi)/n}$ and solve for n.)

14. Suppose a 95% confidence interval for the proportion of people having a rare genetic condition is to be constructed and must be accurate to within .01. A preliminary sample provides a value of $p = .01$. Calculate the smallest sample size n, using both equation (8.5) and equation (8.6), and compare the results. Why are they so different? Which is more appropriate?

8.5 Summary of Confidence Intervals

In this chapter, we learned how to construct confidence intervals for a variety of situations that affect all aspects of our lives. The procedures used to solve such problems are relatively straightforward and very similar to one another. In fact, the greatest difficulty that many students have with such problems lies in distinguishing one type of estimation situation from another. Once you identify the type of problem, it is usually easy to complete the solution.

In this section we summarize all the confidence interval situations covered so far to make it easier for you to identify which procedure applies for a given problem. This information is arranged in Table 8.1.

TABLE 8.1

Situation	Sample Statistics	Conditions	Standard Deviation	Confidence Interval
Mean μ, large sample	$n, \bar{x}, s$	$n > 30$	$\dfrac{\sigma}{\sqrt{n}} \approx \dfrac{s}{\sqrt{n}}$	$\bar{x} \pm z \sigma_{\bar{x}}$ or $\bar{x} \pm z\left(\dfrac{s}{\sqrt{n}}\right)$
Mean μ, small sample	$n, \bar{x}, s$	$n \leq 30$ Population is nearly normal	$\dfrac{s}{\sqrt{n}}$	$\bar{x} \pm t\left(\dfrac{s}{\sqrt{n}}\right)$
Proportion π	n, x or n, p	$np > 10$ $n(1-p) > 10$	$\sqrt{\dfrac{p(1-p)}{n}}$	$p \pm z \sigma_p$ or $p \pm z \sqrt{\dfrac{p(1-p)}{n}}$

To distinguish one type of situation from another, you must read each question very carefully. To begin, you must distinguish between confidence intervals and other types of problems such as probability problems. If a question involves estimation, it almost always has the phrase *confidence interval* or the word *estimate* in it. But if a question asks for the probability that something happens, then it is not a confidence interval situation—you must solve it as a probability problem.

Further, you must distinguish between confidence intervals for means and for proportions. If a problem involves means (averages), then the sample statistics needed are a sample size n, a sample mean $\bar{x}$, and most likely a sample standard deviation s. (Occasionally, you may be given the raw data and have to calculate these quantities.) If the problem involves proportions, then the sample statistics needed are a sample size n and either a sample proportion p (as a decimal or a percentage) or the number x of "successes" out of n cases. The type of sample statistics given in a problem virtually dictates the type of situation. To see this, examine the way that different types of sample statistics for the data are displayed in Table 8.1.

Next, if a question asks for a confidence interval for means, then you must distinguish between large and small samples. Whenever you identify the sample size n, be sure to stop and note whether it is larger than 30.

As a final strategy, we suggest the following. When you face a series of problems that cover a variety of cases, such as those at the end of this section or on a test, you may find it easier to distinguish between them if you read through all the questions first to make an initial identification of each one *before* you solve any of them. When you first look at such a varied series of questions, you bring to it a certain perspective or overview which makes it easier to perform this identification. However, once you become involved in solving any single problem, you often lose much of that overall perspective in the process of doing the calculations. The different questions may then start to look all the same.

Exercise Set 8.5

Applying the Concepts

1. A study conducted by the U.S. Center for Disease Control found that, of 64 cruise ships tested, 34 did not achieve the minimum level of health standards. Determine a 90% confidence interval for the proportion of all cruise ships that do not meet these standards.
2. A survey of 64 randomly selected faculty at a major university finds that the mean salary is $44,500 with a standard deviation of $5800. Determine a 95% confidence interval for the mean salary of all faculty at this college.
3. A cruise line company finds that 78% of all the cabins on its fleet are sold for a sample of 30 ocean cruises one winter. Find a 90% confidence interval for the proportion of all cabins sold on all its ships that winter.

4. In a sample of 25 baseball bats, the mean weight is 32.23 ounces with a standard deviation of .4 ounce. Find a 98% confidence interval for the mean weight of all such bats.

5. A survey finds that, in a random sample of 36 cars waiting to pay a bridge toll to enter a city during the morning rush hour, the average waiting time to get to the toll booth is 16 minutes with a standard deviation of 3 minutes. Find a 90% confidence interval for the mean waiting time for all cars waiting for the toll.

6. A dental experiment involves coating patients' teeth with a special compound which is intended to reduce formation of plaque and so reduce the number of cavities. The compound is applied to the teeth of a sample group of 16 volunteers. After 3 years, these patients developed a mean of 3.2 cavities with a standard deviation of 1.4. Determine a 95% confidence interval for the mean number of cavities developed by all similar patients using this compound for 3 years.

7. Suppose the group using the dental compound in Exercise 6 is compared to a control group of 60 patients who do not use the compound. Over the same 3-year period, the control group has a mean of 4.7 cavities with a standard deviation of 1.1 cavities. Find a 95% confidence interval for the mean number of cavities for all similar individuals who do not use the compound for 3 years. Do the two intervals overlap? What does this indicate to you?

8. A medical study finds that, in a random sample of 250 people selected from a certain demographic group in a particular city, 18 have been exposed to the AIDS virus. Find a 99% confidence interval for the proportion of all people in this group who have been exposed to the AIDS virus.

9. A study is conducted to compare the estimated costs of repairs to a VCR in nine different electronics repair shops. The results, in dollars, are

$$75, 110, 85, 90, 125, 100, 92, 115, 126$$

Determine a 90% confidence interval for the mean cost of repair of this VCR in all such shops in the area.

10. A government agency puts out a call for bids on paper for office supplies. A sample of 20 of the bids submitted by various companies are studied and found to have a mean of \$32 per case with a standard deviation of \$2.20. Find a 99% confidence interval for the mean cost of a case of such paper from all possible suppliers.

11. A telephone survey of 150 homes in a certain area finds that 16 have their radios tuned to a certain daytime radio program. Find a 90% confidence interval for the proportion of all households listening to this station.

12. A tire manufacturer is testing the stopping distance for a new tread design. To do this, a test car is driven at a speed of 50 miles per hour when the driver slams on the brakes. These stopping distances (in feet) are recorded:

$$120, 142, 131, 155, 126, 135, 143$$

Determine a 98% confidence interval for the mean stopping distance for this tread design.

13. A waitress in an expensive restaurant is enrolled in a statistics course, so she decides to apply some statistical methods to her work. She finds that in a random sample of 35 couples dining in this restaurant, the mean tip she receives is $12 with a standard deviation of $3.50. Find a 90% confidence interval for the mean of all tips she receives from couples dining there.
14. The waitress in Exercise 13 is also keeping track of the amounts that diners are paying for their meals and how they are paying. She finds that, in a random sample of 70 parties, 24 pay with a certain credit card. Find a 95% confidence interval for the proportion of all customers at the restaurant who pay with this credit card.
15. The waitress in Exercises 13 and 14 finds that the mean amount charged by the 24 customers using the credit card is $78 with a standard deviation of $34. Find a 95% confidence interval for the mean amount of the bill for all credit card patrons of this restaurant.

Student Projects

1. Continue the investigation you began for your projects in Chapters 2 and 3. You have already collected a set of data and found values for the sample mean and standard deviation. Now use these results to construct 90%, 95% and 99% confidence intervals for the mean μ of the underlying population from which your sample was drawn. Recall that the initial instructions asked you to select a random sample with at least 25 data values. If you selected a small sample, be careful how you proceed.

   After you complete your calculations, you should incorporate the results into a formal statistical project report. The report should include

   - A statement of the topic being studied, including identification of the *specific* underlying population under investigation
   - The source of the data
   - A discussion of how the data were collected and why you believe that it is a random sample
   - A list of the data
   - All the statistical results you previously calculated
   - A discussion of whether the data appear to come from a roughly normal population
   - The results you obtain for the various confidence intervals
   - A discussion of any surprises you may have noted in connection with collecting or using the data for estimation

2. Conduct a survey to obtain a set of data with which you will construct a confidence interval for proportions. For example, you might try to estimate
   (a) The proportion of students at your school who smoke

(b) The proportion of students on the Dean's list
(c) The proportion of people in a community who watch a particular television program
(d) The proportion of adults who support some political or social position, for instance, stricter gun controls, capital punishment, euthanasia, abortion at the request of the pregnant woman, a stronger national defense stand, or an increase in taxes to reduce the national deficit.

To do this project, prepare a statement on the issue you are studying. State it in such a fashion that respondents have two possible answers such as: yes or no, agree or disagree, etc. Poll a minimum of 50 randomly selected individuals to obtain their responses to your question. Use these data to construct 90%, 95%, and 99% confidence intervals for the proportion of all people in the population under study who fall into this category.

Incorporate your results in a formal statistical project report. The report should include

- A statement of the topic being studied, including identification of the *specific* underlying population under investigation
- The source of the data
- A discussion of how the data were collected and why you believe that it is a random sample
- A list of the data
- All the statistical results you calculated, including the different confidence intervals
- A discussion of any surprises you may have noted in connection with collecting or using the data for estimation

CHAPTER 8 SUMMARY

You use **confidence intervals** to predict or estimate the value of a population parameter such as μ or π based on data from a sample.

To estimate the population mean μ when $n > 30$ and σ is known, you use the sample statistics n and $\bar{x}$ to construct a confidence interval

$$\bar{x} \pm z\sigma_{\bar{x}} = [\bar{x} - z\sigma_{\bar{x}}, \bar{x} + z\sigma_{\bar{x}}]$$

where the value of z used depends on the degree of confidence desired.

If σ is not known and the population is roughly normal, you use

$$\sigma_{\bar{x}} = \frac{\sigma}{\sqrt{n}} \approx \frac{s}{\sqrt{n}}$$

If the sample size $n < 30$, then the confidence interval is based on a t-distribution with $n - 1$ degrees of freedom and is

$$\bar{x} \pm t\left(\frac{s}{\sqrt{n}}\right) = \left[\bar{x} - t\left(\frac{s}{\sqrt{n}}\right), \bar{x} + t\left(\frac{s}{\sqrt{n}}\right)\right]$$

provided that the underlying population is approximately normal. The values of t are found in Table IV.

To estimate the population proportion π, use sample statistics n, x, and p to construct the confidence interval

$$p \pm z\sigma_p = [p - z\sigma_p, p + z\sigma_p]$$

where

$$\sigma_p = \sqrt{\frac{\pi(1-\pi)}{n}} \approx \sqrt{\frac{p(1-p)}{n}}$$

provided that

$$np > 10 \quad \text{and} \quad n(1-p) > 10$$

The **level of confidence** is the percentage equivalent of the probability that the confidence interval contains the population parameter being estimated. If you construct a confidence interval with a given level of confidence, then the larger the sample size n, the shorter the confidence interval will be and vice versa (all other things being equal). The higher the level of confidence, the larger the confidence interval, and vice versa (all other things being equal).

To find the sample size n needed to achieve a certain level of accuracy, let E be the maximum acceptable error. The corresponding sample size n is for means:

$$n = \left(\frac{z_0 s}{E}\right)^2$$

for proportions:

$$n = \frac{z_0^2 p(1-p)}{E^2}$$

or

$$n = \left(\frac{\frac{1}{2}z_0}{E}\right)^2$$

9 Hypothesis Testing

In a study conducted by USA TODAY, one out of every four properly addressed letters was delivered late by the U.S. Postal Service. Even worse, one-fifth of those late letters were severely late—three days or more behind the Postal Service's delivery guarantees.

Depending on distance, the Postal Service claims that all first class mail will be delivered within the continental United States in one to three days. However, in this study of 1000 letters mailed to and from various points in the country, the letters arrived an average of half a day late.

On the bright side, the study found that 65% of all the letters sent arrived on time and, in fact, 10% were delivered earlier than the Postal Service guaranteed. But, of the 1000 letters sent, 10 were still undelivered after more than a month. Further, 25% of correctly addressed items arrived at least a day behind schedule. 82% of those letters with intentionally incorrect ZIP codes were late.

The US Postal Service claims that it will provide:

- Next-day service within the same metropolitan area
- Two-day service within 600 miles
- Three-day service beyond 600 miles

To test these claims, the above study conducted by *USA TODAY* had people mail 1000 letters, some within the local city, others within the region and others across the country.

How important is it for the mail to be on time? According to the American Bankers Association's postal services committee, businesses would lose more than $1 billion a year in lost interest alone if all mail arrived just one day late. Consumers would pay an extra $135 million a year in interest on credit cards and other debt payments. The Postal Service plans to spend $4.7 billion on automation from 1991 to 1995 to dramatically improve service.

But are these expectations realistic? Do you think that it is feasible for the Postal Service to be able to deliver every single item mailed between every possible pair of towns in the country within 3 days considering the volume of hundreds of millions of items processed each day? Would you expect different times on different days of the year or in different locations within the same city or state? Might there be a difference depending on whether an item was mailed from the corner mailbox or at the post office or, in some areas, in the individual's home mail box? For that matter, should the focus be on the *mean delivery time* or should we be concerned with the *worst possible* incidents, the outliers? If it is the latter, then the problem reduces to controlling the spread rather than the mean of the delivery times. Is this possible? Is it realistic to expect that any process can always be improved? If so, how do we test and prove that there actually has been improvement?

In this chapter, we will develop statistical methods for conducting tests of claims that apply to a wide variety of important situations that affect all aspects of our lives. These include economic, political, medical and technological applications.

9.1 Hypothesis Tests for Means

In Chapter 8 we studied one of the two main branches of inferential statistics, the estimation of population parameters such as the mean and the proportion of successes. We now introduce the second branch, *hypothesis testing*, which uses sample data to make decisions about an underlying population. In this section we study hypothesis tests for the mean μ of a population. In a later section, we consider hypothesis tests for the population proportion π.

To illustrate, suppose Goodstone Tires advertises that a particular line of tires lasts an average of at least 30,000 miles. A consumer testing organization tests 100 of these tires and finds that they last an average of 20,000 miles under normal driving conditions. What conclusion can we draw from these data? The mean for this sample ($\bar{x} = 20,000$) is so far below the claimed mean for the population ($\mu = 30,000$) that it almost certainly indicates that the claim is vastly inflated. As a consequence, we would likely conclude that the original claim was wrong and therefore would reject it out of hand. It is just too unreasonable to obtain a sample from such a population which has a mean $\bar{x}$ that is this far away from the population mean μ.

Suppose instead that the mean life for the 100 tires is $\bar{x} = 25,000$ rather than 20,000 miles. This is still far below the claimed mean μ. Again, we would probably conclude that the true mean lifetime μ of all these tires was not likely to be as high as 30,000 miles based on the sample data. However, we might not be quite as certain as in the first case. It does seem unlikely that the sample mean could be this far below the true mean based solely on the random variation present between different elements in a population.

Now suppose that the mean lifetime for the 100 tires is $\bar{x} = 28,000$ miles. For these data, we might be unwilling to reject the original claim of 30,000 miles out of hand. The figure 28,000 seems relatively close to the claimed mean of 30,000 and might simply be attributable to our having selected a sample of tires that contained a few bad ones. As a consequence, we might be tempted to reserve judgment in this case.

Next, suppose the mean for the 100 tires in the sample is $\bar{x} = 29,500$ miles. This value is close to the claimed value for the mean of all such tires. As a result, we would probably conclude that these data support the original claim.

Finally, suppose the mean for the 100 tires is $\bar{x} = 31,000$ miles. In this case, we would conclude that the claim was fully justified, since we seemingly are getting more mileage from these tires than the manufacturer advertises.

In making any of these decisions, keep in mind that there is always an element of risk. There is random variation among samples reflecting the variation in the population. Thus, no matter how unlikely the sample data seem, it is always possible that we drew a particularly unrepresentative sample. Alternatively, we might have selected a particularly good sample and have therefore overstated the results.

Moreover, the element of risk goes deeper than just the possibility of making a mistaken judgment. Often there are penalties associated with such an error. For instance, suppose we decide that Goodstone's claim of 30,000 miles is too high and publicly call the company a pack of liars. If the company has data which can prove that its claim is correct, then we can expect Goodstone and its legal department to react with a rather large lawsuit. Before making any such judgments, we must be able to assess the likelihood of our being wrong.

The above ideas form the basis for hypothesis testing. Notice how the sample data are used. If the value for the mean $\bar{x}$ of the sample is unrealistically far from the claimed mean μ of the population, then this evidence suggests that the claimed value for μ is not correct. If the value of the sample mean $\bar{x}$ is relatively close to the value claimed for μ, then it seems to support the claim. However, there is obviously a gray area in between these two extremes.

To study this gray area more carefully, we must put the above ideas into a more rigorous context. First, there is a claim or statement about the mean of a population. This is known as the *null hypothesis* and is denoted by H_0. For the above illustration, we write

$$H_0: \mu = 30,000$$

In addition, we have an *alternate hypothesis*, denoted by H_a, which states our suspicion about the population mean μ. In the tire example, we might suspect that the actual mean μ is less than the claimed value of 30,000. In this case, we write the alternate hypothesis as

$$H_a: \mu < 30,000$$

Note that this says that we suspect the true mean μ is less than 30,000. It does not specify any particular alternate value for μ.

Depending on the particular situation, there are several other possibilities that are appropriate for the alternate hypothesis. We will encounter cases where we believe the true mean μ for some population is actually larger than the value claimed. In such a case, the alternate hypothesis takes the form

$$H_a: \mu > 30{,}000$$

In addition, there are cases where we simply do not believe that the claimed value for μ is correct. We make no indication of whether the claim is too high or too low; we merely contend that the true value for μ is *different* from the claim, and we therefore write

$$H_a: \mu \neq 30{,}000$$

Realize that, in any given problem, there is just one alternate hypothesis. The choice of which of the three possible forms is appropriate for H_a must be made based on the context; that is, based on the reason the study was conducted in the first place.

The first two possibilities for an alternate hypothesis

$$H_a: \mu < 30{,}000$$

and

$$H_a: \mu > 30{,}000$$

are known as *one-tailed tests*. The third possibility

$$H_a: \mu \neq 30{,}000$$

is known as a *two-tailed test*. The reasons for these names will become apparent later.

The null hypothesis is

$$H_0: \mu = \mu_0$$

The alternate hypothesis is

or

$$H_a: \mu < \mu_0$$

$$H_a: \mu > \mu_0$$ } one-tailed test

or

$$H_a: \mu \neq \mu_0$$ two-tailed test

In any hypothesis-testing situation on the mean, we will have a set of sample data relating to some randomly selected sample. In turn, these data give us the values for n, $\bar{x}$ and s.

Our decision in any hypothesis-testing situation for the mean depends on how far the sample mean $\bar{x}$ is from the population mean μ claimed in H_0. If it is too far away to be accounted for by random variation, we *reject* the null hypothesis H_0 and therefore accept the alternate hypothesis H_a in its place. If

$\bar{x}$ falls relatively close to the value for μ, then we *fail to reject* the null hypothesis H_0.

In making these decisions, we can think of the null hypothesis as being on trial. The sample data collected represent the evidence being presented. If the evidence is sufficiently strong, it will convict the null hypothesis and so we reject H_0. If the data are weak, then they do not convict the null hypothesis, and so we fail to reject H_0. Moreover, the basis for our judicial system is the principle that the accused is considered innocent until proved guilty. A similar principle holds in hypothesis testing: *The null hypothesis is considered correct until proved otherwise.* In other words, we assume that the claimed value for μ is true and work with it throughout a problem.

Now let's see how the above ideas are applied to the problem involving the tires. Suppose that the claim is again $\mu = 30{,}000$ miles and that the sample of 100 tires lasted an average of $\bar{x} = 25{,}000$ miles with a sample standard deviation $s = 5000$ miles. Suppose further that tire life is approximately normally distributed. The null and alternate hypotheses are

$$H_0: \mu = 30{,}000$$

$$H_a: \mu < 30{,}000$$

The sample data give

$$n = 100$$

$$\bar{x} = 25{,}000$$

$$s = 5000$$

The question now becomes, How likely is it for the mean of a sample of size $n = 100$ to be this far below the population mean μ? To answer this question, we must interpret the sample mean $\bar{x}$ in terms of its associated z value. We know that the distribution of sample means has parameters

$$\mu_{\bar{x}} = \mu = 30{,}000$$

(since we assume that the null hypothesis is correct) and

$$\sigma_{\bar{x}} = \frac{\sigma}{\sqrt{n}}$$

Because σ is not known, we use

$$\frac{s}{\sqrt{n}} = \frac{5000}{\sqrt{100}} = \frac{5000}{10} = 500$$

in place of $\sigma_{\bar{x}}$, as we did with confidence intervals for the mean. Further, since the sample size $n = 100 > 30$, we can use a normal distribution. Consequently, the z value corresponding to $\bar{x} = 25{,}000$ is

$$z = \frac{\bar{x} - \mu_{\bar{x}}}{\sigma_{\bar{x}}}$$

$$= \frac{25{,}000 - 30{,}000}{500}$$

$$= \frac{-5000}{500}$$

$$= -10$$

That is, the sample mean $\bar{x} = 25{,}000$ miles actually lies 10 standard deviations below the claimed mean $\mu = 30{,}000$ miles. As we have seen, it is virtually impossible for such an outcome to arise purely by chance. As a result, we reject the null hypothesis in this case. We therefore accept the alternate hypothesis that the true mean μ is less than 30,000 miles. However, we make no commitment about its specific value. In particular, we cannot conclude that μ must be 25,000.

Incidentally, note that we used the claimed value $\mu = 30{,}000$ in the calculation of the z value corresponding to $\bar{x}$. In the process of conducting a hypothesis test, we always assume that the null hypothesis is correct.

Suppose now that we rework the above problem with just one change—instead of the sample mean $\bar{x} = 25{,}000$, let's use $\bar{x} = 29{,}500$. The null and alternate hypotheses are still

$$H_0: \mu = 30{,}000$$

and

$$H_a: \mu < 30{,}000$$

The sample data values now give

$$n = 100$$
$$\bar{x} = 29{,}500$$
$$s = 5000$$

and we again use

$$\frac{s}{\sqrt{n}} = \frac{5000}{\sqrt{100}} = 500$$

Therefore, corresponding to $\bar{x} = 29{,}500$, we obtain

$$z = \frac{29{,}500 - 30{,}000}{500}$$

$$= \frac{-500}{500} = -1$$

Based on this result, the evidence seems to support the manufacturer's claim since a sample mean falling 1 standard deviation away from the population mean is quite reasonable. Consequently, we fail to reject the null hypothesis under these circumstances. We still suspect that μ may be less than 30,000 (that is why we went to the trouble of testing 100 tires). However, there is simply insufficient evidence from this sample to support our suspicion. Thus, the fact that we "fail to reject" a null hypothesis does not mean that we automatically accept it as valid.

Let's consider one other case where the sample mean $\bar{x} = 29{,}000$. The only changes are in the corresponding value for z and the associated decision we make based on it. Thus, we find that

$$z = \frac{29{,}000 - 30{,}000}{500}$$

$$= \frac{-1000}{500} = -2$$

With this result, it is not immediately apparent whether we should reject the null hypothesis. It is not entirely unreasonable for a sample mean to fall 2 standard deviations below the population mean μ, though it is not likely. The decision in this case highlights the element of risk we mentioned earlier. We must consider in detail how to assess the types of errors that can arise and see how to anticipate how large an error we are willing to accept in making any such judgment.

Whenever we make a judgment in a hypothesis-testing situation, we can make either the correct decision or an incorrect one. A correct decision arises in two ways:

1. Reject a null hypothesis which is wrong   *or*
2. Fail to reject a null hypothesis which is valid

Similarly, there are two ways to make an incorrect decision:

1. Reject a null hypothesis which is valid   *or*
2. Fail to reject a null hypothesis which is wrong

We introduce the following terminology for these two types of errors:

> A **Type I error** is the error we make when we wrongly reject a valid null hypothesis.
>
> A **Type II error** is the error we make in failing to reject an incorrect null hypothesis.

The four cases that can occur are:

	Fail to Reject H_0	Reject H_0
H_0 true	Correct decision	Type I error
H_0 false	Type II error	Correct decision

As we mentioned, a hypothesis test is analogous to a courtroom trial, and the same four types of outcomes can occur. Two involve correct decisions: A guilty person can be convicted, and an innocent person can be acquitted. However, two errors can be made: An innocent person can be convicted (Type I error), and a guilty person can be acquitted (Type II error). Ideally, we would like to minimize the chance of making both types of errors, but this is generally impossible in practice. When we reduce the chance of making a Type I error, we increase the probability of making a Type II error, and vice versa.

In U.S. society, we consider convicting an innocent person a much more serious mistake than acquitting a guilty person. The same is true in hypothesis testing, where we attempt to keep Type I errors acceptably small. We decide in advance on the level of risk that we are willing to accept, and we denote this *risk level* or *significance level* by the Greek letter alpha, α. It represents the probability of making a Type I error. Typically, $\alpha = .10, .05, .02$ or $.01$ or their percentage equivalents 10%, 5%, 2% or 1%.

Essentially, we reject a null hypothesis if the sample data are too unlikely to have occurred purely by chance. We use the value of α to quantify just what we mean by the phrase *too unlikely*. To see how this works, suppose we take $\alpha = .05$ in the Goodstone tire situation. If the evidence is to "convict" the null hypothesis, then the sample mean $\bar{x}$ must fall sufficiently below the claimed mean of $\mu = 30{,}000$. In particular, we focus on the 5% least likely set of possible values for $\bar{x}$. This corresponds to the bottom 5% of the normal population for the distribution of sample means. We identify the bottom 5% of this population and consider any sample mean $\bar{x}$ that falls into this region as being too unlikely to have occurred purely by chance. (See Figure 9.1.) In turn, the fact that $\bar{x}$ has its value in this range suggests that μ is not likely to be as large as 30,000, but rather is less than 30,000. Such evidence leads us to reject the null hypothesis.

In general, the significance level α indicates which portion of the sampling population is considered too unlikely to arise solely by chance. For instance, if $\alpha = .02$ instead, then we identify the least likely (or the bottom) 2% of the population. If the sample mean $\bar{x}$ falls into this region, then we reject

FIGURE 9.1

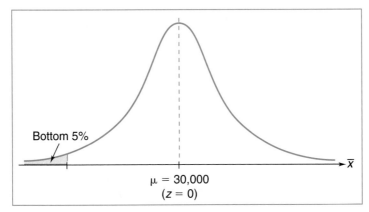

the null hypothesis for μ. Of course, there is a potential error involved in this decision. A sample mean can fall in the bottom 5% or the bottom 2% or even the bottom 1% of the normal population for the distribution of sample means. The chance of each of these is relatively small. But it can still happen, and in such a case we would commit a Type I error by wrongly rejecting a correct null hypothesis. The probability of making such an error is .05, .02 or .01, respectively.

Clearly, the choice for α is extremely important since we can come to different conclusions depending on its value. We note that the value for α must be selected ahead of time. It cannot be chosen for convenience after the z value corresponding to $\bar{x}$ is calculated.

> The associated set of z values corresponding to the value of α is called the **rejection region** or the **critical region**.
>
> The cutoff values for z which define the rejection region are known as the **critical values** for z. See Figure 9.2.

We now consider the third case in the tire problem where $\bar{x} = 29,000$. If $\alpha = .02$, then the bottom or least likely 2% of the population corresponds to z values less than $z = -2.33$. See Figure 9.3. Thus, $z = -2.33$ is the critical value for this hypothesis test at the $\alpha = .02$ significance level. The rejection region consists of all values to the left of $z = -2.33$. We previously found that the z value corresponding to the sample mean $\bar{x} = 29,000$ is $z = -2$. Since this value does not fall in the rejection region, we fail to reject the null hypothesis $\mu = 30,000$ at the $\alpha = .02$ level of significance.

On the other hand, if we use $\alpha = .05$ instead of .02, then the bottom 5% of the normal population lies to the left of the critical value $z = -1.64$. This determines a different rejection region, as shown in Figure 9.4. Corresponding to $\bar{x} = 29,000$, we still have $z = -2$, but this time we reject the null hypothesis at the $\alpha = .05$ level of significance.

We now apply all these ideas in the following examples.

FIGURE 9.2

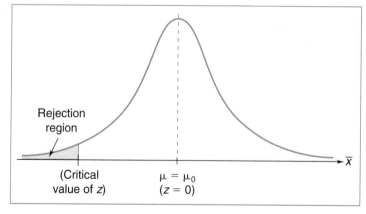

FIGURE 9.3

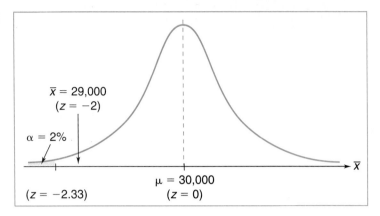

EXAMPLE 9.1 A college claims that its average class size is 35 students. A random sample of 64 classes has a mean size of 37 students with a standard deviation of 6 students. Test at the $\alpha = .05$ level of significance if the claimed value is too low.

SOLUTION The null hypothesis is

$$H_0: \mu = 35$$

Since we suspect that the claim is too low and that the true mean is actually greater than 35, the alternate hypothesis is

$$H_a: \mu > 35$$

The sample data give

$$n = 64$$
$$\bar{x} = 37$$
$$s = 6$$

FIGURE 9.4

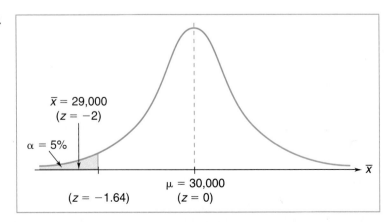

so that

$$\frac{s}{\sqrt{n}} = \frac{6}{\sqrt{64}} = \frac{6}{8} = .75$$

Since $n = 64 > 30$, the distribution of sample means is approximately normal. To reject the null hypothesis, the mean $\bar{x}$ for the sample must fall within the top 5% of the population, as shown in Figure 9.5. This corresponds to a critical value for z of 1.64. Thus, the rejection region consists of all z values greater than 1.64. Note that this is all predetermined before we consider the sample data.

For the sample in question, we find that

$$z = \frac{\bar{x} - \mu_{\bar{x}}}{s/\sqrt{n}}$$

$$= \frac{37 - 35}{.75}$$

$$= \frac{2}{.75} = 2.67$$

Since a z value of 2.67 is beyond the critical value of $z = 1.64$, we reject the null hypothesis. That is, we conclude that the true mean class size is likely more than 35. ∎

EXAMPLE 9.2 A random sample of 100 one-quart containers of milk contain a mean of 31.9 ounces with a standard deviation of .4 ounce. Test if the containers are being underfilled at the $\alpha = .01$ level of significance.

SOLUTION The null hypothesis is

$$H_0: \mu = 32$$

FIGURE 9.5

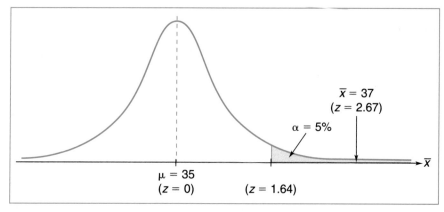

while the alternate hypothesis is

$$H_a: \mu < 32$$

since we suspect that the containers actually contain less than the indicated amount of milk. The sample statistics are

$$n = 100$$
$$\bar{x} = 31.9$$
$$s = .4$$

so that

$$\frac{s}{\sqrt{n}} = \frac{.4}{\sqrt{100}}$$

$$= \frac{.4}{10} = .04$$

Since $n = 100 > 30$, the distribution of sample means is approximately normal. Thus, to test this null hypothesis, we consider the 1% least likely portion of the normal population. As shown in Figure 9.6, the rejection region lies to the left of the critical value $z = -2.33$.

For the sample mean $\bar{x} = 31.9$, we find that

$$z = \frac{31.9 - 32}{.04}$$

$$= \frac{-.1}{.04}$$

$$= -2.50$$

Since this is less than the critical value of $z = -2.33$, we reject the null hypothesis and conclude that the containers are apparently being underfilled. ∎

FIGURE 9.6

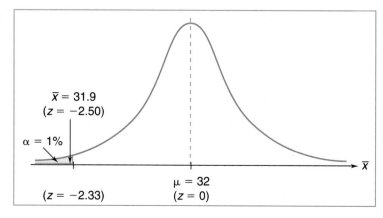

EXAMPLE 9.3 A marketing research group reports that the typical supermarket shopper uses an average of $1.40 in manufacturer "cents-off" coupons per week. A sample of 50 randomly selected shoppers use an average of $1.54 in coupons with a standard deviation of $.62. At the 2% level of significance, test if the report is correct.

SOLUTION The null hypothesis is

$$H_0: \mu = 1.40$$

In this example, we are asked to test if the claim is correct. This does not suggest that it is either too high or too low; we simply want to know whether the mean actually is $1.40. Therefore, the alternate hypothesis is

$$H_a: \mu \neq 1.40$$

The sample data give

$$n = 50$$
$$\bar{x} = 1.54$$
$$s = .62$$

so that

$$\frac{s}{\sqrt{n}} = \frac{.62}{\sqrt{50}}$$

$$= \frac{.62}{7.07} = .088$$

Since $n = 50 > 30$, we again use a normal distribution. Thus, to reject the null hypothesis, the sample mean must be either sufficiently below the claimed value for μ or above it. That is, we reject H_0 if the sample mean falls either in the extreme lower end of the population or in the extreme upper end. Since $\alpha = .02$, this represents the least likely 2% of the normal population. We are using both tails of the population, so that we consider the least likely 1% of the population in *each* tail, as shown in Figure 9.7. The corresponding critical values for z are $z = 2.33$ and $z = -2.33$, and so the rejection region consists of two parts, one in each tail.

For the sample mean $\bar{x} = 1.54$, we find that

$$z = \frac{1.54 - 1.40}{.088}$$

$$= \frac{.14}{.088} = 1.59$$

which does not fall inside the rejection region. As a result, we fail to reject the null hypothesis. That is, there is insufficient evidence to conclude that the average amount is different from $1.40. ∎

FIGURE 9.7

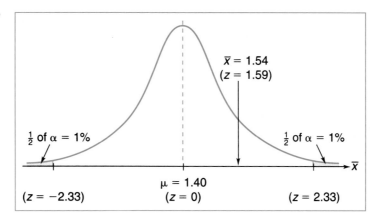

Notice that Examples 9.1 and 9.2 involved alternate hypotheses

$$H_a: \mu > 35$$

and

$$H_a: \mu < 32$$

while Example 9.3 used

$$H_a: \mu \neq 1.40$$

When we introduced alternate hypotheses, we discussed terminology about one- and two-tailed tests. In retrospect, the reason for this terminology should now be obvious. If the alternate hypothesis involves an inequality, < or >, then it is a one-tailed test because only one tail of the normal distribution is used for rejecting the null hypothesis. Moreover, if the inequality is <, then it is the left tail. If the inequality is >, then it is the right tail. When we use $\neq$, it is a two-tailed test and one-half of the value for α is apportioned to each tail, as in Example 9.3. See Figure 9.8.

In many applications involving hypothesis testing, problems have a somewhat different orientation from the examples above. We often have to test whether a new procedure produces a significant difference or a significant improvement over a standard approach. In such cases, you can think of the null hypothesis as standing for the status quo—whatever was the case will remain the case, so there is no change. The alternate hypothesis represents our belief that there has been a change. We demonstrate this in the following situation.

EXAMPLE 9.4 A company has a computer system that can process 1200 bills per hour. A new system is tested which processes an average of 1260 bills per hour with a standard deviation of 215 bills in a sample of 40 hours. Test if the new system is significantly better than the old one at the 5% level of significance.

FIGURE 9.8

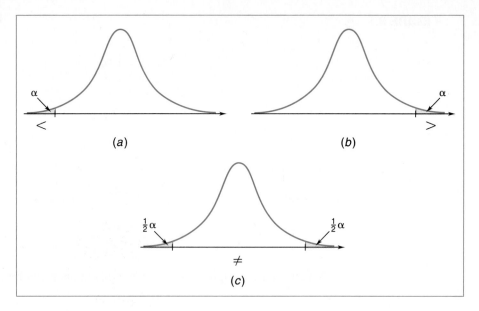

SOLUTION The null hypothesis states that there is no significant improvement in the average number of bills processed each hour, so that

$$H_0: \mu = 1200$$

while

$$H_a: \mu > 1200$$

(Notice that "improvement" suggests a higher mean and hence a one-tailed test.) The sample data yield

$$n = 40$$
$$\bar{x} = 1260$$
$$s = 215$$

so that

$$\frac{s}{\sqrt{n}} = \frac{215}{\sqrt{40}}$$
$$= \frac{215}{6.32} = 34$$

Since $n = 40 > 30$, the distribution of sample means is approximately normal. We locate the top 5% of the population corresponding to a critical value of

FIGURE 9.9

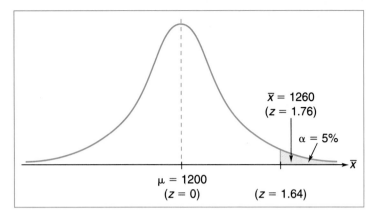

$z = 1.64$, as shown in Figure 9.9. For the sample mean $\bar{x} = 1260$, we find

$$z = \frac{1260 - 1200}{34}$$

$$= \frac{60}{34} = 1.76$$

and so we reject the null hypothesis. That is, we conclude that the new system represents an improvement over the old system at the $\alpha = .05$ level of significance.

Before we leave this example, it is worth noting that the z value for the sample mean $z = 1.76$ is just slightly greater than the critical value of 1.64. As a result, the decision to reject H_0 is a close call. In fact, had we used a lower level of significance, say $\alpha = .025$, then the critical value would be $z = 1.96$ and we would fail to reject the null hypothesis. This illustrates another situation where the choice of α becomes very important—totally different decisions can be made depending on its value.

Finally, we summarize the necessary steps for conducting a hypothesis test:

1. State the null and alternate hypotheses.
2. State the significance level α.
3. List the statistics from the sample data.
4. Decide on the sampling distribution—z or t.
5. Find the critical value(s) for the rejection region.
6. Calculate the test statistic for the sample data.
7. Come to a decision.

This outline applies not only to all the examples we have done in this section, but also to many other varieties of hypothesis tests throughout the rest of this chapter.

EXAMPLE 9.5 A study is conducted to test the claim that the US Postal Service delivers letters between points on the east and west coasts within 3 days. A sample of $n = 54$ letters are sent and the following delivery times, in days, are observed:

1, 1, 2, 2, 2, 2, 2, 2, 3, 3, 3, 3, 3, 3, 3, 3, 3, 3, 3, 3, 3, 3, 3, 3,
3, 3, 3, 4, 4, 4, 4, 4, 4, 4, 4, 4, 4, 4, 4, 4, 4, 4, 4, 5, 5, 5, 5, 5, 5,
6, 6, 6, 6, 8, 9.

Can you conclude that the Postal Service achieves its guarantee of a maximum 3-day delivery at the 5% level of significance?

SOLUTION The null and alternate hypotheses are:

$$H_0: \mu = 3$$
$$H_a: \mu > 3$$

since the test is one-tailed to the right.

We begin by analyzing the data using Program 1: Statistical Analysis of Data. It gives the histogram shown in Figure 9.10a based on 9 classes, 1, 2, 3, ..., 9, as well as the associated mean $\bar{x} = 3.74$ and standard deviation $s = 1.51$ for this sample. We notice from the histogram that the sample is not normal; it is clearly skewed to the right. Therefore, we would not conclude that the underlying population is normal either.

Using the sample data,

$$n = 54$$
$$\bar{x} = 3.74$$
$$s = 1.51$$

we find that

$$\frac{s}{\sqrt{n}} = \frac{1.51}{\sqrt{54}} = .205$$

Further, since $n = 54 > 30$, the distribution of sample means is approximately normal. Therefore, corresponding to the sample mean $\bar{x} = 3.74$, we find that

$$z = \frac{3.74 - 3}{.205} = 3.61$$

compared to the critical value of $z = 1.64$ as shown in Figure 9.10b. Consequently, we reject the null hypotheses at the $\alpha = .05$ level of significance. The mean delivery time for letters between the two coasts seems to be greater than the 3 days claimed. ∎

FIGURE 9.10a

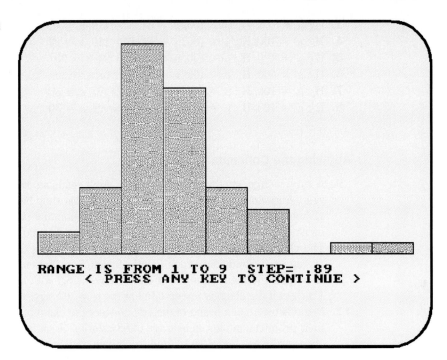

FIGURE 9.10b

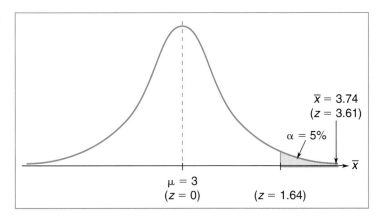

Exercise Set 9.1

Mastering the Techniques

In the following problems, use the given sample data to test the null and alternate hypotheses at a significance level of $\alpha = .05$.

1. $H_0: \mu = 100$, $H_a: \mu > 100$, $n = 64$, $\bar{x} = 105$, $s = 40$
2. $H_0: \mu = 100$, $H_a: \mu > 100$, $n = 64$, $\bar{x} = 105$, $s = 20$

3. $H_0: \mu = 100$, $H_a: \mu > 100$, $n = 64$, $\bar{x} = 110$, $s = 40$
4. $H_0: \mu = 100$, $H_a: \mu > 100$, $n = 64$, $\bar{x} = 110$, $s = 20$
5. $H_0: \mu = 100$, $H_a: \mu < 100$, $n = 36$, $\bar{x} = 94$, $s = 30$
6. $H_0: \mu = 100$, $H_a: \mu < 100$, $n = 36$, $\bar{x} = 88$, $s = 30$
7. $H_0: \mu = 100$, $H_a: \mu \neq 100$, $n = 144$, $\bar{x} = 97$, $s = 30$
8. $H_0: \mu = 100$, $H_a: \mu \neq 100$, $n = 144$, $\bar{x} = 94$, $s = 30$

Applying the Concepts

9. A survey suggests that the typical U.S. adult reads an average of 10 books per year. A random sample of 136 adults shows that they have read an average of 12 books in the last year with a standard deviation of 9 books. Test if the stated mean of 10 is too low at the $\alpha = .05$ level of significance.
10. The chamber of commerce for a particular city claims that the locality has an average of 4 cloudy days each month. A sample of 36 randomly selected months show that the mean number of cloudy days is 5.2 with a standard deviation of 1.8. Test if the claim is understated at the $\alpha = .02$ significance level.
11. Manufacturers of a brand of dietetic cheesecake claim that a 1-ounce serving of their product contains an average of 88 calories. A sample of 36 servings of this cheesecake are tested and found to contain a mean of 90 calories with a standard deviation of 4 calories. Test if the true mean is higher at the 5% level of significance.
12. A government agency claims that the mean gas mileage for a certain model car is 28 miles per gallon. A sample of 64 drivers of such cars report a mean of 26.8 miles per gallon with a standard deviation of 6 miles per gallon. Do the data indicate that the government rating is too high at the $\alpha = .01$ level of significance?
13. A study suggests that the mean reading level of all adults in the United States is equivalent to grade 9. A sample of 400 adults have a mean reading level of grade 9.2 with a standard deviation of 2.4. Test if the claim is correct at the 5% level of significance.
14. The U.S. government claims that the mean family income in a certain state is $32,000 per year. A random sample of 120 families have a mean income of $33,600 with a standard deviation of $4500. Do the data indicate that the claim is incorrect at the 2% significance level?
15. The mean SAT verbal score in a certain school district is 430. A group of 60 randomly selected students are put into a special SAT preparatory course, and their mean score is 462 with a standard deviation of 120. Does the special course result in any improvement in the SAT scores at the $\alpha = .05$ level of significance?
16. A study indicates that the typical person has had a mean of 4.4 tooth cavities filled by the time he or she is 18 years old. A new treatment is developed and used on a random sample of 384 children. The results show that the mean number of cavities they develop is 3.9 with a standard deviation of 1.3. Does this new treatment represent an improvement at the 1% level of significance?

17. At a large department store, the average amount of all sales is $78. On a special holiday promotion day, the total purchase amount of 80 different randomly selected sales has a mean of $92 with a standard deviation of $62. Test if the mean sale amount on a promotion day is higher than on a regular day at the $\alpha = .10$ significance level.
18. A study shows that the mean time needed to drive 1 mile across a downtown area of a particular city during rush hour is 18 minutes. A new timing pattern is tried for the traffic lights, and a sample of 50 cars take an average of 17 minutes with a standard deviation of 4.8 minutes to traverse this stretch. Does the new pattern produce an improvement at the $\alpha = .02$ significance level?

9.2 Hypothesis Testing Using *P*-Values

The technique we developed in Section 9.1 for testing a hypothesis about a population mean is known as the *classical* or *traditional approach*. In this section, we develop an alternative method for hypothesis testing called the *probability value* or *P-value approach*.

Consider a null hypothesis

$$H_0: \mu = \mu_0$$

where μ_0 is some claimed value for μ. Suppose that the alternate hypothesis is

$$H_a: \mu > \mu_0$$

and that the level of significance is α. The sample data provide values for n, $\bar{x}$ and s. Suppose that $\bar{x}$ is sufficiently larger than the claimed value for μ, or equivalently that the z value corresponding to $\bar{x}$ is larger than the critical value for z. We believe that such an eventuality is just too unlikely to occur by chance, and we therefore reject the null hypothesis. This argument is the basis for the classical approach to hypothesis testing.

We now consider this situation from a somewhat different perspective. When we reject a null hypothesis, it is because the sample mean $\bar{x}$ is unlikely to have occurred purely by chance. We consequently ask, What is the probability that such an occurrence will happen by chance? If the probability is small, in some sense, then we reject the null hypothesis. If the probability is relatively large, so that the occurrence of such a sample mean is not unlikely, then we fail to reject the null hypothesis.

To illustrate this approach, suppose we again use the Goodstone tire example of Section 9.1 with

$$H_0: \mu = 30,000$$
$$H_a: \mu < 30,000$$

and sample data that give

$$n = 100$$
$$\bar{x} = 29{,}000$$
$$s = 4000$$

so that

$$\frac{s}{\sqrt{n}} = \frac{4000}{\sqrt{100}} = 400$$

If we use this quantity in place of $\sigma_{\bar{x}}$, the corresponding z value is

$$z = \frac{29{,}000 - 30{,}000}{400} = -2.50$$

We now consider the question, How unlikely is it to obtain this value or worse by chance when the null hypothesis is true? In other words, what is the probability that the mean $\bar{x}$ for a sample of 100 tires is 29,000 miles or less when the actual mean is 30,000 miles? Symbolically, this is

$$P(\bar{x} \le 29{,}000)$$

If this probability is sufficiently small so that the outcome is very unlikely, then we reject the null hypothesis. The key to this decision lies in calculating the corresponding probability.

We can calculate this probability by using the techniques of Section 7.2 since the distribution of sample means is approximately normal. Corresponding to $\bar{x} = 29{,}000$, we found that $z = -2.50$, and so we must find

$$P(z \le -2.50)$$

As seen from Figure 9.11, this is $.5000 - .4938 = .0062$.

This value for the probability is known as the *P-value* associated with the sample mean $\bar{x}$. The P-value can be used in one of two ways. Some statisticians

FIGURE 9.11

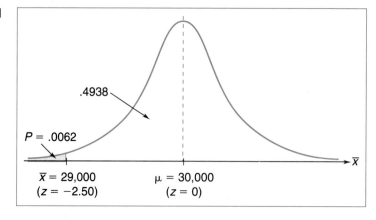

prefer to quote the *P*-value associated with the sample mean and leave the decision on whether to reject the null hypothesis to the reader's discretion. When doing this, statisticians use the following general guidelines:

If $P > .05$, there is insufficient evidence to reject the null hypothesis.

If $.01 < P < .05$, there is sufficient evidence to reject the null hypothesis.

If $P < .01$, there is very strong evidence to reject the null hypothesis.

The other approach is to use the *P*-value in conjunction with a significance level α. We adopt this latter approach here and illustrate it in the context of the tire example.

Suppose that the significance level α is .05. Since the *P*-value we calculated, $P = .0062$, is smaller than the significance level $\alpha = .0500$, the outcome for this sample mean is very unlikely. This is equivalent to the z value falling in the rejection region, and so we reject the null hypothesis.

In general, given the significance level α, we calculate the *P*-value associated with the sample mean and compare these two values:

If $P < \alpha$, then we reject the null hypothesis.

If $P > \alpha$, then we fail to reject the null hypothesis.

We illustrate these ideas in the following examples.

EXAMPLE 9.6   A used-car price guide claims that the average price for a certain used-car model is $4700. A random sample of 36 used-car dealers in a particular geographic region are surveyed and found to be asking an average of $5040 with a standard deviation of $780 for this car. Use *P*-values to test if the average price for this model is higher in this region compared to the national average at the $\alpha = .02$ level of significance.

SOLUTION The null hypothesis is

$$H_0: \mu = 4700$$

while the alternate hypothesis is

$$H_a: \mu > 4700$$

The sample data give

$$n = 36$$
$$\bar{x} = 5040$$
$$s = 780$$

so that

$$\frac{s}{\sqrt{n}} = \frac{780}{\sqrt{36}} = 130$$

Since $n = 36 > 30$, we know that the distribution of sample means is approximately normal.

We now calculate the probability that a sample mean $\bar{x}$ is at least 5040:

$$P(\bar{x} \geq 5040)$$

The corresponding z value is

$$z = \frac{5040 - 4700}{130} = 2.62$$

Consequently, the probability

$$P(z \geq 2.62) = .5000 - .4956 = .0044$$

as seen in Figure 9.12, and so the associated P-value is

$$P = .0044$$

Since this value is smaller than the significance level $\alpha = .02 = .0200$, we reject the null hypothesis and conclude that the mean price for the car in this region is higher than the national average of $4700. ∎

EXAMPLE 9.7 A dairy industry study concludes that school-age children consume an average of 19.4 ounces of milk per day. In a survey of 140 randomly chosen children, it is found that they drink an average of 18.5 ounces with a standard deviation of 6.8 ounces. Use P-values to test if the claim is too high at the 5% level of significance.

FIGURE 9.12

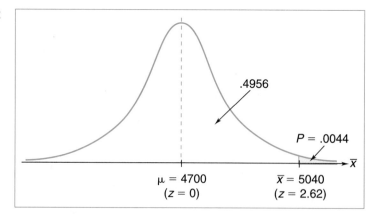

SOLUTION The null and alternate hypotheses are

$$H_0: \mu = 19.4$$
$$H_a: \mu < 19.4$$

The sample data give

$$n = 140$$
$$\bar{x} = 18.5$$
$$s = 6.8$$

so that

$$\frac{s}{\sqrt{n}} = \frac{6.8}{\sqrt{140}} = .575$$

Since $n = 140 > 30$, we can use a normal distribution while finding

$$P(\bar{x} \leq 18.5)$$

Corresponding to $\bar{x} = 18.5$, we find that

$$z = \frac{18.5 - 19.4}{.575} = -1.57$$

From Figure 9.13, we see that

$$P = P(\bar{x} \leq 18.5) = .5000 - .4418 = .0582$$

Since this P-value is greater than the level of significance $\alpha = .0500$, we fail to reject the null hypothesis. There is insufficient evidence to support the suspicion that the claim is too high. ∎

FIGURE 9.13

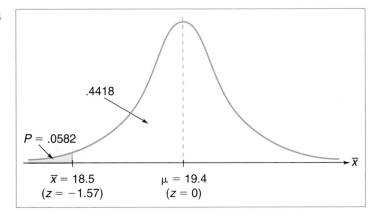

Notice that the *P*-value for a hypothesis test is always a positive quantity (since it represents a probability) regardless of whether $\bar{x}$ is above or below the claimed population mean μ (or whether z is positive or negative).

We need a slight modification of the above approach to handle a two-tailed hypothesis test. In such a case, recall that in the classical approach, we take one-half of the significance level α and apportion it to each tail. In the probability value approach, we essentially reverse this. We are looking for the probability of obtaining a value for the sample mean $\bar{x}$ which is at least that far away from the claimed population mean μ. Since such an unlikely event can occur on either side of μ with a two-tailed test, we must consider the area in both tails. See Figure 9.14. The *P*-value associated with a given sample mean $\bar{x}$ is the area in both tails or, more simply, *twice* the area in either tail.

We illustrate this in the following example.

EXAMPLE 9.8 A certain type of tree is known to grow an average of 11 inches per year under normal conditions. An experimental insecticide is used on a sample of 48 of these trees, and the researcher is interested in seeing if it has any effect on the annual growth of the trees. She finds that with the insecticide, the trees grew an average of 10.3 inches with a standard deviation of 2.3 inches. Use *P*-values to test if the insecticide has a significant effect on the average growth of these trees at the 5% significance level.

SOLUTION The null hypothesis is

$$H_0: \mu = 11$$

There is no indication that we are to test for an increase or decrease in growth; rather, we want to know whether there is a change due to the insecticide. Therefore, we perform a two-tailed test with

$$H_a: \mu \neq 11$$

FIGURE 9.14

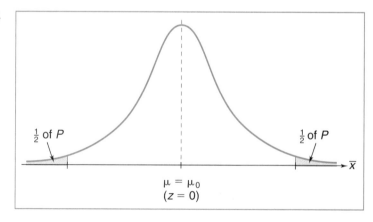

The sample data give

$$n = 48$$
$$\bar{x} = 10.3$$
$$s = 2.3$$

so that

$$\frac{s}{\sqrt{n}} = \frac{2.3}{\sqrt{48}} = .332$$

Since $n = 48 > 30$, we use a normal distribution to find the necessary probabilities.

To find the P-value associated with $\bar{x} = 10.3$, we calculate

$$P(\bar{x} \leq 10.3)$$

Corresponding to $\bar{x} = 10.3$, we find that

$$z = \frac{10.3 - 11}{.332} = -2.11$$

and so, as shown in Figure 9.15,

$$P(\bar{x} \leq 10.3) = .5000 - .4826 = .0174$$

Since this is a two-tailed test, we obtain the P-value by doubling this probability to account for the two tails:

$$P = 2(.0174) = .0348$$

When we compare this P-value to the level of significance, we see that

$$P = .0348 < \alpha = .0500$$

and so we reject the null hypothesis. The insecticide does have a significant effect on the growth of the trees. ∎

FIGURE 9.15

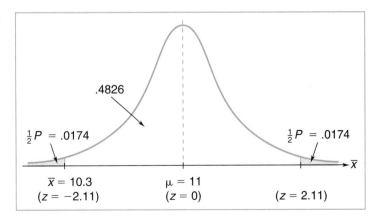

Exercise Set 9.2

Mastering the Techniques

1–8. Repeat Exercises 1 to 8 from Section 9.1, using *P*-values.

Applying the Concepts

9–18. Repeat Exercises 9 to 18 from Section 9.1, using *P*-values.

9.3 Hypothesis Tests for Means Using Small Samples

All the hypothesis testing situations we have considered so far involve tests on the population mean μ and are based on a set of data having a sample size $n > 30$. We also assume that the underlying population is roughly normal in all examples and exercises. Under these circumstances, the distribution of sample means is approximately normal so that we can work with the associated z values.

We also can have a data set with a sample size of 30 or less. In this case, we cannot use a normal distribution, but rather we must work with an appropriate t-distribution with $n - 1$ degrees of freedom, provided that the underlying population is approximately normal. (We assume this to be the case in all following examples.) Otherwise, some of the nonparametric tests we will develop in Chapter 13 may apply.

Fortunately, our previous methods remain essentially the same except for the relatively minor modification of using t values instead of z values, as we will see in the examples below.

EXAMPLE 9.9 A company claims that the mean selling price for a certain type of imported sports car is $42,000. A survey of 16 randomly selected owners of such cars shows that they actually paid a mean of $44,200 with a standard deviation of $6000 for their cars. Test if the company's claim is too low at the $\alpha = .05$ level of significance.

SOLUTION The null hypothesis is

$$H_0: \mu = 42{,}000$$

and the alternate hypothesis is

$$H_a: \mu > 42{,}000$$

The sample data give

$$n = 16$$
$$\bar{x} = 44{,}200$$
$$s = 6000$$

so that

$$\frac{s}{\sqrt{n}} = \frac{6000}{\sqrt{16}} = 1500$$

Further, since $n = 16 < 31$, we must use a *t*-distribution with $n - 1 = 15$ degrees of freedom.

The hypothesis test is a one-tailed test and we use the right tail of the *t*-distribution. Since $\alpha = .05$, the rejection region consists of the top 5% of the *t*-distribution. We find from Table IV that the critical value for *t* with 15 degrees of freedom is $t = t_{.05} = 1.753$, as shown in Figure 9.16. For the sample data, we find that

$$t = \frac{\bar{x} - \mu_{\bar{x}}}{s/\sqrt{n}}$$

$$= \frac{44{,}200 - 42{,}000}{1500}$$

$$= \frac{2200}{1500} = 1.47$$

Since this value falls outside the rejection region, we fail to reject the null hypothesis. There is insufficient evidence to conclude that the mean selling price of all such cars is greater than $42,000. ∎

As we indicated before Example 9.9, the procedure for solving this problem is virtually identical to that for problems involving large samples. However, it is essential that you notice whether the sample size *n* is 30 or less. If it is, you must use *t* values instead of *z* values, assuming that all underlying populations are approximately normal.

FIGURE 9.16

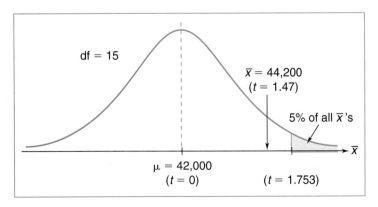

EXAMPLE 9.10 Managers of a diet plan advertise that the mean weight loss for people on their plan is at least 45 pounds in 6 months. A sample of 28 people on this plan lose an average of 35 pounds with a standard deviation of 20 pounds. Test, at the $\alpha = .01$ significance level, if the claim is too high.

SOLUTION The null and alternate hypotheses are

$$H_0: \mu = 45$$
$$H_a: \mu < 45$$

The sample data give

$$n = 28$$
$$\bar{x} = 35$$
$$s = 20$$

so that

$$\frac{s}{\sqrt{n}} = \frac{20}{\sqrt{28}} = \frac{20}{5.29} = 3.78$$

The fact that $n = 28 < 31$ indicates that we must use a *t*-distribution with $n - 1 = 27$ degrees of freedom. In addition, the hypothesis test is a one-tailed test involving the left tail of this *t*-distribution. From Table IV, we find that the critical value for *t* with $\alpha = .01$ and 27 degrees of freedom is $t = t_{.01} = -2.473$, as shown in Figure 9.17. The *t* value for the sample data is found to be

$$t = \frac{35 - 45}{3.78}$$
$$= \frac{-10}{3.78} = -2.646$$

and we therefore reject the null hypothesis. That is, the claim for an average 45 pound weight loss is overstated at the .01 level of significance. ∎

FIGURE 9.17

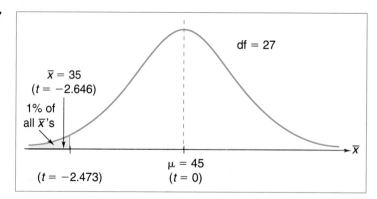

EXAMPLE 9.11

A large company finds that the typical office worker spends an average of 13 minutes per hour on non-work-related activities. On a trial basis, management sets up new enclosed work stations for a randomly selected group of 12 workers and finds that the average number of minutes lost per hour is 11.3 with a standard deviation of 3.7. At the 5% level of significance, can we conclude that the new work stations represent a significant change?

SOLUTION The null and alternative hypotheses are

$$H_0: \mu = 13$$
$$H_a: \mu \neq 13$$

The sample data give

$$n = 12$$
$$\bar{x} = 11.3$$
$$s = 3.7$$

so that

$$\frac{s}{\sqrt{n}} = \frac{3.7}{\sqrt{12}} = \frac{3.7}{3.46} = 1.07$$

Since $n = 12 < 31$, we must use a t-distribution with $n - 1 = 11$ degrees of freedom. Further, this is a two-tailed test with 2.5% of the population in each tail. As a result, we find that the critical values for t are $t = t_{.025} = \pm 2.201$, as shown in Figure 9.18. We find that the t value for the sample mean $\bar{x} = 11.3$ is

$$t = \frac{11.3 - 13}{1.07}$$
$$= \frac{-1.7}{1.07} = -1.589$$

FIGURE 9.18

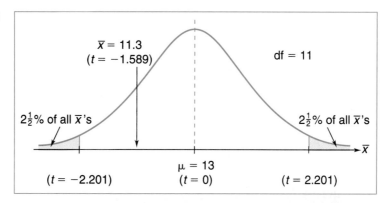

and we therefore fail to reject the null hypothesis. That is, we cannot conclude that the new work stations produce a significant change in the average lost time per worker. ∎

EXAMPLE 9.12 Federal toxic waste standards allow a maximum of 1 part per million (ppm) for the presence of cadmium in waste. Use the data shown in Figure 9.19 to determine whether the garbage on the infamous Long Island garbage barge exceeded this level at the $\alpha = .01$ significance level.

SOLUTION The null and alternate hypotheses are

$$H_0: \mu = 1$$
$$H_a: \mu > 1$$

Rather than being given values for $\bar{x}$ and s as in preceding examples, we must calculate these quantities from the raw measurements

$$20.88,\ 5.93,\ 19.33,\ 22.42,\ 0.0,\ 13.72$$

After a little effort, we obtain,

$$n = 6$$
$$\bar{x} = 13.71$$
$$s = 9.026$$

so that

$$\frac{s}{\sqrt{n}} = \frac{9.026}{\sqrt{6}} = \frac{9.026}{2.449} = 3.685$$

Since $n = 6 < 31$ and we assume that the underlying population is approximately normal, we must use a t-distribution with $n - 1 = 5$ degrees of freedom.

This is a one-tailed test with 1% of the population in the right tail, so that the critical value for t with 5 degrees of freedom is $t = t_{.01} = 3.365$. See Figure

FIGURE 9.19

What's in the Garbage

Amount of cadmium and lead, in parts per million, that was found in samples of garbage from the barge, and the federal government's maximum allowable levels for those materials

	Sample						Maximum
	1	2	3	4	5	6	allowable
Cadmium	20.88	5.93	19.33	22.42	0	13.72	1.00
Lead	35.60	32.40	38.00	44.00	36.40	37.28	5.00

SOURCE: Volumetric Techniques Ltd.

FIGURE 9.20

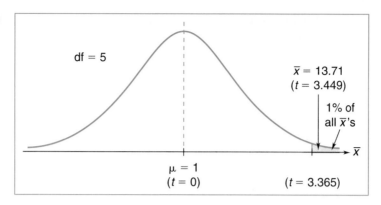

9.20. Corresponding to the sample mean $\bar{x}$, we find that

$$t = \frac{13.71 - 1}{3.685}$$

$$= \frac{12.71}{3.685} = 3.449$$

and therefore we reject the null hypothesis. The level of cadmium present seems to exceed the federal limitation at the $\alpha = .01$ level of significance. ∎

While the outcome of Example 9.12 is not surprising, it is actually much closer than we might expect just from looking at the data. In fact, it is really impossible to depend solely on intuition in such matters. The formal methods we have developed for handling such situations are essential for making these statistical decisions.

We can also use the *P*-value approach when performing hypothesis testing based on small samples. To do so requires that we find the probability associated with a given sample mean $\bar{x}$ and hence with the corresponding *t* value. However, because the *t*-distribution tables we commonly use present only a summary of particular values, it is rare that we can determine a probability precisely. We illustrate this in the following example.

EXAMPLE 9.13 The mean survival time for patients with a particular form of cancer is 5 years. A new treatment is tried on 20 patients with this disease, and their average survival time is 5.7 years with a standard deviation of 1.2 years. Use *P*-values to test if the new treatment constitutes a significant improvement at the $\alpha = .01$ level of significance.

SOLUTION The null and alternate hypotheses are

$$H_0: \mu = 5$$
$$H_a: \mu > 5$$

The sample data give

$$n = 20$$
$$\bar{x} = 5.7$$
$$s = 1.2$$

so that

$$\frac{s}{\sqrt{n}} = \frac{1.2}{\sqrt{20}} = \frac{1.2}{4.47} = .27$$

Since $n = 20 < 31$, we must use a t-distribution with $n - 1 = 19$ degrees of freedom to find

$$P(\bar{x} \geq 5.7)$$

The t value corresponding to $\bar{x} = 5.7$ is

$$t = \frac{5.7 - 5}{.27}$$

$$= \frac{.7}{.27} = 2.593$$

so that we must find

$$P(t \geq 2.593)$$

From Table IV with 19 degrees of freedom, we see that $t = 2.593$ does not occur precisely. Rather, it lies between $t_{.01} = 2.539$ and $t_{.005} = 2.861$. See Figure 9.21. Therefore, the associated P-value for the sample mean is between .005 and .01, but cannot be found more exactly from this table. However, we can conclude that this P-value is less than the level of significance $\alpha = .01$ and seems to constitute a significant improvement in terms of mean survival time at the 1% significance level. ∎

FIGURE 9.21

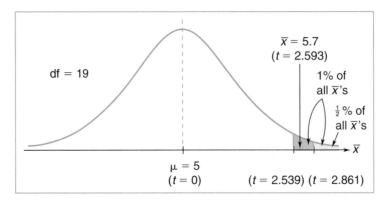

Computer Corner

We have included two programs on hypothesis testing on your disk. Program 28: Hypothesis Tests: Means/Proportions is designed to help you develop your skills at identifying the various features and steps involved in solving hypothesis-testing problems. The program first asks if the hypothesis test is for means or for proportions. At this point, you should choose means. (We discuss hypothesis testing for proportions in the next section.) The program then asks for the population mean μ for the null hypothesis. For instance, if $H_0: \mu = 120$, then you would enter 120. You must also indicate if it is a one-tailed or a two-tailed test; if it is one-tailed, you must indicate which tail. The program then asks for the risk level α: your choices are 10, 5, or 1 percent. Finally, you must supply sample data values for n, $\bar{x}$ and s.

The program also asks whether to use a normal distribution or a t-distribution, and you must make the correct determination based on the sample size. The program next asks you to supply the appropriate critical value, either z or t, from the corresponding table. You have three attempts to determine the critical value of z or t for the appropriate hypothesis test at your selected significance level. If you cannot provide the correct answer, the program will give it to you after three wrong answers.

The program then displays your null hypothesis and sample data and asks you to calculate the test statistic, either z or t, based on the sample data. The program will analyze an incorrect response and offer advice on how to correct it. After three incorrect responses, the program supplies the correct answer. Finally you are asked whether you fail to reject or reject the null hypothesis. The program responds by letting you know if you have answered correctly.

When the analysis and decisions are complete, the program draws the graph of the appropriate normal or t-distribution curve for the sampling distribution. It draws tick marks for all appropriate significance levels and draws one or two vertical lines to indicate the location of the critical value(s) for z or t. Finally, it indicates where the sample mean $\bar{x}$ falls so that you check visually whether it is in the rejection region. When the display is finished, all the pertinent information is summarized on an accompanying text screen.

From the text screen, you can either return to the graph by pressing any key other than the space bar or go on to a menu screen for a variety of choices by pressing the space bar. The choices allow you to change the significance level α, the sample data, the alternate hypothesis, and so forth.

We urge you to use this program as a tutorial if you are having any difficulty in solving hypothesis-testing problems.

The second program available on hypothesis testing is Program 29: Hypothesis Testing Simulation. You may select any of four underlying populations. The program asks whether the test is a one- or a two-tailed test and, if it is a one-tailed test, which tail. The program then asks for the level of significance α you desire, and you can select 10, 5, or 1 percent. The program then draws the sampling distribution and a vertical line or lines corresponding to the critical value(s).

The program next randomly generates 100 different random samples of size 36 from the indicated population. It calculates the mean $\bar{x}$ of each of these samples and shows its location as a vertical line. It also keeps track of the number of sample means that fall inside the rejection region and displays this number and the corresponding proportion at the end of the run. In this way, you can see that if $\alpha = 10\%$, say, then approximately 10% of the sample means fall in the rejection region. Alternatively, if $\alpha = 1\%$, then only about 1 of the 100 sample means falls in the rejection region. A typical output is shown in Figure 9.22 for $\alpha = 5\%$.

Notice that even when the sample mean falls in the rejection region, it is usually fairly close to the critical values indicated by the tall vertical lines. It is very unlikely for a sample mean to fall very far from the population mean.

We suggest that you experiment with the ideas underlying hypothesis testing by running this program with a variety of choices for the underlying population and with different levels of significance α. It should provide you with a better feel for what could happen in practice if repeated sampling were performed.

FIGURE 9.22

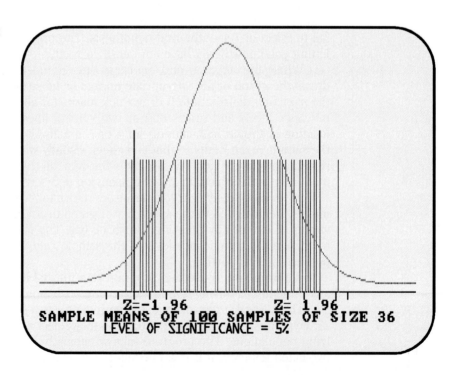

Minitab Methods

You can also use Minitab to perform a hypothesis test on the population mean μ. Suppose that we enter a set of data values for a sample in column C1. The Minitab command **TTEST** starts the hypothesis test procedure, where you must provide the claimed value for the population mean μ and the column in which the data have been stored. This command must then be followed by a subcommand, **ALTERNATIVE**, which provides the alternate hypothesis. The actual format for the command is as follows:

```
TTEST 160 C1
ALTERNATIVE  1     (for a one-tailed test, upper tail)
ALTERNATIVE -1     (for a one-tailed test, lower tail)
ALTERNATIVE  0     (for a two-tailed test)
```

A more intelligible format for such a command to perform a one-tailed test on the null hypothesis that $\mu = 160$, using the data in column C1, might be

```
TTEST of mu = 160 using data in C1
ALTERNATIVE is higher, one-tailed = 1
```

Minitab will then print out the sample size, the sample mean, the sample standard deviation, the standard error of the mean (SE MEAN), the t value corresponding to the sample mean, and the associated probability or P-value. Once you have this information, you will have to compare the calculated value for t to the critical value in the t-distribution table corresponding to the desired significance level α and the appropriate number of degrees of freedom. A sample Minitab session is shown in Figure 9.23.

Exercise Set 9.3

Mastering the Techniques

In the following exercises, use the given sample data to test each hypothesis at a significance level of $\alpha = .05$.

1. H_0: $\mu = 100$, H_a: $\mu > 100$, $n = 16$, $\bar{x} = 115$, $s = 40$
2. H_0: $\mu = 100$, H_a: $\mu > 100$, $n = 16$, $\bar{x} = 115$, $s = 20$
3. H_0: $\mu = 100$, H_a: $\mu > 100$, $n = 25$, $\bar{x} = 115$, $s = 40$
4. H_0: $\mu = 100$, H_a: $\mu > 100$, $n = 25$, $\bar{x} = 115$, $s = 20$
5. H_0: $\mu = 100$, H_a: $\mu < 100$, $n = 9$, $\bar{x} = 84$, $s = 30$
6. H_0: $\mu = 100$, H_a: $\mu < 100$, $n = 9$, $\bar{x} = 78$, $s = 30$
7. H_0: $\mu = 100$, H_a: $\mu \neq 100$, $n = 14$, $\bar{x} = 87$, $s = 30$
8. H_0: $\mu = 100$, H_a: $\mu \neq 100$, $n = 14$, $\bar{x} = 76$, $s = 30$

FIGURE 9.23

```
MTB > PRINT C1
WEIGHT
   136   137   157   144   190   164   147   150   136   163   148
   174   211   169   148   184   163   144   130   181   156   147
   170   148   182   159   140   137   122   158

MTB > MEAN C1
         MEAN =       156.50
MTB > TTEST of mu = 160 using data in C1;
SUBC> ALTERNATIVE is lower, one tailed = -1.

TEST OF MU = 160.000 VS MU L.T. 160.000
                N       MEAN      STDEV    SE MEAN        T    P VALUE
WEIGHT         30     156.500     19.837      3.622    -0.97      0.17
MTB >
MTB > TTEST of mu = 160 using data in C1;
SUBC> ALTERNATIVE is two tailed, = 0.

TEST OF MU = 160.000 VS MU N.E. 160.000
                N       MEAN      STDEV    SE MEAN        T    P VALUE
WEIGHT         30     156.500     19.837      3.622    -0.97      0.34
```

In each of the following exercises, use the *P*-value approach to test each hypothesis at the 5% significance level.

9. $H_0: \mu = 100$, $H_a: \mu > 100$, $n = 16$, $\bar{x} = 115$, $s = 40$
10. $H_0: \mu = 100$, $H_a: \mu > 100$, $n = 25$, $\bar{x} = 115$, $s = 20$
11. $H_0: \mu = 100$, $H_a: \mu < 100$, $n = 9$, $\bar{x} = 84$, $s = 30$
12. $H_0: \mu = 100$, $H_a: \mu \neq 100$, $n = 14$, $\bar{x} = 87$, $s = 30$

Applying the Concepts

13. A special high-tension cable is rated to withstand a mean weight of 1800 pounds without breaking. A sample of 16 such cables are tested and break under a mean load of 1740 pounds with a standard deviation of 60 pounds. Test if the claim is overstated at the $\alpha = .05$ significance level.
14. The state police claim that there are an average of 142 accidents per week on state roads. In a random sample of 9 weeks, records indicate that there was a mean of 160 accidents with a standard deviation of 24. Test if the claim is too low at the $\alpha = .05$ level of significance.
15. A large engineering company claims that the mean salary for its engineering staff is $48,000. A prospective job applicant contacts a random sample of 12 engineers working for the company and finds that their mean salary is $45,850 with a standard deviation of $6300. Test if the company's claim is too high at the $\alpha = .025$ significance level.
16. State health regulations require that the level of a certain chemical contaminant

in drinking water not exceed 4 parts per million (ppm). A sample of drinking water in 20 homes in a certain community shows the chemical present at a mean level of 4.12 ppm with a standard deviation of .58. Do these data indicate that the health limitations are being exceeded at the $\alpha = .01$ level of significance?

17. At a professional football team tryout, a punter claims that he can kick a football an average of 43 yards from the line of scrimmage. The coach allows him 10 kicks, which have a mean of 40.6 yards with a standard deviation of 3.8 yards. Based on these data, is the kicker's claim overstated at the 5% level of significance?

18. A nationwide advertising campaign for a major hamburger chain claims that a customer can get a full meal (large hamburger, French fries and a soft drink) for under $4. A random sample of 15 of their restaurants in different parts of the country show that the mean cost for such a meal is $4.11 with a standard deviation of $.43. Test if the claim is valid at the 5% significance level.

19. The planes used on a certain route seat 186 passengers. The airline claims that it sets aside and sells an average of 42 seats per flight for a special low-fare promotion. A random sample of 22 such flights indicate that the mean number of discount seats sold was 37.2 with a standard deviation of 7.4. Use *P*-values to test if the airline's claim is valid at the 2% significance level.

20. A company feels that its employees are taking advantage of a liberal policy on sick leave. The union counters and claims that the mean number of sick days used by the employees each year is 8.4. The company checks the records of 25 randomly selected employees and finds that the average number of sick days used during the previous year is 9.7 with a standard deviation of 2.8. Use *P*-values to test if the company's suspicion is correct at the 5% level of significance.

21. A bicyclist has set a goal of biking an average of at least 5 miles per day. She checks the odometer reading on her bike on 8 randomly selected days and finds the following distances, in miles:

$$5.3, \ 4.5, \ 4.8, \ 5.1, \ 4.3, \ 4.8, \ 4.9, \ 4.7$$

Can she conclude that she has achieved her goal at the $\alpha = 5\%$ level of significance?

22. An SAT preparatory course advertises that it will increase a student's score by an average of 150 points. A group of 7 students who have taken the course are randomly selected, and the point increases in their scores are

$$140, \ 170, \ 110, \ 180, \ 120, \ 120, \ 140$$

Test if the course meets its advertised claims at the $\alpha = 5\%$ level of significance.

9.4 Hypothesis Tests for Proportions

In the last three sections, we tested hypotheses about the mean μ of a population. We now turn our attention to a comparable technique for testing claims regarding a population proportion π.

To illustrate, suppose that a cereal producer claims that two-thirds of all children prefer their Rice Crunchies to its major competitor, Rice Flakies. A random sample of 100 children taste-test these cereals, and only 40 of the 100, or 40%, prefer Rice Crunchies. This certainly seems to contradict the claim. On the other hand, if 63 of the 100 children, or 63%, prefer Rice Crunchies, then this seemingly supports the company's claim.

As this example indicates, the ideas involving hypothesis testing for proportions are very similar to those for means. We will develop a method to test such hypotheses by using a relatively minor modification of the procedure we previously used for means.

To begin, the null hypothesis now concerns the population proportion π rather than the population mean μ. Thus, in the cereal illustration, we take the null hypothesis to be

$$H_0: \pi = \tfrac{2}{3} = .667$$

Depending on the context, the alternate hypothesis might be written in any of the forms:

$$H_a: \pi < .667$$

$$H_a: \pi > .667$$

or

$$H_a: \pi \neq .667$$

Analogous to hypothesis tests on the mean μ, there is a level of significance α, which takes on the typical values of .10, .05, .02 or .01. Moreover, the hypothesis test is either one-tailed or two-tailed, depending on the alternate hypothesis, and this determines how α is to be apportioned.

Our decision as to whether we reject the null hypothesis on π is based on a set of sample data. This consists of either n (the sample size) and p (the sample proportion) or n and x (the number of successes) from which we find $p = x/n$.

This sample proportion p that we use in making our decision is just one element from the distribution of sample proportions. As we mentioned in Section 8.3,

---

The distribution of sample proportions is approximately normal provided that both

$$n\pi \geq 5 \quad \text{and} \quad n(1 - \pi) \geq 5$$

This distribution has mean

$$\mu_p = \pi$$

and standard deviation

$$\sigma_p = \sqrt{\frac{\pi(1 - \pi)}{n}}$$

SECTION 9.4 • Hypothesis Tests for Proportions

Recall that in Section 8.3 it was necessary to consider an alternative to this formula

$$s_p = \sqrt{\frac{p(1-p)}{n}}$$

because the population proportion π was unknown and, in fact, was the objective when we constructed a confidence interval for π. However, in a hypothesis-testing situation, we assume that the claimed value for π, which is stated in the null hypothesis, is valid until disproved. Thus, since π is known, the correct expression for the standard deviation is σ_p, and it is the first formula which must be used.

With these ideas in mind, let's consider several examples illustrating how the hypothesis-testing procedure is applied to proportions.

EXAMPLE 9.14 A cereal company claims that two-thirds of all children prefer Rice Crunchies to Rice Flakies. In a sample of 100 children, 55 prefer Rice Crunchies. Test if the company's claim is overstated at the $\alpha = .05$ level of significance.

SOLUTION The null hypothesis is

$$H_0: \pi = .667$$

and the alternate hypothesis is

$$H_a: \pi < .667$$

which results in a one-tailed test. The sample data give

$$n = 100$$
$$x = 55$$

so that

$$p = \frac{55}{100} = .55$$

Since

$$n\pi = 100(.667) = 66.7 \geq 5$$

and

$$n(1 - \pi) = 100(.333) = 33.3 \geq 5$$

the distribution of sample proportions is approximately normal with mean $\mu_p = \pi = .667$ and standard deviation

$$\sigma_p = \sqrt{\frac{\pi(1-\pi)}{n}}$$
$$= \sqrt{\frac{.667(.333)}{100}}$$
$$= \sqrt{.002221}$$
$$= .047$$

FIGURE 9.24

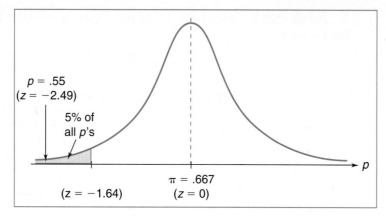

This is a one-tailed test with 5% of the normal population in the left tail, so that the critical value for z is $z = -1.64$, as seen in Figure 9.24. By comparison, the z value for the sample proportion p is

$$z = \frac{p - \pi}{\sigma_p} = \frac{.55 - .667}{.047} = \frac{-.117}{.047} = -2.49$$

and we therefore reject the null hypothesis. That is, based on these sample data, we conclude that the manufacturer's claim is overstated at the $\alpha = .05$ significance level. ∎

As we indicated above, the procedure involved in solving hypothesis-testing problems for proportions is virtually identical to that for hypothesis tests for means.

CAUTION You must use the hypothesized value for π, not the sample proportion p, in calculating the standard deviation σ_p for such problems.

EXAMPLE 9.15 A political party claims that 45% of the voters in an election district prefer its candidate. A sample of 200 voters include 80 who prefer this candidate. Test if the claim is valid at the 5% significance level.

SOLUTION The null hypothesis is

$$H_0: \pi = .45$$

and the alternate hypothesis is

$$H_a: \pi \neq .45$$

since there is no indication of whether we suspect the claim is too high or too low. The sample data give

$$n = 200$$
$$x = 80$$

and so

$$p = \frac{80}{200} = .40$$

Since

$$n\pi = 200(.45) = 90 \geq 5 \quad \text{and} \quad n(1 - \pi) = 200(.55) = 110 \geq 5$$

we conclude that the distribution of sample proportions is approximately normal with mean $\mu_p = \pi = .45$ and standard deviation

$$\sigma_p = \sqrt{\frac{.45(.55)}{200}}$$
$$= \sqrt{.001238} = .035$$

Since this is a two-tailed test with $\alpha = .05$, $2\frac{1}{2}\%$ of the population must lie in each tail, and so the critical values for z are $z = 1.96$ and $z = -1.96$, as shown in Figure 9.25. In comparison, the z value for the sample proportion $p = .40$ is

$$z = \frac{.40 - .45}{.035}$$
$$= \frac{-.05}{.035} = -1.43$$

We therefore fail to reject the null hypothesis. The party's claim might be valid. ∎

FIGURE 9.25

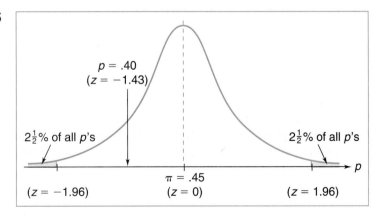

EXAMPLE 9.16 An electric utility survey indicates that 18% of all households in a community own personal computers. A separate study of 80 families with school-age children in this community finds that 22 of them own computers. Test whether the proportion of families with school-age children owning computers is higher than in the general population in that area. Use $\alpha = .02$.

SOLUTION As we have mentioned previously, the null hypothesis often stands for the status quo or the fact that there is no significant difference. Thus, we use

$$H_0: \pi = .18$$
$$H_a: \pi > .18$$

The sample data give

$$n = 80$$
$$x = 22$$

so that the sample proportion is

$$p = \frac{22}{80} = .275$$

Furthermore, since

$$n\pi = 80(.18) = 14.4 \geq 5 \quad \text{and} \quad n(1 - \pi) = 80(.82) = 65.6 \geq 5,$$

we can use a normal distribution with mean $\mu_p = \pi = .18$ and standard deviation

$$\sigma_p = \sqrt{\frac{(.18)(.82)}{80}}$$
$$= \sqrt{.001845} = .043$$

Since $\alpha = .02$ and the hypothesis test is a one-tailed test, the critical value for z is $z = 2.05$, as shown in Figure 9.26. The z value for the sample proportion $p = .275$ is

$$z = \frac{.275 - .18}{.043}$$
$$= \frac{.095}{.043} = 2.21$$

and so we reject the null hypothesis. The proportion of households with school-age children that own computers seems higher than the reported proportion of all families in the area. ∎

We can also apply the *P*-value approach to hypothesis tests on the population proportion π, as illustrated in the following example.

FIGURE 9.26

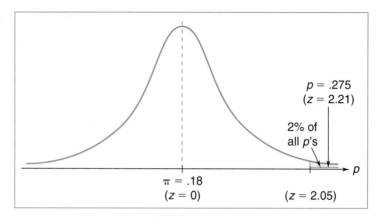

EXAMPLE 9.17

A state's Department of Motor Vehicles claims that 62% of all people who take a driver's test pass. Out of a random sample of 130 prospective drivers at one particular test site, 68 pass the driver's test. Use P-values to determine if the passing rate at this location is lower than the statewide percentage at the $\alpha = .02$ level of significance.

SOLUTION The null and alternate hypotheses are

$$H_0: \pi = .62$$
$$H_a: \pi < .62$$

The sample data give

$$n = 130$$
$$x = 68$$

so that

$$p = \frac{68}{130} = .523$$

Further, since

$$n\pi = 130(.62) = 80.6 \geq 5 \quad \text{and} \quad n(1 - \pi) = 130(.38) = 49.4 \geq 5$$

we can use a normal distribution with mean $\mu_p = \pi = .62$ and standard deviation

$$\sigma_p = \sqrt{\frac{.62(.38)}{130}}$$
$$= \sqrt{.001823} = .043$$

The P-value corresponding to the sample proportion $p = .523$ is equal to the

FIGURE 9.27

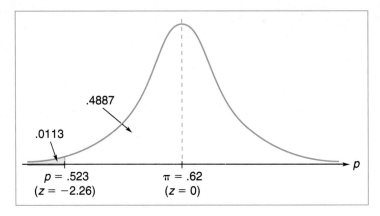

probability of obtaining such a sample proportion or worse, namely,

$$P = P(p \leq .523)$$

Since the z value corresponding to this sample proportion is

$$z = \frac{.523 - .62}{.043} = \frac{-.097}{.043} = -2.26$$

we find that

$$P = P(z \leq -2.26)$$

and, from Figure 9.27, we see that this is equal to .0113. Since this P-value, $P = .0113$, is less than $\alpha = .02$, we reject the null hypothesis. The percentage of people who pass the driver's test at this location seems to be lower than the statewide proportion. ∎

Computer Corner

At the end of the last section, we described Program 28: Hypothesis Tests: Means/Proportions. This program allows you to choose between hypothesis tests for means and proportions. Now select the proportions option. The program asks for the null hypothesis on π. For instance, if H_0: $\pi = .25$, then you would enter .25. It next asks whether the test is one- or two-tailed and, if one, which tail. The program asks you to select a risk level α (10, 5, or 1 percent) and the sample size n and sample proportion p. It then asks *you* to look up and enter the critical value(s) for z from the normal distribution table. The program then displays your null hypothesis and sample data and asks *you* to calculate the z value corresponding to the sample proportion p. Finally, you must enter *your* decision as to whether to reject the null hypothesis. At each stage, as it leads you through the analysis needed to solve the problem, the program attempts to identify any errors you may make.

When the analysis is complete, the program transfers to a graphical display of the normal distribution curve for the distribution of sample proportions. It also shows all the possible significance levels for α as tick marks, the critical value(s), and the location of the sample proportion p. After the graphical display, a text screen summarizes all the results including the P-value associated with the sample proportion p.

Exercise Set 9.4

Mastering the Techniques

In the following Exercises, use the given sample data to test the stated hypothesis on the population proportion π at a significance level of $\alpha = .05$.

1. H_0: $\pi = .50$, H_a: $\pi > .50$, $n = 100$, $x = 55$
2. H_0: $\pi = .50$, H_a: $\pi > .50$, $n = 100$, $x = 62$
3. H_0: $\pi = .50$, H_a: $\pi < .50$, $n = 64$, $x = 28$
4. H_0: $\pi = .50$, H_a: $\pi < .50$, $n = 64$, $x = 24$
5. H_0: $\pi = .50$, H_a: $\pi \neq .50$, $n = 60$, $x = 35$
6. H_0: $\pi = .50$, H_a: $\pi \neq .50$, $n = 60$, $x = 23$
7. Repeat Exercise 1, using P-values.
8. Repeat Exercise 5, using P-values.

Applying the Concepts

9. A vocational training institute claims that only 12% of the people who enter the program fail to complete it. A random sample of 100 people who started the program includes 18 who did not complete it. Test if the claim is too low at the $\alpha = .05$ level of significance.
10. A bank credit card department claims that 60% of all cardholders do not pay their bills on time. In a random sample of 200 cardholders, 136 are behind in their bills. Test if the claim is too low at the $\alpha = .02$ significance level.
11. A company claims that 40% of all people prefer its product to the competing brand. In a sample of 400 consumers, it was found that 125 preferred this particular brand. Test if the claim is too high at the 5% level of significance.
12. A computer chip manufacturer claims that only 0.5% of its chips are defective. In a sample of 1000 such chips, 7 were found to be defective. At the 2% significance level, test if the claim is overstated.
13. A medical study claims that 12% of the people in a certain group are susceptible to a particular disease. A random sample of 200 people in such a group contain 21 who are susceptible. Test if the study results are valid at the $\alpha = .01$ significance level.
14. An industry report claims that 70% of all new-car buyers purchase 5-year extended warranties. A sample of 60 new-car buyers include 26 people who did not purchase such a warranty. Test if the claim is accurate at the $\alpha = .02$ significance level.

15. A weight reduction program claims that 75% of the people on the plan lose a minimum of 40 pounds. In a group of 120 people who participate, 82 lost 40 pounds or more. Test if the claim is too high at the $\alpha = .05$ significance level.
16. A newspaper claims that it reaches 82% of all households in a certain community. A random sample of 250 families in that area contain 184 households that buy the paper. Test if the claim is correct at the $\alpha = .10$ level of significance.
17. A drug rehabilitation center claims that at most 22% of its patients who are certified as drug-free suffer a relapse within 2 years. A study of 35 randomly selected graduates of the program shows that 10 have gone back to drugs within 2 years. Test if the claim is overstated at the 1% significance level.
18. A college claims that at least 42% of all its students receive some form of scholarship assistance. A sample of 75 randomly selected students contain 24 who are not on scholarship. Test if the claim is valid at the 5% level of significance.
19. A government agency claims that at least 35% of all managerial employees are female. A random sample of 92 managers contains 22 women. Use P-values to test if the claim is accurate at the $\alpha = .025$ level of significance.
20. A large corporation sets up a promotion plan which guarantees that 25% of all promotions will go to members of minority groups. A government agency monitors the implementation of the plan and finds that, in a random sample of 200 members who were promoted, 39 were minority employees. Use P-values to test, at the $\alpha = .02$ level of significance, if the plan is being implemented correctly.

9.5 Differences of Means: Hypothesis Tests and Estimation

Up to this point, our work on hypothesis testing has been limited to testing claims about a single population parameter, either μ or π, based on the data from a single sample. A further modification of the methods developed previously permits us to compare two different groups to see if they represent different populations or if they are essentially equivalent with respect to a particular parameter.

Suppose we want to compare major league baseball in 1993 to that in 1970. Specifically, suppose we want to find out if there is any difference in the batting averages of current players compared to those of a generation ago. To do this, we might take a random sample of batting averages of full-time players in 1993 and find the mean $\bar{x}_1 = 253$, say. We might similarly take the mean batting average $\bar{x}_2 = 267$ of a sample of players from 1970. Based on the values for these two sample means, we would like to be able to infer whether the two generations of players have comparable averages or are significantly different.

To do this, we must formalize some of the ideas involved. First, we have two different populations with potentially different means μ_1 and μ_2. These two populations have standard deviations σ_1 and σ_2. We seek to determine whether

$$\mu_1 = \mu_2$$

or
$$\mu_1 \neq \mu_2$$
(so that they are different populations). The statement $\mu_1 = \mu_2$ can be rewritten in the equivalent form $\mu_1 - \mu_2 = 0$. Statistically, it is much easier to work with the *difference of means* than with the equality of means. Thus, we will study whether
$$\mu_1 - \mu_2 = 0$$
The null hypothesis in this situation therefore is
$$H_0: \mu_1 - \mu_2 = 0$$
and asserts that there is no difference or change between the two groups with respect to their means. The alternate hypothesis is often
$$H_a: \mu_1 - \mu_2 \neq 0$$
so that we have a two-tailed test. Situations sometimes arise involving a one-tailed test where the alternate hypothesis is either $\mu_1 - \mu_2 > 0$ or $\mu_1 - \mu_2 < 0$. In such an instance, we suspect that the mean of one group is greater than or smaller than the mean of the other.

To perform such a hypothesis test, we need two separate independent random samples, one drawn from each population. Thus, the sample data give

$n_1, \bar{x}_1, s_1$ from the first group

$n_2, \bar{x}_2, s_2$ from the second group

We consider the difference between the means of these two samples:
$$\bar{x}_1 - \bar{x}_2$$
The set of all such differences of sample means from the two populations forms a new sampling distribution known as the *distribution of differences of sample means*.

> The distribution of differences of sample means is approximately normal with mean
> $$\mu_{\bar{x}_1 - \bar{x}_2} = \mu_1 - \mu_2$$
> and standard deviation
> $$\sigma_{\bar{x}_1 - \bar{x}_2} = \sqrt{\frac{\sigma_1^2}{n_1} + \frac{\sigma_2^2}{n_2}} \tag{9.1}$$
> provided that both
> $$n_1 > 30 \quad \text{and} \quad n_2 > 30$$

When we work with this normal distribution for differences of sample means, the z values are calculated according to the formula

$$z = \frac{(\bar{x}_1 - \bar{x}_2) - (\mu_1 - \mu_2)}{\sigma_{\bar{x}_1 - \bar{x}_2}}$$

Since the null hypothesis asserts that $\mu_1 - \mu_2 = 0$, this formula reduces to

$$z = \frac{\bar{x}_1 - \bar{x}_2}{\sigma_{\bar{x}_1 - \bar{x}_2}}$$

Moreover, it is usually unrealistic to expect that we know the values for σ_1 and σ_2 when we don't know the population means μ_1 and μ_2. Therefore, if both n_1 and n_2 are larger than 30, we can use the values for s_1 and s_2 from the two samples in place of σ_1 and σ_2, respectively. The standard deviation for this sampling distribution is therefore approximated by

$$\sigma_{\bar{x}_1 - \bar{x}_2} \approx \sqrt{\frac{s_1^2}{n_1} + \frac{s_2^2}{n_2}} \tag{9.2}$$

provided that both n_1 and n_2 are greater than 30.

We now apply these ideas to study hypothesis tests for the difference of means.

EXAMPLE 9.18 A study is conducted to compare the batting averages of players in 1970 and 1993. A random sample of 35 players from 1970 have a mean batting average of 267 with a standard deviation of 27. A random sample of 40 players from 1993 have a mean batting average of 255 with a standard deviation of 30. Test whether there is a difference in batting averages between the players in the two seasons at the $\alpha = .05$ level of significance.

SOLUTION The null hypothesis contends that there is no difference between the means for the two seasons and hence is

$$H_0: \mu_1 - \mu_2 = 0$$

while the alternate hypothesis is

$$H_a: \mu_1 - \mu_2 \neq 0$$

The sample data give

1970	1993
$n_1 = 35$	$n_2 = 40$
$\bar{x}_1 = 267$	$\bar{x}_2 = 255$
$s_1 = 27$	$s_2 = 30$

SECTION 9.5 • Differences of Means: Hypothesis Tests and Estimation

Since both samples are large, we are assured that the sampling distribution for the difference of sample means is approximately normally distributed and so we can use

$$\sigma_{\bar{x}_1 - \bar{x}_2} \approx \sqrt{\frac{s_1^2}{n_1} + \frac{s_2^2}{n_2}}$$

$$= \sqrt{\frac{27^2}{35} + \frac{30^2}{40}}$$

$$= \sqrt{\frac{729}{35} + \frac{900}{40}}$$

$$= \sqrt{20.83 + 22.5}$$

$$= \sqrt{43.33}$$

$$= 6.58$$

Since this is a two-tailed test at the $\alpha = .05$ significance level, we want 2.5% of the normal population in each tail, and as shown in Figure 9.28, the corresponding critical values are $z = 1.96$ and $z = -1.96$. We calculate the z value corresponding to the difference in the sample means

$$z = \frac{\bar{x}_1 - \bar{x}_2}{\sqrt{s_1^2/n_1 + s_2^2/n_2}}$$

$$= \frac{267 - 255}{6.58}$$

$$= \frac{12}{6.58}$$

$$= 1.82$$

FIGURE 9.28

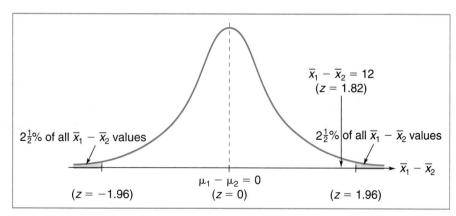

Since this does not fall in the rejection region, we fail to reject the null hypothesis. We cannot conclude that there is a difference between the mean batting averages in the two seasons at the $\alpha = .05$ level of significance. ∎

Before we look at other examples, a word of caution is necessary. When you evaluate the expression

$$\sqrt{\frac{s_1^2}{n_1} + \frac{s_2^2}{n_2}}$$

you should be very careful of the order in which you perform the arithmetic operations. There are a number of potentially tempting errors that you might make, such as combining fractions without first finding a common denominator or taking the square root of the individual terms without first combining them. The best advice we can give is to evaluate each term individually (as decimals rather than as fractions), add them, and take the square root as the last step of the process.

EXAMPLE 9.19 A study is made comparing the prices asked for existing one-family homes in two adjacent communities. In College Heights, the mean asking price for a random sample of 50 homes is $142,000 with a standard deviation of $30,000. In University Gardens, the mean asking price for a random sample of 35 homes is $168,000 with a standard deviation of $40,000. Test whether there is a difference in the mean asking prices for homes in these two areas at the $\alpha = .02$ level of significance.

SOLUTION The null hypothesis is

$$H_0: \mu_1 - \mu_2 = 0$$

while the alternate hypothesis is

$$H_a: \mu_1 - \mu_2 \neq 0$$

The sample data give

College Heights	University Gardens
$n_1 = 50$	$n_2 = 35$
$\bar{x}_1 = 142{,}000$	$\bar{x}_2 = 168{,}000$
$s_1 = 30{,}000$	$s_2 = 40{,}000$

Due to the sizes of the numbers involved, it is convenient to express the values for the sample means and the sample standard deviations in terms of thousands of dollars. Thus, we use

$n_1 = 50$	$n_2 = 35$
$\bar{x}_1 = 142$	$\bar{x}_2 = 168$
$s_1 = 30$	$s_2 = 40$

Further, since both sample sizes are greater than 30, we can use a normal distribution with

$$\sigma_{\bar{x}_1 - \bar{x}_2} \approx \sqrt{\frac{s_1^2}{n_1} + \frac{s_2^2}{n_2}}$$

$$= \sqrt{\frac{30^2}{50} + \frac{40^2}{35}}$$

$$= \sqrt{\frac{900}{50} + \frac{1600}{35}}$$

$$= \sqrt{18 + 45.714}$$

$$= \sqrt{63.714}$$

$$= 7.98$$

Since this is a two-tailed test with $\alpha = .02$, we want 1% of the normal population in each tail, and this corresponds to critical values of $z = 2.33$ and $z = -2.33$, as shown in Figure 9.29. Moreover, corresponding to the difference of the two sample means, we find that

$$z = \frac{\bar{x}_1 - \bar{x}_2}{\sqrt{s_1^2/n_1 + s_2^2/n_2}}$$

$$= \frac{142 - 168}{7.98}$$

$$= \frac{-26}{7.98}$$

$$= -3.26$$

and we therefore reject the null hypothesis. There is a difference between the mean asking prices for homes in these two communities at the $\alpha = .02$ significance level. (In fact, since $z = -3.26$ is negative, we have evidence that the average price is lower in College Heights.) ∎

FIGURE 9.29

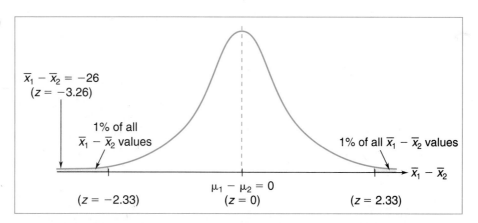

EXAMPLE 9.20 A study is made comparing wages paid to women and men holding comparable jobs in a large company. A random sample of 100 women are paid a mean hourly wage of $7.23 with a standard deviation of $1.64 while a random sample of 75 men are paid a mean hourly wage of $8.06 with a standard deviation of $1.85. Do these data constitute "proof" that, on the average, women are paid less than men at the 5% level of significance?

SOLUTION The phrasing of this question clearly indicates that this is a one-tailed test. In such a case, we must take special care when setting up the alternate hypothesis. We let μ_W refer to the mean wage of the women and μ_M refer to the mean wage of the men. The null hypothesis says that there is no difference between them and so

$$H_0: \mu_W - \mu_M = 0$$

Our suspicion is that women are paid less than men, that is, $\mu_W < \mu_M$ or equivalently $\mu_W - \mu_M < 0$, so the alternate hypothesis is

$$H_a: \mu_W - \mu_M < 0$$

The sample data give

Women	Men
$n_W = 100$	$n_M = 75$
$\bar{x}_W = 7.23$	$\bar{x}_M = 8.06$
$s_W = 1.64$	$s_M = 1.85$

Since both n_W and n_M are greater than 30, we use a normal distribution with

$$\sigma_{\bar{x}_W - \bar{x}_M} \approx \sqrt{\frac{s_W^2}{n_W} + \frac{s_M^2}{n_M}}$$

$$= \sqrt{\frac{1.64^2}{100} + \frac{1.85^2}{75}}$$

$$= \sqrt{.0269 + .0456}$$

$$= \sqrt{.0725}$$

$$= .27$$

Since this is a one-tailed test with $\alpha = .05$, the critical value for z is $z = -1.64$, as shown in Figure 9.30. Corresponding to the two sample means, we find that

$$z = \frac{7.23 - 8.06}{.27}$$

$$= \frac{-.83}{.27}$$

$$= -3.07$$

FIGURE 9.30

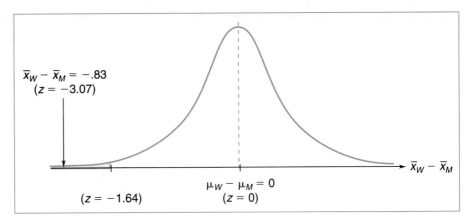

and therefore we reject the null hypothesis. That is, women are seemingly paid lower wages, on the average, than men in this organization at the $\alpha = .05$ level of significance. ∎

EXAMPLE 9.21 A student newspaper at a college is conducting a study on the sleeping habits of the students. As part of this study, it collects the following sets of data on the number of hours per night that male and female students sleep:

Males: 7, 5.5, 9, 6.5, 6, 7, 7.5, 8, 7, 6, 6, 8, 7, 6, 10, 7, 6, 6.5, 7, 7, 5, 6, 8.5, 8, 4, 7, 6, 7, 7.5, 7, 8, 8, 4.5, 7, 6, 8, 5, 6.5, 9, 7, 5.5, 6.5, 6

Females: 8, 7, 8.5, 6, 6, 8, 6.5, 7, 7, 6, 7.5, 8, 9, 6.5, 8, 7, 7, 8, 5.5, 8, 9, 6, 7, 8, 7, 9.5, 6, 7.5, 8, 8, 6, 8.5, 8, 7.5, 6, 7.5, 8, 9, 5, 8, 7, 7, 8, 7.5, 7, 8.5

Based on this set of data, can you conclude that there is any difference in the mean number of hours that students sleep each night based on their gender at the 5% level of significance?

SOLUTION We begin by analyzing the data using Program 1: Statistical Analysis of Data. In Figure 9.31a, we show the associated box plots and observe that there appears to be a difference between the two groups of sleep times. The females have a higher median of 7.5 hours per night compared with the men, whose median is 7 hours. Further, there is more spread in the box plot for the men than for the women.

In Figure 9.31b, we show the associated histograms based on the two sets of data and observe that, while the histogram for the men appears to be approximately normal in shape, that for the women is skewed. We now test to see if there is indeed a difference in the means.

The null and alternate hypotheses are

$$H_0: \mu_M - \mu_F = 0$$
$$H_a: \mu_M - \mu_F \neq 0$$

FIGURE 9.31a

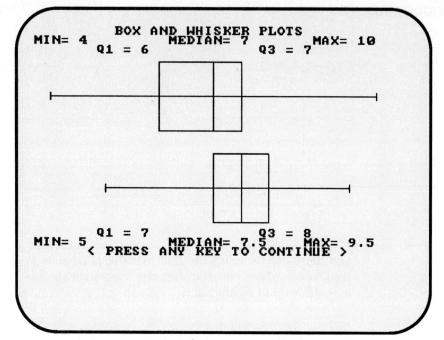

FIGURE 9.31b

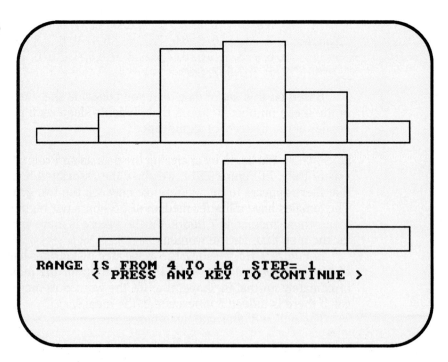

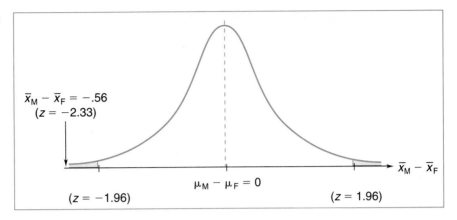

FIGURE 9.31c

The program also provides us with the summary statistics:

$$n_M = 43 \qquad n_F = 46$$
$$\bar{x}_M = 6.81 \qquad \bar{x}_F = 7.37$$
$$s_M = 1.22 \qquad s_F = 1.02$$

We then calculate

$$\sigma_{\bar{x}_M - \bar{x}_F} \approx \sqrt{\frac{(1.22)^2}{43} + \frac{(1.02)^2}{46}} = .24$$

Since both n_M and n_F are greater than 30, the distribution of differences of sample means is approximately normal and so, for the sample data, we find

$$z = \frac{\bar{x}_M - \bar{x}_F}{\sigma_{\bar{x}_M - \bar{x}_F}} = \frac{6.81 - 7.37}{.24}$$
$$= -2.33$$

Hence, as shown in Figure 9.31c, we reject the null hypothesis at the .05 level of significance. There appears to be a difference in the mean number of hours that male and female students sleep each night at this college. ∎

The procedures followed so far in this section have all involved the use of the normal distribution. This is appropriate only if the sample sizes n_1 and n_2 are both larger than 30. Alternative procedures are available when either or both sample sizes are small.

The simplest such procedure requires that the two approximately normal underlying populations have equal variances

$$\sigma_1^2 = \sigma_2^2$$

In this case, the distribution of differences of sample means follows a t-distribution with

$$n_1 + n_2 - 2 \text{ degrees of freedom}$$

The standard deviation for this sampling distribution is approximately

$$\sqrt{\left(\frac{1}{n_1} + \frac{1}{n_2}\right)\left[\frac{(n_1 - 1)s_1^2 + (n_2 - 1)s_2^2}{n_1 + n_2 - 2}\right]} \tag{9.3}$$

This formula certainly will intimidate many of you by its apparent complexity. However, when you have a set of values for n_1, s_1, n_2, and s_2 and substitute them into the formula in the appropriate places, the resulting arithmetic is reasonably simple, as you will see in the example below.

Before proceeding to the example, we consider where some of these expressions come from. Notice that when we are sampling from the first population, there are $n_1 - 1$ degrees of freedom. Similarly, when we are sampling from the second population, there are an additional $n_2 - 1$ degrees of freedom. Algebraically, there are a total of

$$(n_1 - 1) + (n_2 - 1) = n_1 + n_2 - 2$$

degrees of freedom for the two samples.

Further, the term

$$\frac{(n_1 - 1)s_1^2 + (n_2 - 1)s_2^2}{n_1 + n_2 - 2}$$

arises because of the assumption that the two populations have equal variances. To estimate this common variance, it is necessary to combine or *pool* the information from both samples to get one single estimate of the unknown variance σ^2. We call this *pooled variance* s^2_{pool}. It turns out to be equal to the above expression. When this common term is factored out of equation (9.1), we obtain equation (9.3).

We now illustrate how to apply these ideas on hypothesis tests for the difference of means when the sample sizes are small.

EXAMPLE 9.22 The U.S. government has contracted with two companies to produce prototypes for a new battery to be used to power a series of communications satellites. Based on the performance of the two, NASA will select one company to produce all the batteries to be used for the satellite. Company *A* has built and tested 10 models which had a mean effective lifetime of 4.8 years with a standard deviation of 1.1 years. Company *B* has built and tested 12 models with a mean lifetime of 4.3 years with a standard deviation of .9 year. Assume that the underlying populations are approximately normal and have equal variances. Test if there is a difference between the average lifetimes of the two batteries at the $\alpha = .01$ significance level.

SECTION 9.5 • Differences of Means: Hypothesis Tests and Estimation

SOLUTION The null hypothesis is
$$H_0: \mu_1 - \mu_2 = 0$$
while the alternate hypothesis is
$$H_a: \mu_1 - \mu_2 \neq 0$$
The sample data give

Company A	Company B
$n_1 = 10$	$n_2 = 12$
$\bar{x}_1 = 4.8$	$\bar{x}_1 = 4.3$
$s_1 = 1.1$	$s_1 = 0.9$

Since n_1 and n_2 are both less than 30 and we have assumed that the underlying populations are approximately normal with equal variances, the sampling distribution for the difference of sample means is a *t*-distribution with
$$n_1 + n_2 - 2 = 10 + 12 - 2 = 20$$
degrees of freedom. The standard deviation is

$$\sqrt{\left(\frac{1}{n_1} + \frac{1}{n_2}\right)\left[\frac{(n_1-1)s_1^2 + (n_2-1)s_2^2}{n_1 + n_2 - 2}\right]}$$

$$= \sqrt{\left(\frac{1}{10} + \frac{1}{12}\right)\left[\frac{(10-1)(1.1)^2 + (12-1)(.9)^2}{10 + 12 - 2}\right]}$$

$$= \sqrt{(.183)\left[\frac{9(1.21) + 11(.81)}{20}\right]}$$

$$= \sqrt{(.183)\left(\frac{19.80}{20}\right)}$$

$$= \sqrt{.18117} = .426$$

Furthermore, since this is a two-tailed test, the critical values for *t* with 20 degrees of freedom at the $\alpha = .01$ level of significance are $t = -2.845$ and $t = 2.845$, as shown in Figure 9.32. For the difference of sample means, we find that

$$t = \frac{4.8 - 4.3}{.426}$$

$$= \frac{.5}{.426} = 1.174$$

and so we fail to reject the null hypothesis. There may not be a difference in the mean lifetimes of the two types of batteries at the 1% level of significance. Consequently, NASA would base its choice of contractor on other considerations, such as price, promptness of delivery or history of reliability. ∎

FIGURE 9.32

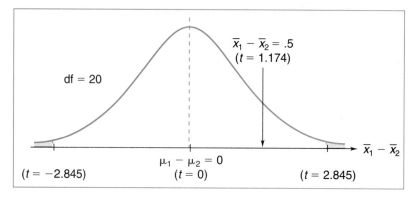

In all the applications of the difference-of-means test that we consider, if the sample sizes are small, then we assume that the underlying populations are approximately normal and that their variances are equal. We note that there is another statistical test which can be used to assess whether variances are indeed equal, but in almost all practical cases, it is not useful and so we do not go into it here. Finally, if the above conditions fail to hold, then some of the nonparametric tests that we develop in Chapter 13 may apply.

We note that the difference-of-means test, while an extremely powerful tool for comparing the means of two potentially different populations, does have some limitations. It is applicable only when the two samples are independent of each other. If there is some linkage between the sample data, then the test does not apply. For instance, if we want to compare the weights of people before and after they go on a certain diet plan, then we are dealing with a set of dependent measurements. Similarly, if we want to compare the prices of a set of 50 items in two different stores, the samples are not independent. We consider a different test that applies to such cases in the next section.

We can also use the ideas in this section to construct confidence intervals to estimate the difference between the population means of two groups based on two samples drawn from them. We use either a z statistic or a t statistic, as appropriate, and formula (9.1) or (9.3), respectively.

> The resulting confidence intervals for the difference of sample means are
>
> $$(\bar{x}_1 - \bar{x}_2) \pm z \sqrt{\frac{s_1^2}{n_1} + \frac{s_2^2}{n_2}}$$
>
> if n_1 and n_2 are both greater than 30 or
>
> $$(\bar{x}_1 - \bar{x}_2) \pm t \sqrt{\left(\frac{1}{n_1} + \frac{1}{n_2}\right)\left[\frac{(n_1 - 1)s_1^2 + (n_2 - 1)s_2^2}{n_1 + n_2 - 2}\right]}$$

if either n_1 or n_2 is less than 30 and if the underlying populations are approximately normal with equal variances.

We illustrate these ideas in the following example.

EXAMPLE 9.23 A study is conducted comparing average starting salaries offered to new B.A. recipients at two universities. A sample of 42 students from one school are offered an average of $1360 per month with a standard deviation of $320 while a sample of 48 students from the other school are offered an average of $1320 with a standard deviation of $375. Construct a 95% confidence interval for the difference in the mean starting salaries.

SOLUTION The sample data give

$$n_1 = 42 \qquad n_2 = 48$$
$$\bar{x}_1 = 1360 \qquad \bar{x}_2 = 1320$$
$$s_1 = 320 \qquad s_2 = 375$$

Since both sample sizes are greater than 30, we can use a normal distribution with

$$\sigma_{\bar{x}_1 - \bar{x}_2} \approx \sqrt{\frac{s_1^2}{n_1} + \frac{s_2^2}{n_2}}$$

$$= \sqrt{\frac{320^2}{42} + \frac{375^2}{48}}$$

$$= \sqrt{2438.1 + 2929.7}$$

$$= \sqrt{5367.8}$$

$$= 73.27$$

Therefore, the corresponding 95% confidence interval using $z = 1.96$ is

$$(1360 - 1320) \pm 1.96(73.27)$$
$$40 \pm 143.60$$

or

$$[-103.60, 183.60]$$

See Figure 9.33. That is, with 95% confidence, we conclude that the difference in the mean salaries offered to graduates of the two universities is between −$103.60 and $183.60. ∎

FIGURE 9.33

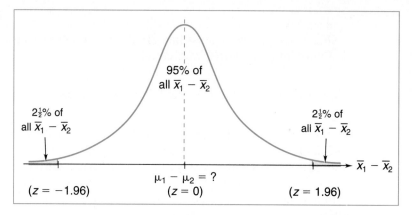

Computer Corner

Program 30: Dist'n of Difference of Means allows you to investigate the distribution of differences of sample means. You may select two populations from a choice of four different underlying populations: a normal population, a skewed population, a uniform population, and a U-shaped population. You can select any two of them or the same one twice. The program asks for the sample size for the samples from each population and the number of such samples you desire. The program then displays the histograms for the two underlying populations you selected. After you press any key, it generates repeated random samples of the indicated sizes from the two populations, calculates the sample mean for each sample, and graphs the difference between these sample means. The resulting distribution allows you to explore the effects that different underlying populations, different sample sizes, and different numbers of samples have on the sampling distribution. For example, what is the apparent shape of the distribution if one sample size is small, say $n_1 = 5$?

When the graph is complete, an accompanying text screen displays the numerical results for the samples in that run. This includes the mean and standard deviation for the differences in all the sample means generated during the run. The program also displays the theoretical values for the difference of means $\mu_1 - \mu_2$ and the appropriate standard deviation, depending on whether the sampling population is normal or t.

Minitab Methods

You can also use Minitab to conduct hypothesis tests on two sets of independent data. Suppose the data are stored in columns C1 and C2 and you want to perform a difference-of-means test on them. The appropriate Minitab command is **TWOSAMPLE** along with the associated subcommand **ALTERNATIVE**. The format for these commands might be

> **TWOSAMPLE** test at the **95%** level for data in **C1** and **C2**
> **ALTERNATIVE** = 0

for a two-tailed test. A shorthand version of the command is simply

> **TWO SAMPLE 95 C1 C2**

Minitab then responds with a table giving, for each of the two sets of data, the sample size, sample mean, sample standard deviation, and standard error of the mean. This is followed by a 95% confidence interval for the difference in the two means as well as the results of the t-test for two independent samples consisting of the t value corresponding to $\bar{x}_1 - \bar{x}_2$, the P-value, and the total number of degrees of freedom. You then must compare this t value to the appropriate critical value from the t-distribution table if either n_1 or n_2 is less than 31; otherwise, you can use the critical value from the normal distribution table.

A sample Minitab session is shown in Figure 9.34.

FIGURE 9.34

```
MTB > DESCRIBE C1-C2
              N      MEAN    MEDIAN   TRMEAN   STDEV   SEMEAN
WEIGHTS1     30    156.50    153.00   155.46   19.84    3.62
WEIGHTS2     30    158.00    157.00   157.92   19.37    3.54

             MIN      MAX       Q1       Q3
WEIGHTS1   122.00   211.00   143.00   169.25
WEIGHTS2   122.00   194.00   143.00   174.00

MTB > TWOSAMPLE test @ 95% level of C1 vs C2;
SUBC> ALTERNATIVE is two tailed = 0.

TWOSAMPLE T FOR WEIGHTS1 VS WEIGHTS2
              N      MEAN    STDEV   SE MEAN
WEIGHTS1     30     156.5     19.8      3.6
WEIGHTS2     30     158.0     19.4      3.5

95 PCT CI FOR MU WEIGHTS1 - MU WEIGHTS2: (-11.6, 8.6)
TTEST MU WEIGHTS1 = MU WEIGHTS2 (VS NE): T = -0.30  P=0.77  DF= 57
```

Exercise Set 9.5

Mastering the Techniques

Test each of the following hypotheses at the $\alpha = .05$ level of significance using the given sets of data.

1. $H_0: \mu_1 - \mu_2 = 0$, $H_a: \mu_1 - \mu_2 \neq 0$, $n_1 = 50$, $\bar{x}_1 = 102$, $s_1 = 20$, $n_2 = 32$, $\bar{x}_2 = 97$, $s_2 = 16$
2. $H_0: \mu_1 - \mu_2 = 0$, $H_a: \mu_1 - \mu_2 \neq 0$, $n_1 = 40$, $\bar{x}_1 = 77$, $s_1 = 15$, $n_2 = 50$, $\bar{x}_2 = 71$, $s_2 = 18$
3. $H_0: \mu_1 - \mu_2 = 0$, $H_a: \mu_1 - \mu_2 \neq 0$, $n_1 = 60$, $\bar{x}_1 = 23$, $s_1 = 24$, $n_2 = 80$, $\bar{x}_2 = 30$, $s_2 = 26$
4. $H_0: \mu_1 - \mu_2 = 0$, $H_a: \mu_1 - \mu_2 > 0$, $n_1 = 50$, $\bar{x}_1 = 105$, $s_1 = 20$, $n_2 = 32$, $\bar{x}_2 = 98$, $s_2 = 16$
5. $H_0: \mu_1 - \mu_2 = 0$, $H_a: \mu_1 - \mu_2 < 0$, $n_1 = 25$, $\bar{x}_1 = 20$, $s_1 = 6$, $n_2 = 35$, $\bar{x}_2 = 25$, $s_2 = 8$
6. $H_0: \mu_1 - \mu_2 = 0$, $H_a: \mu_1 - \mu_2 > 0$, $n_1 = 20$, $\bar{x}_1 = 150$, $s_1 = 20$, $n_2 = 42$, $\bar{x}_2 = 133$, $s_2 = 18$

Applying the Concepts

7. A study is conducted comparing the mean length of hospitalization of men versus women. A random sample of 50 men were hospitalized an average of 5.3 days with a standard deviation of 2.1 days. A random sample of 40 women were hospitalized an average of 6.2 days with a standard deviation of 1.8 days. Test if there is a difference at the $\alpha = .05$ significance level.
8. A company conducts a study comparing the mean number of sick days that employees take off depending on whether they have been employed more than 5 years or less than 5 years. In a random sample of 50 long-term employees, the mean number of sick days in a year was 8.2 with a standard deviation of 4.6; in a sample of 35 short-term employees, the mean was 9.8 days with a standard deviation of 5.2 days. Test if there is a difference at the $\alpha = .02$ level of significance.
9. A scientist conducts a study comparing the effects of two different types of fertilizer on the yield of a particular variety of tomato plant. A random sample of 60 plants using Greenthumb fertilizer had a mean yield of 32.2 tomatoes with a standard deviation of 8.5. A random sample of 72 plants using Supergro fertilizer had a mean yield of 28.4 tomatoes with a standard deviation of 9.3. Test if there is a difference in yield at the 5% significance level.
10. A waitress at a yacht club restaurant conducts a study to compare the average amount of tips she receives per person on a $20 dinner from sailboaters versus motorboaters. In a random sample of 32 sailboaters, she receives a mean of $3.80 with a standard deviation of $1.22. In a random sample of 36 motorboaters, she receives a mean of $4.42 with a standard deviation of $1.04. Test if there is a difference in the average tip at the 5% level of significance.
11. A trucking company wants to know if there is a difference in the mean number of trucks that are out of service in the winter versus the summer. In a random sample of 35 winter days, there are a mean of 16.6 trucks out of service with a

standard deviation of 7.1. In a random sample of 32 summer days, a mean of 12.4 trucks are out of service with a standard deviation of 5.8. Test if there is a difference at the $\alpha = .02$ level of significance.

12. Two businesses sell Christmas trees in a certain town. One company wants to advertise that it carries the tallest trees. To prove this, the employees conduct a survey. On their own lot, a random sample of 40 trees have a mean height of 7.4 feet with a standard deviation of 1.6 feet. At the competitor's lot, a random sample of 32 trees have a mean height of 6.8 feet with a standard deviation of 1.4 feet. At the 2% significance level, is this claim valid?

13. A doctor maintains an office in each of two different communities and wishes to determine if there is a difference in the average number of patients he treats in the two locations each day. He randomly selects 10 days at office A and finds that he treats a mean of 26 patients each day with a standard deviation of 4 patients. In a random sample of 8 days at office B, he finds that he treats a mean of 21 patients with a standard deviation of 5 patients. Test if there is a difference between the average number of patients treated at the two offices at the 5% level of significance.

14. A physics professor wants to determine if there is a difference in test results between her morning and afternoon sections of the same course. In a random sample of 9 students from the morning class, she finds that the mean score on a test is 78.4 with a standard deviation of 18.6. In a random sample of 8 students in the afternoon class, she finds that the mean score on the comparable test is 73.8 with a standard deviation of 20.5. Is there a difference in the test results at the 5% level of significance?

15. Use the information from Exercise 7 to construct a 98% confidence interval for the difference in means for length of hospitalization for women and men.

16. Use the information from Exercise 9 to construct a 95% confidence interval for the difference in means for tomato production based on the two types of fertilizer.

Computer Applications

17. Use Program 2: Random Number Generator to generate two random samples of size 35, one from the National League and the other from the American League, from the list of baseball averages provided in Appendix I. Use Program 1: Statistical Analysis of Data to compute the mean and standard deviation for each sample, and then compare the two. Is there a difference in the batting averages between the two leagues at the 5% level of significance?

18. Use Program 2: Random Number Generator to select a random sample of at least 30 stock prices from the list provided, and then find the mean and standard deviation. Repeat this procedure with a different random sample of at least 30 stock prices found in a comparable stock market listing in a current newspaper. Test if there is a difference in the mean stock prices between the two dates at the 2% level of significance.

19. Test if there is a difference between the grade-point averages of male and female students at your school. Select random samples of 30 students of each gender, collect their GPAs, and conduct the appropriate statistical analysis at the $\alpha = .05$ level of significance.

9.6 The Paired-Data Test

In the last section, we saw how useful the difference-of-means test is for comparing the means of two different *independent* groups to determine whether there is a difference between them. However, the test does not apply if the two samples are linked or paired in some fashion; that is, if they are *dependent*. We now consider this situation to see how a minor modification of ideas we developed previously can be applied when the samples are not independent of each other.

Suppose that 10 people are randomly selected from a large group of individuals who go on a special diet. Their weights *before* and *after* this 6-week weight loss program are given in Table 9.1. We wish to determine if the diet results in a significant weight loss or if the apparent differences in weight for each person are attributable to random variation.

To determine if the diet is effective, first we consider the difference between all pairs of sample data. We do this by calculating the differences $d = x - y$ representing the net weight loss for each individual, as shown in Table 9.1.

Before we proceed, several essential points must be noted. First, the observations are dependent; we are measuring the same randomly selected individual's weight twice, before and after the diet. Second, we are not concerned with the means of the weights, either before or after. Rather, we consider the change or difference in each individual's weight. Third, since we study only the *change* in weight for each person, we are effectively considering a single population consisting of differences in weights before and after the diet.

If the diet is not effective, then we expect that the positive and negative entries among the d's would tend to cancel each other out. Thus, if the mean of all these values is 0, then there is no difference between weights before and after the diet. However, if the diet is effective, then we expect a prepon-

TABLE 9.1

Before x	After y	Difference $d = x - y$
190	185	5
202	197	5
177	185	−8
160	152	8
225	205	20
180	184	−4
196	185	11
208	200	8
185	187	−2
177	170	7

derance of positive entries. Therefore, if the mean of these *d* values is positive, there is a significant weight loss. The problem consequently reduces to considering the mean of *all* the possible differences $d = x - y$ for all members of the population to see whether the mean is actually zero.

To formalize these ideas, we introduce the following notation. The mean for all the possible *d*'s in the population is denoted by μ_d, and their standard deviation is σ_d. The null hypothesis is then formulated as

$$H_0: \mu_d = 0$$

Since we want to show that the diet is successful, we want to test whether the mean of all the differences is actually positive to indicate a positive weight loss. The alternate hypothesis is therefore

$$H_a: \mu_d > 0$$

Further, we assume that the distribution of all the differences is approximately normal in all examples and problems in which the number of data pairs *n* is "small" (under 31). This transforms the problem to a standard hypothesis test on the single mean μ_d. We solve it in the usual fashion discussed in Sections 9.1 to 9.3, with several slight modifications. Previously, the sample data provided values for n, $\bar{x}$ and s. In the weight loss example, we have a sample consisting of $n = 10$ differences. From it, we have the sample size n and can calculate the sample mean $\bar{d}$ of the *d*'s and the corresponding sample standard deviation s_d for the *d*'s. We perform these calculations as in Chapter 3; they are shown in Table 9.2.

Therefore, the mean of the differences is $\bar{d} = 5$, and the sample variance is

$$s_d^2 = \frac{578}{9} = 64.222$$

TABLE 9.2

d	$d - \bar{d}$	$(d - \bar{d})^2$
5	0	0
5	0	0
−8	−13	169
8	3	9
20	15	225
−4	−9	81
11	6	36
8	3	9
−2	−7	49
7	2	4
$\Sigma d = 50$		578

$$\bar{d} = \frac{50}{10} = 5$$

so that the sample standard deviation is

$$s_d = \sqrt{64.222} = 8.01$$

We approximate the standard deviation for the sampling distribution with

$$\frac{s_d}{\sqrt{n}} = \frac{8.01}{\sqrt{10}} = \frac{8.01}{3.16} = 2.53$$

Notice that n represents the number of pairs of entries or differences (in this case 10, since we are studying a sample of 10 differences). It does not represent the total number of measurements (20) originally taken.

We now test the null hypothesis at the $\alpha = .05$ level of significance. Since we assume that the d's are approximately normally distributed, we can use a t-distribution with $n - 1 = 9$ degrees of freedom. For this one-tailed test with 5% of the population in the right-hand tail, the critical value for t with 9 degrees of freedom is $t = t_{.05} = 1.833$, as shown in Figure 9.35. The t value corresponding to the sample data is

$$t = \frac{\bar{d} - \mu_d}{s_d/\sqrt{n}}$$

$$= \frac{5 - 0}{2.53} = 1.976$$

and we consequently reject the null hypothesis. The diet plan does result in weight loss at the $\alpha = .05$ level of significance.

> The procedure we used is known as the **paired-data test** or the **paired t-test**. It is based on the test statistic
>
> $$t = \frac{\bar{d} - \mu_d}{s_d/\sqrt{n}}$$

FIGURE 9.35

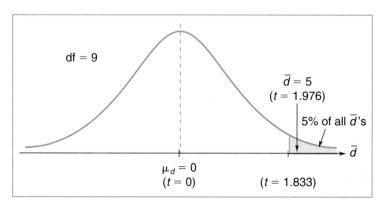

We note that the sample standard deviation s_d can also be calculated by using the equivalent form of the calculating formula from Section 3.2:

$$s_d = \sqrt{\frac{n\Sigma d^2 - (\Sigma d)^2}{n(n-1)}}$$

EXAMPLE 9.24 A study is conducted to compare the pricing at two competing food stores. Twelve common grocery items are chosen at random, and the price of each is noted at the two stores, as follows:

Item	1	2	3	4	5	6	7	8	9	10	11	12
Store A	.89	.59	1.29	1.50	2.49	.65	.99	1.99	2.25	.50	1.99	1.79
Store B	.95	.55	1.49	1.69	2.39	.79	.99	1.79	2.39	.59	2.19	1.99

Test whether there is a mean difference in the prices at the two stores at the $\alpha = .02$ level of significance.

SOLUTION The null hypothesis is

$$H_0: \mu_d = 0$$

while the alternate hypothesis is

$$H_a: \mu_d \neq 0$$

since the phrasing of the question does not indicate which store's prices are higher or lower. From the sample data, we construct a table of differences of respective prices, as shown in Table 9.3.

TABLE 9.3

Item	$d = x - y$	$d - \bar{d}$	$(d - \bar{d})^2$
1	−.06	.013	.0002
2	.04	.113	.0128
3	−.20	−.127	.0161
4	−.19	−.117	.0137
5	.10	.173	.0299
6	−.14	−.067	.0045
7	0	.073	.0053
8	.20	.273	.0745
9	−.14	−.067	.0045
10	−.09	−.017	.0003
11	−.20	−.127	.0161
12	−.20	−.127	.0161
	$\Sigma d = -.88$		$\Sigma(d - \bar{d})^2 = .1944$

$$\bar{d} = \frac{-.88}{12} = -.073$$

From this table, we find that the mean of the sample differences is

$$\bar{d} = \frac{-.88}{12} = -.073$$

and the sample variance is

$$s_d^2 = \frac{.1944}{11} = .0177$$

Therefore, the sample standard deviation is

$$s_d = \sqrt{.0177} = .133$$

Consequently, we use

$$\frac{s_d}{\sqrt{n}} = \frac{.133}{\sqrt{12}} = \frac{.133}{3.464} = .038$$

Since we assume that the population of differences is approximately normal, we have a t-distribution with $n - 1 = 11$ degrees of freedom. Further, since $\alpha = .02$ and this is a two-tailed test, we want 1% of the population in each tail. Therefore, the critical values for t with 11 degrees of freedom are $t = t_{.01} = \pm 2.718$, as shown in Figure 9.36. For the sample data, we find that

$$t = \frac{-.073 - 0}{.038} = -1.921$$

We therefore cannot reject the null hypothesis. There is insufficient evidence to indicate any difference between the average prices at the two stores at the $\alpha = .02$ significance level. ∎

If the sample data consist of more than 30 pairs of values, then it is possible to use a normal distribution to approximate the t-distribution.

Finally, the paired data test is designed to determine whether there is any difference between two sets of measurements on the same individuals or

FIGURE 9.36

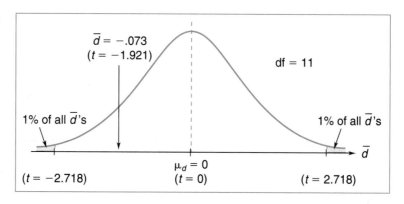

items. For instance, in Example 9.22, we asked if the average prices are different at the two stores. However, sometimes there are other considerations. Often one store will advertise that it is the cheapest store in town. This might suggest that the mean price at that store is lower than that of any competitor. The store might achieve this distinction by simply pricing every item in the store one cent less than the competition. Certainly, the claim is true, but this does not likely represent a *significant* difference in price in any practical sense. If anything, what it represents is a *consistent* difference, which is not the same. In Section 13.2, we consider ways of testing whether there is a consistent difference in sets of paired data.

Minitab Methods

You can use Minitab to perform a paired *t*-test for any pair of dependent sets of data. Suppose the two sets of values are stored in columns C5 and C6. To perform the paired *t*-test, we must consider the differences between the two pairs of entries. To do this, use the Minitab command

```
LET C8 = C6 – C5
```

which calculates the difference between each pair of data values and stores the results in column C8. You can then perform a regular *t*-test on the column of differences by using the command

```
TTEST 0 C8
```

In response to this, Minitab displays the *t* value associated with the set of differences, as well as the sample size, sample mean, sample standard deviation, standard error of the mean, and *P*-value. You then must compare this *t* value to the appropriate critical value from the *t*-distribution table. See Figure 9.37.

Exercise Set 9.6

Mastering the Techniques

For each of the following, test the given hypothesis at the $\alpha = .05$ level of significance using the given data.

1. $H_0: \mu_d = 0$, $H_a: \mu_d \neq 0$, $n = 25$, $\bar{d} = 6$, $s_d = 20$
2. $H_0: \mu_d = 0$, $H_a: \mu_d \neq 0$, $n = 16$, $\bar{d} = 10$, $s_d = 12$
3. $H_0: \mu_d = 0$, $H_a: \mu_d > 0$, $n = 25$, $\bar{d} = 10$, $s_d = 15$
4. $H_0: \mu_d = 0$, $H_a: \mu_d < 0$, $n = 25$, $\bar{d} = -8$, $s_d = 12$

FIGURE 9.37

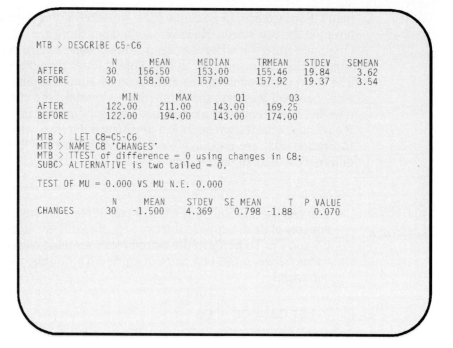

Applying the Concepts

5. The following prices were noted for a particular camera model in the newspaper advertisements for eight different camera stores before and after a major holiday:

Store	1	2	3	4	5	6	7	8
Before	95	109	99	98	105	99	109	102
After	98	105	99	99	109	105	115	110

Test whether the stores, on the average, tend to lower prices before the holiday at the $\alpha = .05$ level of significance.

6. A police department performs an experiment to assess the effects of an obvious radar trap on the speeds of cars. Ten cars are randomly selected on a highway, and their speeds are measured (in miles per hour) just before a radar trap comes into view and just after they pass the trap. The following speeds are recorded:

Car	1	2	3	4	5	6	7	8	9	10
Before	59	52	62	61	59	56	55	62	64	60
After	54	55	65	57	55	65	58	62	54	56

Test if there is a change in car speed because of the presence of the speed traps at the $\alpha = .02$ significance level.

7. The grade-point averages (GPA) of a randomly selected group of 7 college students after their sophomore year are compared to their GPAs after their fresh-

man year. The results are as follows:

Student	1	2	3	4	5	6	7
Freshman	2.50	2.78	3.15	3.62	2.15	2.84	2.61
Sophomore	2.75	2.62	3.34	3.45	2.55	2.58	2.70

Test if there is a change in the GPA from one year to the next at the $\alpha = .05$ level of significance.

8. A group of high school students registered for a special SAT mathematics preparatory course offered in their school. They took a sample SAT the first day and then took another the last day. The scores were as follows:

Student	1	2	3	4	5	6	7	8	9
Before	540	460	520	580	670	590	640	490	530
After	570	510	530	550	640	610	660	520	530

Test whether the course resulted in an improvement in the SAT scores at the 1% significance level.

9.7 Differences of Proportions: Hypothesis Tests and Confidence Intervals

We considered the problem of testing whether there is any difference between the means μ_1 and μ_2 of two populations in Section 9.5. We now turn to a similar procedure for testing for a difference between proportions π_1 and π_2 of two populations.

Suppose a political poll is conducted with the following results: 42% of the men surveyed prefer a certain candidate and 46% of the women surveyed prefer the same candidate. Is there a difference in the proportions of all men and women voters who support this person? The method we use to perform such a test is a minor modification of our previous methods for hypothesis testing.

In general, suppose we have two populations with population proportions π_1 and π_2. We wish to determine whether $\pi_1 = \pi_2$, or equivalently whether $\pi_1 - \pi_2 = 0$. The null hypothesis for this *difference-of-proportions test* is

$$H_0: \pi_1 - \pi_2 = 0$$

The alternate hypothesis will often be

$$H_a: \pi_1 - \pi_2 \neq 0$$

leading to a two-tailed test. In some cases, we require a one-tailed test, and so the alternate hypothesis is either

$$H_a: \pi_1 - \pi_2 > 0$$

or

$$H_a: \pi_1 - \pi_2 < 0$$

Our decision to reject the null hypothesis or reserve judgment on it is based on two sets of sample data. We take a random sample from each group and the sample data usually provide

n_1 and x_1 from group 1

n_2 and x_2 from group 2

In turn, these give rise to the sample proportions

$$p_1 = \frac{x_1}{n_1} \quad \text{for group 1}$$

and

$$p_2 = \frac{x_2}{n_2} \quad \text{for group 2}$$

Realize that each of these samples is just one possible sample drawn from the underlying population. When we consider all possible sample proportions and construct their differences $p_1 - p_2$, we obtain the distribution of differences of sample proportions.

The **distribution of differences of sample proportions** is approximately normal with mean

$$\mu_{p_1 - p_2} = \pi_1 - \pi_2$$

and standard deviation

$$\sigma_{p_1 - p_2} = \sqrt{\frac{\pi_1(1 - \pi_1)}{n_1} + \frac{\pi_2(1 - \pi_2)}{n_2}} \quad (9.4)$$

provided that

$$n_1 \pi_1 \geq 5 \quad n_1(1 - \pi_1) \geq 5$$
$$n_2 \pi_2 \geq 5 \quad n_2(1 - \pi_2) \geq 5$$

Realistically, we do not know the values for π_1 and π_2. Therefore, we must estimate their values based on the sample data. In the situations we are considering here, the null hypothesis asserts that the two population parameters are equal, say

$$\pi_1 = \pi_2 = \pi$$

Consequently, we must estimate this common proportion π. The best way to do this is to *pool* all the available data from the two samples and so we use

$$\bar{p} = \frac{x_1 + x_2}{n_1 + n_2}$$

as the best estimate for the common proportion π. That is, we total the number of successes $x_1 + x_2$ from both samples and divide this quantity by the total number sampled $n_1 + n_2$. The result is a weighted average which reflects the individual sizes of the two sample groups.

We now simplify Equation (9.4) as follows. We first introduce π as the common value for both π_1 and π_2 and then use $\bar{p}$ in place of π to obtain

$$\sigma_{p_1-p_2} \approx \sqrt{\bar{p}(1-\bar{p})\left(\frac{1}{n_1} + \frac{1}{n_2}\right)}$$

provided that

$$n\bar{p} \geq 10 \quad \text{and} \quad n(1-\bar{p}) \geq 10$$

where $n = n_1 + n_2$. The test statistic for z is then

$$z = \frac{p_1 - p_2}{\sqrt{\bar{p}(1-\bar{p})\left(\frac{1}{n_1} + \frac{1}{n_2}\right)}}$$

We illustrate the technique in the following examples.

EXAMPLE 9.25 In a political poll, 42 out of 100 randomly selected men surveyed preferred candidate Smith. Also 92 out of 200 women preferred Smith. Test whether there is any difference in the proportions of men and women who prefer Smith at the $\alpha = .05$ level of significance.

SOLUTION The null hypothesis is

$$H_0: \pi_1 - \pi_2 = 0$$

while the alternate hypothesis is

$$H_a: \pi_1 - \pi_2 \neq 0$$

The sample data give

Men	Women
$n_1 = 100$	$n_2 = 200$
$x_1 = 42$	$x_2 = 92$
$p_1 = \frac{42}{100} = .42$	$p_2 = \frac{92}{200} = .46$

We now calculate the pooled proportion

$$\bar{p} = \frac{42 + 92}{100 + 200}$$

$$= \frac{134}{300}$$

$$= .447$$

Notice that the value for $\bar{p}$ is a weighted average of .42 and .46. It comes out closer to .46 because that corresponds to the larger sample group. Moreover, we see that

$$n\bar{p} = 300(.447) = 134.1 \geq 10$$

and

$$n(1 - \bar{p}) = 300(1 - .447)$$
$$= 300(.553) = 165.9 \geq 10$$

so that the sampling distribution for the difference of proportions is approximately normally distributed.

Using this value for $\bar{p}$, we then find that

$$\sigma_{p_1 - p_2} \approx \sqrt{\bar{p}(1 - \bar{p})\left(\frac{1}{n_1} + \frac{1}{n_2}\right)}$$
$$= \sqrt{(.447)(.553)\left(\tfrac{1}{100} + \tfrac{1}{200}\right)}$$
$$= \sqrt{(.447)(.553)(.01 + .005)}$$
$$= \sqrt{.0037} = .061$$

As seen from Figure 9.38, we use a two-tailed test with $\alpha = .05$, so that the associated critical values are $z = \pm 1.96$. Using the two sample proportions, we find that

$$z = \frac{p_1 - p_2}{\sqrt{\bar{p}(1 - \bar{p})\left(\frac{1}{n_1} + \frac{1}{n_2}\right)}}$$
$$= \frac{.42 - .46}{.061}$$
$$= \frac{-.04}{.061} = -.66$$

FIGURE 9.38

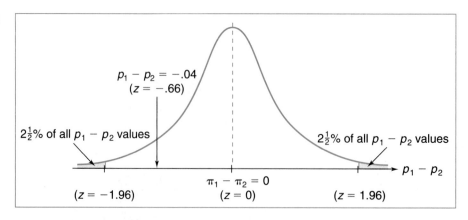

and therefore we cannot reject the null hypothesis. We cannot conclude that there is any difference in the proportions of men and women who support this candidate. (The apparent difference is not large enough to be statistically significant.) ∎

EXAMPLE 9.26 A company is considering introducing a new version of its Zippi Cola soft drink by changing the formula. It first conducts a series of taste tests comparing Zippi to the leading brand of cola. In the first test based on the original formula of Zippi, 120 out of 500 people who tried it preferred Zippi. After the Zippi formula was changed, the test was repeated on a new group of 1000 tasters and 300 of them preferred the new Zippi to the leading brand. Test whether the new formula represents an improvement in the proportion of people who prefer Zippi to the leading brand at the 2% significance level.

SOLUTION To test if there is an improvement in the proportion of people who prefer the new formula, π_2, to those who prefer the old formula, π_1, we set up the null hypothesis

$$H_0: \pi_1 - \pi_2 = 0$$

The alternate hypothesis is based on $\pi_1 < \pi_2$, so that

$$H_a: \pi_1 - \pi_2 < 0$$

The sample data give

Original	New
$n_1 = 500$	$n_2 = 1000$
$x_1 = 120$	$x_2 = 300$
$p_1 = \frac{120}{500} = .24$	$p_2 = \frac{300}{1000} = .30$

Therefore, the pooled proportion is

$$\bar{p} = \frac{120 + 300}{500 + 1000}$$

$$= \frac{420}{1500}$$

$$= .28$$

Further, we find that

$$n\bar{p} = 1500(.28) = 420 \geq 10$$

and

$$n(1 - \bar{p}) = 1500(.72) = 1080 \geq 10$$

so that the sampling distribution is approximately normal. Thus, we use

$$\sqrt{\bar{p}(1-\bar{p})\left(\frac{1}{n_1}+\frac{1}{n_2}\right)} = \sqrt{(.28)(.72)\left(\frac{1}{500}+\frac{1}{1000}\right)}$$
$$= \sqrt{(.28)(.72)(.002+.001)}$$
$$= \sqrt{.000605} = .025$$

For a one-tailed test at the $\alpha = .02$ level of significance, the critical value is $z = -2.05$, as shown in Figure 9.39. Corresponding to the sample data, we find

$$z = \frac{.24 - .30}{.025}$$
$$= \frac{-.06}{.025}$$
$$= -2.40$$

and hence we reject the null hypothesis. The proportion of test tasters who prefer the new version of Zippi to the leading brand is higher than the proportion who prefer the original version. Consequently, the company may decide to use the new formula to increase its share of the market. ∎

EXAMPLE 9.27 According to the medical study reported in Figure 9.40, the use of aspirin can significantly reduce the chances of heart attacks in men. In the study, 104 out of 11,037 male doctors using aspirin had heart attacks while 189 out of 11,034 using a placebo had heart attacks. Perform a hypothesis test to justify the use of the phrase *significantly reduce* in this statement. Use an $\alpha = .01$ level of significance.

FIGURE 9.39

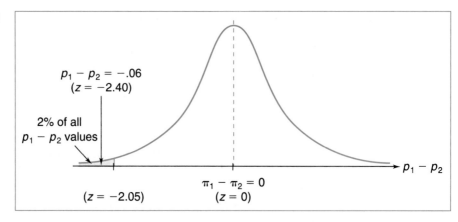

FIGURE 9.40

Heart Attack Risk Cut by Aspirin Use

In a study of more than 22,000 male physicians conducted by the Harvard Medical School, half took an ordinary aspirin tablet every second day and half took a placebo. The study was halted after five years when the benefits of using aspirin became apparent. Shown below are the number of heart attacks suffered by those in the study during the five year period.

	Aspirin	Placebo
Fatal	5	18
Nonfatal	99	171
Total	104	189
Sample Size	11037	11034

Source: New England Journal of Medicine

SOLUTION We must test if the use of aspirin produced a significantly lower proportion of heart attacks. The null and alternate hypotheses are

$$H_0: \pi_1 - \pi_2 = 0$$

and

$$H_a: \pi_1 - \pi_2 < 0$$

where π_1 represents the proportion of men having heart attacks while taking aspirin and π_2 represents the proportion having heart attacks without aspirin (that is, while taking the placebo).

The sample data give

With aspirin: $n_1 = 11{,}037$ $x_1 = 104$

Without aspirin: $n_2 = 11{,}034$ $x_2 = 189$

so that the sample proportions are

$$p_1 = \frac{104}{11{,}037} = .0094$$

$$p_2 = \frac{189}{11{,}034} = .0171$$

Moreover, the pooled proportion is

$$\bar{p} = \frac{104 + 189}{11{,}037 + 11{,}034}$$

$$= \frac{293}{22{,}071} = .0133$$

so that

$$1 - \bar{p} = .9867$$

Since $n\bar{p}$ and $n(1 - \bar{p})$ are both considerably larger than 10, we can use a normal distribution. Further, we use

$$\sqrt{\bar{p}(1 - \bar{p})\left(\frac{1}{n_1} + \frac{1}{n_2}\right)} = \sqrt{(.0133)(.9867)\left(\frac{1}{11{,}037} + \frac{1}{11{,}034}\right)}$$

$$= \sqrt{.0000024}$$

$$= .00155$$

Since this is a one-tailed test with $\alpha = .01$, the critical value is $z = -2.33$, as shown in Figure 9.41. Corresponding to the sample proportions p_1 and p_2, we find that

$$z = \frac{.0094 - .0171}{.00155}$$

$$= \frac{-.0077}{.00155} = -4.97$$

As a result, we reject the null hypothesis and conclude that the use of aspirin does make a significant difference in the proportion of male doctors who get heart attacks at the $\alpha = .01$ level of significance. ∎

FIGURE 9.41

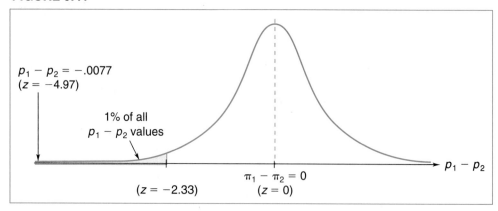

The results of such a study apply only to the population being sampled—in this case, male physicians. The conclusion does not necessarily apply to any other population such as male lawyers or female physicians or all adults in general. Needless to say, comparable studies involving other population groups are currently underway.

Phrases such as *studies have shown* or *significantly increase or decrease* are frequently used in advertising claims for many products. The type of analysis we just did is typical of the statistical procedures used to justify such statements. We illustrate the details of such a study in Figure 9.42 for the drug Seldane which is intended to provide relief for allergy sufferers.

We can also modify the ideas in this section to construct a confidence interval to estimate the difference in proportions for two populations based on samples drawn from each. The resulting confidence interval is

$$(\hat{p}_1 - \hat{p}_2) \pm z \sqrt{\bar{p}(1-\bar{p})\left(\frac{1}{n_1} + \frac{1}{n_2}\right)}$$

provided that the sample sizes are larger enough. We illustrate this in the following example.

FIGURE 9.42 Experience from clinical studies, including both controlled and uncontrolled studies involving more than 2,400 patients who received Seldane, provides information on adverse experience incidence for periods of a few days up to six months. The usual dose in these studies was 60 mg twice daily, but in a small number of patients, the dose was as low as 20 mg twice a day, or as high as 600 mg daily.

In controlled clinical studies using the recommended dose of 60 mg b.i.d., the incidence of reported adverse effects in patients receiving Seldane was similar to that reported in patients receiving placebo. (See Table below.)

ADVERSE EVENTS REPORTED IN CLINICAL TRIALS

Adverse Event	Percent of Patients Reporting				
	Controlled Studies*			All Clinical Studies**	
	Seldane N=781	Placebo N=665	Control N=626***	Seldane N=2462	Placebo N=1478
Central Nervous System					
Drowsiness	9.0	8.1	18.1	8.5	8.2
Headache	6.3	7.4	3.8	15.8	11.2
Fatigue	2.9	0.9	5.8	4.5	3.0
Dizziness	1.4	1.1	1.0	1.5	1.2
Nervousness	0.9	0.2	0.6	1.7	1.0
Weakness	0.9	0.6	0.2	0.6	0.5
Appetite Increase	0.6	0.0	0.0	0.5	0.0
Gastrointestinal System					
Gastrointestinal Distress (Abdominal distress, Nausea, Vomiting, Change in Bowel habits)	4.6	3.0	2.7	7.6	5.4
Eye, Ear, Nose, and Throat					
Dry Mouth/Nose/Throat	2.3	1.8	3.5	4.8	3.1
Cough	0.9	0.2	0.5	2.5	1.7
Sore Throat	0.5	0.3	0.5	3.2	1.6
Epistaxis	0.0	0.8	0.2	0.7	0.4
Skin					
Eruption (including rash and urticaria) or itching	1.0	1.7	1.4	1.6	2.0

*Duration of treatment in "CONTROLLED STUDIES" was usually 7-14 DAYS.
**Duration of treatment in "ALL CLINICAL STUDIES" was up to 6 months.
***CONTROL DRUGS: Chlorpheniramine (291 patients), d-Chlorpheniramine (189 patients), Clemastine (146 patients).

EXAMPLE 9.28 A study is conducted to compare the proportions of satisfied new-car buyers for domestic and foreign models. In a sample of 600 buyers of new domestic cars, 483 reported they were highly satisfied with their cars. In a sample of 450 buyers of new foreign cars, 352 indicated high satisfaction. Construct a 95% confidence interval for the difference in the two proportions.

SOLUTION The sample data give

$$n_1 = 600 \qquad n_2 = 450$$
$$x_1 = 483 \qquad x_2 = 352$$
$$p_1 = \frac{483}{600} = .805 \qquad p_2 = \frac{352}{450} = .782$$

Since n_1 and n_2 are both large, we can use a normal distribution with pooled proportion

$$\bar{p} = \frac{483 + 352}{600 + 450} = \frac{835}{1050} = .795$$

so that

$$\sigma_{p_1 - p_2} \approx \sqrt{(.795)(.205)\left(\tfrac{1}{600} + \tfrac{1}{450}\right)}$$
$$= \sqrt{(.795)(.205)(.0039)}$$
$$= \sqrt{.0006356} = .025$$

Therefore, as shown in Figure 9.43, the 95% confidence interval for the difference in proportions is

$$(.805 - .782) \pm 1.96(.025) = .023 \pm .049$$

or

$$[-.026, .072]$$

FIGURE 9.43

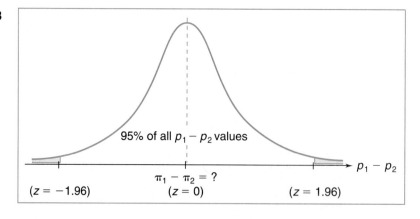

Computer Corner

Program 31: Dist'n of Diff'ce of Proportions allows you to explore some of the properties of the distribution of the differences of sample proportions. You can define two different binomial populations by entering values for n_1 and π_1 for the first and n_2 and π_2 for the second. (You can use the same population twice, if you prefer.) You also have to provide the number of different sets of samples you desire.

The program then generates a random binomial sample of size n_1 drawn from the first population and a second random binomial sample of size n_2 drawn from the second population. It calculates the sample proportions of successes p_1 and p_2 and then calculates and graphs the difference in proportions $p_1 - p_2$. In this way, the program simulates the sampling distribution for the difference of sample proportions. A typical output from the program is shown in Figure 9.44. Does the distribution appear to be roughly normal?

After the graph is complete, an accompanying text screen displays the actual numerical results and compares them to the theoretical results.

We encourage you to experiment with this program to see the effects of different values for n_1 and n_2 or for π_1 and π_2 on the distribution of the differences of sample proportions. For example, suppose you use the same value for π_1 and π_2. Will p_1 and p_2 always be the same?

FIGURE 9.44

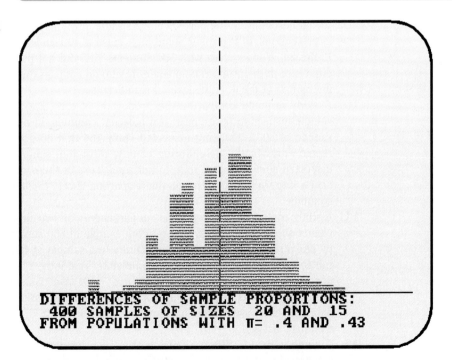

Exercise Set 9.7

Mastering the Techniques

Test each of the following hypotheses at the $\alpha = .05$ level of significance using the given values.

1. $H_0: \pi_1 - \pi_2 = 0$, $H_a: \pi_1 - \pi_2 \neq 0$, $n_1 = 100$, $x_1 = 55$, $n_2 = 100$, $x_2 = 50$
2. $H_0: \pi_1 - \pi_2 = 0$, $H_a: \pi_1 - \pi_2 \neq 0$, $n_1 = 100$, $x_1 = 63$, $n_2 = 150$, $x_2 = 57$
3. $H_0: \pi_1 - \pi_2 = 0$, $H_a: \pi_1 - \pi_2 \neq 0$, $n_1 = 60$, $x_1 = 22$, $n_2 = 80$, $x_2 = 30$
4. $H_0: \pi_1 - \pi_2 = 0$, $H_a: \pi_1 - \pi_2 > 0$, $n_1 = 250$, $x_1 = 125$, $n_2 = 400$, $x_2 = 195$
5. $H_0: \pi_1 - \pi_2 = 0$, $H_a: \pi_1 - \pi_2 < 0$, $n_1 = 80$, $x_1 = 32$, $n_2 = 80$, $x_2 = 40$
6. $H_0: \pi_1 - \pi_2 = 0$, $H_a: \pi_1 - \pi_2 > 0$, $n_1 = 480$, $x_1 = 125$, $n_2 = 500$, $x_2 = 120$

Applying the Concepts

7. In a pre-election poll, 40 of 100 prospective voters favor a certain candidate. In a follow-up poll one week later, 68 of 150 different people prefer this candidate. Does this represent a change in the proportion of voters who support this candidate at the $\alpha = .05$ level of significance?
8. Two new synthetic proteins are being tested to compare their effectiveness in preventing frost damage to citrus trees. In a test of protein A on 200 orange trees, 120 trees were pronounced as not having suffered frost damage during a cold spell. In a simultaneous test of protein B on 100 different, nearby trees, 51 trees were pronounced as not having suffered damage. Test if there is a difference in the proportions at the $\alpha = .05$ significance level.
9. Two different professors are giving the same introductory psychology course to large sections. In Professor Steinberg's class of 400 students, 80 did not pass. In Professor Wong's class with 500 students, 125 did not pass. Test if there is a difference in the proportion of students who do not pass this course with the two instructors at the $\alpha = .05$ level of significance.
10. A study is made comparing the proportion of cars that fail a state-mandated vehicle inspection program in two adjacent counties. In one county, a random sample of 300 cars being inspected show 164 that do not pass without some repair work. In the neighboring county, a random sample of 500 cars being inspected show 284 cars that do not pass. Test, at the $\alpha = .05$ level of significance, if there is a difference in the proportion of cars that fail the test in the two counties.
11. A study compares the proportions of passengers on two airlines who issue complaints about their flights. In a random sample of 600 comment cards received by one airline, 322 contained complaints. In a sample of 480 cards received by the second airline, 227 contained complaints. Test if there is a lower proportion of complaints at the second airline at the $\alpha = .01$ level of significance. (Note that the results of such a study may not adequately reflect all passengers since the data used are based solely on those passengers who bothered to respond on a voluntary basis.)
12. A study is made of the proportion of women versus men who earn academic scholarships at a large university. In a sample of 250 women, 145 have scholar-

ships. In a sample of 320 men, 150 have scholarships. Can you conclude that the proportion of women with scholarships is higher than that for men at the 5% level of significance?

13. A study is conducted comparing the proportions of high school students who drop out of school prior to graduation in two different communities in the same state. In the first school district, out of a random sample of 600 students studied, 90 dropped out. In the second school district, 48 out of a random sample of 400 students dropped out. Can you conclude that the proportions of dropouts in the two districts are different at the 2% significance level?

14. A city's transportation department is studying the effects of an increase in fare for mass transit. Prior to the fare increase, a random sample of 400 commuters in the city contain 248 who use public transportation (as opposed to cars) to commute to and from work. After the fare increase, a random sample of 480 commuters contain 265 who use public transportation. Can you conclude that the fare increase resulted in a decline in the proportion of commuters who use mass transit at the 5% level of significance?

15. Use the sample information in Exercise 7 to construct a 95% confidence interval for the difference in the proportions of voters who support the candidate.

16. Use the sample information in Exercise 9 to construct a 98% confidence interval for the difference in the proportions of students who do not pass the introductory psychology course with the two professors.

9.8 Summary of Hypothesis Testing

In each section in this chapter, we have introduced a hypothesis test for a different type of situation. The procedures used to test the various types of hypotheses are quite straightforward and very similar to one another. In fact, the greatest difficulty that many students encounter when solving such problems lies in distinguishing one type of situation from another. Once students identify the type of problem, the solution is usually easy to complete.

In this section, we summarize all the hypothesis test situations covered so far to make it easier for you to identify which procedure applies in a given problem. The interrelationships between the cases are shown in Figure 9.45 and the essential information is arranged in Table 9.4. The key to determining which type of situation a particular problem falls into depends most importantly on a careful reading of the problem itself.

Does the question refer to one or two sets of data? If one set, does it refer to means (averages) or to proportions?

If it refers to means, then the values given include a sample size n, a sample mean $\bar{x}$, and most likely a sample standard deviation s.

If it refers to proportions, then the values given include a sample size n and either a sample proportion p (as a decimal or a percentage) or the number of successes x out of n trials.

If there are two sets of data, are they independent or dependent? If they are independent sets of sample data, then you must conduct a difference-of-means or difference-of-proportions test. If the two sets of data

FIGURE 9.45

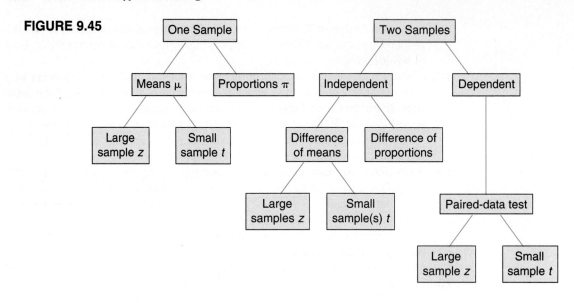

TABLE 9.4

Situation	Sample Statistics	Conditions	Standard Deviation	Test Statistic
Mean μ, large sample	$n, \bar{x}, s$	$n > 30$	$\dfrac{s}{\sqrt{n}}$	$z = \dfrac{\bar{x} - \mu}{s/\sqrt{n}}$
Mean μ, small sample	$n, \bar{x}, s$	$n \leq 30$ Population nearly normal	$\dfrac{s}{\sqrt{n}}$	$t = \dfrac{\bar{x} - \mu}{s/\sqrt{n}}$
Proportion	$n, x,$ or n, p	$n\pi \geq 5$ $n(1 - \pi) \geq 5$	$\sqrt{\dfrac{\pi(1 - \pi)}{n}}$	$z = \dfrac{p - \pi}{\sqrt{\pi(1 - \pi)/n}}$
Difference of means, large samples	$n_1, \bar{x}_1, s_1$ $n_2, \bar{x}_2, s_2$	$n_1 > 30, n_2 > 30$	$\sqrt{\dfrac{s_1^2}{n_1} + \dfrac{s_2^2}{n_2}}$	$z = \dfrac{\bar{x}_1 - \bar{x}_2}{\sqrt{s_1^2/n_1 + s_2^2/n_2}}$
Difference of means, small samples	$n_1, \bar{x}_1, s_1$ $n_2, \bar{x}_2, s_2$	$n_1 < 30$ or $n_2 < 30$ Populations nearly normal, equal variances	$\sqrt{\left(\dfrac{1}{n_1} + \dfrac{1}{n_2}\right)\left[\dfrac{(n_1 - 1)s_1^2 + (n_2 - 1)s_2^2}{n_1 + n_2 - 2}\right]}$	$z = \dfrac{\bar{x}_1 - \bar{x}_2}{sd}$
Paired data	$x_1, x_2, \ldots, x_n$ $y_1, y_2, \ldots, y_n$	Population nearly normal	$\dfrac{s_d}{\sqrt{n}} = \dfrac{\sqrt{\Sigma(d - \bar{d})^2/(n - 1)}}{\sqrt{n}}$	$t = \dfrac{\bar{d}}{\left(\dfrac{s_d}{\sqrt{n}}\right)}$
Difference of proportions	n_1, x_1 n_2, x_2	$n = n_1 + n_2$ $\bar{p} = \dfrac{x_1 + x_2}{n_1 + n_2}$ $\bar{q} = 1 - \bar{p}$ $n\bar{p} \geq 10$ $n\bar{q} \geq 10$	$\sqrt{\bar{p}\,\bar{q}\left(\dfrac{1}{n_1} + \dfrac{1}{n_2}\right)}$	$z = \dfrac{p_1 - p_2}{\sqrt{\bar{p}\,\bar{q}\left(\dfrac{1}{n_1} + \dfrac{1}{n_2}\right)}}$

are dependent, such as before and after values, then you must use a paired-data test.

Keep in mind that the type of sample information given in a problem virtually dictates the type of problem.

Furthermore, you should be extremely alert to the phrasing of the final question so that you can determine if the hypothesis test is one- or two-tailed. Phrases such as *too high, overstated, too low, increase, decrease,* and so forth all indicate a one-tailed test. In all such instances, ask yourself what kind of data would convict the null hypothesis. If you suspect the claim is too high, use < and the left tail. If you suspect the claim is too low, use > and the right tail. Alternatively, phrases such as *different from, valid, correct,* and so forth all suggest a two-tailed test. In all cases, use common sense.

Finally, when you face a series of problems that cover a variety of cases (such as those at the end of this section or on a test), we suggest the following strategy. It is much easier to distinguish among types of situations if you read through all the problems first and make an initial identification of each question *before* you attempt to solve any of them. When you first look at such a varied series of questions, you have a certain perspective which makes it easier to perform this identification. Once you become immersed in any single problem, you may lose much of that overall perspective in the process of doing the calculations, and the different questions may start to look all the same.

Exercise Set 9.8

Applying the Concepts

1. The official unemployment rate for a certain state is 7%. A particular community in the state believes that it has a higher unemployment rate. To determine if this is so, the local government selects a random sample of 500 people between the ages of 18 and 65 years and finds that 41 of them are unemployed. Test if the actual proportion of unemployed persons in this community is higher than the claimed percentage at the $\alpha = .05$ level of significance.

2. During contract negotiations, a college claims that the mean salary paid to its professors is $42,000 a year. The union selects a random sample of 36 faculty members and finds that the mean salary is $38,500 with a standard deviation of $6900. Test whether the administration's claim is too high at the $\alpha = .02$ level of significance.

3. To justify raising airfares, an airline claims that only 64% of all the seats on its flights are sold. A sample of 50 of its flights are randomly selected, and it is found that 72% of the seats are filled. Test whether the claimed proportion of sold seats is too low at the $\alpha = .05$ level of significance.

4. A correctly inflated basketball, when dropped from a height of 10 feet, is supposed to bounce back to a mean height of 6 feet. A sample of 25 basketballs from a new manufacturer are randomly selected and subjected to this bounce test. They bounce back to an average height of 5.6 feet with a standard deviation of .8 foot. Test if these basketballs differ from the bounce standard at the 5% significance level.

5. A company requires that circular holes with radius 0.2 inch be drilled through a certain type of metal plate. A new laser drill is tested by having it drill 100 such holes in a plate. The mean radius of these holes is .207 inch with a standard deviation of .024 inch. Test, at the 1% level of significance, whether the laser drill can be expected to produce holes of the desired radius.

6. Suppose the laser drill in Exercise 5 is being compared to a standard mechanical drill. The mechanical drill makes 120 holes in the same metal plate, and these holes have a mean radius of .197 inch with a standard deviation of .084 inch. Test whether the two types of drill produce different radii at the $\alpha = .05$ level of significance.

7. A study claims that, in a certain demographic group in a particular city, 32% of all people have been exposed to the AIDS virus. A random sample of 300 people in this group is tested, and 110 of them test positive for the virus. Determine if the claim is correct for this population group at the $\alpha = .02$ significance level.

8. A study is being conducted to compare the relative proportions of individuals from a certain demographic group in two different cities who have been exposed to the AIDS virus. In the first city, 110 out of a sample of 300 people in this group test positive. In the second city, 56 out of 175 people in the group test positive. Is there a difference in the proportions of people in these two cities in this group who have been exposed to AIDS at the $\alpha = .01$ significance level?

9. The Air Force orders 15 prototypes of a new air-to-ground missile to be built. These missiles are designed to land within 4 feet of a target. In a series of tests of these missiles, they landed a mean distance of 5.2 feet from the target with a standard deviation of 2.1 feet. Based on these data, should the Air Force accept these missiles as meeting the indicated standards? Test this at the $\alpha = .01$ level of significance.

10. The missiles in Exercise 9 are being compared to a different set of 20 prototypes developed by a competing company. In a comparable series of tests, the second set of missiles landed a mean distance of 4.3 feet from the target with a standard deviation of 2.4 feet. Test if there is a difference between them at the 5% level of significance.

11. A tire manufacturer is testing two new tread designs in terms of stopping distance. To do this, the company uses two test cars driving side by side at the same speed. Both cars have automatic braking systems so that both sets of brakes engage simultaneously on a signal. The stopping distances in feet are recorded as follows:

Car A	120	142	131	155	126	135	143
Car B	134	152	122	163	131	122	170

Test whether there is a difference in stopping distance between the two tread designs at the $\alpha = .05$ level of significance.

12. A company which does a large volume of business by mail decides to test whether there is a difference in mail delivery between those items brought to a post office as compared to those put in a corner mailbox. Out of 100 letters sent to customers in the same city which were mailed from the post office on a par-

ticular day, 92 were received the following day. Out of 250 letters that were sent from the corner mailbox, 217 were received the following day. Test if there is a difference in the proportions of letters that are delivered within 1 day based on the point of mailing at the 5% level of significance.

Student Projects

Conduct one of the following statistical studies:

1. The mean height for adult U.S. males is 5 feet 10 inches. Select a random sample of at least 30 adult men either on your campus or in the neighboring community, and use it to test whether the belief that $\mu = 70$ inches is accurate for the population sampled.
2. The mean height for adult U.S. females is 5 feet $4\frac{1}{2}$ inches. Select a random sample of at least 30 adult women either on your campus or in the neighboring community, and test if the heights of the women in this population differ significantly from those in the U.S. population.
3. Research the proportion of registered voters in your state or community who are registered with a particular party. (The local Board of Elections or the local Democratic or Republican party headquarters should be able to supply you with this information.) Then select a random sample of at least 50 adults either on your campus or in the neighboring community, and poll them as to their political affiliation. Use these data to conduct a hypothesis test to determine whether the claimed proportion is correct.
4. Continue the investigation you began as a project in Chapters 2 and 3. Select a new random sample of at least 30, and conduct a difference-of-means test to determine if there is any difference (change) in the means since the first sample was taken.
5. Select a pair of random samples from two related groups, and use the statistical data to conduct a difference-of-means test to determine if there is a difference between the means of the two groups. For example, you might use
   (a) Batting averages in the American League versus in the National League or batting averages this year versus those in some prior year (say 1970)
   (b) The acceleration time needed for cars to go from 0 to 50 miles per hour for current models compared to models in some previous year
   (c) The amounts paid by customers in a store on a charge card versus the amounts paid by personal check
   (d) The age of admission to hospital of men versus women
   (e) The length of hospitalization of smokers versus nonsmokers with similar diseases
   (f) The number of hours that women sleep per night versus the number of hours that men sleep

(g) The prices of homes or rents for apartments in two neighboring communities

6. Select two random samples consisting of at least 50 people in each group to conduct a difference-of-proportions test. For example, you might use
   (a) The proportion of men smokers versus women smokers
   (b) The proportion of student smokers versus nonstudent smokers
   (c) The proportion of women versus men (or students versus nonstudents) who agree with a stand on some social or political issue, such as abortion, gun control, defense policies, or candidate preference

7. Select a random sample of at least 30 items in a store such as a supermarket and note their prices. Find the prices of the same items at a different store, and conduct a paired data test to determine whether there is a mean difference in the prices at the two stores. (You may want to avoid "sale" items.)

For whichever of these projects you choose, prepare a formal statistical project report. The report should include

- A statement of the topic being studied.
- The source of the data.
- A discussion of how the data were collected and why you believe that it is a random sample.
- A statement of the underlying assumptions for the hypothesis test you use and why you think they are fulfilled.
- A statement of the null and alternate hypotheses and an appropriate level of significance α you will use in performing the hypothesis test.
- A list of the data and any frequency distributions and graphical displays you feel are needed.
- All the statistical calculations needed to analyze the data and to perform the hypothesis test.
- A discussion of any surprises you may have noted in connection with collecting and organizing the data or with the results of the hypothesis test. In particular, if you find that there is some difference, then you should attempt to explain it or account for it.

CHAPTER 9 SUMMARY

Hypothesis testing is used to make decisions about a population parameter. The **null hypothesis** H_0 asserts a claim or supposed fact about the population parameter being tested. The **alternate hypothesis** H_a asserts your suspicion about the claim. The test may be **one-tailed** (the claim is too high or too low) or **two-tailed** (the claim is wrong). You use a set of sample data to make a decision as to whether to reject the null hypothesis.

In making your decision, there is always an element of risk or **significance level** α which represents the least likely set of possible values for the sample statistic. This determines the **critical value(s)**, either z or t, thus forming the **rejection region** associated with the test. If the value for the **test statistic** falls in the rejection region, you conclude that this is too unlikely to have happened by chance alone, and so you **reject** the null hypothesis. If the test statistic does not fall into the rejection region, then you conclude that the sample evidence is not strong enough to disprove the null hypothesis, so you **fail to reject** H_0. Failure to reject the null hypothesis does not mean that you accept it. Specific details on the various cases considered in this chapter are summarized in Section 9.8.

To perform a hypothesis test, use the following steps:

1. State the null and alternate hypotheses.
2. State the level of significance α.
3. List the statistics from the sample data: n, $\bar{x}$, and s for a single mean; n, x, and p for a single proportion; etc.
4. Decide on the sampling distribution (z or t).
5. Find the critical value(s) to determine the rejection region.
6. Calculate the test statistic for the sample data.
7. Make a decision.

The *P*-**value** or **probability value** represents the probability of obtaining the sample value or worse by chance alone if the null hypothesis is true.

If you have two dependent samples, you must use the **paired-data test** instead of the **difference-of-means test**.

You can also construct confidence intervals for the difference of means or the difference of proportions.

10 Correlation and Regression Analysis

A consumer advocacy group has challenged the city's plan to raise subway and bus fares next year. The group, United Straphangers, claims that any fare increase will be counterproductive. They point to past fare increases which had the effect of reducing the number of daily commuters using mass transportation. They conclude that any future hikes in the fare will further diminish the ridership, and so actually reduce the income to the transit system. In turn, this will increase the number of commuters who use cars to go to work in the city and so worsen the city's air pollution, parking and driving problems.

The argument used here is based on the fact that two quantities are apparently related and that a change in one produces a consequent change in the other. But how do we know that there is indeed a relationship between two quantities? Further, if there is a relationship, how can we determine precisely what the relationship is? It is not adequate merely to decide that commuter ridership decreases when fares go up. If we are to make an intelligent argument, then we must have a way of knowing exactly what the relationship is in an algebraic sense so that we can predict precisely what the most likely result of a change of fare will be. It is not enough to argue in generalities.

For example, will *any* increase in fare, large or small, lead to a decrease in ridership? Can we predict how large the drop in ridership will be? Is it evident that the decrease in ridership due to a fare increase will be so large that it will wipe out all increased revenue? Certainly, if the drop in ridership is very large, this will be the case. But what if there is only a relatively small drop? In that case, the argument is specious. We can only make intelligent decisions on such matters if we know the exact relationship between the two quantities.

Moreover, even if there is a formula to express the relationship between two such quantities, how far can we extend it? That is, such a relationship may suggest that if fares are raised high enough, there will be a negative ridership. Does that make sense?

Finally, just because two quantities are related, does that really mean there is necessarily a cause-and-effect relationship? For instance, if that were the case, we could conclude that reducing the fare would increase the ridership dramatically. Will this necessarily happen or might there be other factors, such as convenience and safety concerns, that might keep commuters in their cars regardless of the fare?

In this chapter, we will consider ways to determine whether two quantities are in fact related. If they are, we will learn how to determine a formula to express that relationship mathematically and how to use that formula to predict the value of one quantity depending on the value of the other.

10.1 Introduction to Correlation and Linear Regression

Some of the most useful statistical tools are the ones which allow us to make comparisons between sets of data. In Chapter 9, we studied the difference-of-means test, the paired-data test and the difference-of-proportions test. In each case, we essentially compared two measurements of the same quantity. In a baseball context, we would use the difference-of-means test to compare the average number of home runs hit by players in one league to those hit by players in the other league. To do this, we use two independent samples consisting of the number of home runs per player. Alternatively, we would use the paired-data test to compare the number of home runs hit by the same player during two consecutive seasons. In this case, there are two dependent measurements on each player.

In this chapter, we consider situations in which there are two distinct, but possibly related, quantities for each individual. In particular, we consider ways in which we can determine whether the two different quantities are related to each other and, if so, how we can find a relationship and use it in a predictive manner. For instance, we might want to determine if there is any relationship between the number of home runs hit by a player and the number of RBIs he has or the number of times he strikes out. Thus, we have measurements on two distinct quantities involving each player in a sample. Are they related? If so, can we use the relationship to predict an individual player's performance in one category based on the other? The methods we use in solving such problems are known as correlation and regression analysis.

To illustrate these ideas, suppose that a college seeks to use a student's high school average (HSA) to predict his or her grade-point average (GPA) after the freshman year in college. Thus, there are two measurements for each student involved, an HSA and a GPA. First the college must determine if there is any relationship between HSAs and GPAs; if there is, then the college must find a formula expressing this relationship. Finally, the college will be able to use the formula to predict the most likely GPA for any given entering student based on his or her HSA.

> In any correlation and regression analysis situation, there are three stages:
>
> 1. Determine whether the two quantities are related and the degree of relationship, if any.
> 2. Assuming that there is a high degree of relationship, find a formula expressing the relationship.
> 3. Use this relationship to predict the most likely value of the second variable corresponding to any given value of the first variable.

Each of these interrelated ideas involves lengthy development and the use of what appear at first to be quite complicated formulas. We therefore present an overview of the associated concepts in this section before proceeding to the details in upcoming sections.

To illustrate these ideas, let's look at a particular example. Suppose we collect the following data on a group of randomly selected students in a college:

Student	HSA	GPA
1	80	2.4
2	85	2.8
3	88	3.3
4	90	3.1
5	95	3.7
6	92	3.0
7	82	2.5
8	75	2.3
9	78	2.8
10	85	3.1

For the most part, it seems that students with solid HSAs get high GPAs, while students with low HSAs get relatively lower GPAs. However, to gain a better perspective on just how these two sets of values are interrelated, it is generally preferable to display the values graphically. We show the result of graphing these 10 points in Figure 10.1. This is known as a *scattergram* or *scatterplot* of the data. From this picture, it is much easier to see how the points are arranged. In fact, by looking at the scattergram, it is clear that the 10 points more or less lie along a straight line. Of course, it is also obvious that no single straight line will pass through all 10 points.

Whenever a set of points in a scattergram appear to lie in a linear pattern such as this, we suspect that there is indeed a linear relationship between the two variables. What we need is a way of measuring the degree of relationship or *correlation* between the two quantities. The measure we use is known as the *correlation coefficient* and is denoted by r. This quantity has a numerical value between -1 and 1.

FIGURE 10.1

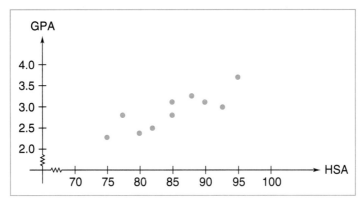

If the value for *r* comes out close to 1, then there is a strong *positive correlation* between the two quantities. (In Figure 10.2a, notice that the points are tightly clustered in a linear pattern.) On the other hand, if the value for *r* is close to −1, then there is a strong *negative correlation* between the variables. (In Figure 10.2b, notice that the points are tightly clustered about a line having negative slope.) If *r* is close to 0, then there is little correlation between *x* and *y* and they are unrelated. (In Figure 10.2c, the points are spread out seemingly randomly throughout the scattergram.) If *r* is positive but moderately small, then there is a slight positive correlation between *x* and *y*. (In Figure 10.2d, the points tend to fall in a linear pattern, but they are not tightly clustered.) Finally, if the correlation coefficient is somewhat negative, there is a slight negative correlation. (In Figure 10.2e, the points fall in a downward-trending linear pattern, but are not tightly clustered.)

Studies have shown that, in general, there is a high degree of positive correlation between the time a student devotes to studying statistics and the grades earned on the exams. Further, there is a high degree of negative correlation between the literacy rate in a country and the infant mortality rate. In general, the higher the literacy rate, the lower the infant mortality rate; the lower the literacy rate, the higher the infant mortality rate. Finally, there is no correlation between the number of keys that a person carries and the number of credit cards that he or she owns.

In the next section, we will see how to calculate the value for the correlation coefficient for any set of paired data. Moreover, we will also see that it is important to be able to determine whether the level of correlation is significant in a statistical sense or if there is really no meaningful correlation.

Suppose, then, that there is a high degree of correlation between two variables *x* and *y*. This suggests that there is a linear relationship between them. However, it is evident that we cannot connect all the points in a scattergram, such as the one shown in Figure 10.1, with a single straight line. Obviously, we can draw a great number of different lines which seem to be "close to" all the points in the scattergram, as shown in Figure 10.3. We want

FIGURE 10.2

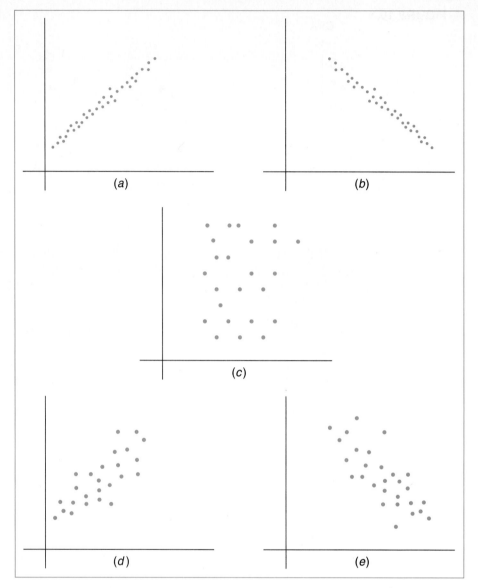

to find *the* one particular line which comes closest, in some sense, to *all* the points in the scattergram.

It can be shown, by using calculus, that there always is one single line which is better than any other possible line. It is based on the idea that the sum of the squares of the vertical distances between every data point and this line is a minimum. See Figure 10.4. This particular line is known as the *regression line* or the *least-squares line*. We will see how to determine this line in a fairly simple way (certainly without the use of calculus!) in Section 10.3.

FIGURE 10.3

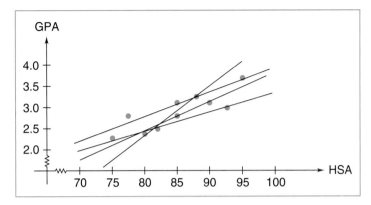

Incidentally, we note that the calculation of both the correlation coefficient and the regression line can be accomplished particularly easily by using one of the programs on your disk as well as with Minitab. We also discuss this in the next section. They can also be found with many calculators having statistical capabilities.

Finally, having found the equation of the regression line which is the best fit to a set of data, we want to use it to make predictions about individual members of the underlying population. For instance, if we have the equation for the line relating GPAs to HSAs, then we can estimate the most likely GPA for any college freshman based on her or his high school record. However, as we have seen previously when we considered confidence interval estimates for either a population mean or a proportion, there is always some degree of uncertainty involved in the prediction. Thus, we will also have to consider this problem of developing an appropriate estimate for the predictions. We do this in Section 10.4.

10.2 Correlation

In Section 10.1 we introduced the idea of studying data based on two possibly related quantities x and y. In any such instance, it is first essential to determine whether there is any degree of correlation between the two variables. Often we may suspect that such a relationship exists. The methods developed in

FIGURE 10.4

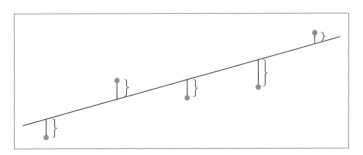

this section will allow us to verify that the suspicion is indeed correct or to find that it is incorrect. Sometimes we may even find that a high correlation exists between quantities which we would not expect to be related.

We use the correlation coefficient to measure the degree of relationship between two quantities. Let $\bar{x}$ and $\bar{y}$ represent the respective means of the x and y values from the sample data. The *correlation coefficient r* is then defined as

$$r = \frac{\Sigma(x - \bar{x})(y - \bar{y})}{\sqrt{\Sigma(x - \bar{x})^2} \sqrt{\Sigma(y - \bar{y})^2}} \tag{10.1}$$

Using algebra we can reduce this to a considerably easier formula for calculating r, namely,

$$r = \frac{n(\Sigma xy) - (\Sigma x)(\Sigma y)}{\sqrt{n(\Sigma x^2) - (\Sigma x)^2} \sqrt{n(\Sigma y^2) - (\Sigma y)^2}} \tag{10.2}$$

This formula is not as complicated as it looks at first glance if you look at each piece carefully. First, n represents the number of data pairs. In our example on student HSAs and GPAs, $n = 10$. Further, we let x represent a student's HSA and y represent the same student's GPA. The formula for r involves both Σx and Σy, the sums of the x values and y values, respectively, which are both very simple to calculate. The only pieces remaining in the above formula for r are Σxy, Σx^2, and Σy^2. The first, Σxy, involves calculating the product of each x and y value and then adding them. The second term, Σx^2, involves squaring each x value and then summing them. The third term, Σy^2, similarly involves squaring each y value and then summing them.

Keep in mind the following:

> The correlation coefficient r is always between -1 and 1.

If you ever get an answer outside this range, you have made an error in your calculations.

The simplest way to calculate the correlation coefficient by hand is to work in tabular form, much as we did in computing a standard deviation. In fact, as you will see below, the complexity of the work is no greater than in calculating the standard deviation. To do this work in a table, we require separate columns for each of the terms that we will eventually add: Σx, Σy, Σxy, Σx^2, and Σy^2. We illustrate this process in the following examples.

EXAMPLE 10.1 Determine the correlation coefficient for the data in Section 10.1 relating the HSA to the GPA.

SOLUTION We begin by setting up a table with the headings mentioned above, and we record the 10 sets of scores in the first two columns, as shown in Table 10.1. We then fill in the balance of the entries by forming the products xy and the squares x^2 and y^2 of these data values. For instance, in the first row

TABLE 10.1

x	y	xy	x^2	y^2
80	2.4	192.0	6,400	5.76
85	2.8	238.0	7,225	7.84
88	3.3	290.4	7,744	10.89
90	3.1	279.0	8,100	9.61
95	3.7	351.5	9,025	13.69
92	3.0	276.0	8,464	9.00
82	2.5	205.0	6,724	6.25
75	2.3	172.5	5,625	5.29
78	2.8	218.4	6,084	7.84
85	3.1	263.5	7,225	9.61
$\Sigma x = 850$	$\Sigma y = 29.0$	$\Sigma xy = 2{,}486.3$	$\Sigma x^2 = 72{,}617$	$\Sigma y^2 = 85.78$

where $x = 80$ and $y = 2.4$, the product is $xy = 80(2.4) = 192$, $x^2 = 6400$, and $y^2 = 5.76$. We finally sum the entries in each column, as shown.

Using the totals from these columns, we have

$$n(\Sigma xy) - (\Sigma x)(\Sigma y) = 10(2486.3) - (850)(29)$$
$$= 24{,}863 - 24{,}650$$
$$= 213$$

$$n(\Sigma x^2) - (\Sigma x)^2 = 10(72{,}617) - (850)^2$$
$$= 726{,}170 - 722{,}500$$
$$= 3670$$

and

$$n(\Sigma y^2) - (\Sigma y)^2 = 10(85.78) - (29)^2$$
$$= 857.8 - 841$$
$$= 16.8$$

Consequently, the correlation coefficient r is

$$r = \frac{n(\Sigma xy) - (\Sigma x)(\Sigma y)}{\sqrt{n(\Sigma x^2) - (\Sigma x)^2} \sqrt{n(\Sigma y^2) - (\Sigma y)^2}}$$

$$= \frac{213}{\sqrt{3670}\sqrt{16.8}}$$

$$= \frac{213}{(60.58)(4.10)}$$

$$= \frac{213}{248.38}$$

$$= .858$$

which suggests a fairly strong positive correlation between the HSA and GPA based on the data. ∎

Before we go on, let's look at some of the values in the calculation above. Notice that while most of the terms in the formula for the correlation coefficient r are extremely large (for instance, 726,170 and 722,500), they are close in size. Thus, when we perform the necessary subtractions, the resulting values are relatively small (for instance, $726{,}170 - 722{,}500 = 3670$). This is very typical of what happens whenever you calculate the correlation coefficient. Thus, if the terms you calculate by hand are not close in size, either in the numerator or within one of the square roots in the denominator, you should suspect an error in your work.

Moreover, recall that you can only take the square root of a positive quantity or zero. Therefore, if you ever get a negative term for either of the square roots in the denominator, you have made an error.

EXAMPLE 10.2 A study is conducted to determine if there is a relationship between the weight of a car and its gas mileage. The following set of values is obtained and listed in the form (weight in pounds, miles per gallon):

(2800, 19), (2650, 23), (2500, 27), (2450, 25), (2200, 32), (2300, 26), (2500, 22), (2600, 18)

Determine the correlation coefficient for these data.

SOLUTION We begin by drawing the scatterplot for the set of data, as shown in Figure 10.5. There certainly appears to be a linear relationship between the two quantities, and the overall slope is negative. However, this makes sense since we would expect gas mileage to decrease as the weight of a car increases.

To calculate the correlation coefficient, we organize the data entries in a table, as shown in Table 10.2, and then fill in the additional values corresponding to the product xy and the squared terms x^2 and y^2. We therefore find that

FIGURE 10.5

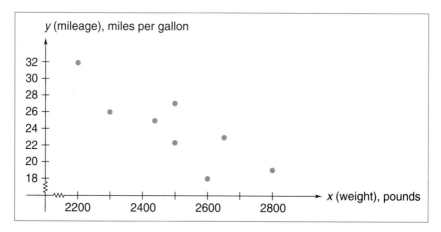

TABLE 10.2

x	y	xy	x^2	y^2
2,800	19	53,200	7,840,000	361
2,650	23	60,950	7,022,500	529
2,500	27	67,500	6,250,000	729
2,450	25	61,250	6,002,500	625
2,200	32	70,400	4,840,000	1,024
2,300	26	59,800	5,290,000	676
2,500	22	55,000	6,250,000	484
2,600	18	46,800	6,760,000	324
$\Sigma x = 20,000$	$\Sigma y = 192$	$\Sigma xy = 474,900$	$\Sigma x^2 = 50,255,000$	$\Sigma y^2 = 4,752$

$$n(\Sigma xy) - (\Sigma x)(\Sigma y) = 8(474,900) - (20,000)(192)$$
$$= 3,799,200 - 3,840,000$$
$$= -40,800$$

$$n(\Sigma x^2) - (\Sigma x)^2 = 8(50,255,000) - (20,000)^2$$
$$= 402,040,000 - 400,000,000$$
$$= 2,040,000$$

and

$$n(\Sigma y^2) - (\Sigma y)^2 = 8(4752) - (192)^2$$
$$= 38,016 - 36,864$$
$$= 1152$$

Hence, the correlation coefficient is

$$r = \frac{-40,800}{\sqrt{2,040,000}\,\sqrt{1152}}$$
$$= \frac{-40,800}{(1428.29)(33.94)}$$
$$= \frac{-40,800}{48,476.2}$$
$$= -.842$$

which suggests a strong negative correlation between the weight of a car and its gas mileage. ∎

The complexity of performing the above calculations makes this type of computation ideal for computer use. In fact, it is unrealistic to do this type of calculation by hand, especially if the data set is any larger than the ones we used in the above examples. For that reason, we now digress to consider how to utilize the computer to perform such calculations. We assume that you will use the computer from now on.

Computer Corner

Program 32: Linear Regression Analysis is designed to perform a complete regression and correlation analysis for any set of paired data (x, y). Thus, it includes several features which we have not fully explored yet.

To use the program, you must input your data pairs in the form

$$X, Y$$

where you are prompted first for x and then for y. Press Enter after each entry. Continue in this manner until you have entered all your data pairs. To indicate that you have finished, enter the value 9999 for x, and again for y after your last data entry. Once the data have been entered, the program asks if you wish to edit the values, in case you made some errors that you want to correct.

The program draws the scatterplot for your data and then calculates and draws the regression line based on your entries, as illustrated in Figure 10.6. The program also prints the equation of the regression line in slope-intercept form:

$$y = mx + b$$

Be careful to read the printed information regarding the range of values for x and y used in the display. We discuss this in detail in Section 10.3.

After the graph is complete, you can transfer to an accompanying text screen by pressing any key. In addition to reprinting the equation of the regression line, the program prints the value for the correlation coefficient r. It also prints the value for an error estimate, called the *standard error*, involved in using the regression line for predictions. We discuss this quantity in Section 10.4.

From the text screen, you can transfer back to the graph by pressing any key other than the space bar. The space bar will take you to the menu screen, where you will be presented with a variety of choices, including entering a new set of data, editing the present set of data (adding, changing or deleting points), having the results printed (just be certain the printer is On), or saving all results as a disk file (be sure that you do not use a previously used name for this file, or else the old data will be erased). We suggest that you experiment with any set of data to see the effects of making changes in it. For instance, if you have any points that lie relatively far from the regression line, see what happens to the value for r if you change the coordinates of those points to move them closer to the line.

FIGURE 10.6a

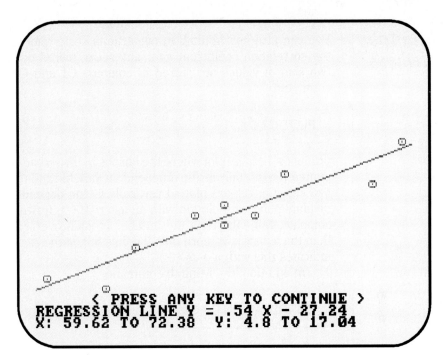

```
        < PRESS ANY KEY TO CONTINUE >
REGRESSION LINE Y = .54 X - 27.24
X: 59.62 TO 72.38    Y: 4.8 TO 17.04
```

FIGURE 10.6b

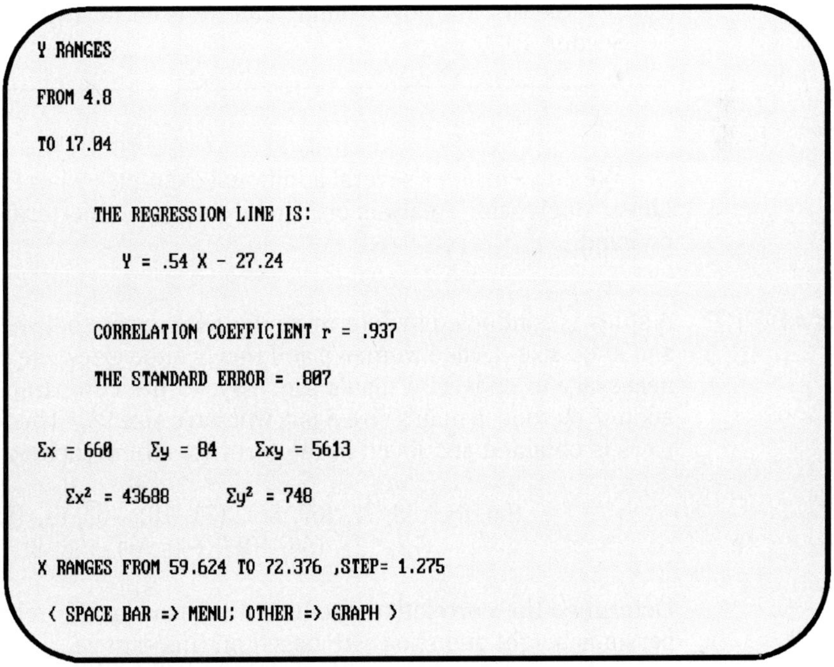

```
Y RANGES

FROM 4.8

TO 17.04


      THE REGRESSION LINE IS:

         Y = .54 X - 27.24


      CORRELATION COEFFICIENT r = .937

      THE STANDARD ERROR = .807

Σx = 660    Σy = 84    Σxy = 5613

   Σx² = 43688    Σy² = 748


X RANGES FROM 59.624 TO 72.376 ,STEP= 1.275

< SPACE BAR => MENU; OTHER => GRAPH >
```

Minitab Methods

You can also use Minitab to construct a scatterplot and to calculate the correlation coefficient r for any set of paired data. Suppose the two sets of values are stored in columns C1 and C2. The Minitab command

> PLOT C2 C1

produces a scatterplot where the entries in the *second* column, C1, are plotted horizontally as the independent variable and the entries in the *first* column, C2, are plotted vertically as the dependent variable. Individual points are plotted with an asterisk (*). If two or more points coincide, Minitab prints the number of such overlaps, say 4, rather than the asterisk. If more than 9 points fall on the same spot, Minitab indicates this with a plus (+).

In addition, the Minitab command

> CORRELATION C2 C1

causes the value for the correlation coefficient r to be calculated and displayed. As with the **PLOT** command, the first column listed, C2, is interpreted as containing the dependent variable, and the second column listed, C1, contains the independent variable. We illustrate both commands in the sample Minitab session in Figure 10.7.

We discuss how Minitab can be used for regression analysis in Section 10.3.

We now consider several additional examples where we use Program 32: Linear Regression Analysis or Minitab instead of performing the calculations by hand.

EXAMPLE 10.3 A study is conducted to determine the relationship between a person's height and shoe size. (Since women's and men's shoe sizes use different scales, it is necessary to convert a man's size, say, to the comparable woman's size by adding $1\frac{1}{2}$; thus, a man's size 8 is a woman's size $9\frac{1}{2}$.) The following set of data pairs is obtained and listed in the form (height in inches, shoe size):

$$(66, 9), (63, 7), (67, 8\tfrac{1}{2}), (71, 10), (62, 6), (65, 8\tfrac{1}{2}),$$
$$(72, 12), (68, 10\tfrac{1}{2}), (60, 5\tfrac{1}{2}), (66, 8)$$

Determine the correlation coefficient measuring the relationship between a person's height and shoe size based on this sample.

FIGURE 10.7

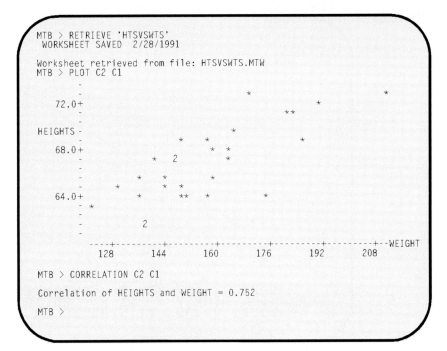

SOLUTION We first enter these data pairs in the program and examine the scatterplot shown in Figure 10.8. There certainly seems to be a linear relationship between the two sets of measurements. The program also provides the value for the correlation coefficient

$$r = .951$$

which suggests a very high degree of positive correlation between a person's height and her or his shoe size. ∎

EXAMPLE 10.4 Consider the first three digits of a person's telephone number (the exchange) and the first three digits of that person's social security number. A typical set of such data in the form (telephone exchange, social security digits) is as follows:

(532, 081), (665, 115), (243, 092), (728, 125),
(663, 097), (243, 134), (368, 130), (665, 084),
(665, 103), (584, 088), (665, 104), (243, 065)

We show the scatterplot for this set of data in Figure 10.9. From the diagram, we decide that the points do not seem to fall into any linear relationship, but rather appear to be spread randomly.

At first thought, there is absolutely no reason why these two quantities should be related. Nevertheless, there is always the possibility that some unanticipated relationship may exist, even between such apparently unrelated

FIGURE 10.8*a*

FIGURE 10.8*b*

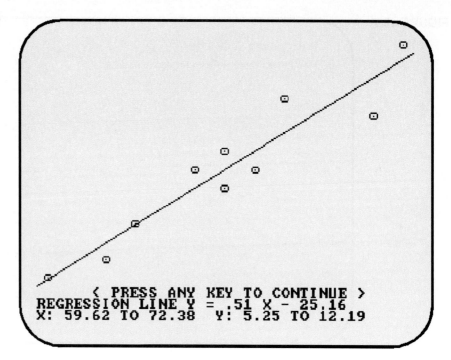

FIGURE 10.9

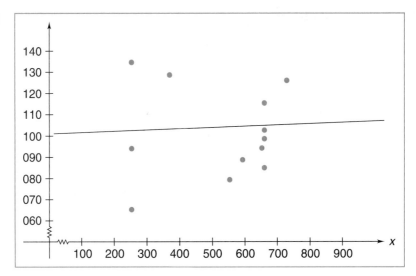

quantities as these. To determine if such a relationship does exist, we calculate the correlation coefficient using Program 32: Linear Regression Analysis. It yields a value for r of .057, which tends to verify our original conjecture that there is virtually no relationship between a person's telephone exchange and social security number. ∎

Many of the conclusions we drew in the preceding examples were quite obvious. In Example 10.3, the value obtained for the correlation coefficient r was .951, and this certainly suggests a very strong positive correlation between the two variables in the underlying population of heights versus shoe sizes of all people. In Example 10.4, the value for r was .057, which is so close to 0 that it suggests no significant relationship between the two quantities (a person's telephone number and social security number). However, in Examples 10.1 and 10.2, the values for r were .858 and $-.842$, respectively. Both values are reasonably close to 1 and -1, respectively, but are they close enough to justify our concluding that there is a high level of correlation between the two variables in each case? In other words, there obviously must be a point, possibly at $r = .85$ or $r = .75$ or $r = .60$ or $r = .48$, at which the degree of relationship between x and y is no longer considered high. The question is, What is the appropriate value for r that separates a high degree of correlation from an insignificant degree of correlation?

Before we answer this question completely, you should realize that the result must depend on the sample size n. If n is very small, then there is considerable opportunity for variation in the sample data. As a result, a high level of correlation requires that the value for r be very close to either 1 or -1. But if the sample size is large, then far less variation is likely, so that a smaller value for r would suffice to indicate a high level of correlation.

Furthermore, to answer this question, we must interpret the ideas in a somewhat broader context. The decision we wish to make hinges on whether there is correlation between two quantities x and y in the original population. The value we calculate for the correlation coefficient r is based on just one sample drawn from that underlying population. Therefore, we really want to ask, In the underlying population, is there any correlation between the two variables? We answer this question by using the data, namely the value for r, based on the single sample. If we consider all members of the original bivariate population, then we could theoretically calculate a correlation coefficient for the population; we denote this value by ρ (the Greek letter rho). If there is no correlation between these two quantities in the population, then $\rho = 0$. Otherwise, ρ has some nonzero value. We use the value for r from the single random sample as an estimate for the population parameter ρ and perform a hypothesis test to see whether there is any correlation between the variables. In this hypothesis test, the null hypothesis typically asserts that there is no correlation,

$$H_0: \rho = 0$$

while the alternate hypothesis states that

$$H_a: \rho \neq 0$$

so that we have a two-tailed test.

If we suspect that there is only positive correlation or only negative correlation, then we could use a one-tailed test and the corresponding alternate hypothesis would be $H_a: \rho > 0$ or $H_a: \rho < 0$, respectively.

The sample statistic we use to make a decision on whether to reject the null hypothesis is the sample correlation coefficient r that we calculate for the single sample.

To make this decision, we must compare the sample value r to an appropriate critical value based on the sampling distribution of the correlation coefficient. These critical values for r have been calculated and are displayed in Table V. They give us the cutoff points for an appropriate level of significance α. Notice that these critical values for r depend on the sample size n. The number of degrees of freedom (df) for the r statistic is 2 less than the sample size n:

$$df = n - 2$$

As the number of pairs of points used increases, the critical value for r decreases.

To illustrate how to use Table V, suppose we work with the results of Example 10.1, comparing HSAs to GPAs. We found $r = .858$ based on a sample size of $n = 10$. With a level of significance $\alpha = .05$, the corresponding entry from Table V is $r = .632$. This represents the *minimum* acceptable value of r which indicates a degree of correlation. Since our value $r = .858$ is considerably larger than this critical value, we conclude that there is correlation between the HSA and GPA at the $\alpha = .05$ level of significance based on this set of data. However, if our calculated value for r had been less than this critical value of .632, say $r = .528$, then we could not conclude that there was any correlation between the two quantities.

Furthermore, a similar analysis holds when the calculated value for r comes out negative. Because of symmetry, all we need do is to find the appropriate critical value of r from Table V and reverse its sign. If the value of r that we calculate for the data falls between this negative critical value and -1, we conclude that there is negative correlation. But if our calculated value for r falls between this negative critical value and 0, we cannot conclude that there is any correlation. For instance, suppose the calculated value for $r = -.723$. If the associated critical value is $-.632$, say, then we reject the null hypothesis that $\rho = 0$ and so conclude that there is negative correlation. If the critical value is $-.511$, then we cannot reject the null hypothesis and so cannot conclude that there is any correlation.

EXAMPLE 10.5 Use the results of Example 10.2 to determine if there is any correlation between the weight of a car and its gas mileage at the $\alpha = .05$ level of significance.

SOLUTION We begin with the null and alternate hypotheses:

$$H_0: \rho = 0$$
$$H_a: \rho \neq 0$$

Based on the set of data used in Example 10.2, we found that the sample correlation coefficient $r = -.842$. Since the number of data pairs is $n = 8$, we use the critical value $r = .707$ from the table but now reverse its sign to get $-.707$. For us to conclude that there is any correlation, the calculated value for r must be more negative than $-.707$; that is, it must be between $-.707$ and -1. Since we found $r = -.842$ for the sample, we reject the null hypothesis and so conclude that there is negative correlation between the two variables, car weight and gas mileage, at the 5% level of significance. ∎

One of the most common errors made with statistics involves an incorrect interpretation of correlation. Just because we establish that two quantities have correlation does not necessarily indicate that there is a cause-and-effect relationship between them. For instance, several studies have shown that there is high positive correlation between teacher salaries in the schools and the amount of alcohol consumed by the students. We certainly cannot conclude that either one causes the other. (It is unreasonable to think that lowering teachers' salaries would lower the students' drinking level; it is even more foolish to think that if students stopped drinking, then teacher salaries would drop.) Rather, we should realize that there are other pertinent factors in force which may be the cause of both occurrences. For example, the school might be in a very affluent area where the residents can afford taxes to pay high salaries and the students have lots of spending money to afford alcohol.

Computer Corner

We have included Program 33: Correlation Simulation on your disk to let you experiment with some of the ideas behind correlation analysis. The program includes a built-in population consisting of data pairs. This population has a correlation coefficient $\rho = .88$. The program allows you to see the effects of drawing different random samples from this bivariate population and graphing the correlation coefficient r for each sample. To use the program, you must enter the sample size you desire (a minimum of 3 and a maximum of 30) and the number of samples you want. The program first displays a vertical line corresponding to the correlation coefficient for the underlying population $\rho = .88$. It then selects the indicated random samples, calculates the correlation coefficient for each, and graphs the value for r. We include a sample output from the program in Figure 10.10. You may notice that this sampling distribution is not particularly bell-shaped (even when the sample size n is relatively large) and is typically skewed to one side of the population correlation coefficient ρ.

When the graph is complete, you can transfer to a text screen which displays some numerical results of the simulation. In particular, it gives the mean and standard deviation for all the sample correlation coefficients generated. We suggest that you experiment with this program by choosing different sample sizes to see the effects on the sampling distribution for the correlation coefficient.

FIGURE 10.10

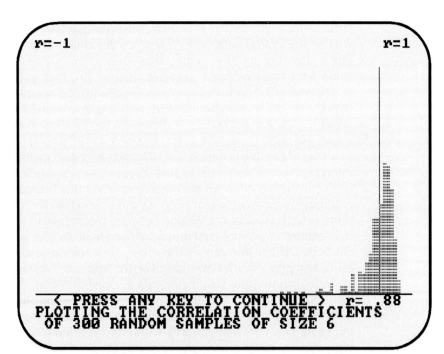

Exercise Set 10.2

Mastering the Techniques

Draw the scattergram and calculate (by hand) the correlation coefficient for the following sets of data:

1.
x	2	5	7	10	11
y	10	20	35	50	65

2.
x	5	12	15	18	25
y	30	50	45	65	90

3.
x	5	8	10	12	15	22
y	50	42	40	35	33	28

4.
x	2	10	15	18	27	33	35
y	80	66	60	52	47	53	27

5–8. For each of the sets of data in Exercises 1 through 4, determine if there is any correlation between the two variables at the 5% level of significance.

Applying the Concepts

9. A small company is interested in analyzing the effects of advertising on its sales. Over a 5-month period, it finds the following results

x	5	8	10	15	22
y	6	15	20	30	39

where x represents the money spent on advertising (in hundreds of dollars) and y represents the total sales (in thousands of dollars). Use these data to determine the correlation coefficient. Is there any correlation between sales and advertising at the 5% significance level?

10. A supermarket owner is studying how the average waiting time y in minutes for customer checkout depends on the number x of checkout clerks working. The results are

x	3	4	5	5	6	7
y	9	6	6	4	2	1

Use these data to determine the correlation coefficient. Is there any correlation between waiting time and the number of clerks on duty at the 5% level of significance?

11. A college collects the following set of data on the number of credits y that a randomly selected group of students carry and the number of hours x that they work during the week:

x	20	25	30	50	20	23
y	12	13	12	15	16	16

Find the correlation coefficient. Is there any correlation between the two quantities at the $\alpha = .05$ level of significance?

12. The college in Exercise 11 uses the above information on the number of hours x that students work to study the relationship between time spent on a job with y, the grade-point average. The results are as follows:

x	20	25	30	50	20	23
y	3.4	3.0	2.8	2.4	2.9	2.9

Find the correlation coefficient for these data. Does it indicate any correlation between the number of hours worked by students and their GPAs at the 5% level of significance?

13. A government study on energy conservation is conducted to determine the relationship between the price of home heating oil x (in cents) and the mean number of gallons y used per month during January in a variety of different communities with similar climates. The results are as follows:

x	75	80	86	90	95	98	106
y	120	125	114	110	112	106	97

Find the correlation coefficient that relates oil use to cost. Is there any correlation between the two at the 1% significance level?

14. A TV network is concerned about the high cost of producing many of its programs. It therefore conducts a study to relate the production costs for 30 minutes of programming x (in hundreds of thousands of dollars) to the ratings y that the program gets in a national ratings survey. The results are

x	1.2	1.6	1.8	2.5	2.7	3.0	3.5	4.4
y	3.3	3.9	5.7	4.2	4.5	8.2	6.1	4.6

Find the correlation coefficient for the ratings and the production costs. Is there a correlation between the two at the 1% level of significance?

15. Show that formula (10.2) for the correlation coefficient r is equivalent to the defining formula (10.1) by expanding (10.1) algebraically.

Computer Applications

16. From the data on major league baseball results in Appendix I, select a random sample of 25 players in either league and record each player's batting average and number of home runs. Use Program 32: Linear Regression Analysis to find

the correlation coefficient. Test whether there is any correlation between these two variables at the 5% level of significance.
17. Repeat Exercise 16 with a comparison between the player's number of home runs and number of RBIs. Is there any correlation between them at the 5% level of significance?
18. Select a random sample of 25 stocks from the stock lists in Appendix III and record the closing price of each stock and the volume (number of shares sold) in hundreds, given in the second column of the list. Use the program to determine the correlation coefficient. Test whether there is any correlation between the two at the 5% level of significance.
19. Select a random sample of 30 cities from the weather listings in Appendix II. Record the high and low temperatures for the given day, and use the program to determine the correlation coefficient. Test whether there is any correlation between high and low temperatures at the 5% level of significance.

10.3 Linear Regression

In Section 10.2, we introduced the correlation coefficient as a measure of the degree of relationship between two variables. Suppose we have two quantities which have nonzero correlation, so that we suspect that they are related in a linear fashion. We now seek an algebraic formula to express this linear relationship. Once we have such an expression, we can use it to make predictions about the values for the second variable by using the first. Of course, this does not imply that there is a cause-and-effect relationship between the two quantities.

To illustrate these ideas, we again consider the example involving the apparent relationship between a student's high school average (HSA) and grade-point average (GPA) after the freshman year of college. There are two random variables, HSA and GPA. We saw in Section 10.2 that there is a high degree of correlation between these two quantities. To determine the best linear relationship between them, we once more examine the set of sample data pairs:

HSA	GPA
80	2.4
85	2.8
88	3.3
90	3.1
95	3.7
92	3.0
82	2.5
75	2.3
78	2.8
85	3.1

FIGURE 10.11

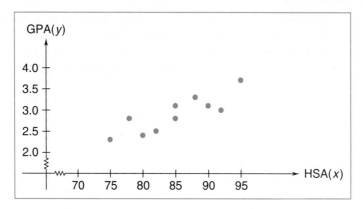

We also consider the associated scatterplot shown in Figure 10.11. The data points more or less lie along a straight line, which also suggests that there is a linear relationship between the two quantities. Despite this, clearly we cannot connect all 10 points with a single straight line. In fact, we can draw a great many different lines which all seem to be close to the points in the scatterplot, as shown in Figure 10.12. Our objective is to find the equation of the one particular line which comes closest to *all* the points in the scattergram. This line of "best fit" is known as the *regression line* or the *least-squares line*.

The most common way to find the line of best fit for the given data points in the scatterplot is to find the line with the property that the distances from the individual points in the scattergram to the line are as small as possible. We interpret the distance from a point to a line in the vertical sense, so that we want to minimize the sum of the vertical distances between the points and the still unknown regression line. However, there is a complication here if we consider only vertical distances; some points are above the line and so the vertical distances are positive, while others are below the line and the vertical distances are negative. Thus, we run into the same type of difficulty we faced when trying to develop a measure for the spread of a set of data in Chapter

FIGURE 10.12

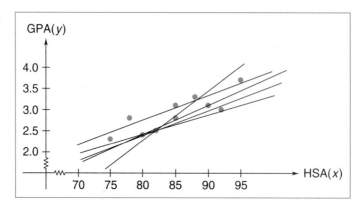

FIGURE 10.13

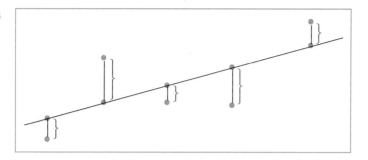

3. When we introduced the standard deviation, we found it was necessary to consider the sum of the squares, $\Sigma(x - \mu)^2$, of the individual deviations $x - \mu$.

In a comparable way, it is now necessary to consider the sum of the squares of the vertical distances from the points in the scatterplot to the desired line. We illustrate this in Figure 10.13. Fortunately, it turns out (from calculus) that for any set of points, a unique line, called the *regression line*, is determined which minimizes this sum of the squares.

Before we discuss this regression line in detail, we want to recall some ideas from algebra regarding the equation of a line. The most common approach to expressing the equation of a line involves using the *slope-intercept form* for the line

$$y = mx + b$$

In this equation, m is the slope (which measures the angle of inclination of the line), and b is the y-intercept (which indicates the height where the line crosses the y-axis). When the slope m is positive, the line is rising toward the right; when m is negative, the line is falling toward the right. See Figure 10.14. As a reference, when the slope $m = 1$, the line is inclined at a 45° angle. When m is larger than 1, the angle is steeper than 45°. When m is smaller than 1

FIGURE 10.14

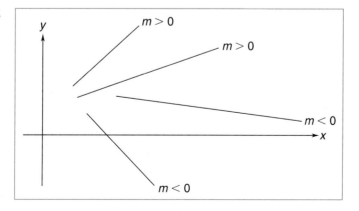

FIGURE 10.15

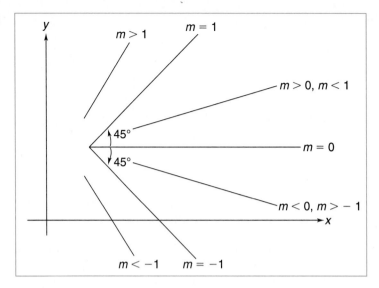

but still positive, the angle is flatter than 45°. Similarly, when $m = -1$, the angle is inclined downward at a 45° angle. Further, if the slope is 0, then the line is horizontal. See Figure 10.15.

An alternative format for writing the equation of a line is the *point-slope form*, which highlights the slope m of the line and one particular point $P(x_0, y_0)$ that the line passes through. This form for the equation is

$$y - y_0 = m(x - x_0)$$

For example, the line passing through the point $P(5, 3)$ with slope $m = 2$ has equation

$$y - 3 = 2(x - 5)$$

With these ideas in mind, let's return to the question of finding the regression line corresponding to a set of data. Although we do not go into the derivation here, it turns out that, for any set of data pairs, the corresponding regression line always passes through the point $P(\bar{x}, \bar{y})$, where $\bar{x}$ represents the mean of the x values and $\bar{y}$ represents the mean of the y values. Thus, if there are n pairs of values, then

$$\bar{x} = \frac{\Sigma x}{n} \quad \text{and} \quad \bar{y} = \frac{\Sigma y}{n}$$

Therefore, by using the point-slope formula for the equation of a line, the equation of the regression line is

$$y - \bar{y} = m(x - \bar{x})$$

Furthermore, the value for the slope m is given by

$$m = \frac{n(\Sigma xy) - (\Sigma x)(\Sigma y)}{n(\Sigma x^2) - (\Sigma x)^2}$$

This formula is not as complicated as it looks at first. In fact, it is remarkably similar to the formula we used to calculate the correlation coefficient

$$r = \frac{n(\Sigma xy) - (\Sigma x)(\Sigma y)}{\sqrt{n(\Sigma x^2) - (\Sigma x)^2} \sqrt{n(\Sigma y^2) - (\Sigma y)^2}} \tag{10.3}$$

In particular, the numerator in the formula for the slope

$$n(\Sigma xy) - (\Sigma x)(\Sigma y)$$

is identical to the numerator for r. Further, the denominator in the slope formula

$$n(\Sigma x^2) - (\Sigma x)^2$$

is precisely the same term that occurs inside the first square root in the denominator of the formula for r. Therefore, it is a relatively simple matter to calculate the value for m, particularly if you have already calculated the value for r.

We illustrate how to use these results by applying them to the original problem involving HSAs and GPAs. Consider Table 10.3, which we reproduce from Section 10.2, when we calculated the correlation coefficient $r = 0.858$ based on these $n = 10$ data pairs.

To obtain the equation of the regression line, we begin by finding the point, $P(\bar{x}, \bar{y})$, that the line passes through. Clearly,

$$\bar{x} = \frac{\Sigma x}{n} = \frac{850}{10} = 85$$

and

$$\bar{y} = \frac{\Sigma y}{n} = \frac{29}{10} = 2.9$$

so the point is $P(85, 2.9)$. Next we find the slope m of the regression line using the totals of the columns in Table 10.3. In particular,

$$n(\Sigma xy) - (\Sigma x)(\Sigma y) = 10(2486.3) - (850)(29)$$
$$= 24{,}863 - 24{,}650$$
$$= 213$$

which we calculated in Section 10.2. Similarly,

$$n(\Sigma x^2) - (\Sigma x)^2 = 10(72{,}617) - (850)^2$$
$$= 726{,}170 - 722{,}500$$
$$= 3670$$

TABLE 10.3

x	y	xy	x^2	y^2
80	2.4	192.0	6,400	5.76
85	2.8	238.0	7,225	7.84
88	3.3	290.4	7,744	10.89
90	3.1	279.0	8,100	9.61
95	3.7	351.5	9,025	13.69
92	3.0	276.0	8,464	9.00
82	2.5	205.0	6,724	6.25
75	2.3	172.5	5,625	5.29
78	2.8	218.4	6,084	7.84
85	3.1	263.5	7,225	9.61
$\Sigma x = 850$	$\Sigma y = 29.0$	$\Sigma xy = 2{,}486.3$	$\Sigma x^2 = 72{,}617$	$\Sigma y^2 = 85.78$

which we have already calculated as well. Consequently, the slope m is given by

$$m = \frac{n(\Sigma xy) - (\Sigma x)(\Sigma y)}{n(\Sigma x^2) - (\Sigma x)^2}$$

$$= \frac{213}{3670}$$

$$= .058$$

Consequently, the equation of the regression line for this set of data is

$$y - \bar{y} = m(x - \bar{x})$$

or

$$y - 2.9 = .058(x - 85)$$

To transform this equation to the equivalent slope-intercept form for the equation of a line, we merely expand the above equation and collect like terms. Thus,

$$y - 2.9 = .058x - 4.93$$

so that

$$y = .058x - 4.93 + 2.9$$
$$= .058x - 2.03$$

Having obtained the above formula for the regression line, we want to consider what it represents. In this situation, x represents a student's HSA and y his or her GPA. Our primary goal is to use the regression equation in a predictive sense. This works as follows: Suppose a particular student has an

HSA of $x = 93$. We would like to predict his or her GPA after the freshman year in college based on this HSA. The regression equation is

$$y = .058x - 2.03$$

Thus, when $x = 93$, this equation becomes

$$y = .058(93) - 2.03$$
$$= 5.394 - 2.03$$
$$= 3.364 \approx 3.36$$

This value is taken as the best estimate for the GPA of the student. We note that it is certainly a reasonable prediction based on the initial set of data.

Similarly, if a student has an HSA of $x = 78$, then the best estimate for the GPA is

$$y = .058(78) - 2.03$$
$$= 4.524 - 2.03$$
$$= 2.494 \approx 2.49$$

which is again reasonable.

Keep in mind, though, that we usually cannot use a regression equation in a predictive manner if the values for x are taken well outside the interval given in the original data. For instance, our regression equation for GPA is based on a set of HSAs between 75 and 95. It would not necessarily be at all accurate to use it for a student with a high school average of $x = 60$, say. We cannot assume that the linear relationship continues to hold outside the range of the original data.

Moreover, it is unreasonable to expect that every student with an HSA of 93 will end up with the same GPA of 3.36. Rather, the regression prediction y gives us the most likely point estimate or average value for the GPA for any student with that particular HSA. However, this value must be interpreted much as the value for the sample mean $\bar{x}$ is interpreted when we are trying to estimate a population mean μ. That is, for a given value of x, the predicted value for y is just the center of an interval of likely values (a confidence interval) for the GPA. Thus, we can think of the prediction as being the center of a range of values above and below each point (x, y) on the regression line. The result is a band of values centered along the regression line, as shown in Figure 10.16. When the number of data pairs n is sufficiently large, the spread about the predicted y value on the regression line is approximately normal, for each value of x. We discuss the entire notion of using the regression equation in a predictive manner in Sections 10.4 and 10.5. As you will see, the concepts and methods used are slightly more sophisticated than those in this section.

From a practical point of view, finding the equation of the regression line for a set of data pairs is usually done in the same context as calculating the correlation coefficient for those data. We normally start by calculating r. If it turns out that, based on this sample, there is correlation between the two

FIGURE 10.16

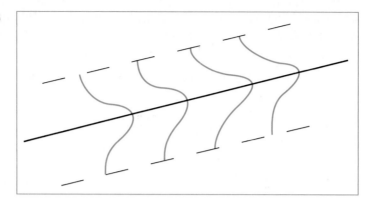

variables, then it makes sense to proceed to calculate the equation of the regression line based on the data. However, if there is no correlation between the two, then there is no point in determining the equation of the regression line, especially if you are doing the calculations by hand. All the values needed to determine the point $P(\bar{x}, \bar{y})$ and the slope m for the regression line are already determined in the process of calculating the value for r. There is no need to reproduce the table of values or to recalculate the expressions in the numerator and denominator of m.

EXAMPLE 10.6 A study is conducted to determine the relationship between the weight of a car in pounds and its gas mileage in miles per gallon. The following set of data pairs is obtained:

(2800, 19), (2650, 23), (2500, 27), (2450, 25), (2200, 32), (2300, 26), (2500, 22), (2600, 18)

Find the equation of the regression line based on this sample.

SOLUTION In Example 10.2 we found a correlation coefficient of $r = -0.842$, which indicated a high degree of negative correlation between the two quantities based on the critical value in Table V. Further, the scatterplot (Figure 10.17) falls into a linear pattern and seems to be slanted downward. Therefore, we should expect the slope of the regression line to be negative.

We perform all the appropriate calculations for the regression equation using Table 10.4. (Actually, we simply delete the last column of entries from the table used in Example 10.2.) Since there are $n = 8$ data pairs, we immediately find that

$$\bar{x} = \frac{20{,}000}{8} = 2500 \quad \text{and} \quad \bar{y} = \frac{192}{8} = 24$$

Further, either by using the values we previously calculated in Section 10.2

FIGURE 10.17

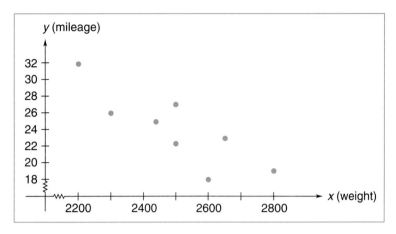

or from the appropriate subtotals in Table 10.4, we find that

$$n(\Sigma xy) - (\Sigma x)(\Sigma y) = -40{,}800$$
$$n(\Sigma x^2) - (\Sigma x)^2 = 2{,}040{,}000$$

Therefore, the slope m of the regression line is

$$m = \frac{-40{,}800}{2{,}040{,}000} = -.02$$

As a result, the equation of the regression line is

$$y - 24 = -.02(x - 2500)$$

or equivalently

$$y = -.02x + 74 \qquad \blacksquare$$

The fact that the slope m turned out negative simply means that there is an inverse relationship between the two quantities. The greater the weight of a car, in general, the lower the gas mileage, and vice versa.

TABLE 10.4

x	y	xy	x^2
2,800	19	53,200	7,840,000
2,650	23	60,950	7,022,500
2,500	27	67,500	6,250,000
2,450	25	61,250	6,002,500
2,200	32	70,400	4,840,000
2,300	26	59,800	5,290,000
2,500	22	55,000	6,250,000
2,600	18	46,800	6,760,000
$\Sigma x = 20{,}000$	$\Sigma y = 192$	$\Sigma xy = 474{,}900$	$\Sigma x^2 = 50{,}255{,}000$

In Example 10.6 obviously a variety of other factors might have been taken into account. For instance, a detailed study would also consider the age of the car, the condition of the engine, the habits of the driver, the type of driving, and so forth. It is rather unrealistic to consider just a single factor, the weight of the car, in determining the gas mileage. We discuss some extensions of regression ideas that apply to such situations in a later section.

Moreover, as with correlation analysis, the level of computation involved in finding the equation of a regression line from the raw data is so great that it is natural to use computers or special statistical calculators. With this in mind, we assume that in the future you will use Program 32: Linear Regression Analysis or Minitab to calculate the equation of the regression line, as well as the correlation coefficient, based on a set of data pairs.

Minitab Methods

To apply Minitab to determine the equation of the regression line based on a set of paired data stored in columns C1 and C2, use the command

> **REGRESS C2 1 C1**

or more fully

> **REGRESS C2** on **1** variable in **C1**

Notice that the column containing the values for the dependent variable, C2, comes first and the column for the values of the independent variable, C1, is listed last. In response to this command, Minitab performs a complete regression analysis based on the two sets of data.

Before you use it, however, we strongly recommend that you first enter the command

> **BRIEF** level of output = **1**

or simply

> **BRIEF 1**

Otherwise, Minitab will present you with a seemingly overwhelming display of information that is well beyond the scope of this course. Even with the **BRIEF 1** level of output, Minitab will print the equation of the regression line (in terms of the names given to the two variables) as well as several tables of additional information. We illustrate this in the sample Minitab session in Figure 10.18.

FIGURE 10.18

```
MTB > RETRIEVE 'HTSVSWTS'
   WORKSHEET SAVED  2/28/1991

Worksheet retrieved from file: HTSVSWTS.MTW
MTB > BRIEF level of output = 1
MTB >
MTB > REGRESS analysis of C2 on 1 variable in C1

The regression equation is
HEIGHTS = 48.5 + 0.118 WEIGHT

Predictor      Coef      Stdev    t-ratio        p
Constant     48.502      3.069      15.80    0.000
WEIGHT      0.11756    0.01946       6.04    0.000

s = 2.079    R-sq = 56.6%    R-sq(adj) = 55.0%

Analysis of Variance
SOURCE        DF       SS        MS         F        p
Regression     1   157.70    157.70     36.49    0.000
Error         28   121.00      4.32
Total         29   278.70

MTB >
```

EXAMPLE 10.7 A study is conducted to determine the relationship between a person's height and shoe size. The following set of data pairs is obtained:

$(66, 9), (63, 7), (67, 8\frac{1}{2}), (71, 10), (62, 6), (65, 8\frac{1}{2}),$
$(72, 12), (68, 10\frac{1}{2}), (60, 5\frac{1}{2}), (66, 8)$

Find the equation of the regression line relating height to shoe size based on this sample. Use it to predict the most likely shoe size for a person who is 70 inches tall and for a person who is 61 inches tall.

SOLUTION We enter the data pairs into Program 32: Linear Regression Analysis, and it produces the scatterplot and the graph of the associated regression line in Figure 10.19. In addition, the program gives us the equation of the regression line

$$y = .51x - 25.16$$

for a person's shoe size y as a function of height x.

Using this regression equation, we find that the estimated shoe size corresponding to $x = 70$ is

$$y = .51(70) - 25.16$$
$$= 35.7 - 25.16 = 10.54 \approx 10\frac{1}{2}$$

FIGURE 10.19

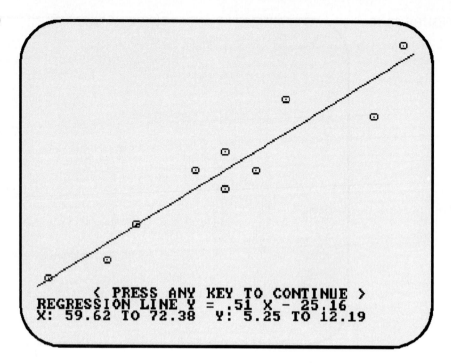

Similarly, if $x = 61$, then the corresponding shoe size is estimated to be

$$y = .51(61) - 25.16$$
$$= 31.11 - 25.16 = 5.95 \approx 6$$

■

Exercise Set 10.3

Mastering the Techniques

Draw the scattergram and calculate by hand the regression equation for the following sets of data:

1.

x	2	5	7	10	11
y	10	20	35	50	65

2.

x	5	12	15	18	25
y	30	50	45	65	90

3.

x	5	8	10	12	15	22
y	50	42	40	35	33	28

4.

x	2	10	15	18	27	33	35
y	80	66	60	52	47	53	27

Applying the Concepts

For each of the following, answer the given question provided that there is correlation between the indicated variables based on the given sample data pairs. Refer to your answers to the comparable exercises in Exercise Set 10.2.

5. A small company is interested in analyzing the effects of advertising on its sales. Over a 5-month period, it finds the following results

x	5	8	10	15	22
y	6	15	20	30	39

where x represents the money spent on advertising (in hundreds of dollars) and y represents the total sales (in thousands of dollars). Use these data to determine the linear regression equation for sales as a function of advertising.

6. A supermarket owner is studying how the average waiting time y in minutes for customer checkout depends on the number x of checkout clerks working. The results are

x	3	4	5	5	6	7
y	9	6	6	4	2	1

Use these data to determine the linear regression equation for waiting time as a function of the number of clerks on duty.

7. A college collects the following set of data on the number of credits y that a randomly selected group of students carry and the number of hours x they work during the week.

x	20	25	30	50	20	23
y	12	13	12	15	16	16

Find the equation for the regression line based on these data.

8. The college in Exercise 7 uses the above information on the number of hours x that students work to study the relationship between time spent on a job with grade-point average y. The results are:

x	20	25	30	50	20	23
y	3.4	3.0	2.8	2.4	2.9	2.9

Find the equation for the regression line based on these data.

9. A U.S. government study on energy conservation is conducted to explore the relationship between the price of home heating oil x (in cents) and the mean number of gallons y used per month during January in a variety of different communities with similar climates. The results are

x	75	80	86	90	95	98	106
y	120	125	114	110	112	106	97

Find the linear regression equation that relates oil use to cost.

10. A TV network is concerned about the high cost of producing many of its programs. It therefore conducts a study to relate the production costs for 30 minutes of programming x (in hundreds of thousands of dollars) to the ratings y that the program gets in some national ratings survey. The results are

x	1.2	1.6	1.8	2.5	2.7	3.0	3.5	4.4
y	3.3	3.9	5.7	4.2	4.5	8.2	6.1	4.6

Find the linear regression equation relating the ratings to the production costs.

11. Show algebraically that the y-intercept term b in the regression formula $y = mx + b$ can also be calculated as

$$b = \bar{y} - m\bar{x} = \frac{(\Sigma x^2)(\Sigma y) - (\Sigma x)(\Sigma xy)}{n(\Sigma x^2) - (\Sigma x)^2}$$

10.4 The Standard Error of the Estimate

There are three components in any regression-correlation analysis study. The first is the determination of the correlation coefficient r. Unless the value of r for the sample data indicates that there is some correlation, either positive or negative, between the two variables x and y, there is no point in performing any further analysis. The results would have no statistical validity or significance.

Once we know that there is correlation between the variables, we can proceed to the second component, the determination of the equation of the regression line. Our purpose in finding this equation is to use it in a predictive manner to estimate the value of the independent variable y corresponding to any given value of x.

However, we should not be content with merely making a prediction for the value of y. Even though this value for y may be the best point estimate based on any given value of x, there is always a certain degree of error in any such estimate. As a consequence, the third component in any regression-correlation analysis is measuring the maximum likely error in the estimate.

The estimate for the dependent variable is some particular value of y, say y_p, determined from the regression equation based on the particular value of x. We can view this predicted value as a point estimate for the actual value of y. For instance in the last section, we found that the regression equation relating a student's GPA (y) to his or her HSA (x) was

$$y = .058x - 2.03$$

Then, if a student had an HSA of $x = 93$, we predicted a most likely GPA of $y_p = 3.36$. However, a better prediction for y should actually be a confidence interval of the form

$$y_p \pm \text{error estimate}$$

This is analogous to what we did earlier in our study of confidence intervals where we estimated a population mean μ or a population proportion π using interval estimates.

To see how this might be accomplished, let's consider the entire process from the beginning. We start with a set of sample data pairs (x_1, y_1), (x_2, y_2), ..., (x_n, y_n). These points represent *known* or *observed* values. Based on these points, we calculate the equation of the regression line

$$y - \bar{y} = m(x - \bar{x})$$

or

$$y = mx + b$$

Anticipating some notation we will need later, we write $b_1 = m$ and $b_0 = b$, so that the regression equation becomes

$$y = b_1 x + b_0$$

or preferably

$$y = b_0 + b_1 x$$

Using this regression equation, we can predict the value of y corresponding to *any* given value of x.

We focus on the known values from the original data pairs to develop a way of estimating the error in any prediction. That is, the regression equation provides predictions about the values for $y_1, y_2, \ldots, y_n$. Since we know the actual values, it is possible to compare how accurately the predictions match the known results. Thus, for each of the n values of x given in the original data set, we calculate the corresponding n values predicted by the regression equation and compare them to the actual values observed for y.

In Figure 10.20, we indicate the original set of points (marked as y's) and the graph of the regression equation. The predicted values y_p all lie *on* the regression line. Consequently, the error or difference between the actual observed or known value y and the predicted value y_p is just the difference in height between each data point and the regression line. This is simply $e = y - y_p$. What we seek is a way of measuring the total error in this set of estimates.

FIGURE 10.20

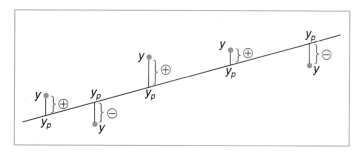

The problem is quite similar to the one we faced previously when we tried to calculate the total variation in a set of data and were led to develop the equation for the standard deviation. In that derivation, we introduced the square of the differences $(x - \mu)^2$ because the individual deviations $x - \mu$ were either positive or negative and we had to find a way to avoid the fact that they canceled one another. The same is true in this situation. As you can see from Figure 10.20, some of the differences $y - y_p$ are positive while others are negative. If we consider just the total of the differences, these error terms will likewise partially cancel. Therefore, as with the standard deviation calculation, we square each of the differences $(y - y_p)^2$ and find their sum

$$\Sigma(y - y_p)^2$$

We might now be tempted to average this quantity over the original number of data points by dividing this sum by the number of data pairs n. Recall, though, that when we calculate the standard deviation for a sample, it is necessary to divide the sum of squares by $n - 1$ instead of n since we lose a degree of freedom by using the mean $\bar{x}$ from the sample. The quantity $n - 1$ represents the number of degrees of freedom. In this case, we have two samples, one of x values and the other of y values. Thus, since we use the two means $\bar{x}$ and $\bar{y}$, we lose 2 degrees of freedom. As a result, there are only $n - 2$ degrees of freedom present. Consequently, we must divide the sum of squares by $n - 2$. Finally, as with the standard deviation, we take the square root of the result. Therefore, the estimate we will use for the error in any prediction based on the regression equation is given by

$$s_e = \sqrt{\frac{\Sigma(y - y_p)^2}{n - 2}}$$

This quantity is known as the *standard error of the estimate.*

We illustrate how to calculate it in the following example.

EXAMPLE 10.8 Determine the value of the standard error of the estimate s_e based on the data relating high school average (HSA) to grade-point average (GPA) from Sections 10.2 and 10.3.

SOLUTION As with all the other quantities involved in regression and correlation analysis, the best way to perform this calculation by hand is with a table. It is usually simplest to extend the table used to calculate m and r by adding several columns, one for the predicted value y_p, a second for the difference $y - y_p$, and a third for the square of this difference $(y - y_p)^2$. In the interests of clarity, however, we display only those columns we need in Table 10.5.

Recall that the regression equation we obtained for this set of data in Section 10.3 is

$$y = .058x - 2.03 \tag{10.4}$$

TABLE 10.5

x	y	y_p	$y - y_p$	$(y - y_p)^2$
80	2.4	2.61	−.21	.0441
85	2.8	2.90	−.10	.0100
88	3.3	3.07	.23	.0529
90	3.1	3.19	−.09	.0008
95	3.7	3.48	.22	.0484
92	3.0	3.31	−.31	.0961
82	2.5	2.73	−.23	.0529
75	2.3	2.32	−.02	.0004
78	2.8	2.49	.31	.0961
85	3.1	2.90	−.10	.0100
$\Sigma x = 850$	$\Sigma y = 29.0$			$\Sigma(y - y_p)^2 = .4017$

Consider the first row in Table 10.5 with $x = 80$ and the actual observed value of $y = 2.4$. We use the regression equation (10.4) to calculate the predicted value y_p based on $x = 80$, namely,

$$y_p = .058(80) - 2.03$$
$$= 4.64 - 2.03 = 2.61$$

compared to the actual value of $y = 2.4$. Consequently, the error or difference

$$y - y_p = 2.4 - 2.61 = -.21$$

is recorded in the following column, and the square of this

$$(y - y_p)^2 = (-.21)^2 = .0441$$

is recorded in the last column. In the same way, we use each of the other given values for x to find the corresponding predicted value y_p and compare it to the observed value y. The full results are shown in the table, from which we see that

$$\Sigma(y - y_p)^2 = .4017$$

Since there are $n = 10$ data pairs, there are $n - 2 = 8$ degrees of freedom. Hence, the standard error of the estimate is

$$s_e = \sqrt{\frac{\Sigma(y - y_p)^2}{n - 2}}$$

$$= \sqrt{\frac{.4017}{8}}$$

$$= \sqrt{.0502}$$

$$= .22$$

∎

The standard error of the estimate plays a role comparable to the standard deviation for a sampling distribution. For instance, suppose a student enters college with an HSA of 93. The regression equation (10.4) predicts a GPA of 3.36 as the most likely value. However, a more meaningful prediction for the GPA would be

$$3.36 \pm \text{error estimate}$$

where the error estimate is based on the standard error of the estimate s_e. We consider the problem of determining this error estimate, and hence constructing an analog of a confidence interval for estimating the values for the y_p predictions, in the next section.

Computer Corner

We mentioned previously that Program 32: Linear Regression Analysis also calculates the value for the standard error of the estimate s_e. It is displayed along with the value for the correlation coefficient on the text screen that follows the graphics page. Therefore, this program essentially provides you with a means of performing all the detailed computations involved in any linear regression and correlation analysis problem for any set of paired data. Moreover, Minitab also prints out the value for the standard error of the estimate as part of the output associated with the **REGRESS** command.

In addition, you can get a very interesting perspective on regression analysis and its relationship to the standard error by using another program on your disk. Program 34: Regression Simulation provides a simulation of regression analysis. To this point, we have started with a set of data and used it to calculate the correlation coefficient, the equation of the regression line, and now the standard error of the estimate. However, from a broader point of view, this set of data actually represents just a single random sample of n pairs of values drawn from a larger population consisting of all possible pairs of values for x and y. Thus, if we draw a different sample, we will end up with a different regression line (as well as different values for r and s_e). The question is, Just how different are these regression lines likely to be?

We use Program 34: Regression Simulation to explore this question. The program contains a built-in bivariate population and asks for the sample size you desire. Suppose we start with a very small sample size, say $n = 3$. The program draws the regression line corresponding to the entire population of data pairs. It then draws two parallel lines representing plus and minus the standard error of the estimate for the underlying population. Since we know that the regression line always passes through the point corresponding to the means of the x

and y values, the program also highlights the point (μ_x, μ_y) on the regression line. It then selects a random sample of size $n = 3$ from the population of data pairs, calculates the regression line, draws its graph, and displays the equation of the sample regression line. After each such line is drawn, you are asked if you want another regression line. If you enter anything other than an N for NO, a further line will be displayed.

In Figure 10.21 we show the results of the first few sample regression lines constructed on a typical run of the simulation. Notice that most of the lines shown lie relatively close to the central (lighter) line representing the regression line for the population. However, at least one goes off at a rather sharp angle. Clearly, it corresponds to a set of three data points from the population which are not typical of the population. Figure 10.22 shows the results of many more such samples. While some regression lines are drastically different from the population line, the majority are fairly close to it.

If we change to a sample size of $n = 12$, as shown in Figure 10.23, then some changes become evident. First, virtually all the sample regression lines lie quite close to the population regression line. As the sample size increases, the possible variation decreases, so that it is very unlikely you will obtain a sample regression line that goes off at a sharp angle.

FIGURE 10.21

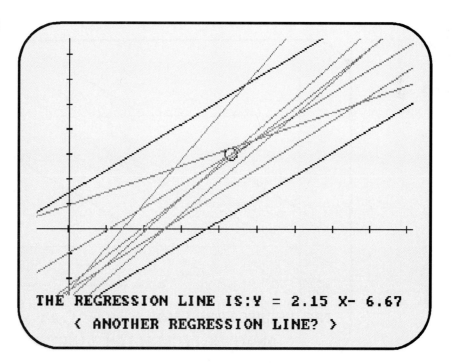

Second, at least for the portion of the graph displayed on the computer screen, all these lines remain within the two bands determined by the standard error of the estimate for the population regression equation. (Of course, unless the lines are parallel, we know that if the sample regression lines are extended far enough—off the computer screen—they will eventually cross the two bands. However, when using regression analysis, we are usually concerned with values of x between or at least close to the original data entries for the independent variable x.)

Third, most of the sample lines pass very close to the point (μ_x, μ_y) on the population line. In retrospect, this makes sense. When we take a sample of x values, their mean $\bar{x}$ is usually fairly close to the population mean μ, provided that the sample size is large enough. In this case, each sample regression line is based on a collection of x values and y values. Each sample regression line passes through the point $(\bar{x}, \bar{y})$. Since we expect $\bar{x}$ to be fairly close to μ_x and $\bar{y}$ to be fairly close to μ_y, it makes sense that the sample regression lines all pass relatively close to the point (μ_x, μ_y).

Based on these observations, we ask you to anticipate what is likely to happen if you take a still larger sample size, say $n = 20$ or $n = 30$. Try the program yourself with these values and see if your predictions are indeed correct.

FIGURE 10.22

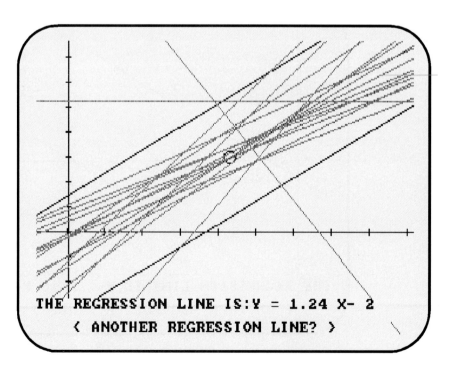

FIGURE 10.23

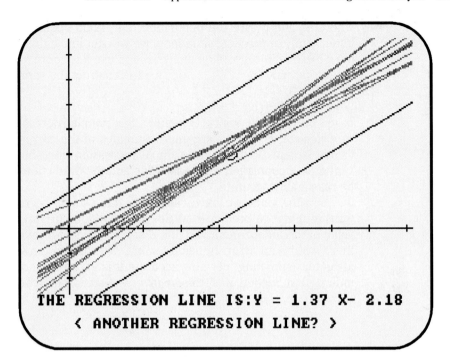

Exercise Set 10.4

Mastering the Techniques

1–4. For each of Exercises 1 through 4 in Section 10.2, calculate by hand the standard error of the estimate.

Applying the Concepts

5–10. For each of the situations in Exercises 9 through 14 of Section 10.2, calculate the standard error of the estimate.

Computer Applications

11–14. For each of the situations you studied in Exercises 16 through 19 of Section 10.2, determine the value for the standard error of the estimate.

10.5 Applications of Statistical Inference in Regression Analysis

We now turn our attention to some slightly more sophisticated applications of statistical inference in regression analysis. Suppose we have a bivariate population composed of x and y values with a population regression equation

$$y = \beta_0 + \beta_1 x$$

where β_0 and β_1 are the population regression coefficients (β is the Greek letter beta). In practical situations, we will not know this relationship. In fact, our objective in regression analysis is to use a random sample from the population $(x_1, y_1), (x_2, y_2), \ldots, (x_n, y_n)$ to construct the sample regression line

$$y = b_0 + b_1 x$$

based on the data points. Hopefully, the sample regression coefficients that we calculate, b_0 and b_1, are good estimates of the corresponding population values β_0 and β_1. If so, the sample regression line should be close to the (unknown) population regression line and the predictions based on it should be reasonably accurate.

Further, once we have calculated this sample regression equation, we want to use it to predict the value for y_p which corresponds to any desired value of the independent variable x, say x_0. Obviously, in any such prediction, there will be an error, hopefully small. We therefore need to develop some ideas for estimating just how accurate this approximation actually is. As we indicated in Section 10.4, this involves the standard error of the estimate

$$s_e = \sqrt{\frac{\Sigma(y - y_p)^2}{n - 2}}$$

There are several distinct questions that we must address in this context. The first concerns the value of the regression coefficient β_1, which is the slope of the regression line. If β_1 is 0, then the prediction equation reduces to

$$y = \beta_0 + 0x = \beta_0$$

which is the equation of a horizontal line. Since this is a constant, we would obtain the same value of y no matter what value of $x = x_0$ we chose. Hence our first question: Is $\beta_1 = 0$? We can answer this by performing a hypothesis test on the regression coefficient β_1. The null hypothesis states that β_1 is 0, so that

$$H_0: \beta_1 = 0$$

while the alternate hypothesis states

$$H_a: \beta_1 \neq 0$$

If we reject the null hypothesis, then we can apply regression analysis to predict y on the basis of x. The corresponding sampling distribution for b_1 consists of the slopes of all possible regression lines based on different samples of size n drawn from the original bivariate population.

The sampling distribution of b_1 is approximately normal provided that the sample size n is large enough. This distribution has a standard deviation given by

$$\sigma_{b_1} = \frac{s_e}{\sqrt{\Sigma x^2 - \frac{(\Sigma x)^2}{n}}}$$

where s_e is the standard error of the estimate.

Usually the criterion on n for normality is expressed as $n - 2 > 30$. In the cases we consider here, we use relatively small sample sizes. The hypothesis test is then a t-test with $n - 2$ degrees of freedom provided that the underlying distribution is approximately normal. We assume that the appropriate conditions are fulfilled in all subsequent examples and exercises.

The test statistic we use in the hypothesis test is the value of b_1 for the slope of the regression line based on the sample data. We illustrate the approach in the following example.

EXAMPLE 10.9 Test whether the slope of the regression line based on the sample data for the HSAs and GPAs from Sections 10.1 through 10.4 is nonzero at the 5% level of significance.

SOLUTION For this hypothesis test, the null and alternate hypotheses are

$$H_0: \beta_1 = 0$$
$$H_a: \beta_1 \neq 0$$

We have already found that the slope of the sample regression line is $b_1 = .058$, and the standard error of the estimate for this set of data pairs is $s_e = .22$. For this set of data, $\Sigma x = 850$ and $\Sigma x^2 = 72{,}617$, so that

$$\sigma_{b_1} = \frac{s_e}{\sqrt{\Sigma x^2 - \frac{(\Sigma x)^2}{n}}}$$

$$= \frac{.22}{\sqrt{72{,}617 - \frac{(850)^2}{10}}}$$

$$= \frac{.22}{\sqrt{367}}$$

$$= .01148$$

Since we have $n = 10$ data pairs, we use a t-distribution with $n - 2 = 8$ degrees of freedom for the hypothesis test. At the 5% level of significance, the critical values for t are ± 2.306. For the sample data, we find that

$$t = \frac{b_1 - \beta_1}{\sigma_{b_1}}$$

$$= \frac{.058 - 0}{.01148}$$

$$= 5.052$$

FIGURE 10.24

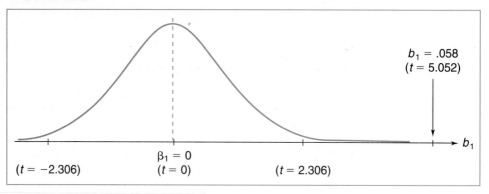

Since this falls in the rejection region, as shown in Figure 10.24, we reject the null hypothesis and so conclude that the regression coefficient β_1 is not zero based on the 10 sample data pairs at the 5% level of significance. ∎

Once we have established that $\beta_1 \neq 0$, we can use the sample regression line on a predictive basis. Thus, for each value of $x = x_0$, we calculate the predicted value

$$y_p = b_0 + b_1 x$$

However, this value y_p is only a point estimate for the correct value of y corresponding to x_0. It is more meaningful to construct the prediction as a range of possible values for y. That is, we want to construct an interval estimate centered at a particular point y_p within which we believe that the true value for y will fall with a fairly high degree of confidence. The process is very similar to what we did previously when constructing confidence intervals for the population mean or proportion. Therefore, the second question we must consider is, How do we construct a confidence interval for the predicted values of y given a value of $x = x_0$?

To answer this question, we use the sampling distribution of the predicted values for y, y_p, corresponding to a given value of $x = x_0$. We assume that the distribution of the y's is approximately normal for any given x and that the variances (and hence the standard deviations) of the distribution of the y values are the same at all values of x. Under these conditions:

> The sampling distribution of the y_p values is approximately normal provided the sample size n is large enough. It has a standard deviation
>
> $$\sigma_{y_p} = s_e \sqrt{1 + \frac{1}{n} + \frac{(x_0 - \bar{x})^2}{\Sigma x^2 - (\Sigma x)^2/n}}$$
>
> where $\bar{x}$ is the mean of the observed x values.

That is, if we take repeated samples from the underlying bivariate population, find the regression line based on each sample (recall the discussion in the Computer Corner in Section 10.4), and calculate the predicted value of y for that same x_0 using each possible regression line, then the sampling distribution of y_p is approximately normal.

We saw in the computer investigation of the different sample regression lines that the sample size n plays an important role in the level of accuracy. As n increases, the sample regression lines remain closer to the population regression line over longer intervals of x values. Nevertheless, the sample regression line almost always diverges from the population line. Therefore, when the value selected for x_0 is close to the mean $\bar{x}$, the sample and the population regression lines will usually be close and the corresponding approximation y_p will usually be quite accurate. However, if x_0 is chosen relatively far from $\bar{x}$, the level of accuracy of the approximation decreases. To account for this, we multiply s_e by the term involving the square root in the above formula for σ_{y_p}. The effect is to increase the error estimate term and so widen the confidence interval the farther that x_0 is from the mean $\bar{x}$. That is, it adjusts the size of the possible error needed to achieve a given level of confidence in the prediction depending on how far x_0 is from $\bar{x}$.

We can picture this graphically by thinking of a series of vertical confidence intervals centered along the sample regression line, one for each possible value of x_0, as shown in Figure 10.25. The closer that x_0 is to the mean $\bar{x}$, the shorter the confidence interval necessary to achieve a given level of confidence in the prediction. Therefore, the confidence intervals for values of x_0 relatively far from $\bar{x}$ will be longer than those for values of x_0 close to $\bar{x}$. This produces a pair of bands, one lying above the regression line, the other lying below it. For each given value of x_0, the most likely values predicted for y_p

FIGURE 10.25

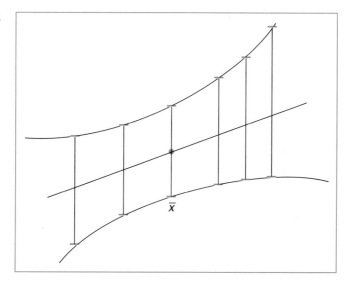

514 CHAPTER 10 • Correlation and Regression Analysis

will lie vertically within these bands. We note that the confidence intervals so produced are often called *prediction intervals*.

Moreover, as with hypothesis tests on the slope, β_1, we consider only situations with relatively small sample sizes, and so the sampling distribution is a *t*-distribution with $n - 2$ degrees of freedom. The sample data that we use consist of the predicted value y_p corresponding to any desired x_0. We illustrate this procedure in the following example.

EXAMPLE 10.10 Construct 95% confidence intervals for the predicted GPAs of two students, one having an HSA of 84 and the other having an HSA of 95, based on the data we have been using.

SOLUTION The regression equation we use is

$$y = .058x - 2.03$$

Corresponding to an HSA of $x_0 = 84$, we find the point estimate $y_p = 2.84$. Further, we have already found that the standard error of the estimate is $s_e = .22$. In addition, $\Sigma x = 850$, $\Sigma x^2 = 72{,}617$, and $\bar{x} = 85$. Therefore,

$$\sigma_{y_p} = s_e \sqrt{1 + \frac{1}{n} + \frac{(x_0 - \bar{x})^2}{\Sigma x^2 - (\Sigma x)^2/n}}$$

$$= .22 \sqrt{1 + \frac{1}{10} + \frac{(84 - 85)^2}{72{,}617 - (850)^2/10}}$$

$$= .22 \sqrt{1 + .1 + \frac{1}{367}}$$

$$= .22 \sqrt{1.1027}$$

$$= .231$$

Furthermore, since $n = 10$, we use a *t*-distribution with $n - 2 = 8$ degrees of freedom. For a 95% confidence interval, the corresponding value of *t* is $t = 2.306$. Consequently, the 95% confidence interval for the predicted GPA for a student with an HSA of $x_0 = 84$ is

$$y_p \pm t\sigma_{y_p}$$

$$2.84 \pm 2.306(.231)$$

$$2.84 \pm .53$$

or the interval

$$[2.31, 3.37]$$

See Figure 10.26.

For the student with an HSA of $x_0 = 95$, the corresponding point estimate is $y_p = 3.48$. The only other thing that changes in the above solution is the value for σ_{y_p} where the term $(x_0 - \bar{x})^2$ now involves $x_0 = 95$ rather than 84.

FIGURE 10.26

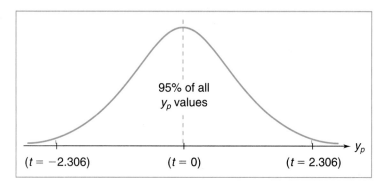

Consequently, we find that

$$\sigma_{y_p} = s_e \sqrt{1 + \frac{1}{n} + \frac{(x_0 - \bar{x})^2}{\Sigma x^2 - (\Sigma x)^2/n}}$$

$$= .22 \sqrt{1 + \frac{1}{10} + \frac{(95 - 85)^2}{72{,}617 - (850)^2/10}}$$

$$= .22 \sqrt{1 + .1 + \frac{100}{367}}$$

$$= .22 \sqrt{1.372}$$

$$= .258$$

Since we still want to construct a 95% confidence interval based on a sample of $n = 10$ data pairs, the value for t is again $t = 2.306$, and so the corresponding confidence interval is now

$$y_p \pm t\sigma_{y_p}$$

$$3.48 \pm 2.306(.258)$$

$$3.48 \pm .59$$

or the interval

$$[2.89, 4.07]$$

See Figure 10.26 also. ■

Notice that the width of the second confidence interval is somewhat larger than that of the first interval. This merely reinforces the point we made previously that the farther the value of x_0 is taken from the mean $\bar{x}$, the larger the confidence interval must be to achieve the desired degree of confidence. Further, we mentioned earlier that regression predictions based on values of x_0 beyond the original interval for the data should be avoided. In the second case here, the value $x_0 = 95$ is at the upper limit of the original data values, and the corresponding interval extends up to 4.07, beyond the usual upper limit for GPAs.

Exercise Set 10.5

Mastering the Techniques

1–4. For each of Exercises 1 through 4 in Sections 10.2 through 10.4, can you conclude that the slope β_1 of the regression line is nonzero at the 5% level of significance?

For each of the Exercises 1 through 4 in Sections 10.2 through 10.4, construct a 95% confidence interval for the predictions y_p based on the indicated values of $x = x_0$:

5. $x_0 = 4; x_0 = 8$
6. $x_0 = 8; x_0 = 16$
7. $x_0 = 10; x_0 = 20$
8. $x_0 = 12; x_0 = 30$

Applying the Concepts

9. For the situation in Exercise 9 of Section 10.2, can you conclude that the slope β_1 of the regression line is nonzero at the $\alpha = .05$ level of significance?
10. Repeat Exercise 9 based on the situation in Exercise 11 in Section 10.2.
11. For the situation in Exercise 9 in Section 10.2, construct a 95% confidence interval for the predictions based on $x_0 = 6; x_0 = 12$
12. For the situation in Exercise 11 of Section 10.2, construct a 90% confidence interval for the predictions based on $x_0 = 12; x_0 = 14$.

Computer Applications

13–16. For each of the situations you studied in Exercises 16 through 19 of Section 10.2, can you conclude that the slope β_1 of the regression line is nonzero at the $\alpha = .05$ level of significance?

10.6 Extensions of the Regression Concept

Regression analysis is one of the most powerful statistical tools available for a wide variety of applications for making predictions based on a set of data involving several distinct random variables. However, the ideas we discussed in previous sections represent only an introduction to the subject. In this section, we briefly indicate a number of directions in which the type of regression analysis we already considered can be extended.

The development of the linear regression equation in Section 10.3 was based on two assumptions:

1. There is some correlation between the two variables.
2. The set of data pairs, when graphed in a scattergram, lies more or less along a straight line.

In this case, it makes sense to speak of the regression line that best fits the data. Clearly, from the examples and problems we have encountered up to this point, this is a very reasonable assumption for a wide variety of sets of data. However, in many other cases, the data set is not linear when displayed in a scattergram. Several such sets of data are shown in Figure 10.27. For each of these sets, it would be unrealistic to try to use a straight line to approximate the data points.

One extension of the regression concept is to introduce the idea of *nonlinear regression*. Instead of constructing a regression line, we construct a *regression curve* which best fits the data set. For the first set of data, displayed in Figure 10.27a, the pattern might suggest a parabola as the corresponding regression curve, and so the equation for it would take the form

$$y = Ax^2 + Bx + C$$

The problem then becomes one of determining the best values for A, B and C to produce that parabola which best fits the data points in the least-squares sense.

FIGURE 10.27

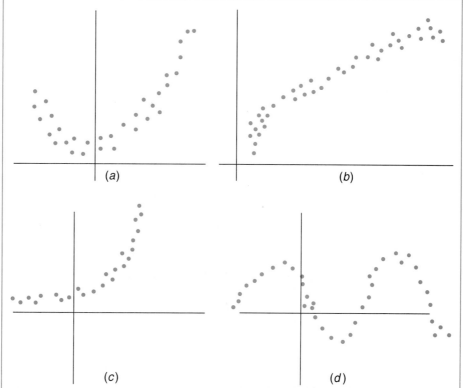

It is also possible that a higher degree polynomial, say
$$y = Ax^3 + Bx^2 + Cx + D$$
or
$$y = Ax^4 + Bx^3 + Cx^2 + Dx + E$$
might be needed to fit a set of data.

Alternatively, it is often necessary to use some other type of mathematical function besides a polynomial to determine the best fit. The second set of data, shown in Figure 10.27b, might suggest using a logarithmic function or a square root function for the regression equation. The third data set, in Figure 10.27c, might suggest using some type of exponential function. The data might represent the values associated with population growth in some geographic region or in some species and are therefore exponential. The fourth set of data, in Figure 10.27d, would likely involve a trigonometric function such as $\sin x$ or $\cos x$ to reflect the periodic nature of the data. This set of data might arise from a series of measurements of temperature in a certain locality over several years. Obviously, there will be fluctuation depending on the season, as shown in the data set. Computer programs are widely available to implement the necessary procedures for constructing the corresponding regression equations based on a variety of assumptions regarding the shape of the set of data points.

We illustrate some of these ideas on nonlinear regression in the following situation. Let us try to find a regression curve that best fits the following set of points:

$$(-1, 12), (0, 5), (1, 1), (2, -1), (3, 2), (4, 7), (5, 14)$$

We begin by drawing the scattergram for this set of data, as shown in Figure 10.28, which was generated by Program 32: Linear Regression Analysis. The equation of the regression line is

$$y = .39x + 4.93$$

From the diagram, it is obvious that the regression *line* does not provide a good fit to the data points. Moreover, the corresponding correlation coefficient is $r = .15$. As we might expect, this value for the correlation coefficient is very small, and using Table V, we find that it does not indicate the presence of any linear correlation between x and y. Further, the standard error of the estimate $s_e = 6.117$ is quite large and so also tends to indicate that the regression line is not a good fit to the data.

From the scattergram, it is apparent that, rather than a linear pattern, the data seem to fall more into a parabolic pattern which can best be fitted with a quadratic curve of the form

$$y = Ax^2 + Bx + C$$

Without going into the details, we simply indicate that the best choice for the coefficients is $A = 1.488$, $B = -5.560$, and $C = 4.929$, so that the regression

FIGURE 10.28

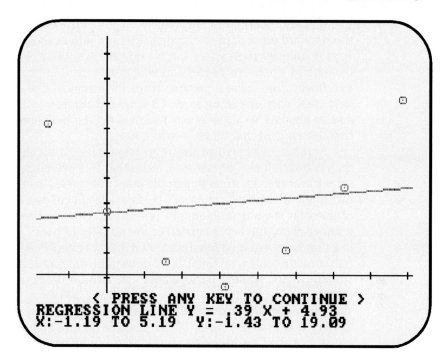

equation based on least squares is given by

$$y = 1.488x^2 - 5.560x + 4.929$$

The graph of this *quadratic regression equation* is shown superimposed over the scattergram for the data set in Figure 10.29. We see that it represents a considerably better fit to the data points than the linear regression equation did.

FIGURE 10.29

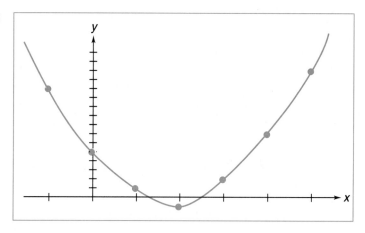

It is also possible to introduce versions of the correlation coefficient and the standard error of the estimate in this and other cases involving polynomial (and nonlinear) regression, but we do not do so here. Further, we point out that such a nonlinear regression equation is used in precisely the same way as a linear regression equation. Thus, for instance, if we want to predict the best value corresponding to $x = 3.4$, we substitute $x = 3.4$ into the equation and so obtain $y = 3.226$ as our estimate for the best predicted value, having first determined that there is correlation.

Another extension of the regression concept involves the use of more than two variables. For instance, in Section 10.1, we discussed the use of high school averages in predicting grade-point averages. However, many other factors might also affect a student's performance in college. We suggested some of these in several problems in previous sections—they might include the number of credits being carried or the number of hours that a student works in a non-school-related job. In addition, many schools use a placement test, in either mathematics or English, to determine an appropriate course for each student. Thus, student performance can also be predicted by the results on this test.

One approach that could be used to make predictions of academic performance is to isolate each of the possible factors and develop a regression equation based on it alone. Thus, there might be one regression equation based on high school average, another based on score on an admissions test, a third based on scores on the SAT or ACT, a fourth based on number of hours worked, and so forth. However, if this approach were used, we would have a whole battery of predictions, many of which would undoubtedly contradict one another.

A far better approach is to develop a notion called *multivariate linear regression analysis* or simply *multivariate regression analysis*. Basically, this involves considering a single variable, say y, which depends on a set of m different variables, often denoted by $x_1, x_2, \ldots, x_m$. For instance, these might be a student's high school average x_1, SAT score x_2, score on an admissions or placement exam x_3, the number of hours worked x_4, and so forth.

It is possible to extend the notion of the correlation coefficient to develop a measure for the degree of correlation between the quantity y and all the individual variables. It is also possible to study the degree of correlation between y and each of the x values individually. If there is correlation between the variables and if the relationship between them is linear, then it is possible to relate y to the other variables by using a linear equation of the form

$$y = A_1 x_1 + A_2 x_2 + \cdots + A_m x_m + B$$

All the coefficients, $A_1, A_2, \ldots, A_m$ and B must be estimated based on the set of measurements for a sample group. Once this has been done, the equation is used as a predictor. Further, it is also possible to extend the idea of the standard error to reflect the error of such an estimate. However, all these ideas are beyond the scope of this book, and we will not go into them.

We have used the term *multivariate linear regression analysis* to describe the type of regression equation used because the equation $y = A_1x_1 + A_2x_2 + \cdots + A_mx_m + B$ is linear in each of the x's. There is a relatively simple geometric interpretation possible for this as well. When we study simple linear regression, the regression equation is represented as a line in the plane. If we perform a multivariate linear regression analysis with two independent variables, then the resulting regression equation has the form $y = A_1x_1 + A_2x_2 + B$. Mathematically, the graph of such an equation is a plane in three-dimensional space, so that we sometimes talk about the *regression plane* instead of the *regression line* when there are two independent variables. It is very difficult to give a comparable geometric perspective to the regression equation when there are three or more independent variables.

We illustrate some of these ideas in the following example.

EXAMPLE 10.11 A college desires to predict student performance in terms of a grade-point average y after the freshman year in college based on the student's high school average x_1, total SAT score x_2, and score on an academic placement exam x_3. To do this, the values for a randomly selected group of 5 students are used and subjected to a multivariate regression analysis. These scores (x_1, x_2, x_3, y) are used:

Student 1:   (80, 980, 50, 2.90)
Student 2:   (85, 1060, 60, 2.83)
Student 3:   (90, 1250, 55, 3.25)
Student 4:   (94, 1320, 66, 3.50)
Student 5:   (88, 1080, 46, 3.05)

SOLUTION There are three independent variables x_1, x_2, and x_3 and a dependent variable y. Further, we are given five sets of data, one for each student. Using a computer program such as Minitab, we find that the multivariate correlation coefficient is $R = .95$ (which is quite large) and the multivariate linear regression equation is

$$y = -.003x_1 + .002x_2 - .006x_3 + 1.256$$

We use this as a predictor as follows. Suppose a student enters the college with a high school average of 84 (x_1), 1030 on the SAT (x_2), and a score of 57 (x_3) on the placement exam. When these values are substituted into the multivariate regression equation, the predicted value of $y = 2.722$ is produced. If we compare this prediction to the data on the original five students, we see that the result is certainly reasonable. ■

Minitab Methods

You can use Minitab to perform a multivariate regression analysis on a number of different sets of data. Suppose you have three sets of data that you want to store in columns C1, C2, and C3. It might be easier to use the Minitab command **READ C1 - C3** to enter the three sets of data rather than the **SET** command we have used until now. The **SET** command is used to enter all the data in a single column at a time. If there were three (or more) sets of data, then you would have to enter all the values for column C1, then all the values for column C2, and so forth. With the types of multivariate data we use in regression analysis, it is usually easier to enter all the values associated with each individual separately, that is, to enter the data row by row rather than column by column. The **READ** command allows this.

To conduct the regression analysis, use the Minitab commands

> **BRIEF = 1**
> **REGRESS C3** on **2** sets of data in columns **C1** and **C2**

or more simply,

> **REGRESS C3 2 C1-C2**

Minitab then responds with the associated regression equation as well as considerably more details that we will not concern ourselves with here. We illustrate a sample Minitab session in Figure 10.30.

FIGURE 10.30

```
MTB > NAME C1 'AGE'
MTB > NAME C2 'WEIGHT'
MTB > NAME C3 'HEIGHT'
MTB > READ C1-C3
      5 ROWS READ
MTB > END
MTB > BRIEF 1
MTB > REGRESS C3 on 2 sets of data in columns C1 and C2

The regression equation is
HEIGHT = 23.8 + 10.0 AGE - 0.923 WEIGHT

Predictor       Coef      Stdev    t-ratio        p
Constant       23.75      10.39       2.29    0.150
AGE            10.050      4.159      2.42    0.137
WEIGHT        -0.9234      0.5971    -1.55    0.262

s = 3.885    R-sq = 89.3%    R-sq(adj) = 78.5%

Analysis of Variance

SOURCE          DF         SS         MS        F        p
Regression       2     251.02     125.51     8.32    0.107
Error            2      30.18      15.09
Total            4     281.20

MTB >
```

Exercise Set 10.6

Computer Applications

1. Use multivariate regression analysis with Minitab to perform a study comparing the number of RBIs for a baseball player depending on the number of hits he has and the number of home runs he hits. In particular, select a random sample of 30 players from the data set in Appendix I. Enter the values for the number of hits, the number of home runs, and the number of RBIs as the input data and have the program produce the multivariate regression equation for you. Use this equation as a predictor based on the number of hits and the number of home runs for several other players you randomly select.
2. Repeat Exercise 1 to study the multivariate regression equation that arises to relate the price of a stock to the sales volume (given in hundreds) and the change in price that day.

Student Projects

Select a random sample consisting of at least 30 pairs of values for two possibly related quantities, and perform a full regression and correlation analysis on them. For example, you might want to study the relationship, if any, between:

1. Batting averages and RBIs
2. The closing price of a stock and the volume (number of shares traded) on a given day
3. The weight of a car (often listed on the automobile registration card) and its gas mileage
4. The age of a car and its value (price on a used-car dealer's lot, perhaps)
5. Monthly electric bill and monthly telephone bill
6. A student's GPA and the number of hours per week that she or he works
7. The temperature and the amount of rain in 24-hour periods in a particular city
8. A man's height and his shirt size
9. A woman's weight and her pulse rate
10. The asking price for homes in the classified ads and the number of rooms in the house

Once you have collected the sample data, you should

(a) Calculate the correlation coefficient r and test if there is any nonzero correlation between the two quantities you are studying.
(b) Calculate the equation of the regression line.
(c) Calculate the standard error of the estimate.
(d) Test if the slope of the regression line is nonzero.
(e) Construct 95% confidence intervals for the predicted values y_p corresponding to at least three values x_0 that you select for the first variable.

You should incorporate the results of your study into a formal statistical project report. The report should include

- A statement of the topic being studied.
- The source of the data.
- A discussion of how the data were collected and why you believe that it is a random sample.
- A list of the data, a scatterplot, and possibly some other tables and/or diagrams to display the data.
- All the statistical calculations including the means for the two sets of data, the correlation coefficient, the equation of the regression line, and the standard error of the estimate. This might involve hand calculations or output directly from the computer by using appropriate programs on your disk.
- A discussion as to whether there is correlation between the two variables. Be sure to include a rationale for the level of significance you select.
- Your conclusions and a discussion of your results.
- A discussion of any surprises you may have noted in connection with collecting and organizing the data or with the results of the study in general.

CHAPTER 10 SUMMARY

Correlation and **regression analysis** is used

1. To determine whether two quantities are related to each other and what the degree of relationship is, if any
2. If there is correlation, to determine a formula expressing the relationship
3. To apply the relationship to predict the most likely value of the second variable corresponding to any given value of the first variable

The data can be displayed in a **scattergram**.

The **correlation coefficient r** is

$$r = \frac{n(\Sigma xy) - (\Sigma x)(\Sigma y)}{\sqrt{n(\Sigma x^2) - (\Sigma x)^2} \sqrt{n(\Sigma y^2) - (\Sigma y)^2}}$$

The value for the correlation coefficient is always between -1 and 1.

You can test if the population correlation coefficient ρ is nonzero by using a hypothesis test based on the sample correlation coefficient r. Use Table V to find the critical value for r.

Realize that high correlation between two variables does not indicate a cause-and-effect relationship between them. Other factors may exist, causing the results you observe.

The equation of the **regression line** is

$$y - \bar{y} = m(x - \bar{x})$$

where

$$\bar{x} = \frac{\Sigma x}{n} \qquad \bar{y} = \frac{\Sigma y}{n}$$

and

$$m = \frac{n(\Sigma xy) - (\Sigma x)(\Sigma y)}{\sqrt{n(\Sigma x^2) - (\Sigma x)^2}}$$

This can be rewritten as

$$y = b_0 + b_1 x$$

The **standard error of the estimate** measures the error in any prediction based on the regression equation. It is given by

$$s_e = \sqrt{\frac{\Sigma(y - y_p)^2}{n - 2}}$$

where y_p is the value of y predicted by the regression equation for each value of x in the original data set.

You can test if the regression coefficient β_1 in the population regression equation

$$y = \beta_0 + \beta_1 x$$

is nonzero by using a *t*-test with $n - 2$ degrees of freedom.

The best interval estimate for the value of y corresponding to any desired value of x_0 using the sample regression equation is

$$y_p \pm t\sigma_{y_p}$$

where y_p is the value obtained from the regression equation,

$$\sigma_{y_p} = s_e \sqrt{1 + \frac{1}{n} + \frac{(x_0 - \bar{x})^2}{\Sigma x^2 - \frac{(\Sigma x)^2}{n}}}$$

and the t value used is based on $n - 2$ degrees of freedom.

Review Exercises
The following table lists the world records (in minutes and seconds) set in the mile run during the twentieth century. Use these values to answer the following questions:

x (Year)	y (Time)	
1911	4:15.4	(John Paul Jones, U.S.)
1913	4:14.6	(John Paul Jones, U.S.)
1915	4:12.6	(Norman Taber, U.S.)
1923	4:10.4	(Paavo Nurmi, Finland)
1931	4:09.2	(Jules Ladoumegue, France)
1933	4:07.6	(Jack Lovelock, New Zealand)
1934	4:06.8	(Glen Cunningham, U.S.)
1937	4:06.4	(Sidney Wooderson, Great Britain)
1942	4:06.2	(Gunder Haegg, Sweden)
1942	4:06.2	(Arne Andersson, Sweden)
1942	4:04.6	(Gunder Haegg, Sweden)
1943	4:02.6	(Arne Andersson, Sweden)
1944	4:01.6	(Arne Andersson, Sweden)
1945	4:01.4	(Gunder Haegg, Sweden)
1954	3:59.4	(Roger Bannister, Great Britain)
1954	3:58.0	(John Landy, Australia)
1957	3:57.2	(Derek Ibbotson, Great Britain)
1958	3:54.5	(Herb Elliott, Australia)
1962	3:54.4	(Peter Snell, New Zealand)
1964	3:54.1	(Peter Snell, New Zealand)
1965	3:53.6	(Michel Jazy, France)
1966	3:51.3	(Jim Ryun, U.S.)
1967	3:51.1	(Jim Ryun, U.S.)
1975	3:51.0	(Filbert Bayi, Tanzania)
1975	3:49.4	(John Walker, New Zealand)
1979	3:49.0	(Sebastian Coe, Great Britain)
1980	3:48.9	(Steve Ovett, Great Britain)
1981	3:48.8	(Sebastian Coe, Great Britain)
1981	3:48.7	(Steve Ovett, Great Britain)
1981	3:47.6	(Sebastian Coe, Great Britain)
1985	3:46.5	(Steve Cram, Great Britain)

1. Determine the correlation coefficient for the world-record time for the mile (in seconds) and the year in which the record was set. (Convert the times into seconds.)
2. Determine if there is correlation between the two at the $\alpha = .05$ level of significance.
3. Determine the linear regression equation to predict record time in the race as a function of the year.
4. Find the standard error of the estimate based on this regression equation.
5. Can you conclude that the slope of the regression line is nonzero at the $\alpha = .05$ level of significance?
6. What is your best estimate for the world record in the mile in the year 2000?

7. Estimate the year in which the world record may be 3 minutes, 30 seconds.
8. Estimate the world record for the mile run in the year 2050 based on this equation.
9. Based on this regression equation, estimate the world record for the mile in the year 1492. Is this a reasonable value? Explain your answer.
10. Based on this regression equation, estimate the world record for the mile in the year 2500. Is this a reasonable value? Explain your answer.

The Chi-Square Distribution

According to a major study released today, there are more high school and college graduates concentrated in the west than in any other part of the United States. Nationwide, approximately 76.9% of all adults over 25 years of age have high school diplomas. 81% of those adults living in the west have high school diplomas while only 73% of those in the south have completed high school. In the northeast and midwest, the percentages are 78% and 79%, respectively.

In a breakdown by state rather than by region, the highest high school graduation rate is in Utah and Washington State, where 88% of all adults have diplomas. They are followed by Alaska, Wyoming, Minnesota, Nevada and Oregon. The lowest rate is in Alabama, with 63% of all adults having graduated from high school. Among metropolitan areas, Seattle has the highest percentage of high school graduates, over 90%.

Much the same pattern exists with college graduates. The west again leads with 25% of all adults over age 25 having graduated from college. This is followed by the northeast with 23% and the midwest and south, each with 19%.

This study is based on a random sample of 5400 people across the country.

In a study such as this, the researchers have gathered data on educational achievements of residents in each of the 50 states, as well as in other geographical areas. When comparing the results in the different states, there are 50 proportions, one for each state. We should therefore expect that approximately half these sample proportions will naturally fall above the nationwide average and approximately half will fall below it.

In New York, several newspapers used the above study to conclude that New Yorkers fell below the national average for completing high school. Their implication is that the data indicate that New York's educational system is at fault. Is this justified? Might the difference be explainable by the fact that over the last decade, more new professional level jobs have been created in the west, so there has been a movement of better educated people in that direction? More importantly, in the fine print, we find that the high school graduation rate in New York is 76.7%, which *is* lower than the nationwide average of 76.9%. But is this really a significant difference?

What is more important is to know how far above or below the nationwide average a particular result falls. Is the result for any one state significantly better or worse than the results for the other 49 states? For that matter, are the percentages for the 50 states essentially the same, so that any apparent differences in the sample proportions simply represent the kind of variation we would expect from one sample to another? How do we determine if there are any significant differences among such a large number of sample proportions? That is, do the apparent differences among the sample proportions reflect an actual difference in the true proportion of people in each state who have completed high school or who have completed college?

In this chapter, we will develop a relatively simple method for testing whether there is indeed a difference among the set of underlying population proportions on the basis of a set of sample proportions.

11.1 Introduction to Chi-Square Analysis

In Section 9.6, we studied situations involving proportions from two populations to see whether they are equal or if there is a difference between them. For example, we may wish to determine whether a person's sex influences whether he or she is a smoker. More specifically, we ask if there is a difference in the proportion of men and women who smoke. We answered this question in Section 9.6 using a difference-of-proportions test based on two sets of sample data.

However, there are many comparable situations involving more than two categories. For instance, we might want to know if there is a difference in the proportions of people who smoke depending on their age, where we might consider four different age groups: under 20, 20 to 35, 36 to 50, and over 50 years. For each of these four age categories, there is a corresponding proportion of the population who smoke. We would like to determine whether the proportion of people in each age group in the population who smoke is the same. To do this, we take a random sample from the population. In that sample, suppose we find that 18% of those under age 20 smoke, that 21% of those between 20 and 35 years old smoke, that 13% of those from 36 to 50 years old smoke, and that 24% of those over age 50 smoke. We now want to determine if, in the entire population, the same proportion of people in each age group smoke or if there is any difference among the proportions. We use these percentages from the sample to test this hypothesis.

Before we consider how to approach this problem, we recall how we set up the hypothesis test used in the difference-of-proportions test. In such a case, we have two population proportions π_1 and π_2, and the null hypothesis asserts that they are equal, so that

$$H_0: \pi_1 = \pi_2$$

or equivalently,

$$H_0: \pi_1 - \pi_2 = 0$$

We then perform the difference-of-proportions test studied in Section 9.6.

Unfortunately, this cannot be done when there are three or more categories which result in three or more population proportions $\pi_1, \pi_2, \ldots, \pi_m$ for some number m.

To see how we must change the format when there are more than two categories, we return to the above situation where we seek to determine patterns of smoking based on four different age groups. We have four population proportions π_1, π_2, π_3 and π_4, representing the percentage of smokers in each age group in the population. We want to determine whether the four proportions are equal. Thus, in the hypothesis test for this example, the null hypothesis is

$$H_0: \pi_1 = \pi_2 = \pi_3 = \pi_4$$

To treat such a case involving three or more categories, we have to develop a new statistical procedure known as *chi-square analysis*.

We note that these ideas have many different applications. For instance, rather than studying smoking, we might want to conduct a political poll and analyze the support for a particular candidate (for, against or undecided) depending on the respondent's

- Sex (male or female)
- Religion (Protestant, Catholic, Jewish, Muslem, Hindu, Buddhist or other)
- Income
- Level of education

or many other possible characteristics. In each case, there are a number of different categories. Our goal is to determine if the proportions in each group in the population are the same or if the apparent differences found in a survey are just due to random variation.

We develop the appropriate methods in the following situation. Suppose Professor Abalone is accused of favoring the men in his classes by seemingly giving them higher grades. To prove or disprove this accusation, data on previous grades that Professor Abalone has given have been collected and are tabulated as follows:

	A or B	C or D	F
Men	120	80	50
Women	100	90	60

From this table, we observe that he has apparently given more high grades to men and more low grades to women. To analyze this more carefully, however, we need to look at the row and column totals for each category and then calculate the corresponding percentages or proportions for each group. Totaling the results, we obtain Table 11.1. We now see that there are a total of 500 students involved, 250 men and 250 women. Further, 220 out of 500, or 44% of all his grades, were A's and B's; 170 out of 500, or 34%, were C's and D's; and 110 out of 500, or 22%, were F's.

TABLE 11.1

	A or B	C or D	F	Total
Men	120	80	50	250
Women	100	90	60	250
	220	170	110	500

However, the distribution based on the sex of the student is somewhat different from this overall distribution. Thus, 120 out of 250, or 48%, of the men received A's and B's while only 100 out of 250, or 40%, of the women did so. At the other extreme, 50 out of 250, or 20%, of the men received F's while 60 out of 250, or 24%, of the women did so. We display these results in Table 11.2. Based on these results, can we conclude that Professor Abalone gives a higher percentage of high grades to men and a higher percentage of low grades to women? Alternatively, is the proportion of grades distributed in each category the same for both sexes or is there a difference based on gender?

TABLE 11.2

	A or B	C or D	F
All students	44%	34%	22%
Men only	48%	32%	20%
Women only	40%	36%	24%

To answer this, we need some additional terminology. First, the display given in Table 11.1 is known as a *contingency table*. Each of the positions in the table, corresponding to each possible category, is known as a *cell*. Moreover, the entries in this table are called the *observed frequencies* and are denoted by f_O. They are the actual numbers observed or recorded in each category from the random sample. The technique we develop will be a hypothesis test to determine whether a factor such as gender makes a significant difference among the proportions involved. As such, there is a null hypothesis which states

$$H_0: \pi_1 = \pi_2 = \cdots = \pi_m$$

or in words,

H_0: There is no difference among the proportions in each category.

The alternate hypothesis states:

H_a: There is a difference among the proportions.

To test this hypothesis, we must determine precisely what it means to say that there is no difference in the proportions, as stated in the null hypothesis. If we look back at the data in the contingency table shown in Table 11.1, we notice that one-half of the students involved are men and one-half are women. Professor Abalone gave 220 A's and B's. If there is no difference in the grade distribution based on gender, as the null hypothesis states, then we expect that one-half of these high grades will go to men and one-half will go to women. That is, if the null hypothesis is true, then 110 men should get A's and B's and 110 women should get A's and B's. This accounts for the total of 220 A's and B's.

By the same line of reasoning, since one-half of the students are men, we expect that one-half of the 170 C's and D's will be given to men. That is, 85 men and similarly 85 women should receive C's and D's. Finally, of the 110 F's given, we expect that 55 should be allotted to men and 55 to women. Thus, our expectation of what should happen ideally, assuming that there is no difference in the proportions based on gender, leads us to produce another table, this one consisting of these *expected frequencies*, which we denote by f_e.

f_e	A or B	C or D	F	Total
Men	110	85	55	250
Women	110	85	55	250
	220	170	110	500

Notice that this table has precisely the same subtotals in each row and each column as the previous table for the observed frequencies.

Rather than using two separate tables, it is customary to combine both the observed and expected frequencies into a single table and to list the expected frequency for each cell in parentheses, as shown:

f_o (f_e)	A or B	C or D	F	Total
Men	120 (110)	80 (85)	50 (55)	250
Women	100 (110)	90 (85)	60 (55)	250
	220	170	110	500

Our objective in performing a hypothesis test on a contingency table is to compare the observed frequencies and the expected frequencies. Keep in mind:

> The observed frequencies list exactly what **did happen** in our sample.
>
> The expected frequencies represent the theoretical expected outcomes of what **should have happened,** assuming the null hypothesis is true.

The question we seek to answer is, How different are these two sets of values? If the observed frequencies are relatively close to the expected frequencies, we can attribute any differences simply to chance variation between samples and so conclude that the null hypothesis may be valid. But if the differences between the two sets of frequencies are large, in some sense, then the differences cannot be attributed solely to chance—there likely is a difference among the proportions in the population, and we would have evidence to reject the null hypothesis.

Statisticians have developed a fairly simple method for measuring the differences between the observed and expected frequencies in such a contingency table. We use a quantity known as the *chi-square* (χ^2) *statistic* to do this. We study how to calculate this measure and to apply it in Section 11.2.

Before we proceed, however, one additional comment is in order. The argument we used to calculate the expected frequencies in the grade distribution versus gender example is actually too simplistic to use in most cases. It worked well here because the subtotals for the two rows resulted in a particularly simple 50-50 breakdown, so we could argue that the total for each column should be allocated in a similar way. However, in most cases, things are generally more complicated, and we need a more effective way to calculate the expected frequencies.

When we found the expected frequency 110 for the top left cell, what we actually did was the following: The proportion of men in the sample group is $\frac{250}{500}$. Therefore, of the 220 A or B grades (which is the column total for that cell), we expect the fraction $\frac{250}{500}$ of the 220 A's and B's to be given to men. That is, the expected frequency for this cell is calculated as

$$\frac{250}{500} \times 220 = 110$$

In a similar fashion, since 250 of the 500 students are women, we expect that

$$\frac{250}{500} \times 220 = 110$$

of these grades should be given to women.

If you examine these numbers, you will see that, to calculate the expected frequency for each cell, we multiply the column total for the cell by the fraction formed by dividing the corresponding row total by the grand total. Thus, since each cell in the table lies in a particular row and a particular column, the expected frequency for any cell is actually given by

$$\text{Expected frequency} = \frac{\text{row total} \times \text{column total}}{\text{grand total}}$$

under the assumption that there are no differences among the proportions in each category.

For instance, to calculate the expected frequency for the first (top left) cell, we simply multiply the row total for that cell (250) by the column total for that cell (220) and divide by the grand total (500). Thus, the expected frequency for the first cell is

$$f_e = \frac{250 \times 220}{500}$$
$$= 110$$

Similarly, for the second cell on the top row, we find that

$$f_e = \frac{\text{row total} \times \text{column total}}{\text{grand total}} = \frac{250 \times 170}{500}$$
$$= 85$$

and so forth for all the remaining entries.

Exercise Set 11.1

Mastering the Techniques

For each of the following contingency tables, construct the corresponding table of expected frequencies under the assumption that there are no differences among the proportions in each category.

1. | 20 | 50 |
    |----|-----|
    | 60 | 150 |

2. | 20 | 50 | 30 |
    |----|-----|-----|
    | 80 | 150 | 170 |

3. | 20 | 60 | 70 |
    |----|----|----|
    | 30 | 50 | 70 |
    | 40 | 60 | 90 |

4. | 25 | 40 | 15 | 60 |
    |----|-----|----|----|
    | 75 | 125 | 50 | 50 |
    | 50 | 35 | 45 | 30 |

Applying the Concepts

5. A consumer research organization conducts a survey of drivers to determine if there is any difference in their choice of brand of American-made car based on their gender. The results are as follows:

	Chrysler	Ford	General Motors
Women	70	80	150
Men	40	60	100

Construct the corresponding table of expected frequencies.

6. The consumer research organization in Exercise 5 conducts a further survey to determine if there is any difference between the proportion of drivers who express satisfaction or dissatisfaction with the performance of their American, Japanese or European cars. These are the results:

	American	Japanese	European
Satisfied	140	120	40
Not satisfied	70	40	20

Construct the corresponding table of expected frequencies.

7. A survey is conducted among workers in New York City to determine if there is any difference between the proportions of women and men who drive to work, take the bus to work, or take the subway to work. The results are as follows:

	Drive	Bus	Subway
Women	25	100	125
Men	75	120	205

Construct the corresponding table of expected frequencies for the proportions using the different modes of commutation based on gender.

8. A study is conducted comparing the proportions of people of different age groups who prefer several major brands of soft drink. These are the results:

	Under Age 25	Ages 25 to 45	Over Age 45
Coca-Cola	300	200	400
Pepsi Cola	400	150	250
Seven-Up	150	100	150

Determine the corresponding expected frequencies for the proportions of people preferring the different brands based on age.

11.2 Contingency Tables and the Chi-Square Distribution

In Section 11.1, we introduced the question of determining whether the sample data in a contingency table indicate that the proportions for different categories in a population are the same or different. We do this by measuring how "close" the set of expected frequencies is to the set of observed frequencies. If the observed frequencies (what did happen) are relatively close to the expected frequencies (what should have happened if there is no difference in the population proportions), we can attribute any apparent differences to chance variation among samples and hence conclude that the underlying population proportions are likely the same. But if the difference between the two sets of frequencies is large, in some sense, then the difference cannot be attributed solely to chance—we conclude that the proportions are not the same.

We now see how to calculate the value for the chi-square statistic which measures how *close* the two sets of frequencies are. We also investigate how to use this measure to decide whether there is any difference in the proportions based on the sample data in the contingency table.

The difference between the observed and expected frequencies in a contingency table is measured by the *chi-square* (χ^2) *statistic*, given by the formula

$$\chi^2 = \sum \frac{(f_o - f_e)^2}{f_e} \tag{11.1}$$

To calculate this statistic for a contingency table, we perform the following steps:

1. Calculate the difference $f_o - f_e$ for each cell.
2. Square the difference $(f_o - f_e)^2$ for each cell.
3. Divide each of the quantities from step 2 by the expected frequency of the corresponding cell

$$\frac{(f_o - f_e)^2}{f_e} \quad \text{for each cell}$$

4. Add the values obtained in step 3 for all cells.

We illustrate these calculations using the example on Professor Abalone's grade distribution compared to the gender of his students from Section 11.1. The extended contingency table of observed and expected frequencies is as follows:

f_o (f_e)	A or B	C or D	F	Total
Men	120 (110)	80 (85)	50 (55)	250
Women	100 (110)	90 (85)	60 (55)	250
	220	170	110	500

The value of the χ^2 statistic can be calculated either in tabular form or from the formula directly. We demonstrate both approaches. First, in tabular form, we organize the data into a single table and consider each cell across one row:

	f_o	f_e	$f_o - f_e$	$(f_o - f_e)^2$	$\frac{(f_o - f_e)^2}{f_e}$
Row 1	120	110	10	100	.909 (= $\frac{100}{110}$)
	80	85	−5	25	.294
	50	55	−5	25	.455
Row 2	90	110	−10	100	.909
	80	85	5	25	.294
	60	55	5	25	.455

Finally, the chi-square statistic is the sum of the entries in the last column

$$\chi^2 = .909 + .294 + .455 + .909 + .294 + .455$$
$$= 3.316$$

Alternatively, we can calculate the chi-square value directly from the formula:

$$\chi^2 = \sum \frac{(f_O - f_e)^2}{f_e}$$

$$= \frac{(120 - 110)^2}{110} + \frac{(80 - 85)^2}{85} + \frac{(50 - 55)^2}{55} + \frac{(100 - 110)^2}{110}$$

$$+ \frac{(90 - 85)^2}{85} + \frac{(60 - 55)^2}{55}$$

$$= \frac{10^2}{110} + \frac{(-5)^2}{85} + \frac{(-5)^2}{55} + \frac{(-10)^2}{110} + \frac{5^2}{85} + \frac{5^2}{55}$$

$$= \frac{100}{110} + \frac{25}{85} + \frac{25}{55} + \frac{100}{110} + \frac{25}{85} + \frac{25}{55}$$

$$= .909 + .294 + .455 + .909 + .294 + .455$$

$$= 3.316$$

Chi-square values play a comparable role in chi-square analysis to what z and t values played in previous situations. The z values refer to a normal distribution; the t values refer to a t-distribution based on $n - 1$ degrees of freedom. In an analogous way, χ^2 values refer to another distribution known as the *chi-square distribution*. As with the t-distribution, the chi-square distribution also depends on the number of degrees of freedom. Therefore, there are many different chi-square distributions. Unlike the normal and t-distributions, however, the chi-square distribution is not symmetric about a center. It is skewed to the right. In Figure 11.1, we show several different chi-square distributions with different degrees of freedom. As the number of degrees of freedom increases, the corresponding chi-square distribution approaches a normal distribution.

To find the number of degrees of freedom for the chi-square distribution associated with a particular contingency table, we take the product of 1 less than the number of rows and 1 less than the number of columns in the contingency table:

$$\text{df} = (\text{number of rows} - 1) \times (\text{number of columns} - 1)$$
$$= (r - 1) \times (c - 1)$$

In the above example, there are 2 rows and 3 columns, so that

$$\text{df} = (2 - 1) \times (3 - 1) = 1 \times 2 = 2$$

FIGURE 11.1

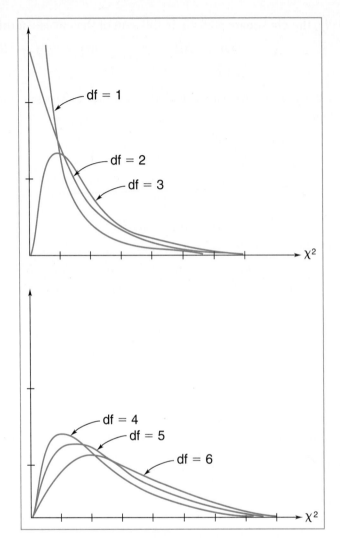

Recall that when we conducted a hypothesis test on the mean μ based on a small sample, we calculated the t value for the data and compared it to the critical value for t listed in a table of values for the t-distributions. In a totally analogous way, we compare the value calculated for the χ^2 statistic based on the contingency table data ($\chi^2 = 3.316$ in the above example) to a critical value for χ^2 given in Table VI showing the critical values for the chi-square distribution. The entries listed represent critical values for a 5% level of significance, $\chi^2_{0.05}$, a 1% level of significance, $\chi^2_{0.01}$, and other levels corresponding to the different numbers of degrees of freedom.

In calculating the value of the χ^2 statistic by using equation (11.1), notice that when the expected and observed frequencies are quite different from each other, the value for χ^2 is large and we then reject the null hypothesis. But when the expected frequencies are close to the observed frequencies, the value for χ^2 is small and we cannot reject the null hypothesis that there is no difference among the population proportions. The evidence is not strong enough. Realize that all hypothesis tests involving contingency tables are one-tailed tests using the tail to the right.

Suppose now that we want to complete the above hypothesis test on the issue of gender versus grades at the 5% level of significance. From Table VI, the critical value we test against, using df = 2, is

$$\chi^2 = 5.991$$

See Figure 11.2. Since the value we calculated for $\chi^2 = 3.316$ is less than the critical value from the table, we conclude that we cannot reject the null hypothesis. There is insufficient evidence to conclude that there is a difference among the proportions of men and women who received each grade. The differences observed can be attributed simply to random variation. The charge of favoritism against Professor Abalone cannot be substantiated based on the evidence from this sample at the 5% level of significance.

On the other hand, if the value we calculated for χ^2 came out larger than the critical value given in Table VI, then we would reject the null hypothesis. Some of the differences between observed and expected frequencies would be "too large" to attribute just to chance. There would likely be a difference among the proportions. Thus, good overall agreement between the observed and the expected frequencies leads to small values for χ^2. Disagreement in some cells yields a large value for χ^2, and this usually leads to our rejecting the null hypothesis.

We illustrate the use of chi-square analysis in a number of additional examples below. As you will see, even though the method is relatively lengthy, it is quite simple to perform once you have used it several times. In addition, it is particularly natural to implement the procedure by computers, as you will see in the Computer Corner at the end of this section.

FIGURE 11.2

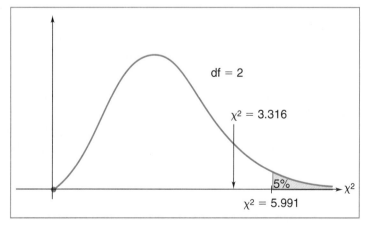

EXAMPLE 11.1 A study is conducted to compare the number of car accidents with the gender of the driver. The results are given in the following contingency table:

f_o	\multicolumn{4}{c}{Accidents}	Total
	0	1	2	3 or more	
Men	100	250	100	50	500
Women	200	300	150	50	700
	300	550	250	100	1200

Test if there is any difference in the proportions of men and women who have had various numbers of car accidents at the $\alpha = .05$ level of significance.

SOLUTION For this problem, the null hypothesis is

H_0: There is no difference among the proportions of men and women having a given number of accidents.

while the alternate hypothesis is

H_a: There is some difference among the proportions.

To perform the chi-square analysis, first we must obtain the expected frequencies. Thus, for the first cell, we find

$$f_e = \frac{\text{Row total} \times \text{column total}}{\text{Grand total}} = \frac{500 \times 300}{1200}$$

$$= 125$$

for the second cell,

$$f_e = \frac{500 \times 550}{1200} = 229.2$$

and so forth across the top row. Similarly, for the first cell on the second row,

$$f_e = \frac{700 \times 300}{1200} = 175$$

and so forth. The table of observed and expected frequencies is as follows:

f_o (f_e)	\multicolumn{4}{c}{Accidents}	Total
	0	1	2	3 or more	
Men	100 (125)	250 (229.2)	100 (104.2)	50 (41.6)	500
Women	200 (175)	300 (320.8)	150 (145.8)	50 (58.4)	700
	300	550	250	100	1200

Incidentally, you probably noticed that many of these calculations are actually unnecessary. Once the first cell of the top row is known, the first cell of the second row is predetermined by the column total of 300. In fact, all the entries of the bottom row will be predetermined by the top row and the individual column totals. Further, once the first three values of the top row are found, then the last entry is automatically predetermined by the row total of 500. Thus, all we actually need to calculate are three entries in this table. All the other entries can then be obtained directly from these three and the appropriate subtotals.

The number of degrees of freedom in this example is given by

$$\text{df} = (r - 1) \times (c - 1) = (2 - 1) \times (4 - 1)$$
$$= 1 \times 3 = 3$$

It is no coincidence that this number 3 is precisely the same as the number of expected frequencies that *had* to be calculated (while all other expected frequencies could then be obtained by subtracting from the row and column totals). The degrees of freedom actually indicate the number of entries in the table of expected frequencies that are "free" in the sense that all others are predetermined by the row and column totals.

We now calculate the χ^2 value for this contingency table:

$$\chi^2 = \frac{(100 - 125)^2}{125} + \frac{(250 - 229.2)^2}{229.2} + \frac{(100 - 104.2)^2}{104.2}$$
$$+ \frac{(50 - 41.6)^2}{41.6} + \frac{(200 - 175)^2}{175}$$
$$+ \frac{(300 - 320.8)^2}{320.8} + \frac{(150 - 145.8)^2}{145.8} + \frac{(50 - 58.4)^2}{58.4}$$
$$= \frac{(-25)^2}{125} + \frac{(20.8)^2}{229.2} + \frac{(-4.2)^2}{104.2} + \frac{(8.4)^2}{41.6} + \frac{25^2}{175} + \frac{(-20.8)^2}{320.8}$$
$$+ \frac{(4.2)^2}{145.8} + \frac{(-8.4)^2}{58.4}$$
$$= \frac{625}{125} + \frac{432.64}{229.2} + \frac{17.64}{104.2} + \frac{70.56}{41.6} + \frac{625}{175}$$
$$+ \frac{432.64}{320.8} + \frac{17.64}{145.8} + \frac{70.56}{58.4}$$
$$= 5 + 1.888 + .169 + 1.696 + 3.571 + 1.349 + .121 + 1.208$$
$$= 15.002$$

Since there are 3 degrees of freedom and we are testing this at the 5% level of significance, we find the critical value of

$$\chi^2 = 7.815$$

from Table VI. Since the χ^2 value calculated for our data, $\chi^2 = 15.002$, is

FIGURE 11.3

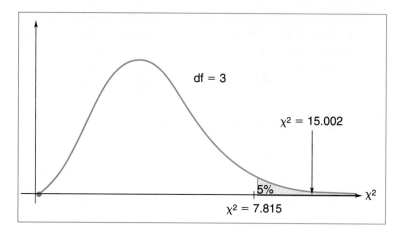

considerably larger than this critical value (see Figure 11.3), we reject the null hypothesis and conclude that there is a difference in the proportions of car accidents based on the gender of the driver. ∎

To perform this type of chi-square analysis for a contingency table and come to a meaningful conclusion, it is recommended that the expected frequencies f_e in each cell be greater than or equal to 5. If any of the expected frequencies is less than 5, then the chi-square distribution may not be a good approximation and the results become invalid. Thus, when working with a contingency table, if any of the expected frequencies is smaller than 5, we combine two or more of the categories into a single category. We illustrate this in the following example.

EXAMPLE 11.2 A survey is conducted among the evening students at a college to determine if there is any difference in student attitudes toward sex education in the schools (the level should be increased, should remain the same, should be decreased, have no opinion) depending on marital status (married, single, divorced or widowed). The results of the survey are as follows:

f_0	Increase	Same	Decrease	No Opinion
Married	100	200	50	30
Single	300	400	80	70
Divorced or widowed	20	30	5	5

Test if there is any difference in the proportions of students with each response depending on their marital status at the 5% level of significance.

SOLUTION To solve this problem, we first calculate the row and column totals and then find the expected frequencies, obtaining

f_o (f_e)	Increase	Same	Decrease	No Opinion	Total
Married	100 (123.7)	200 (185.6)	50 (39.8)	30 (30.9)	380
Single	300 (276.7)	400 (415.1)	80 (89.0)	70 (69.2)	850
Divorced or widowed	20 (19.5)	30 (29.3)	5 (6.3)	5 (4.9)	60
	420	630	135	105	1290

(Note that, due to rounding in the intermediate calculations, the sum of the expected frequencies in each row or column may not always exactly equal the row or column total.)

We note that the last cell in the above table has an expected frequency of 4.9, which is smaller than 5. As a consequence, the results of performing a chi-square analysis on this contingency table may be inaccurate. We therefore combine two categories to eliminate this difficulty. There are several ways we could do this. One possibility is to combine the last column, No Opinion, with one of the other columns, possibly that the level should remain the same. However, these are not necessarily compatible opinions, and this would probably not be advisable.

A second possibility is to combine the last row, divorced or widowed, with one of the other two rows. Actually, this could be done with either of the two categories Married or Single. In fact, the choice as to which one to do might very well depend on the question being addressed. For some issues, we might anticipate that divorced or widowed people would react more similarly to single people. For others, such as the question under consideration here dealing with sex education in the schools, it is probably more intelligent to classify divorced and widowed people with married individuals. As a consequence, we construct a new table of observed frequencies:

f_o	Increase	Same	Decrease	No Opinion	Total
Married, divorced, or widowed	120	230	55	35	440
Single	300	400	80	70	850
	420	630	135	105	1290

The corresponding expanded table including the expected frequencies is as follows:

f_o (f_e)	Increase	Same	Decrease	No Opinion	Total
Married, divorced, or widowed	120 (143.3)	230 (214.9)	55 (46.0)	35 (35.8)	440
Single	300 (276.7)	400 (415.1)	80 (89.0)	70 (69.2)	850
	420	630	135	105	1290

FIGURE 11.4

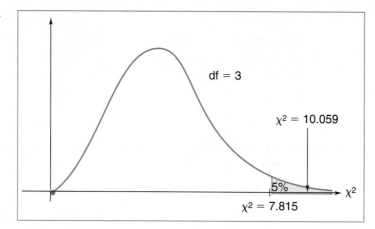

Notice that all the f_e values are now considerably larger than 5.
Using this table of entries, we calculate

$$\chi^2 = \frac{(120 - 143.3)^2}{143.3} + \frac{(230 - 214.9)^2}{214.9} + \frac{(55 - 46)^2}{46} + \frac{(35 - 35.8)^2}{35.8}$$
$$+ \frac{(300 - 276.7)^2}{276.7} + \frac{(400 - 415.1)^2}{415.1} + \frac{(80 - 89)^2}{89} + \frac{(70 - 69.2)^2}{69.2}$$

which eventually reduces to

$$\chi^2 = 10.059$$

The number of degrees of freedom is found from the new, altered table rather than from the original contingency table. We therefore have

$$(r - 1) \times (c - 1) = 1 \times 3 = 3$$

degrees of freedom, so that the critical value for χ^2 at the $\alpha = .05$ level of significance is $\chi^2 = 7.815$. Based on this, we conclude that there is a difference in attitude toward sex education depending on marital status. See Figure 11.4. ∎

Computer Corner

Since chi-square analysis is such a heavily computational procedure, it is natural to perform such calculations on the computer. We have provided Program 35: Chi-Square Analysis. When you run this program, it first asks for the number of rows and the number of columns for your data. The program then asks you to supply the entries for each cell in the contingency table. As you type the entries, they are positioned in the appropriate cell. After all the entries have been completed, the program displays the row and column totals and then the expected frequencies for each cell. The expected frequencies are printed in parentheses within the cell. Finally, the program calculates and displays the value of χ^2 for your data and the number of degrees of freedom. However, you will still need to use the χ^2 table to decide whether any difference exists among the proportions.

Incidentally, you may notice that a difference sometimes occurs between the value for χ^2 given by the program and the value you calculate by hand. In the above examples, we performed the calculations by rounding each number to three decimal places for display purposes. However, the program does not round until the final answer is printed. Since quite a few such roundings take place when you do the calculations by hand, the difference can become relatively large. As a result, you will occasionally notice a fairly large difference between a chi-square value you calculate by hand and one supplied by the program if you round your intermediate results. We recommend that you avoid such intermediate roundings as a matter of course.

The program also allows you to edit the data either to correct an error in entry or to see the effects of changing the entries in one or more cells. Use the left, right, up and down keys to highlight the entry you desire to change, press the space bar to indicate that you want to change that cell, type in the new entry and press Enter. When you are finished changing the entries, type N for "no more changes." The program then displays the results as described above.

We assume hereafter that you will use this program to perform all chi-square analysis computations.

EXAMPLE 11.3

As part of a nationwide survey of public attitudes on various social issues, the question of stricter gun control legislation was considered. Possible responses were: should be Stricter, is OK as is, and is Too Strict now. The results are to be analyzed based on the level of education of the respondents: did not complete high school, high school graduate, college graduate. The results are as follows:

f_0	Stricter	OK	Too Strict
Non-high school graduate	298	311	167
High school graduate	489	578	297
College graduate	503	504	228

Test if level of education makes any difference in terms of opinion on this issue at the $\alpha = .01$ level of significance.

SOLUTION The null and alternate hypotheses are

H_0: There is no difference in the proportions based on level of education.

H_a: There is a difference in the proportions.

FIGURE 11.5

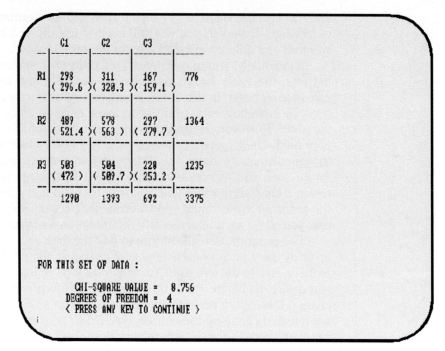

Using Program 35: Chi-Square Analysis, we obtain the results shown in Figure 11.5. In particular, the chi-square statistic is $\chi^2 = 8.756$. Since there are 4 degrees of freedom and we are conducting the test at the 1% level of significance, the critical value is $\chi^2 = 13.277$, as shown in Figure 11.6. Since our sample value is smaller than this critical value, we cannot reject the null hypothesis, and so we conclude that there may not be any difference in the proportions having different opinions on stricter gun control legislation based on level of education. ∎

FIGURE 11.6

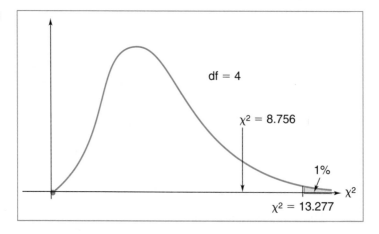

We have also provided Program 36: Chi-Square Distributions to allow you to experiment with the shapes of different chi-square distributions. You can use the program to investigate the shapes of a sequence of chi-square distributions with successively larger numbers of degrees of freedom (much like what is displayed in Figure 11.1). Notice how successive distribution curves seem to become more symmetric and seemingly approach a normal distribution as the number of degrees of freedom increases. The program also allows you to investigate how the chi-square distribution with any desired number of degrees of freedom compares to the approximating normal distribution. You just have to enter the value for the degrees of freedom, and the program draws the graph of the corresponding chi-square distribution and the approximating normal distribution. Based on your investigations, how many degrees of freedom seem to be necessary for the chi-square distribution to be reasonably well approximated by a normal distribution?

Minitab Methods

You can also use Minitab to perform a chi-square analysis for any contingency table. The observed frequency values should be entered into a series of columns, say C1 through C4, by using the **READ** command. Once this is done, all that you need do is to enter the Minitab command

```
CHISQUARE C1-C4
```

and Minitab will respond with a table giving the observed and the expected frequencies, the associated totals, the value for the chi-square statistic (composed of the individual contributions for each cell in the table), and the number of degrees of freedom. You then must compare this calculated value of χ^2 to the critical value found in Table VI corresponding to your desired significance level and the number of degrees of freedom. A sample session is shown in Figure 11.7.

Exercise Set 11.2

Mastering the Techniques

For the contingency tables in Exercises 1 to 4, determine

(a) The number of degrees of freedom
(b) The value of the chi-square statistic χ^2

1. | 20 | 50 |
   | 60 | 150 |

2. | 20 | 50 | 30 |
   | 80 | 150 | 170 |

FIGURE 11.7

```
MTB > READ C1-C4
      3ROWS READ
MTB > END
MTB >
MTB > CHISQUARE C1-C4

Expected counts are printed below observed counts
         C1      C2      C3      C4     Total
   1     34      47      63      38      182
       36.79   42.65   66.42   36.14
   2     26      36      57      42      161
       32.55   37.73   58.75   31.97
   3     53      48      84      31      216
       43.66   50.62   78.83   42.89
 Total  113     131     204     111      559

ChiSq = 0.212 + 0.443 + 0.176 + 0.096 +
        1.316 + 0.079 + 0.052 + 3.147 +
        1.996 + 0.136 + 0.340 + 3.297 = 11.290
DF = 6

MTB >
```

3.	20	60	70	4.	25	40	15	60
	30	50	70		75	125	50	50
	40	60	90		50	35	45	30

Applying the Concepts

5. A consumer research organization conducts a survey of drivers to determine if there is any difference in their choice of brand of American-made car based on their gender. These are the results:

	Chrysler	**Ford**	**General Motors**
Women	70	80	150
Men	40	60	100

Test whether there is any difference in the proportion of drivers who prefer a particular brand based on gender at the $\alpha = .05$ level of significance.

6. The consumer research organization in Exercise 5 conducts a further survey to determine if there is a difference between the proportion of drivers who express satisfaction or dissatisfaction with the performance of their American, Japanese or European cars. These are the results:

	American	Japanese	European
Satisfied	140	120	40
Not satisfied	70	40	20

Test whether there is any difference in the proportions who are satisfied with the different cars at the $\alpha = .05$ level of significance.

7. A survey is conducted among workers in New York City to determine if there is any difference between the proportions of women and men who drive to work, take the bus to work, or take the subway to work. The results are as follows:

	Drive	Bus	Subway
Women	25	100	125
Men	75	120	205

Test whether there is any difference in the proportions using the different modes of commutation based on gender at the $\alpha = .01$ level of significance.

8. A study is conducted comparing the proportions of people of different age groups who prefer several major brands of soft drink, with these results:

	Under Age 25	Ages 25 to 45	Over Age 45
Coca-Cola	300	200	400
Pepsi Cola	400	150	250
Seven-Up	150	100	150

Test whether there is any difference in the proportions of people preferring the different brands based on age at the 5% level of significance.

Computer Applications

Use Program 35: Chi-Square Analysis to answer the following questions.

9. A publisher test-markets a new novel, using four different colors for the dust jacket (red, blue, yellow and green) to determine if there is any difference in the proportions of sales of the different color jackets depending on gender. The results are as follows:

	Red	Blue	Yellow	Green
Women	62	34	71	42
Men	125	223	52	54

Test whether there is any difference at the 1% significance level between the proportions of men and women who are attracted to different colors on the dust jackets.

10. A survey is conducted comparing the proportions of college students having various political leanings based on the geographic location of their schools, with these results:

	Northwest	Northeast	Southwest	Southeast
Democratic	203	504	296	408
Republican	328	307	504	315
Undecided	97	294	192	194

Test whether there is any difference in the proportions of political leanings based on geographic location at the 5% level of significance.

11. A study is conducted asking people in different cities to express their degree of satisfaction with the city in which they live. The results are as follows:

City	Very Happy	Somewhat Happy	Unhappy
San Diego	220	121	63
Chicago	130	207	75
Pittsburgh	84	54	24
Denver	156	95	43
Boston	122	164	73

Test whether there is any difference in the proportions of people who express satisfaction living in each of these five cities at the $\alpha = .05$ level of significance.

11.3 The Goodness-of-Fit Test

Throughout our study of statistics, we have repeatedly talked about random events, random variables, random samples and random numbers. However, we have never directly addressed the question of how we know that something is actually random. For instance, we have often referred to a fair coin or a fair die, but how do we know that a particular coin or die is actually fair? It turns out that a simple variation on the type of chi-square analysis we applied in Section 11.2 allows us to analyze some kinds of data to study randomness and related characteristics.

Let us consider the question of whether a die is fair. To test it, we would probably decide to roll it a large number of times and observe the frequencies of the possible outcomes 1, 2, 3, ..., 6. We would then attempt to decide if the actual outcomes came close to matching what we would expect to get if the die were actually fair. For example, suppose we roll this die 600 times. Since the probability that any particular face occurs is $\frac{1}{6}$ (assuming that the die is in fact fair), we expect one hundred 1s, one hundred 2s, one hundred 3s, ..., one hundred 6s. Of course, it is unrealistic to expect precisely this breakdown in practice, but we expect a set of outcomes reasonably close to this set of theoretical values. Suppose we obtain the following set of *observed frequencies*

Outcome	1	2	3	4	5	6	Total
f_O	106	92	97	105	88	112	600

compared to these *expected frequencies* if the die is fair:

Outcome	1	2	3	4	5	6	Total
f_e	100	100	100	100	100	100	600

Specifically, we wish to determine if the variations from the expected frequencies that occur in the observed frequencies are just the result of chance variation or if they indicate a real departure from the expected values. Clearly, this is very similar to the chi-square analysis we did in Section 11.2. In fact, this decision can be made in the same way with one simple, but important, modification.

In performing a chi-square analysis for a contingency table, it is essential to determine the number of degrees of freedom for the corresponding chi-square distribution. In Section 11.2, we saw that for any contingency table

$$\text{df} = (r - 1) \times (c - 1)$$

which is equal to the number of "free," or independent, entries.

Unfortunately this formula does not apply in the present case, so we must determine the number of degrees of freedom directly. In this case, we are looking at information on 600 rolls of a die with 6 possible outcomes. Once the first 5 entries for the expected frequencies have been determined (for obtaining a 1, a 2, ..., a 5), the last entry is predetermined, given the fact that the total number of rolls is 600. Therefore, there are $6 - 1 = 5$ degrees of freedom.

We now test to determine whether the above die is fair based on the information obtained. To begin, the null hypothesis represents our belief that the die is fair, or equivalently that there is no difference between the observed and the expected frequencies. That is, the null hypothesis states

H_0: The die is fair.

compared to the alternate hypothesis which states

H_a: The die is not fair.

Suppose we test this hypothesis at the $\alpha = .05$ level of significance. The test is essentially identical to the one used in Section 11.2.

We begin by calculating the χ^2 value as

$$\chi^2 = \sum \frac{(f_O - f_e)^2}{f_e}$$

$$= \frac{(106 - 100)^2}{100} + \frac{(92 - 100)^2}{100} + \frac{(97 - 100)^2}{100} + \frac{(105 - 100)^2}{100}$$

$$+ \frac{(88 - 100)^2}{100} + \frac{(112 - 100)^2}{100}$$

$$= \frac{6^2}{100} + \frac{(-8)^2}{100} + \frac{(-3)^2}{100} + \frac{5^2}{100} + \frac{(-12)^2}{100} + \frac{12^2}{100}$$

$$= \frac{36}{100} + \frac{64}{100} + \frac{9}{100} + \frac{25}{100} + \frac{144}{100} + \frac{144}{100}$$

$$= .36 + .64 + .09 + .25 + 1.44 + 1.44 = 4.22$$

Since there are 5 degrees of freedom and $\alpha = .05$, we find the critical value of $\chi^2 = 11.070$ from Table VI. See Figure 11.8. As a result, we cannot reject the null hypothesis. The apparent differences between the observed and expected frequencies for the outcomes on the die are not judged significant. We cannot conclude that the die is unfair.

The type of procedure we just performed is known as a *goodness-of-fit test* since it provides us with a means of determining whether a set of data is a good fit to a theoretical model or prediction. In general, the number of degrees of freedom for any goodness-of-fit test is given by

$$df = \text{number of categories} - 1$$

CAUTION Do not confuse the number of categories (e.g., the number of outcomes on the die) with the number of measurements.

We further illustrate these ideas in the following series of examples.

FIGURE 11.8

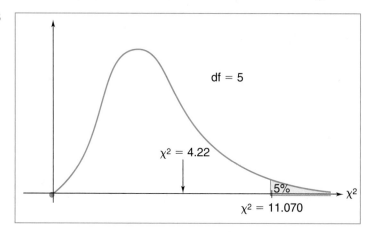

EXAMPLE 11.4

A large company believes that many of its employees are taking advantage of a liberal absence policy by taking off a disproportionate number of Mondays and Fridays. The following set of data, showing the number of employee absences by the day of the week, is collected.

Day	Monday	Tuesday	Wednesday	Thursday	Friday
f_O	57	39	37	54	63

Does this set of data indicate that the company's suspicions are valid? Test at the $\alpha = .05$ level of significance.

SOLUTION The null hypothesis states

H_0: There is no difference in the number of employee absences based on the day of the week.

while the alternate hypothesis is

H_a: There is a difference.

Since the data involve a total of 250 absences, the null hypothesis implies that these 250 should be evenly distributed over the 5 days of the week, and so we would expect 50 absences per day. Thus, the expected frequencies are:

Day	Monday	Tuesday	Wednesday	Thursday	Friday	Total
f_e	50	50	50	50	50	250

Therefore, we find that

$$\chi^2 = \frac{(57-50)^2}{50} + \frac{(39-50)^2}{50} + \frac{(37-50)^2}{50}$$
$$+ \frac{(54-50)^2}{50} + \frac{(63-50)^2}{50}$$
$$= \frac{49}{50} + \frac{121}{50} + \frac{169}{50} + \frac{16}{50} + \frac{169}{50}$$
$$= .98 + 2.42 + 3.38 + .32 + 3.38 = 10.48$$

Since there are $5 - 1 = 4$ degrees of freedom, we compare this to the critical value of $\chi^2 = 9.488$ (using $\alpha = .05$) from Table VI. See Figure 11.9. We therefore reject the null hypothesis and so conclude that there is a difference in absenteeism based on the day of the week. ∎

EXAMPLE 11.5

The random number generator in a computer is being tested to determine if the numbers generated are indeed random. As part of the test, the following 100 one-digit numbers are generated:

FIGURE 11.9

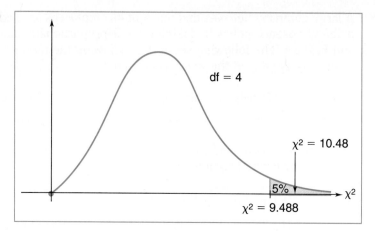

7	2	7	5	4	3	5	2	5	4	5	6	7	0	9	6	9	7	0	4
8	1	3	9	4	6	9	3	4	9	2	7	6	6	5	0	8	1	1	4
9	1	3	9	6	2	5	6	9	8	0	5	4	5	0	7	0	0	9	3
5	7	0	3	7	2	5	8	5	2	9	9	6	1	5	5	9	5	4	8
1	0	2	9	1	2	0	8	8	5	1	5	8	7	0	2	1	1	9	8

Test if they are random at the $\alpha = .05$ level of significance.

SOLUTION The null hypothesis states

H_0: The digits are random.

while the alternate hypothesis is

H_a: The digits are not random.

We begin by counting the number of 0s, 1s, ..., in this set of 100 digits to produce the observed frequencies:

Digit	0	1	2	3	4	5	6	7	8	9
f_O	11	10	9	6	8	16	8	9	9	14

If the digits are indeed random, then we expect an equal number of occurrences of each digit, and so the expected frequencies are

Digit	0	1	2	3	4	5	6	7	8	9
f_e	10	10	10	10	10	10	10	10	10	10

After the usual calculation, we find that the χ^2 statistic for this set of data is $\chi^2 = 8.000$. Since there are $10 - 1 = 9$ degrees of freedom and the level of significance is $\alpha = .05$, we find from Table VI that the critical value is $\chi^2 =$

FIGURE 11.10

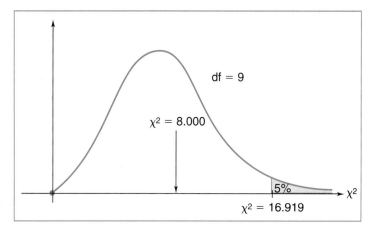

16.919. See Figure 11.10. Consequently, we cannot reject the null hypothesis. There is no difference between the observed values and what we would expect under the assumption that the digits generated are random. ∎

The above situations all involve cases in which the assumption of no difference implies that the full set of observations should be assessed uniformly across all categories. The underlying distribution, in fact, is known as a *uniform distribution*. We can also have occasion to check if our observed data behave according to some other distribution or rule. In the next example, we compare a set of observed data to a binomial distribution to see how good a fit it is.

EXAMPLE 11.6 A doctor claims that a certain drug will increase a woman's likelihood of having boys rather than girls. To prove this claim, he administers the drug to a group of women and then studies 80 of them who have had three children while taking the drug. These are the results:

No. boys	3	2	1	0
Frequency	14	36	24	6

Test at the $\alpha = .05$ level of significance if this drug does indeed have an effect on the sex of the offspring.

SOLUTION The null hypothesis states

H_0: The drug has no effect on the gender of the children.

while the alternate hypothesis states

H_a: The drug does have an effect.

We cannot simply apportion the 80 outcomes as 20, 20, 20 and 20, since the theoretical probabilities of the different outcomes are not the same. In par-

FIGURE 11.11

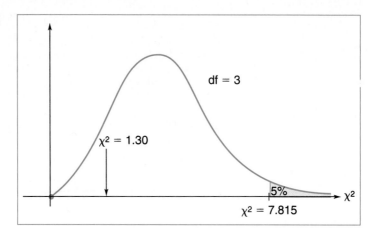

ticular, the number of boys follows a binomial distribution with $n = 3$ and $\pi = .5$ under the null hypothesis. The associated probabilities are

$$P(3 \text{ boys}) = \tfrac{1}{8}$$
$$P(2 \text{ boys}) = \tfrac{3}{8}$$
$$P(1 \text{ boy}) = \tfrac{3}{8}$$
$$P(0 \text{ boys}) = \tfrac{1}{8}$$

As a consequence, the expected frequencies must reflect these probabilities. Based on 80 repetitions, we expect the following number of outcomes for each:

No. boys	3	2	1	0
f_e	10	30	30	10

We now calculate the χ^2 value for this set of data and so obtain $\chi^2 = 1.30$. The critical value of χ^2 based on $4 - 1 = 3$ degrees of freedom and $\alpha = .05$ is $\chi^2 = 7.815$. See Figure 11.11. As a result, we cannot reject the null hypothesis. The data do not validate the claim, and we conclude that the drug has no significant effect on the gender of the children. ∎

Exercise Set 11.3

Applying the Concepts

1. A die is rolled 360 times with the following outcomes:

1	2	3	4	5	6
53	57	56	63	62	69

Based on these data, test whether the die is fair at the $\alpha = .05$ level of significance.

2. A company keeps records on the number of major production defects in its product according to the day of the week. Based on the following set of data, can the company conclude that there is any difference in the number of major defects produced depending on the day of the week at the $\alpha = .01$ level of significance? (Assume that total daily production is constant.)

Monday	Tuesday	Wednesday	Thursday	Friday
230	170	180	190	250

3. A fire department keeps track of the number of false alarms received over the course of a week:

Sunday	Monday	Tuesday	Wednesday	Thursday	Friday	Saturday
22	12	15	8	14	27	35

At the $\alpha = .01$ level of significance, is there any difference in the number of calls received based on the day of the week?

4. A state police department keeps track of the number of fatal automobile accidents throughout the state over the course of a year, with the following results:

Jan.	Feb.	Mar.	Apr.	May	June	July	Aug.	Sept.	Oct.	Nov.	Dec.
50	32	16	4	30	45	35	22	16	16	21	54

On the basis of these data, can the police conclude that the month makes any difference in terms of the number of fatal accidents at the 5% level of significance?

5. Two dice are rolled 360 times with the following results, based on the total dots on the two faces:

2	3	4	5	6	7	8	9	10	11	12
8	15	26	42	50	65	48	44	32	22	8

At the 5% level of significance, can you conclude that the pair of dice is fair?

6. A group of 160 families, each having four children, are studied with regard to the gender of the children with the following results:

No. girls	0	1	2	3	4
No. families	7	33	63	45	12

Test at the 5% level of significance if this breakdown is different from what we would expect based on probability theory.

7. Blood types in the general population are distributed as follows: 45% have type O, 40% have type A, 10% have type B, and 5% have type AB. A group of 200 immigrants from a certain area are tested and found to have the following frequencies for the different blood types:

Type	O	A	B	AB
Frequency	80	72	24	24

On the basis of these results, can we conclude that people from this area have a different distribution of blood types at the $\alpha = .05$ level of significance?

Computer Applications

8. Run Program 5: Coin Flipping Simulation with $n = 3$ coins, and record the numerical results from the text screen. Test these results to determine whether the simulation produces results that agree with the predictions based on probability theory for a binomial process at the $\alpha = .05$ level of significance. Repeat this for a second run.
9. Repeat Exercise 8 using $n = 4$ coins.
10. Run Program 6: Dice Rolling Simulation and record the numerical results for the 360 flips of the two dice. Test these results to determine whether the computer simulation produces results that agree with the theory at the $\alpha = .05$ level of significance.
11. Use Program 2: Random Number Generator to produce a set of 100 random digits between 0 and 9. Test if the digits generated are random at the $\alpha = .05$ level of significance.
12. Use Program 2: Random Number Generator to produce a set of 60 random digits between 1 and 6. (These might be used to simulate rolling a die 60 times.) Test if the digits are randomly generated at the $\alpha = .05$ level of significance.

11.4 The Kolmogorov-Smirnov Test for Normality (Optional)

One of the underlying themes that pervades all of statistics is normality. We have considered probability problems based on normal distributions. We have approximated distributions such as the binomial distribution with a normal distribution. We have used the Central Limit Theorem to ensure that the sampling distribution of the mean is approximately normal, provided the sample size is large enough. We have also tacitly assumed that the underlying distribution or distributions in a variety of hypothesis-testing situations are normal or approximately normal. However, if this condition of normality does not hold, then the results of the associated hypothesis tests may be invalid.

Despite all this, we have never formally addressed the question of determining whether a given population or set of data is indeed normal. Clearly, however, this is an essential question that must be answered if the various statistical procedures we have studied can be applied. Needless to say, statisticians have developed a variety of methods for testing whether a set of data is actually normal.

At a very simplistic level, we could construct a histogram based on the set of data and, by eye, decide whether it appears to fall into a normal pattern.

Such a coarse approach is likely to be adequate if the pattern is clearly not normal in shape, say, if it is highly skewed to one side or if it is U-shaped or if it is roughly uniform in appearance. However, if the histogram is somewhat bell-shaped, then we must be able to decide definitively if it is sufficiently close to normal to allow us to use the desired statistical procedure.

We can approach this question in several ways. We first consider a variation on the goodness-of-fit process from Section 11.3. Suppose we have a set of 200 measurements on some random variable x and find that this sample has a mean $\bar{x}$ and a standard deviation s. Further suppose that the associated histogram is roughly "mound-shaped," so that we suspect that the data are approximately normally distributed. We could compare the histogram based on the data to a theoretical normal distribution curve. Since we don't know the population mean μ or the population standard deviation σ, our best estimate for each is to use the sample values instead. That is, we presume that $\mu = \bar{x}$ and $\sigma = s$ and consider the normal distribution based on μ and σ.

For this (or any) normal distribution, we know several characteristics. First, approximately 34% of the population lies between the mean and 1 standard deviation to the right. Further, about 13.5% of the population lies between 1 and 2 standard deviations above the mean. Also about 2.5% of the population lies more than 2 standard deviations above the mean. Similar percentages hold below the mean. See Figure 11.12.

If we apply the above percentages, we expect that, of the 200 measurements we have, approximately 2.5% (or 5) will have z values below -2, another 13.5% of them (or 27) will have z values between $z = -2$ and $z = -1$, another 34% of them (or 68) will have z values between $z = -1$ and $z = 0$, another 34% of them (or 68) will have z values between $z = 0$ and $z = 1$, and so forth. This gives rise to a set of expected frequencies based on the assumption that the data are indeed normally distributed:

z	< -2	-2 to -1	-1 to 0	0 to 1	1 to 2	> 2
f_e	5	27	68	68	27	5

FIGURE 11.12

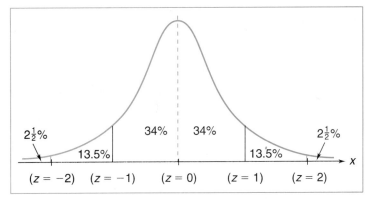

We can then compare these expected frequencies to the observed frequencies for the actual data entries and perform a goodness-of-fit test to see if the assumption of normality is justified.

The problem with this approach is that, for the goodness-of-fit test, each cell should have an expected frequency of at least 5. Thus, it would apply as outlined here only if the data set consisted of at least 200 entries, which is fairly large for our purposes. Moreover, the results would be more accurate if we used more intervals of z values, say from -3 to -2.5, from -2.5 to -2, and so forth. However, in that case, the recommendation that the expected frequency for each cell be at least 5 requires a much larger sample size than $n = 200$.

As an alternative to the goodness-of-fit approach, we consider a method due to the Russian mathematicians Kolmogorov and Smirnov which is very simple to apply and very effective with considerably smaller sets of data. Suppose we have 10 measurements $x_1, x_2, \ldots, x_{10}$ on a random variable x which have sample mean $\bar{x}$ and sample standard deviation s. As before, we assume that the associated normal distribution has parameters $\mu = \bar{x}$ and $\sigma = s$. Using these values, we can calculate the z values $z_1, z_2, \ldots, z_{10}$ corresponding to each of the 10 measurements $x_1, x_2, \ldots, x_{10}$. We assume that the x values and the associated z values are arranged in numerical order. That is, z_1 is the smallest or most negative z value, and z_{10} is the largest or most positive z value.

We next calculate the total area under the normal distribution curve from the extreme left up to each of these z values. We call these probabilities $P_1, P_2, \ldots, P_{10}$. Thus, P_1 is equal to the total area under the normal curve to the left of $z = z_1$, P_2 is the total area under the normal curve to the left of $z = z_2$, and so forth. See Figure 11.13. We compute these values by using the methods from Chapter 5. These probabilities $P_1, P_2, \ldots, P_{10}$ represent values of the total or *cumulative* probability under the normal distribution curve. The graph of this cumulative probability function is shown in Figure 11.14. It starts with a height near 0 at the far left, since there is minimal area under the normal curve below $z = -3$. As we move to the right under the normal curve, the amount of area swept out increases and so this cumulative probability curve also increases. As we move far to the right, the total area swept out under the normal curve increases to a total of 1 square unit, and so the cumulative probability curve rises to a maximum height of 1.

We now construct an expression known as a *step function* based on the fact that there are $n = 10$ data values:

$$S_1 = \tfrac{1}{10} \quad \text{for } z \text{ between } z_1 \text{ and } z_2$$
$$S_2 = \tfrac{2}{10} \quad \text{for } z \text{ between } z_2 \text{ and } z_3$$
$$S_3 = \tfrac{3}{10} \quad \text{for } z \text{ between } z_3 \text{ and } z_4$$
$$\vdots$$
$$S_{10} = \tfrac{10}{10} = 1 \quad \text{for } z \text{ beyond } z_{10}$$

FIGURE 11.13

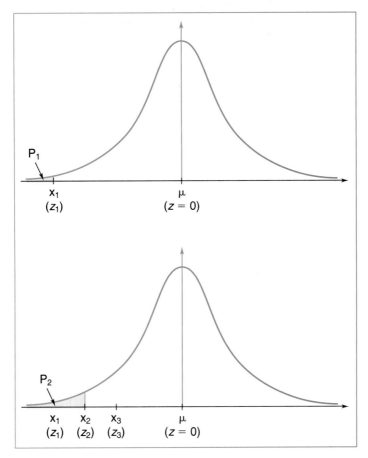

FIGURE 11.14

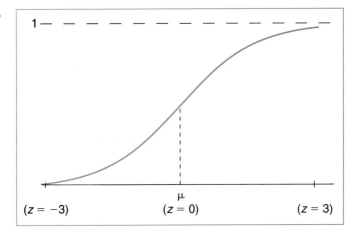

FIGURE 11.15

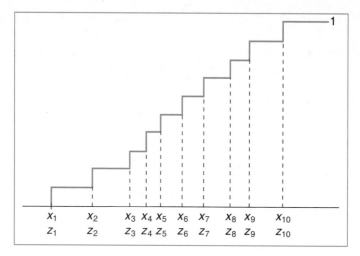

(If there were $n = 40$ data values, then there would be 40 of these expressions, $S_1 = \frac{1}{40}, S_2 = \frac{2}{40}, \ldots, S_{40} = \frac{40}{40} = 1$.) We show the graph of this step function in Figure 11.15.

It turns out that if the cumulative probability curve and the step function are "close" in some sense, then the normal curve is a good fit to the data and we can conclude that the data come from a normally distributed population. To compare them visually, we superimpose the two graphs in Figure 11.16. The Kolmogorov-Smirnov statistic is used to measure how close the two graphs are. Geometrically, this statistic is the maximum difference in height between the curve and either the top or the bottom of any of the steps for any of the z values. That is, at each point z_i, $i = 1, 2, \ldots, 10$ (corresponding to the x values x_i, $i = 1, 2, \ldots, 10$), we compare the normal cumulative proba-

FIGURE 11.16

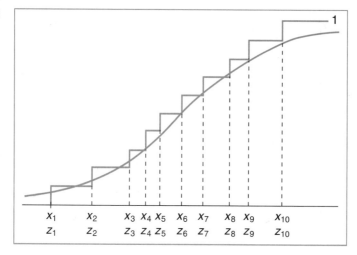

bility value P_i to both S_i (the top of the step) and S_{i-1} (the bottom of the step) and calculate the differences $P_i - S_i$ and $P_i - S_{i-1}$. We then select the largest difference in absolute value across all the z_i values. This is the *Kolmogorov-Smirnov statistic* and is denoted by D_{10} since there are $n = 10$ data values. In general, if there are n data values, then it is denoted by D_n.

Finally, we compare this value for D_n to a corresponding critical value found in Table VII based on n data values. If the test statistic D_n is greater than the critical value from the table, then we conclude that the data are not normally distributed with $\mu = \bar{x}$ and $\sigma = s$. If the test statistic D_n is smaller than the critical value, then we cannot reject the assumption of normality. That is, the data would seem to follow a normal pattern with these parameters.

While the above description may seem somewhat complicated because of the extensive notation used, the actual implementation of the method is quite simple, as we illustrate now.

EXAMPLE 11.7 Determine if the following set of measurements could come from a normal population, using an $\alpha = .05$ level of significance:

$$5, 12, 15, 18, 20, 21, 23, 27, 32, 37$$

SOLUTION We first observe that the given data values are already in numerical order. We find that the sample mean and sample standard deviation for this set of $n = 10$ data values are $\bar{x} = 21$ and $s = 9.428$, respectively. We therefore consider the normal distribution with mean $\mu = 21$ and standard deviation $\sigma = 9.428$. The null and alternate hypotheses are then

H_0: The data come from a normal distribution with $\mu = 21$ and $\sigma = 9.428$.

H_a: The data do not come from this normal distribution.

We now conduct the Kolmogorov-Smirnov test for normality in tabular form, as shown in Table 11.3. Notice that the first column contains the original data entries. The second column consists of the associated z values based on the hypothesis that $\mu = 21$ and $\sigma = 9.428$. Thus, corresponding to the first entry $x = 5$, we find

$$z = \frac{5 - 21}{9.428} = -1.70$$

Furthermore, the area to the left of this z value is found by using the normal distribution table, Table III. Corresponding to $z = -1.70$, we obtain an entry of .4554, which represents the area under the curve from the mean μ where $z = 0$ to $z = -1.70$. Therefore, the area to the left of this is

$$.5000 - .4554 = .0446$$

and this value is recorded as the first entry of the third column. The remaining entries in the second and third columns corresponding to the other data values are obtained in a similar way.

TABLE 11.3

x_i	z_i	P_i	S_i	$P_i - S_i$	$P_i - S_{i-1}$
5	−1.70	.0446	.1	−.0554	
12	−.95	.1711	.2	−.0289	.0711
15	−.64	.2611	.3	−.0389	.0611
18	−.32	.3745	.4	−.0255	.0745
20	−.11	.4562	.5	−.0438	.0562
21	0	.5000	.6	−.1000	0
23	.21	.5832	.7	−.1168	−.0168
27	.64	.7389	.8	−.0611	.0389
32	1.17	.8790	.9	−.0210	.0790
37	1.70	.9554	1.0	−.0446	.0554

The fourth column consists of the values for the step function S_i. Since $n = 10$, they are $S_1 = \frac{1}{10} = .1$, $S_2 = \frac{2}{10} = .2$, and so forth.

The fifth column consists of the differences between the cumulative probability values P_i in the third column and the step function *upper* values S_i in the fourth column. Thus, the first entry is

$$.0446 - .1 = -.0554$$

the second entry is

$$.1711 - .2 = -.0289$$

and so forth.

The entries in the sixth column consist of the differences between the cumulative probability values P_i in the third column and the step function *lower* values S_{i-1} in the *previous* row of the fourth column. Thus, the first entry in the sixth column is blank since there is no previous row to the first row. The second entry in this column is

$$.1711 - .1 = .0711$$

the third entry in the column is

$$.2611 - .2 = .0611$$

and so forth.

We now examine all the entries in the last two columns of differences and select the one which is largest in absolute value. It is $-.1168$, and so the Kolmogorov-Smirnov statistic is

$$D_{10} = .1168$$

To complete the hypothesis test for normality, we compare this value to the critical value for the Kolmogorov-Smirnov statistic in Table VII with $\alpha = .05$. This critical value is .409. Since the test statistic $D_{10} = .1168$ is smaller than the critical value, we cannot reject the null hypothesis. The data may well come from a normal distribution having mean $\mu = 21$ and standard deviation $\sigma = 9.428$. ∎

Notice that the Kolmogorov-Smirnov test does not tell us that the data actually come from a normal distribution. It only lets us conclude that we cannot reject the possibility of normality. However, the same limitation also applies to other tests of normality, such as the goodness-of-fit approach we outlined above.

Note that the Kolmogorov-Smirnov test can also be applied by using your own choice of μ and σ (not just $\mu = \bar{x}$ and $\sigma = s$) if you have some reason to suspect a particular normal distribution. Furthermore, the test can be applied to check whether a set of data comes from some population other than the normal. However, we do not consider such cases here.

Computer Corner

Since the calculations involved in performing a Kolmogorov-Smirnov test for normality are fairly complicated, we have provided Program 37: Kolmogorov-Smirnov Test. To use this program, you must enter your set of data values. The program calculates the sample mean and sample standard deviation and then applies the test to check whether the set of data could come from a normal distribution. The program first produces a graphical display showing both the cumulative normal distribution curve and the associated step function based on the data entries. It then calculates and prints the corresponding Kolmogorov-Smirnov statistic D_n. All that remains is for you to compare this test value to the critical value for D_n found in Table VII based on n data points and the desired level of significance.

Exercise Set 11.4

Mastering the Techniques

Construct one or more histograms for each of the following sets of data to see if the data appear to be normal. Then apply the Kolmogorov-Smirnov test to test whether the data might come from a normally distributed population. In each case, use a 5% level of significance for the test and use the sample statistics $\bar{x}$ and s.

1. 11, 12, 15, 20, 23, 25, 30, 33, 37, 44
2. 21, 21, 22, 23, 24, 25, 26, 26, 27, 27, 27, 28, 28, 28, 28, 29, 29, 29, 30, 31
3. 53, 47, 59, 66, 36, 69, 84, 77, 42, 57, 51, 60, 78, 63, 46, 63, 42, 55, 63, 48, 75, 60, 58, 80, 44, 59, 60, 75, 49, 63
4. 2.8, 3.5, 7.2, 5.8, 6.3, 4.1, 5.7, 8.2, 2.3, 4.4, 7.1, 8.0, 6.8, 5.2, 4.3, 3.0, 3.6, 5.4, 6.3, 6.6, 5.7, 8.2, 4.9, 6.0, 7.2
5. 860, 944, 1126, 905, 840, 980, 1044, 1220, 860, 775, 1005, 875, 890, 905, 930, 1040, 1280, 1025, 975, 1330, 890, 980, 1260, 980, 760

6. 17, 39, 26, 20, 17, 49, 22, 36, 17, 17, 39, 26, 15, 20, 18, 22, 46, 17, 28, 17, 36, 19, 18, 46, 16, 19, 28, 15, 20, 48, 16, 18, 36, 29, 15, 16, 28, 36, 15, 18
7. 304, 221, 242, 154, 259, 271, 189, 253, 311, 225, 247, 151, 209, 314, 278, 190, 195, 315, 222, 185, 317, 288, 250, 190
8. 135, 175, 166, 148, 183, 206, 190, 128, 147, 156, 166, 174, 158, 196, 120, 165, 189, 174, 148, 225, 192, 177, 154, 140, 180, 172

Computer Applications

9–12. Repeat Exercises 5 through 8, using Program 37: Kolmogorov-Smirnov Test.
13. Apply the Kolmogorov-Smirnov test to the data on National League batting averages in Appendix I to determine whether the batting averages are actually normally distributed, as we have assumed.

11.5 The Distribution of Sample Variances: Estimation and Hypothesis Tests (Optional)

Throughout our study of inferential statistics, we have seen how important it is to know the characteristics of the distribution of sample means so that we can compare the mean $\bar{x}$ of one sample to the means of all other comparable samples. Those of you who read Section 7.5 have seen how similar ideas can be considered for measures of central tendency other than the mean which describe a set of sample data. These include the median, the mode and the midrange.

We have also seen how the mean for a set of data is not adequate to give a complete picture of the data. In addition, it is necessary to consider the standard deviation to measure the spread or variation in the data. For instance, in most production processes, it is essential to minimize or at least control the variability in the product being manufactured. This requires a study of either the variance or the standard deviation. Thus, instead of studying the distribution of sample means, we consider the *distribution of sample standard deviations s*. Even though such a topic is valuable, statisticians have discovered that it is more useful to consider the associated *distribution of sample variances* s^2 based on samples of size n drawn from an underlying population having mean μ, standard deviation σ, and hence variance σ^2.

It turns out that

> If the underlying population is normally distributed, then the random variable
> $$\frac{(n-1)s^2}{\sigma^2}$$
> has a χ^2 distribution with $n-1$ degrees of freedom.

FIGURE 11.17

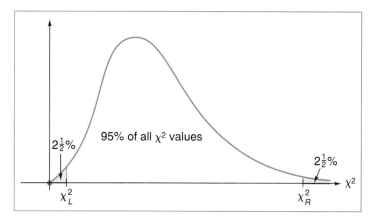

To make use of this fact, we cannot work directly with values of s or even s^2. Instead, we have to work with values of the related statistic

$$\chi^2 = \frac{(n-1)s^2}{\sigma^2}$$

By combining ideas from previous chapters on confidence intervals and hypothesis testing and from the first few sections in this chapter on the chi-square distribution, we are able to develop comparable methods to construct confidence intervals for either σ or σ^2 and to test hypotheses regarding these parameters.

Suppose we wish to construct a 95% confidence interval for the population variance σ^2. This involves locating the "middle" 95% of the distribution of $(n-1)s^2/\sigma^2$ values or, equivalently, locating $2\frac{1}{2}\%$ in each tail of a chi-square distribution. However, unlike either a normal or a t-distribution, a chi-square distribution consists of only positive values and is not symmetric about its mean. See Figure 11.17 above.

As a result, there are two critical values for χ^2—one for the left tail, which we call χ^2_L, and another for the right tail, χ^2_R. These values depend on the number of degrees of freedom $n-1$ as well as on the confidence level we select. For example, for a 95% confidence interval based on $n-1 = 22$ degrees of freedom, we find from Table VI that $\chi^2_L = 10.982$ (corresponding to .025 in the left tail which is equivalent to .975 in the right tail) and $\chi^2_R = 36.781$ (corresponding to .025 in the right tail). See Figure 11.18.

After some algebraic manipulation that we will not go into here, we find the following:

> The **confidence interval** for the population variance σ^2 is
>
> $$\left[\frac{(n-1)s^2}{\chi^2_R}, \frac{(n-1)s^2}{\chi^2_L}\right]$$

The corresponding confidence interval for the population standard deviation σ is

$$\left[\sqrt{\frac{(n-1)s^2}{\chi^2_R}},\ \sqrt{\frac{(n-1)s^2}{\chi^2_L}}\right]$$

EXAMPLE 11.8 A company is producing computer memory boards which must fit into a tight slot inside a computer. A sample of 30 boards are tested, and the connection pins are found to have a mean of .47 millimeter with a standard deviation of .032 millimeter. Construct 95% confidence intervals for the variance and the standard deviation of the pin sizes of all such boards produced. Assume that the sizes of the pin connections are normally distributed.

SOLUTION The sample data yield

$$n = 30$$
$$\bar{x} = 1.47$$
$$s = .032$$

so that

$$s^2 = (.032)^2 = .001024$$

As indicated above, the statistic $(n-1)s^2/\sigma^2$ follows a chi-square distribution with $n - 1 = 29$ degrees of freedom. From Table VI, for a 95% confidence interval, we use $\chi^2_L = 16.047$ and $\chi^2_R = 45.722$, as shown in Figure 11.19. Therefore, a 95% confidence interval for the population variance σ^2 is

$$\left[\frac{(n-1)s^2}{\chi^2_R},\ \frac{(n-1)s^2}{\chi^2_L}\right]$$

FIGURE 11.18

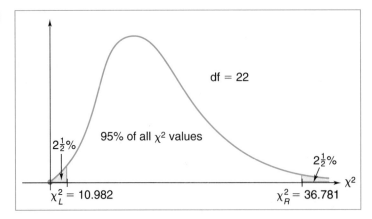

FIGURE 11.19

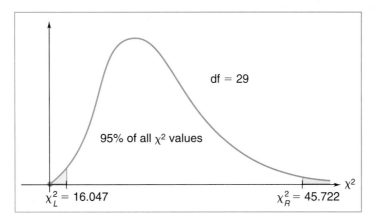

or

$$\left[\frac{(29)(.001024)}{45.722}, \frac{(29)(.001024)}{16.047} \right]$$

or

$$[.000649, .001851]$$

Consequently, the corresponding 95% confidence interval for the population standard deviation σ is

$$[\sqrt{.000649}, \sqrt{.001851}]$$

or

$$[.025, .043]$$

That is, with 95% confidence, we conclude that the standard deviation for the size of all these connection pins is between .025 and .043 millimeter. ■

We use the same ideas to test a hypothesis about the population variance σ^2. The test statistic we use is $\chi^2 = (n-1)s^2/\sigma^2$. If the test is a one-tailed test, then we must locate a single critical value. If it is a two-tailed test, then there are two critical values. We illustrate the technique in the following example.

EXAMPLE 11.9 Nuclear reactors must be taken off line for refueling every few years. If such a plant is out of service even for a relatively short time, the utility must be able to plan ahead to provide adequate power for its customers. In a random sample of 15 refueling operations, it is found that the mean time a reactor is out of service is 3 months with a standard deviation of 1.2 months. Assuming that the refueling time is normally distributed, test if the variance is greater than 1 month at the $\alpha = .05$ level of significance.

SOLUTION The null hypothesis asserts that

$$H_0: \sigma^2 = 1$$

while the alternative hypothesis states

$$H_a: \sigma^2 > 1$$

The sample data yield

$$n = 15$$
$$\bar{x} = 3$$
$$s = 1.2$$

so that the sample variance is

$$s^2 = 1.44$$

Since the underlying population is normally distributed, we know that the statistic $\chi^2 = (n - 1)s^2/\sigma^2$ follows a χ^2 distribution with $n - 1 = 14$ degrees of freedom. Therefore, for this one-tailed test, the critical value is $\chi^2 = 23.685$ at the $\alpha = .05$ level of significance, as shown in Figure 11.20.

For the sample variance $s^2 = 1.44$, we find

$$\chi^2 = \frac{(n - 1)s^2}{\sigma^2}$$

$$= \frac{(14)(1.44)}{1^2}$$

where we use the null hypothesis assertion that $\sigma^2 = 1$. Thus,

$$\chi^2 = 20.160$$

FIGURE 11.20

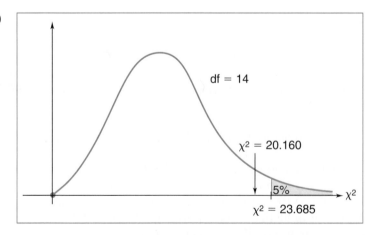

We therefore cannot reject the null hypothesis since this test statistic is smaller than the critical value. The evidence is not strong enough for us to conclude that the population variance in refueling time exceeds 1 month at the 5% level of significance. ∎

Computer Corner

We now explore some of the properties of the sampling distribution of the variance, using Program 38: Distribution of Sample Variances. The program provides the choice of four underlying populations—normal, skewed, uniform, and U-shaped—as well as the choice of the sample size n and the number of desired random samples.

Suppose we select the normal population and request 400 samples of size $n = 4$. For each of the 400 random samples, the program calculates the sample variance s^2 and graphs the multiple of the variance $(n - 1)s^2/\sigma^2$. A set of typical results is shown in Figure 11.21. Since the values for the standard deviation and hence the sample variance are nonnegative, there is a distinct lower limit of 0 to the values that can be achieved for s^2. Thus, the left-hand limit shown on the graph corresponds to a sample variance of 0.

If you examine the graph in Figure 11.21, you will see that the shape of the distribution is approximately that of a chi-square distribution. The distribution seems to rise upward to a "bulge" near the left end and then very slowly tails away to the right.

Furthermore, the mean of the 400 sample variances displayed in Figure 11.21 is 7.26 while their standard deviation is 5.42. The underlying normal population has mean $\mu = 68.07$ and standard deviation $\sigma = 2.76$, so that the population variance is $\sigma^2 = (2.76)^2 = 7.618$. We therefore see that the mean for the 400 sample variances, 7.26, is relatively close to the population variance $\sigma^2 = 7.618$.

If we now change the sample size to $n = 9$ and use 400 such simulations, then we obtain a result such as that shown in Figure 11.22. From this graph, we see much the same chi-square distribution shape, although the bulge does seem more pronounced. It is certainly not at all symmetric. It starts off quite sharply at the left (where the variance is zero), climbs upward toward a peak, and then decays slowly to the right. The corresponding numerical results for this case show a mean for the sample variances of 7.99 with a standard deviation of 3.96.

Before we come to any conclusions about this distribution, we investigate the results of using still larger values for n. In Figures 11.23 and 11.24, we show typical results with $n = 16$ and $n = 36$, respectively. In each case, the distribution of $(n - 1)s^2/\sigma^2$ values has the appearance of a chi-square distribution. However, as the sample size n increases, the shape becomes somewhat normal in appearance.

FIGURE 11.21

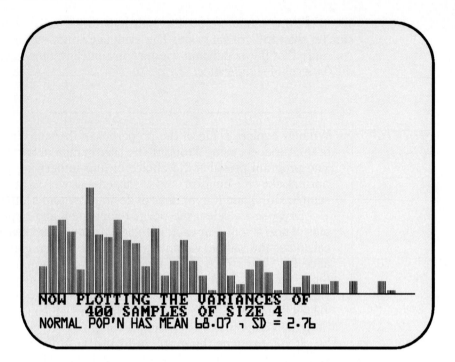

FIGURE 11.22

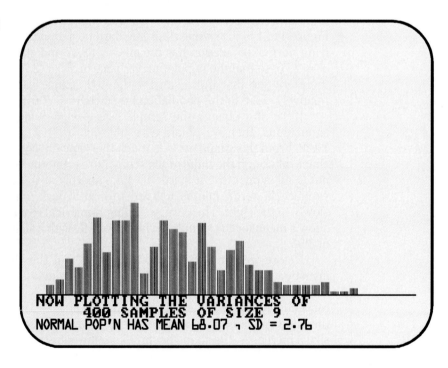

As we have seen in Section 11.2, this is true of the chi-square distribution in general as the number of degrees of freedom increases. Furthermore, in each case, numerically the means of the sample variances s^2 (8.16 and 8.02, respectively) come out fairly close to the population variance $\sigma^2 = 7.618$. However, there does not seem to be a clear-cut pattern for the values of the standard deviation (2.93 and 1.95) for the sample results other than the obvious conclusion that they appear to decrease in size as the sample size n increases.

We summarize the known results for this sampling distribution below:

> The **distribution of sample variances** s^2 based on random samples of size n drawn from a population with standard deviation σ has mean,
>
> $$\mu_v = \sigma^2$$
>
> and standard deviation
>
> $$\sigma_v = \sqrt{\frac{2\sigma^4}{n-1}} \qquad (11.2)$$
>
> If the underlying population is normal, then the distribution of the values of the statistic $(n-1)s^2/\sigma^2$ follows a chi-square distribution with $n-1$ degrees of freedom.

We now conduct a similar study when the underlying population is not normal. In particular, we consider the case of a skewed population. As above, we study 400 samples of sizes $n = 4, 9, 16$ and 36. Typical graphical results are shown in Figures 11.25 to 11.28. Again, these distributions seem to be chi-square in general outline and approach a normal distribution as the sample size increases. As for the numerical results, we first observe that the population has a mean of 71.19 and a standard deviation of 3.29, so that its variance is $\sigma^2 = (3.29)^2 = 10.82$. The means for the sample variances in each case are, respectively, 9.73, 11.48, 11.63 and 11.44, which seems to suggest that the mean of this sampling distribution should be $\sigma^2 = 10.82$. Furthermore, the standard deviations in the four cases are 8.18, 5.96, 4.71 and 2.88, respectively. When we compare these values to the predictions based on formula (11.2), we conclude that they certainly tend to reflect the formula even in a case where the underlying population is not normal.

We leave experimentation with the remaining two underlying populations, the uniform and the U-shaped, for you to do on your own.

FIGURE 11.23

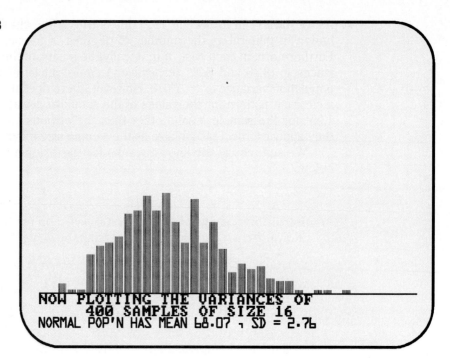

FIGURE 11.24

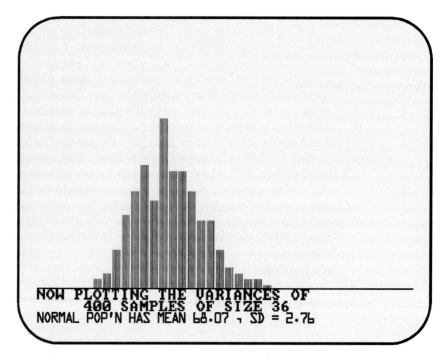

SECTION 11.5 • The Distribution of Sample Variances: Estimation and Hypothesis Tests (Optional) **575**

FIGURE 11.25

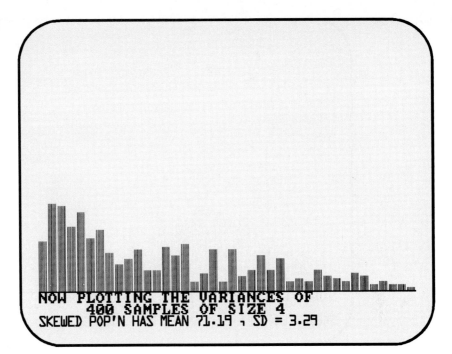

FIGURE 11.26

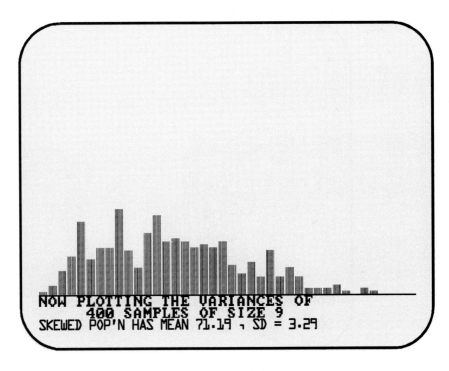

FIGURE 11.27

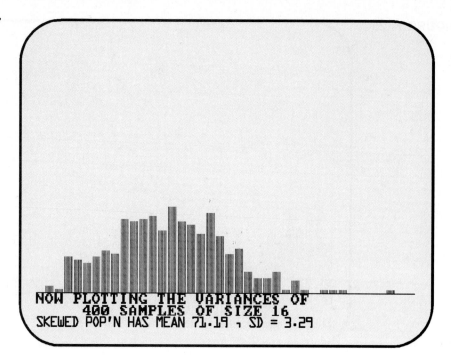

FIGURE 11.28

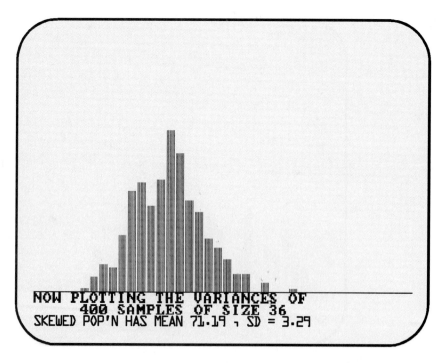

SECTION 11.5 • The Distribution of Sample Variances: Estimation and Hypothesis Tests (Optional)

Exercise Set 11.5

Mastering the Techniques

1. Find the critical values for a 95% confidence interval for the variance based on a sample of size $n = 15$.
2. Find the critical values for a 95% confidence interval for the standard deviation based on a sample of size 25.
3. Find the critical values for a 90% confidence interval for the standard deviation based on a sample of size 25.
4. Find the critical values for a 98% confidence interval for the variance based on a sample of size 30.
5. Find the critical values for a two-tailed hypothesis test on the standard deviation at the $\alpha = .05$ significance level based on a sample of size 25.
6. Find the critical value for a one-tailed (right-tail) hypothesis test on the standard deviation at the 5% level of significance based on a sample of size 18.

Applying the Concepts

7. In a study of the time that school-age children spend watching television, it was found that a randomly selected group of 30 children watch an average of 48 hours per week with a standard deviation of 12.4 hours. Find a 98% confidence interval for the standard deviation of the time all such children spend watching television, assuming that the times are normally distributed.
8. A fast-food restaurant owner is studying the sometimes large spread in waiting time for customers to be served. In a random sample of 25 customers, the mean waiting time for service during dinner hour is 3.4 minutes with a standard deviation of 2.7 minutes. Construct 95% confidence intervals for the variance and the standard deviation of the waiting time per customer. Assume that the waiting times are normally distributed.
9. After a more comprehensive study, the restaurant owner in Exercise 8 concludes that the standard deviation for the waiting time of all customers is 2.75 minutes. The owner then implements a new single-line queue (rather than a separate line for each server) in order to reduce the standard deviation. During a trial implementation, a sample of 30 customers have a mean wait of 3.6 minutes with a standard deviation of 2.32 minutes. Test if the new approach reduces the standard deviation for the waiting time at the 5% level of significance.
10. The same fast-food restaurant owner wants to control the amount of roast beef going into each roast beef sandwich so that the mean weight of the meat is 4 ounces with a standard deviation of .04 ounce. In a random sample of 20 sandwiches, it is found that the mean weight of roast beef is 4.13 ounces with a standard deviation of .056 ounce. Assuming that the weight of roast beef used per sandwich is normally distributed, test if the employees are meeting the management requirement on the standard deviation at the $\alpha = .01$ level of significance.

11. In a random sample of 15 joggers, a nurse finds the following set of pulse rates after a measured 2-mile course:

$$84, 76, 95, 88, 78, 82, 84, 79, 90, 83, 74, 90, 98, 86, 88$$

   (a) Construct a 95% confidence interval for the mean pulse rate of all such joggers.
   (b) Construct a 95% confidence interval for the standard deviation of the pulse rate of all such joggers.
   Assume that all pulse rates are normally distributed.

Computer Applications

12. For both the uniform and the U-shaped populations available with Program 38: Distribution of Sample Variances, repeat the investigations performed in the Computer Corner. In particular, for each population, what can you conclude about the following?
   (a) The shape of the sampling distribution
   (b) The mean of the sampling distribution
   (c) The standard deviation of the sampling distribution

Student Projects

Conduct a chi-square analysis for a contingency table based on a survey you design and administer. You should select a topic of interest to you, preferably some issue of social or political significance. For example, it might be

1. Abortion
2. Gun control
3. Capital punishment
4. Nuclear power plants
5. Imposition of tax surcharges to fund environmental cleanups
6. Allowing children with AIDS to attend public schools
7. The use of affirmative action quotas in hiring

Construct a strong statement regarding the issue you select. It can be either a positive or a negative position on the issue. For instance, under abortion, you might use:

   Abortion should be at the sole discretion of the mother.

or

   Abortion should not be allowed under any circumstances.

Using this statement, conduct a survey to determine people's reaction to the issue. The survey should provide five possible responses: strongly agree, agree, no opinion, disagree, and strongly disagree. In addition, select a *single* demographic category such as gender (female versus male), age (less than 20, 20 to 25, 26 to 35, 36 to 45, over 45), race, religion, marital status, or academic major. You should select a demographic category for which you suspect there may be a difference in responses. For example, if your issue is abortion, then gender, religion, or marital status would be likely choices; race or academic major would probably not be interesting choices.

Conduct the survey on a randomly selected group of at least 100 people, and then perform a complete chi-square analysis on the resulting contingency table to determine whether there is a difference in attitude on the issue based on the demographic category you select.

Incorporate all your results into a formal statistical project report. The report should include

- A statement of the topic being studied
- A statement of the particular population being studied
- The source of the data
- A discussion of how the data were collected and why you believe that it is a random sample
- A list of the data including possibly some tables and/or diagrams to display the data
- All the statistical analysis
- Your conclusions based on this analysis and a rationale for the level of significance you select
- An indication of which groups have disproportionately higher and lower percentages supporting the issue under investigation if you find that there is a difference in the proportions who hold various opinions based on the demographic factor
- A discussion of any surprises you may have noted in connection with collecting and organizing the data or with your conclusions

CHAPTER 11 SUMMARY

Chi-square analysis is used to test if there is any difference among proportions when you have three or more populations.

The **chi-square statistic**

$$\chi^2 = \sum \frac{(f_0 - f_e)^2}{f_e}$$

measures how close the **observed frequencies** f_0 in a **contingency table** are to the **expected frequencies** f_e based on the assumption that there is no difference among the proportions. The expected frequencies are found from

$$\text{Expected frequency} = \frac{\text{row total} \times \text{column total}}{\text{grand total}}$$

The value for the chi-square statistic is compared to the associated critical value based on

$$df = (\text{row} - 1) \times (\text{column} - 1)$$

from Table VI.

The **goodness-of-fit test** is a variation on chi-square analysis which lets you determine if a set of data fits a supposed distribution pattern. If there are n possible outcomes or categories, there are $n - 1$ degrees of freedom for the test.

The **Kolmogorov-Smirnov test** can be used to find out whether a set of data comes from a normal population. The critical values for the test are found in Table VII.

Review Exercises

1. A survey of color preferences for jeans among college students is conducted and the following breakdown on the basis of gender is obtained.

	Whitewashed	Stonewashed	Blue	Other
Women	174	122	80	54
Men	220	182	74	44

Test if there is any difference in the jeans color preferences for women and men college students at the 5% level of significance.

2. A stock broker keeps track of the number of stock purchase orders she receives from her clients during a certain week. The results are as shown.

Monday	Tuesday	Wednesday	Thursday	Friday
41	51	46	54	38

Test if there is any difference in the number of orders received based on the day of the week at the $\alpha = .05$ level of significance.

3. The party affiliations for the registered voters in Appletree County are: 38% Democrats, 41% Republicans, 8% Liberals, 6% Conservatives and 7% Independents. In a recent election, the breakdown of the actual votes cast was:

Democrats	Republicans	Liberals	Conservatives	Independents
4535	5263	974	788	940

At the 5% level of significance, can you conclude that voters followed their party affiliations when voting in this election?

*4. A town recycling center keeps track of the amount of paper that residents bring in. During one day, the following amounts, in pounds, were brought in for recycling:

$$12, 15, 17, 5, 15, 11, 13, 24, 16, 10, 14, 21, 17, 13, 15,$$
$$9, 16, 21, 18, 16, 14, 17, 20, 15, 11, 15$$

Test if these values indicate that the weight of papers recycled per household could be normally distributed at the $\alpha = .05$ level of significance.

---

*Indicates a question based on an optional topic.

Analysis of Variance

According to data from the U.S. census, women consistently earn less than men at every level of education. For example, all women who are college graduates earn an average of $27,344 compared to an average of $42,500 for all men with the same level of education. The average salary for all women high school graduates is only $17,809 compared with an average of $26,750 for all men high school graduates. Thus, the average earnings of female college graduates are more comparable to the average earnings of men with only a high school education.

The gap between women and men actually depends on their ages and gets wider for older adults. For people between 18 and 24 with college education, a woman earns an average of 92 cents for each dollar earned by a man. However, for people between 55 and 64, the gap has widened considerably so that the woman earns an average of 54 cents for each dollar earned by a man. The gaps are just as wide for men and women who are only high school graduates.

Do the results reported in this study suggest to you that there may be discrimination patterns in earnings based on a person's gender? Does the increasingly large gap between men's and women's earnings suggest that men receive more money as they get older than women do? Does the smaller gap that exists at ages 18 to 24 suggest that past inequities are being corrected?

In a study such as this, we are comparing the averages of different groups to see if there are any significant differences among their means. A statistical technique called *analysis of variance*, which we will study in this chapter, allows us to test if there are any differences among the means of more than two groups.

If significant differences exist, they can be used as the basis for legal challenges on the grounds of discrimination. Women's rights advocates would claim that they indicate a clear pattern of discrimination since women consistently earn less than men. Opponents might counter by pointing out that the average salary earned by all women is lowered by the number of women who choose traditionally lower-paid careers (nurses, secretaries, social workers and teachers) rather than the higher paid positions in fields such as medicine, law and business. Moreover, women also tend to take off more time from their careers for family responsibilities and so might accumulate less seniority and hence less pay by a particular age. Consequently, any definitive conclusions require a much more detailed study which reflects additional factors such as the number of years and type of employment.

In this chapter, we will learn how to conduct an Analysis of Variance study to compare the means for more than two groups such as the different age groups or different levels of education.

12.1 Introduction to One-Way Analysis of Variance

In the last chapter, we used chi-square analysis to compare the proportions $\pi_1, \pi_2, \ldots, \pi_m$ from a group of m (three or more) populations to determine if the population proportions are the same:

$$\pi_1 = \pi_2 = \cdots = \pi_m$$

As we have seen throughout our study of statistics, situations involving proportions are duals of comparable situations involving means. Therefore, it should come as no surprise that a related problem often arises: Compare the means of three or more populations.

In particular, suppose we have a series of populations having means $\mu_1, \mu_2, \ldots, \mu_m$. We wish to determine if these population means are all equal:

$$\mu_1 = \mu_2 = \cdots = \mu_m$$

For instance, suppose that three different mathematics professors are teaching a particular course. A student (or a dean) might be interested in knowing if there is any difference among the average grades that each professor's students earn on a uniform final exam. To study this, random samples of four students from each class are selected, and their final-exam grades are listed in Table 12.1

TABLE 12.1

Adams	Bonino	Cohen	
79	71	82	
86	77	68	
94	81	70	
89	83	76	
348	312	296	
$\bar{x}_A = 87$	$\bar{x}_B = 78$	$\bar{x}_C = 74$	(sample means)

Suppose the mean of all final exam grades in Professor Adams' class is denoted by μ_A, the mean in Professor Bonino's class is μ_B, and the mean in Professor Cohen's class is μ_C. The question we wish to answer is: Are these three means essentially equal?

We might be tempted to approach this problem by treating it as a group of three related difference-of-means tests. We could compare μ_A to μ_B, then μ_A to μ_C, and finally μ_B to μ_C. If none of the three null hypotheses were rejected, then we would conclude that all three means might be equal. However, if we could reject the equality of any one pair of means, then we would conclude that a difference exists among the set of three means.

While this approach sounds relatively straightforward, it becomes extremely cumbersome if we are studying more than three different professors. For example, if we were considering 10 professors, then we would have to perform 45 distinct difference-of-means tests. More importantly, the level of significance α represents the chance of rejecting a correct null hypothesis. If we perform a sequence of hypothesis tests each at the 5% level of significance, the total chance of making an incorrect decision does not remain at 5% over the entire process. In fact, it becomes considerably larger than 5%. For instance, with three professors, the chance of an error is as large as 15% if the individual hypothesis tests are independent of each other. If we studied the results for 10 professors, the chance of an error could be as high as 90%. Clearly, this is unacceptable.

We therefore introduce a method known as *ANalysis Of VAriance* (or *ANOVA* for short) which allows us to test all the means simultaneously to see if there is any difference among them.

To understand the logic behind the analysis of variance procedure, consider the two sets of data displayed in Figures 12.1 and 12.2. In each case, we have three measurements on each of three quantities A, B, and C. In the case shown in Figure 12.1, the sample means for the three samples are $\bar{x}_A = 20$, $\bar{x}_B = 25$, and $\bar{x}_C = 30$. Notice that the individual data values in each sample are clustered closely about the corresponding sample mean. Moreover, each

FIGURE 12.1

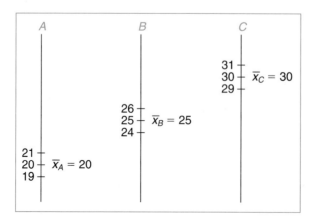

FIGURE 12.2

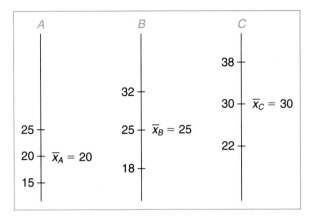

cluster of data values is clearly separated from the other clusters. This suggests that the difference among the three sample means is significant, and so we would likely conclude that the three population means μ_A, μ_B, and μ_C are different.

In Figure 12.2, the three sample means are again $\bar{x}_A = 20$, $\bar{x}_B = 25$, and $\bar{x}_C = 30$. This time, however, the individual sets of data values are spread widely so that they overlap each other. We are no longer quite so confident that the three population means are different. They may be the same because of the large spread or variation within each of the three sets of data.

The analysis of variance procedure is designed to measure this spread and to apportion the total amount of variance present for all the sample data into two components, one corresponding to what happens *between* the different sets of data and the other corresponding to what happens *within* each set of data. If there is little variance within each sample group and a relatively large difference between the sample means (as in Figure 12.1), then we reject the null hypothesis that all the population means are equal. If there is a considerable amount of variance within each sample group and there is relatively little difference between the sample means (as in Figure 12.2), then we cannot conclude that the population means are different.

Before introducing the details on the ANOVA procedure, we discuss some terminology. In any analysis of variance problem, there is a *factor* under study. In our original situation of final exam grades with different professors, the factor is professor. The individual members of the factor are known as *factor levels* or *treatments*. Thus, in our example, the individual professors are the factor levels. In ANOVA, we compare the means of different factor levels A, B, C, ... to see whether

$$\mu_A = \mu_B = \mu_C = \cdots$$

To test this hypothesis, we use a set of sample data on each of the factor levels. In the example, the sample data consist of final exam grades obtained by four randomly selected students in each professor's course.

TABLE 12.2

	Factor Levels	
Adams	**Bonino**	**Cohen**
79	71	82
86	77	68
94	81	70
89	83	76
348	312	296

956 = grand total

Further, the underlying assumptions needed to apply analysis of variance are that the items under investigation (the exam grades here) are approximately normally distributed with equal variances and that the samples are independent.

We begin the ANOVA procedure by considering the given data in Table 12.2, which is a slightly different format from what we gave in Table 12.1 originally.

Clearly, there is an apparent difference among the sample means for the three professors. They are $\bar{x}_A = 87$, $\bar{x}_B = 78$, and $\bar{x}_C = 74$. What we want to determine is how much of this difference can be attributed to the usual chance variation between samples and how much of it can be attributed to an actual difference within each factor level or treatment (the professors). The ANOVA procedure consists of a series of calculations to see how much of this difference can be attributed to the *Factor* and how much can be attributed to randomness or to the *Error*.

The most effective way to do this is to construct a new table, known as an *analysis of variance table* (or simply an *ANOVA table*) in which we display the results of the appropriate calculations and how the differences are apportioned between the Factor and the Error. See Table 12.3. We discuss the meanings of the parts of this table in the following paragraphs.

If you examine this table, you will see that there is room for three different entries under "Sums of Squares," one for the *Sum of Squares due to the Factor*, another for the *Sum of Squares of Error*, and a third for the *Total Sum of Squares*. Similarly, there are *degrees of freedom for the Factor*, *degrees of freedom for the Error*, and the *Total degrees of freedom*. Further, there are two entries known as mean squares, one the *Mean Square for the Factor* and the other

TABLE 12.3

	Sums of Squares	Degrees of Freedom	Mean Squares	F-Ratio
Factor				
Error				
Total				

the *Mean Square for the Error*. Finally, there is a single quantity known as the *F-ratio* whose value determines our decision.

We first indicate how these various quantities are calculated and then show how they are used to decide whether a difference exists among all the population means. We begin by computing the Sum of Squares for the Factor, denoted by SS(Factor) or simply SSF, from the formula

$$\text{SS(Factor)} = \frac{\Sigma(\text{column total})^2}{\text{number of entries per column}} - \frac{(\text{grand total})^2}{\text{total sample size}}$$

From Table 12.2, we see that

1. The column totals are 348, 312, and 296, respectively.
2. There are four entries per column.
3. The grand total is 956.
4. The total sample size is 12.

When we substitute these values into the above formula for the Sum of Squares for the Factor, we obtain

$$\text{SS(Factor)} = \frac{348^2 + 312^2 + 296^2}{4} - \frac{956^2}{12}$$

$$= \frac{121{,}104 + 97{,}344 + 87{,}616}{4} - \frac{913{,}936}{12}$$

$$= \frac{306{,}064}{4} - 76{,}161.33$$

$$= 76{,}516 - 76{,}161.33$$

$$= 354.67$$

Note that many statisticians use the notation SS(treatment) instead of SS(Factor).

The Sum of Squares for the Error, denoted by SS(Error) or SSE, is given by

$$\text{SS(Error)} = \Sigma x^2 - \frac{\Sigma(\text{column total})^2}{\text{number of entries per column}}$$

where Σx^2 represents the Sum of Squares of each of the original 12 entries. Therefore,

$$\text{SS(Error)} = 79^2 + 86^2 + 94^2 + \cdots + 76^2 - \frac{348^2 + 312^2 + 296^2}{4}$$

$$= 6241 + 7396 + 8836 + \cdots + 5776 - 76{,}516$$

$$= 76{,}838 - 76{,}516$$

$$= 322$$

Finally, the Total Sum of Squares, denoted by SS(Total) or SST, is given by

$$\text{SS(Total)} = \sum x^2 - \frac{(\text{grand total})^2}{\text{total sample size}}$$

so that

$$\text{SS(Total)} = 79^2 + 86^2 + \cdots + 76^2 - \frac{956^2}{12}$$

$$= 76{,}838 - 76{,}161.33$$

$$= 676.67$$

If you examine these three values for the different sums of squares—354.67, 322, and 676.67—you will notice that

$$354.67 + 322 = 676.67$$

In general,

$$\text{SS(Total)} = \text{SS(Factor)} + \text{SS(Error)}$$

Consequently, we have to calculate only two of these three expressions. The third can then be obtained immediately from the other two. In fact, it is usually simplest to calculate the Total Sum of Squares and the Sum of Squares due to the Factor and then obtain

$$\text{SS(Error)} = \text{SS(Total)} - \text{SS(Factor)}$$

These three entries make up the first column of the ANOVA table, as shown in Table 12.4.

We now find the degrees of freedom associated with an analysis of variance problem. First, the number of degrees of freedom attributable to the Factor is precisely 1 less than the number of factor levels or treatments (pro-

TABLE 12.4

	Sums of Squares	Degrees of Freedom	Mean Squares	F-Ratio
Factor	354.67			
Error	322			
Total	676.67			

fessors). Equivalently, this is 1 less than the number of columns c in the original table, Table 12.2:

$$df(Factor) = c - 1$$

In our example,

$$df(Factor) = 3 - 1 = 2$$

Further, the total number of degrees of freedom is simply 1 less than the total number of entries in the original data table. However, since the total number of entries is simply the product of the number of rows r and the number of columns c, this reduces to

$$df(Total) = r \times c - 1$$

In our illustration,

$$df(Total) = (4)(3) - 1 = 11$$

As with the sums of squares,

$$df(Total) = df(Factor) + df(Error)$$

so that

$$df(Error) = df(Total) - df(Factor)$$

In our example,

$$df(Error) = 11 - 2 = 9$$

These three entries form the second column of the ANOVA table, as shown in Table 12.5.

Up to this point, we have calculated two quantities attributable to the Factor: the Sum of Squares = 354.67 and the degrees of freedom = 2. Similarly, for the Error, we have the Sum of Squares = 322 and the degrees of freedom = 9. The third column in an ANOVA table, the Mean Squares, con-

TABLE 12.5

	Sums of Squares	Degrees of Freedom	Mean Squares	F-Ratio
Factor	354.67	2		
Error	322	9		
Total	676.67	11		

sists of the ratios of these values. That is, we essentially average the sums of squares over the number of degrees of freedom. Thus, the Mean Square for the Factor is given by

$$MS(Factor) = \frac{SS(Factor)}{df(Factor)}$$

and the Mean Square for the Error is

$$MS(Error) = \frac{SS(Error)}{df(Error)}$$

In our example,

$$MS(Factor) = \frac{354.67}{2} = 177.34$$

and

$$MS(Error) = \frac{322}{9} = 35.78$$

We record these values in the ANOVA table, as shown in Table 12.6.

TABLE 12.6

	Sums of Squares	Degrees of Freedom	Mean Squares	F-Ratio
Factor	354.67	2	177.34	4.96
Error	322	9	35.78	
Total	676.67	11		

We now introduce one more quantity, the F-ratio or F-statistic. The F-statistic is calculated as

$$F = \frac{MS(Factor)}{MS(Error)}$$

In our example, we obtain

$$F = \frac{177.33}{35.78} = 4.96$$

and enter it as the final item in the ANOVA table.

Just as the t-statistic follows a t-distribution and the χ^2 statistic follows a χ^2 distribution, the F-statistic follows an F-distribution. Moreover, as with the

t-distribution and the χ^2 distribution, there are many different F-distributions and they are identified in terms of degrees of freedom. However, it turns out that the F-distribution is based on two different degrees of freedom, one "for the numerator" and one "for the denominator." In analysis of variance situations, the degrees of freedom for the numerator are just the number of degrees of freedom for the Factor, and the degrees of freedom for the denominator are the number of degrees of freedom for the Error. The graphs of a series of F-distributions are shown in Figure 12.3. Notice that these curves are not symmetric. All ANOVA hypothesis tests involving the F-distribution are one-tailed tests; we reject the null hypothesis that the means are equal only if the F-statistic is too large.

The F-statistic is treated in precisely the same way as the χ^2 value is used in a chi-square analysis. That is, we compare it to a critical value found in a table of F values. If the value computed (in this case $F = 4.96$) is larger than the table entry, then we conclude that there is indeed a difference among the population means—the variation is due to a difference in the levels of the factor. But if the calculated value is smaller than the critical value from the table, then we cannot reject the null hypothesis and so there may not be any difference among the means—the variation in the sample data can be explained simply on the basis of random variation between samples.

The critical values for the F-distribution are given in Table VIII. When you examine this table, you will notice that the critical values listed depend on the number of degrees of freedom in the numerator, df(Factor), and the number of degrees of freedom in the denominator, df(Error). Moreover, since analysis of variance is a hypothesis test, there is a significance level α associated with it.

In our example, suppose we use $\alpha = .05$ as the significance level. As we have seen, the degrees of freedom for the numerator and denominator are 2

FIGURE 12.3

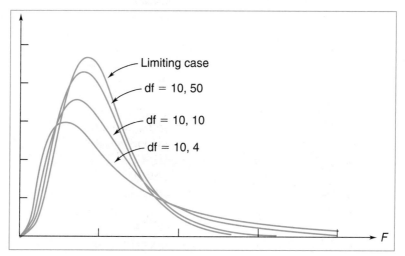

FIGURE 12.4

Degrees of Freedom for Denominator	Degrees of Freedom for Numerator									
	1	2	3	4	5	6	7	8	9	10
1	161	200	216	225	230	234	237	239	241	242
2	18.5	19.0	19.2	19.2	19.3	19.3	19.4	19.4	19.4	19.4
3	10.1	9.55	9.28	9.12	9.01	8.94	8.89	8.85	8.81	8.79
4	7.71	6.94	6.59	6.39	6.26	6.16	6.09	6.04	6.00	5.96
5	6.61	5.79	5.41	5.19	5.05	4.95	4.88	4.82	4.77	4.74
6	5.99	5.14	4.76	4.53	4.39	4.28	4.21	4.15	4.10	4.06
7	5.59	4.74	4.35	4.12	3.97	3.87	3.79	3.73	3.68	3.64
8	5.32	4.46	4.07	3.84	3.69	3.58	3.50	3.44	3.39	3.35
9	5.12	4.26	3.86	3.63	3.48	3.37	3.29	3.23	3.18	3.14
10	4.96	4.10	3.71	3.48	3.33	3.22	3.14	3.07	3.02	2.98
11	4.84	3.98	3.59	3.36	3.20	3.09	3.01	2.95	2.90	2.85
12	4.75	3.89	3.49	3.26	3.11	3.00	2.91	2.85	2.80	2.75
13	4.67	3.81	3.41	3.18	3.03	2.92	2.83	2.77	2.71	2.67
14	4.60	3.74	3.34	3.11	2.96	2.85	2.76	2.70	2.65	2.60
15	4.54	3.68	3.29	3.06	2.90	2.79	2.71	2.64	2.59	2.54

and 9, respectively. Therefore, the critical value for F from Table VIII is $F = 4.26$. See the portion of the table reproduced in Figure 12.4. We compare our calculated value of F to this critical value. Since our value $F = 4.96$ is larger than the table entry $F = 4.26$, we reject the null hypothesis and so conclude that there is a difference among the population means. That is, the average grades given by the three professors (factor levels) are different at the $\alpha = .05$ level of significance.

Admittedly, you may feel that analysis of variance is a very complicated process. In fact, it is usually performed by computer, as we discuss in the next section. However, even if you do it by hand, ANOVA is not that difficult as long as you approach the problem systematically. After you have solved a few problems on your own, you will probably find that the process is not difficult at all. To help in this, we summarize the steps needed to solve any ANOVA problem:

Step 1. Calculate SS(Factor).

$$\text{SS(Factor)} = \frac{\Sigma (\text{column total})^2}{\text{number of entries per column}} - \frac{(\text{grand total})^2}{\text{total sample size}}$$

Step 2. Calculate SS(Total).

$$\text{SS(Total)} = \Sigma x^2 - \frac{(\text{grand total})^2}{\text{total sample size}}$$

Step 3. Find SS(Error).
$$SS(Error) = SS(Total) - SS(Factor)$$
Step 4. Calculate df(Factor).
$$df(Factor) = c - 1$$
Step 5. Calculate df(Total).
$$df(Total) = r \times c - 1$$
Step 6. Find df(Error).
$$df(Error) = df(Total) - df(Factor)$$
Step 7. Calculate MS(Factor).
$$MS(Factor) = \frac{SS(Factor)}{df(Factor)}$$
Step 8. Calculate MS(Error).
$$MS(Error) = \frac{SS(Error)}{df(Error)}$$
Step 9. Calculate the F value.
$$F = \frac{MS(Factor)}{MS(Error)}$$
Step 10. Draw a conclusion. Compare this test statistic to the critical value of F found in the F-distribution table, using the appropriate numbers of degrees of freedom and the appropriate significance level α.

We now illustrate this procedure in the following purely numerical example to assist you in mastering the technique.

EXAMPLE 12.1 Perform an analysis of variance calculation for the data given in Table 12.7 at the $\alpha = .05$ level of significance.

TABLE 12.7

Levels of Factor			
A	B	C	
25	30	40	
25	40	50	
50	60	80	
100	130	170	400 = grand total

We perform the various calculations below and enter all the results into the ANOVA table that follows the work.

Step 1:
$$SS(\text{Factor}) = \frac{\Sigma(\text{column total})^2}{\text{number of entries per column}} - \frac{(\text{grand total})^2}{\text{total sample size}}$$

$$= \frac{100^2 + 130^2 + 170^2}{3} - \frac{400^2}{9}$$

$$= \frac{55{,}800}{3} - \frac{160{,}000}{9}$$

$$= 18{,}600 - 17{,}777.78 = 822.22$$

Step 2:
$$SS(\text{Total}) = \Sigma x^2 - \frac{(\text{grand total})^2}{\text{total sample size}}$$

$$= 25^2 + 25^2 + 50^2 + \cdots + 80^2 - \frac{400^2}{9}$$

$$= 20{,}350 - 17{,}777.78 = 2572.22$$

Step 3:
$$SS(\text{Error}) = SS(\text{Total}) - SS(\text{Factor})$$
$$= 2572.22 - 822.22 = 1750$$

Step 4: $df(\text{Factor}) = c - 1 = 2$

Step 5: $df(\text{Total}) = r \times c - 1 = 8$

Step 6: $df(\text{Error}) = df(\text{Total}) - df(\text{Factor})$
$$= 8 - 2 = 6$$

Step 7:
$$MS(\text{Factor}) = \frac{SS(\text{Factor})}{df(\text{Factor})}$$
$$= \frac{822.22}{2} = 411.11$$

Step 8:
$$MS(\text{Error}) = \frac{SS(\text{Error})}{df(\text{Error})}$$
$$= \frac{1750}{6} = 291.67$$

Step 9:
$$F = \frac{411.11}{291.67} = 1.41$$

We now enter all these values in an ANOVA table.

	Sums of Squares	Degrees of Freedom	Mean Squares	F-Ratio
Factor	822.22	2	411.11	F = 1.41
Error	1750	6	291.67	
Total	2572.22	8		

At the $\alpha = .05$ level of significance with 2 degrees of freedom for the numerator and 6 degrees of freedom for the denominator, the critical value for F from Table VIII is $F = 5.14$. Consequently, since the calculated F-ratio $F = 1.41$ is smaller than the critical value $F = 5.14$, we cannot reject the null hypothesis. We therefore conclude that there may not be any difference among the 3 population means. See Figure 12.5. ■

As we indicated at the beginning of the section, the analysis of variance procedure is designed to apportion the total amount of variance present for all the sample data into two components, one corresponding to what happens *between* the different sets of data and the other corresponding to what happens *within* each set of data. The Mean Square for the Factor measures the amount of variance attributable to the factor (between) while the Mean Square for the Error measures the amount of variance attributable to the error (within). The F-ratio sets these quantities up as a quotient. When F is large, there is a disproportionate amount of variance attributable to the factor and so we reject the null hypothesis. If F is small, the amount of variance attributable to the error is relatively large compared to that attributable to the factor, and so we cannot reject the null hypothesis. The different population means may be equal.

FIGURE 12.5

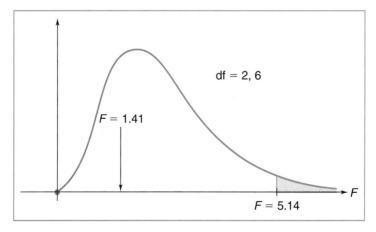

Exercise Set 12.1

Mastering the Techniques

For each of the following sets of data, construct the ANOVA table and compare the F value you calculate to the critical value at the given level of significance.

1. | A | B | C |
    |----|----|----|
    | 15 | 18 | 6 |
    | 10 | 20 | 15 |
    | 15 | 22 | 10 |

    $\alpha = .05$

2. | A | B | C |
    |----|----|----|
    | 12 | 10 | 8 |
    | 14 | 8 | 20 |
    | 9 | 12 | 12 |

    $\alpha = .01$

3. | A | B | C |
    |----|----|----|
    | 10 | 14 | 11 |
    | 12 | 9 | 12 |
    | 15 | 11 | 10 |
    | 13 | 11 | 7 |

    $\alpha = .01$

4. | A | B | C |
    |----|----|----|
    | 22 | 17 | 21 |
    | 19 | 20 | 17 |
    | 21 | 25 | 19 |
    | 18 | 23 | 18 |

    $\alpha = .05$

5. | A | B | C | D |
    |----|---|----|----|
    | 8 | 5 | 7 | 9 |
    | 10 | 9 | 11 | 12 |
    | 9 | 6 | 5 | 5 |
    | 8 | 8 | 7 | 9 |

    $\alpha = .05$

6. | A | B | C | D |
    |----|----|----|----|
    | 20 | 18 | 32 | 24 |
    | 25 | 27 | 25 | 27 |
    | 22 | 30 | 23 | 31 |
    | 23 | 25 | 30 | 28 |

    $\alpha = .01$

12.2 Applications of ANOVA

In Section 12.1, we developed the analysis of variance technique as a method for testing whether any difference exists among a set of three or more population means $\mu_1, \mu_2, \ldots, \mu_m$. As we pointed out, however, the method involved in solving analysis of variance problems is a highly computational procedure. As such, it lends itself naturally to computer usage. In this section, we indicate how the computer can be used to produce an ANOVA table for any set of sample data. We then illustrate a variety of applications of the method as well as some extensions of the ideas beyond what we did in Section 12.1. For example, when we developed the idea of analysis of variance, all samples were taken to be the same size. However, with some minor modifications, we can apply ANOVA to cases where the sample sizes are unequal.

Computer Corner

We have included Program 39: Analysis of Variance as a way of quickly and easily solving the type of ANOVA problems we discussed in Section 12.1. They are known as *one-way analysis of variance* problems.

The program performs a one-way analysis of variance with equal sample sizes. You must enter the number of factor levels or treatments. The program allows between two and five. You must also enter the number of measurements on each factor level for your data. The program allows two to ten entries for each. You then must supply the individual entries in the table of data values in column format for each factor level.

Once you have entered the data values, the program displays the column totals and the grand total for the data. It then transfers to a new text page on which the full ANOVA table is displayed, including the results of all the calculations and the value of the F-statistic. You then must determine the critical value of F from the F-distribution table, Table VIII, corresponding to the appropriate numbers of degrees of freedom and the value of α you desire. Finally, you must decide whether there is any difference among the means.

Minitab Methods

You can also use Minitab to perform a one-way analysis of variance test on any set of data stored in a series of indicated columns, say C1 through C4. The Minitab command

AOVONEWAY C1-C4

causes Minitab to print the complete analysis of variance table based on your data. In addition, it prints a variety of other information related to your data, including a series of graphical 95% confidence intervals for the mean of each set of values. We illustrate this in the sample Minitab session shown in Figure 12.6.

FIGURE 12.6

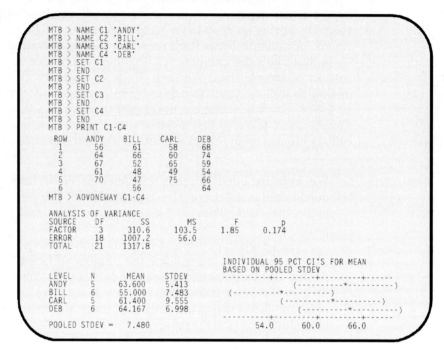

We now consider a number of applications and extensions of the analysis of variance method. In each, we will use the computer to produce the ANOVA table. If you would like to make sure that you can perform this test by hand calculation, you may want to go through these examples using the 10-step method from Section 12.1 to see that you obtain the same table of results.

EXAMPLE 12.2 A research group tests four different gasoline additives A, B, C and D in the same car to determine if there is any difference in gas mileage based on the additive used. A measured amount of gasoline with each additive is used in each test drive. These are the results (in miles per gallon) of five trials using each of the additives:

A	B	C	D
25	28	32	24
23	31	33	24
20	27	30	23
27	28	28	27
20	26	32	22

Determine whether a difference exists in mean gas mileage based on the gasoline additive used at the $\alpha = .05$ level of significance.

SOLUTION The factor levels or treatments are the four gasoline additives being tested. We set up the null and alternate hypotheses as follows:

H_0: There is no difference among the means: $\mu_A = \mu_B = \mu_C = \mu_D$.

H_a: There is a difference among the means.

When we enter the given data values into Program 39: Analysis of Variance, it responds with this ANOVA table:

	Sums of Squares	Degrees of Freedom	Mean Squares	F-Ratio
Factor	205	3	68.33	$F = 13.33$
Error	82	16	5.125	
Total	287	19		

We compare this test value of F to the appropriate critical value from Table VIII. In this case, the number of degrees of freedom for the numerator is 3 and the number of degrees of freedom for the denominator is 16. Further, since the level of significance is $\alpha = .05$, the critical value for F is $F = 3.24$. See Figure 12.7. Since the test value $F = 13.33$ is larger than this critical value, we reject the null hypothesis and conclude that there is a difference in mean gas mileage based on the additive used. ∎

Incidentally, now that we know that a difference likely exists among the population means as a result of the analysis of variance test, in practice we would like to be able to identify precisely where the difference is. Statisticians

FIGURE 12.7

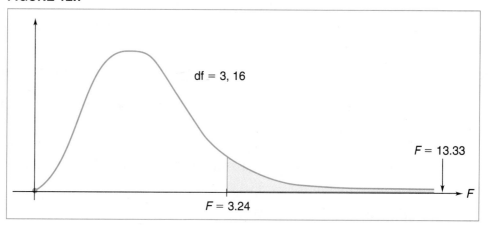

have developed sophisticated methods for performing both pairwise comparisons and more complex combinations of comparisons among the means. In particular, Duncan's method, Tukey's *T*-method, and Scheffé's *S*-method are used for this purpose. Although these methods are beyond the scope of this book, the interested reader can find them in more advanced texts on applied statistics.

A number of extensions of the analysis of variance procedure are possible. First, we considered a relatively simple case where the same number of observations or measurements is made on each treatment. This situation is known as *one-way analysis of variance with equal sample sizes*. It is also possible to consider situations involving different sample sizes for each factor. This is known as *one-way analysis of variance with unequal sample sizes*. We can do this with only a small modification of our previous approach.

In particular, we calculated the Sum of Squares of the Factor as

$$\text{SS(Factor)} = \frac{\Sigma(\text{column total})^2}{\text{number of entries per column}} - \frac{(\text{grand total})^2}{\text{total sample size}}$$

where the first denominator consists of the number of entries in each column. In this formula, we sum the squares of the column totals and then divide the result by the common number of entries per column. Since we now want to allow the possibility of different numbers of measurements in each column (equivalently, different sample sizes for each factor level), we must first divide the square of each column total by the corresponding number of entries in that column and then sum the results. The formula becomes

$$\text{SS(Factor)} = \Sigma \frac{(\text{column total})^2}{\text{number of entries in column}} - \frac{(\text{grand total})^2}{\text{total sample size}}$$

The only other change we must make is in the number of degrees of freedom for the total. It is now 1 less than the total number of data values:

$$\text{df(Total)} = n - 1$$

We illustrate such a case in the following example involving different sample sizes.

EXAMPLE 12.3 A study is conducted comparing the average daily high temperatures in North America, Europe and Asia. Major cities on each continent are randomly selected, and the high temperatures on the chosen day are as follows:

North America
    Chicago: 95   Denver: 73   Fairbanks: 73   Kansas City: 96
    Montreal: 70   Miami: 87   Pittsburgh: 85   Seattle: 80
Europe
    Athens: 95   Geneva: 72   London: 77   Moscow: 86   Rome: 88
    Warsaw: 73

Asia
Beijing: 91   Jerusalem: 88   New Delhi: 94   Tokyo: 77
Hong Kong: 90

Test whether there is any difference in the mean daytime high temperatures on the three continents on the given day at the $\alpha = .01$ level of significance based on these sample data.

SOLUTION The null hypothesis states

H_0: There is no difference among the mean temperatures on each continent on the day is question: $\mu_A = \mu_B = \mu_C$.

while the alternate hypothesis is

H_a: There is a difference in mean temperature.

Note that the sample mean for the eight North American cities is 82.4, for the six European cities is 81.8, and for the five Asian cities is 88.
The resulting ANOVA table is:

	Sums of Squares	Degrees of Freedom	Mean Squares	F-Ratio
Factor	127.40	2	63.70	F = 0.78
Error	1304.70	16	81.54	
Total	1432.10	18		

The critical value of F based on 2 and 16 degrees of freedom for the numerator and denominator, respectively, and $\alpha = .01$ is $F = 6.23$. Since the test statistic $F = .78$ for the given data, we cannot reject the null hypothesis and so conclude that there may be no difference among the average daily temperatures on the three continents in the northern hemisphere on that particular day. See Figure 12.8. ∎

FIGURE 12.8

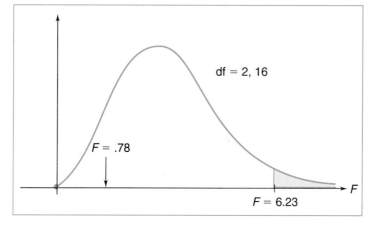

The method we have used for one-way analysis of variance is so called because it involves an analysis based on a single factor. It is also possible to consider two or more factors in such an analysis. For instance, in Section 12.1 we introduced the notion of ANOVA through the problem of the average grades achieved on a final exam by students having three different instructors. The instructor is then a single factor. It is also possible to consider other factors in such a situation that may also be important in determining a student's grade. For instance, factors such as class size or the time of day that a class meets might also be important in terms of the grades received. For such a study which tests different instructors as well as different times of day or different class sizes, an extension of ANOVA known as *two-way analysis of variance* can be used. However, we do not go into this topic here.

Exercise Set 12.2

Mastering the Techniques

1. A farmer is testing the effects of three different fertilizers—A, B, and C—on the yields of tomato plants. He applies the fertilizers and then monitors the number of tomatoes grown on a random sample of four plants for each of the three types of fertilizer. The results in terms of number of tomatoes per plant are as follows:

A	B	C
24	21	16
18	26	22
27	32	19
28	25	17

   Test if there is any difference in the mean number of tomatoes grown per plant based on the type of fertilizer used at the $\alpha = .05$ level of significance.

2. A medical experiment is conducted to study the effects of three different drugs—A, B, and C—in reducing high blood pressure. Each of the drugs is administered to five different randomly selected patients, and the reduction in their diastolic readings after 1 month are measured as follows:

A	B	C
8	10	9
7	11	10
9	14	8
11	12	7
10	13	9

Test if there is any difference in the mean decrease in diastolic reading per patient based on the use of the different drugs at the $\alpha = .01$ significance level.

3. The manager of the shoe department in a large department store is conducting a study of the number of customers who pay by personal check, by store credit card, and by bank credit card. During the course of five different days, she records the following number of purchases made by each method:

Check	Store Card	Bank Card
28	35	33
21	42	38
30	32	31
18	25	42
23	27	29

Test if there is any difference in the mean number of purchases made with each method at the $\alpha = .01$ significance level.

4. A statewide teachers' union is studying the average class size for kindergarten classes in three different school districts covering three major cities in the state. The respondents to a survey provide the following data on such classes:

District A: 18 18 20 19 20

District B: 21 20 22 20 17 21

District C: 24 23 25 23 20 24 25

Determine whether there is any difference in the average class sizes in the different school districts at the $\alpha = .05$ level of significance.

5. A large corporation experiments with different types of background music to assess the effects on the productivity of secretaries in a typing pool. Over the course of several days, it uses soft background music, rock music, classical music, and no music and measures the number of letters typed by random selections of four secretaries in 2-hour periods. The results are as follows:

Soft: 15 18 22 17

Rock: 13 20 16 15

Classical: 15 19 24 18

None: 14 23 17 14

Test if there is any difference among the means for the number of letters typed per 2-hour period based on the type of background music at the $\alpha = .05$ level of significance.

6. The local environmental protection group is responsible for monitoring the level of pollutants at four beaches during the summer to ensure that the beaches are safe for swimmers. At each beach, a variety of different readings are taken. The

results one day are as follows:

Beach A:	15	11	18	16	10	15
Beach B:	9	14	10	15	18	13
Beach C:	18	13	11	14	12	15
Beach D:	12	8	10	14	9	12

Based on these readings, test if there is any difference among the mean pollutant levels at the four beaches at the $\alpha = .01$ level of significance.

Computer Applications

Exercises 7 through 10 refer to the set of data used in Exercise 5 above:

15	18	22	17
13	20	16	15
15	19	24	18
14	23	17	14

7. Double each of the values given in this set of readings, and use Program 39: Analysis of Variance to calculate the value of the F-statistic. How does it compare to the value you found in Exercise 5?
8. Divide each of the values given in this set of readings by 10, and use Program 39: Analysis of Variance to determine the value of F. How does it compare to the previous value?
9. Subtract 13 (the smallest reading) from each of the given values in the set of data, and use Program 39: Analysis of Variance to determine the value of F. How does it compare to the previous value?
10. Based on Exercises 7 through 9, can you devise a scheme for calculating the F-statistic by hand that will minimize the amount of work to be done?

Student Projects

Conduct an analysis of variance for a table of data based on a study you personally perform. Choose a topic of interest to you involving three or more levels of a factor whose means you want to compare. For example, it might be

1. Average class sizes in different academic areas—mathematics, social sciences, humanities, etc.—at your school
2. GPA for students at your school based on academic year—freshman, sophomore, junior, senior
3. Average time needed for different cashiers at a supermarket to ring up and bag customer orders
4. Average number of credits being carried by students based on the academic year

5. Average number of points scored by professional basketball players in each of the NBA's divisions
6. Average number of words per sentence in *The New York Times*, *USA TODAY*, your school newspaper, etc.

Conduct the study by using a randomly selected group of at least 5 measurements for each factor level, and then perform a complete analysis of variance on the results to determine whether there is any difference among the means for the subject you are studying.

You should incorporate all your results into a formal statistical project report. The report should include

- A statement of the topic being studied including the particular population being sampled, the factor, and the different factor levels
- The source of the data
- A discussion of how the data were collected and why you believe that it is a random sample
- A list of the data including possibly some tables and/or diagrams to display the data
- All the statistical analysis
- Your conclusions based on this analysis and a rationale for the level of significance you select
- An indication of which groups or individuals have higher and lower means, if you find that there is a difference among the means
- A discussion of any surprises you may have noted in connection with collecting and organizing the data or with your conclusions

CHAPTER 12 SUMMARY

Analysis of Variance (ANOVA) is used to test whether the means of a set of three or more populations are equal. **One-way analysis of variance** is used to test the means associated with a single **factor**. The factor will have three or more different **factor levels** or **treatments**, one for each of the populations. The sample sizes may be **equal** or **unequal**.

The null hypothesis for ANOVA states that

$$H_0: \mu_1 = \mu_2 = \cdots = \mu_m$$

The analysis of variance procedure apportions the total amount of variance present for all the sample data into two components, one corresponding to what happens **between** the different sets of data and the other corresponding to what happens **within** each set of data. The **Mean Square for the Factor** measures the amount of variance attributable to the factor (between) while the **Mean Square for the Error** measures the amount of variance attributable to the error (within). The **F-ratio** or **F-statistic** is the ratio of the two Mean

Squares. The *F*-statistic follows an **F-distribution** based on two values, the **number of degrees of freedom for the numerator** and the **number of degrees of freedom for the denominator.** The critical values for *F* are found in Table VIII. When the test value for *F* is large, there is a disproportionate amount of variance attributable to the factor and so we reject the null hypothesis. If *F* is small, then the amount of variance attributable to the error is relatively large compared to that attributable to the factor and so we cannot reject the null hypothesis.

13 Nonparametric Statistical Tests

A group of environmental scientists claim that last summer's unusually hot temperatures were not just an aberration but rather part of a long-term global warming pattern. Their study contributes additional evidence to support the theory that the earth is experiencing an overall increase in temperatures due to a phenomenon called the greenhouse effect. The scientists point out that if the theory is indeed true, as more and more environmentalists believe, it will have a significant impact not only on our lifestyles but also on our very ability to maintain life on earth.

What constitutes evidence that the earth is undergoing global warming? How do scientists distinguish between an unusual heat wave one year and a long-term trend? The key is in determining if there is a consistent increase in temperature and this can be tested statistically. Should the comparison be made on a year-by-year basis comparing the average daily high temperature this year to the average of previous years? Should the comparison be made on a day-by-day basis comparing the high temperature on each day to the daily high on the same day 20 or 50 years ago? Should the number of new record highs be compared to the number of records set in previous years? Any such study cannot focus on just a single location, say Minneapolis or Moscow or Melbourne. Rather, it is necessary to compare temperatures in many different locations or to compare regional, continental or even global averages for each day or for each year.

What should scientists look for in these results that would support the global warming hypothesis? For example, using a year-to-year study, regression analysis might be applied. If the regression line has a positive slope, this would suggest a trend of overall rising temperatures. Using a day-by-day comparison, suppose that the daily high temperatures this year set new records on 30 different dates, or 50 different dates, or 100 different dates. Would this suggest that global warming is going on? Suppose that the high (or the low) temperature on different dates this year is consistently higher than the comparable temperature on the same date of other years. Would this constitute proof?

In this chapter, we will develop statistical tools that allow us to make various types of comparisons when the methods we have previously considered do not apply. In particular, one of these methods is ideal for carrying out some of the studies suggested for examining the global warming debate and we will ask you to conduct such a study as a **Student Project** at the end of the chapter.

13.1 Introduction

Our study of inferential statistics has covered a wide variety of hypothesis tests. Most of these methods are based on the assumption that the underlying population is approximately normal or they depend on the Central Limit Theorem to ensure that the corresponding sampling distribution is roughly normal, provided the sample size is large enough. The resulting procedures are called *parametric tests*.

In many cases that arise in practice, we face situations where we cannot assume normality, or even near normality, for the underlying population. Consequently, the parametric tests do not apply, since each of these tests has certain conditions that must be met for it to be valid. As a result, statisticians have developed a series of *nonparametric* or *distribution-free statistical tests* which do *not* depend on the distribution of the underlying population from which we sample. We study some of the more common tests in this chapter.

For the most part, you will find that these nonparametric tests are considerably easier and faster to apply than many of the parametric tests we used previously. One tradeoff, however, is that they sometimes tend to be less powerful than the parametric tests because they do not always make full use of the data available. However, if the conditions for the parametric tests cannot be met, the nonparametric tests are the only alternative.

13.2 The Sign Test

We begin our study of nonparametric statistical tests with a variation on a situation previously encountered. In many applications concerning two populations, we considered the problem of determining whether any difference exists between the means of two groups. For instance, we might want to study the prices at two supermarkets. To do this, we might select a random sample of items sold at both stores and compare the prices on these common products. This would entail using the paired-data test, since there is a direct link between the two dependent samples. This procedure, however, requires that the underlying populations both be approximately normal, and this might not necessarily be the case.

In this instance, we look at a variation on this problem. Often, as part of an advertising campaign, one supermarket will claim that its prices are lower than those of its competitor. Let's see how this can be achieved. The simplest way is for the first supermarket to determine the regular (nonsale) prices of all its competitor's products and then set its own prices precisely one cent lower on every item. In this way, the first supermarket can legitimately claim that its prices are lower. For that matter, it can do this for a majority (but not necessarily all) of the items that are sold and still make the same claim legitimately.

Suppose that for all items sold at both Ace and Bestway, the price at Ace is one cent lower than the corresponding price at Bestway. When we consider the fact that the average price of products sold in supermarkets today is considerably more than a dollar, then most customers would not consider the difference in the mean price between the two stores to be significant in a practical sense. For instance, if a customer purchases 100 different items, then the total bill is likely to be on the order of $150 to $200, while the total savings will be $1.

But this situation does illustrate a different type of difference between two sets of data—a *consistent difference*. In this case, prices at Ace are *consistently lower* than those at Bestway. In this section, we develop a nonparametric statistical test, known as the *sign test*, to determine whether a consistent difference exists between two sets of data in the sense that one set of values is usually smaller than the other.

To see how this is done, let's use the example of the two supermarkets. It is obvious that if all prices at Ace are lower than the corresponding prices at Bestway, then Ace is consistently cheaper than Bestway. Suppose that the prices of 90% of the items at Ace are lower than those at Bestway while 10% are higher at Ace. (We ignore those items where the prices are the same.) Again, we conclude that the preponderance of prices at Ace are lower than at Bestway and so decide that Ace is usually cheaper than Bestway. However, if the percentage of items that are cheaper at Ace is as low as 50%, we would no longer come to this conclusion. Therefore, the key to making such a decision lies with the proportion of prices at Ace found to be lower than those at Bestway. If the proportion is high enough, we conclude that the preponderance of prices at Ace are lower than those at Bestway and so Ace is usually cheaper than Bestway. But if the proportion is not high enough, then we reject the claim that Ace's prices are usually lower than Bestway's. These last two statements, however, should remind you of a simple hypothesis test on proportions, and, in fact, the sign test is precisely a minor variation on it.

At this point, you may want to review the details on such a hypothesis test in Section 8.4.

Suppose the following set of data is obtained on prices for the same 15 items at the two competing supermarkets:

Ace	Bestway
1.29	1.39
1.49	1.55
1.99	1.89
.89	.99
2.45	2.59
.69	.75
.99	.99
1.12	1.09
1.49	1.45
1.59	1.75
2.09	2.39
1.79	1.89
1.59	1.49
.53	.59
3.19	3.39

Before proceeding, we rewrite this table to emphasize the difference in prices between the two stores:

Ace	Bestway	A − B	+/−
1.29	1.39	−.10	−
1.49	1.55	−.06	−
1.99	1.89	.10	+
.89	.99	−.10	−
2.45	2.59	−.14	−
.69	.75	−.06	−
.99	.99	0	0
1.12	1.09	.03	+
1.49	1.45	.04	+
1.59	1.75	−.16	−
2.09	2.39	−.30	−
1.79	1.89	−.10	−
1.59	1.49	.10	+
.53	.59	−.06	−
3.19	3.39	−.20	−

In this set of 15 prices, Ace is cheaper than Bestway for 10 items, and there are 10 minus signs; Bestway is cheaper than Ace for 4 items, so there are 4 plus signs; and their prices match for 1 item, so there is one 0. Based on this, can we conclude that Ace is usually cheaper than Bestway?

It is fairly obvious that the item with the same price at both stores will not affect our decision, so it really reduces to considering the 14 products for which the prices differ. Of these, Ace is cheaper than Bestway for 10 of the 14 items, or 10 of the 14 signs are minus. Either way, this is equivalent to

$\frac{10}{14}$ = 71.4% or a sample proportion of .714. We set this up as a hypothesis test on proportions.

For such a problem, the null hypothesis states that

H_0: There is no consistent difference.

If this is true, then we would expect about 50% of the signs to be plus and 50% to be minus, or equivalently we would presume that π = .5. Notice here that π refers to the proportion of minus signs (or plus signs) out of all the price differences. It does not refer to either of the two populations consisting of store prices under consideration.

Similarly, the alternate hypothesis might be

H_a: There is a consistent difference: $\pi \neq .5$

for a two-tailed test or, more likely in this context,

H_a: Ace tends to be cheaper than Bestway: $\pi > .5$

for a one-tailed test. Further, there is a significance level for the test, say α = .05.

To perform this test of proportions, we first collect the sample data:

$n = 14$ (we discard the "match")

$x = 10$ (number of "successes" out of n)

$p = \dfrac{x}{n} = .714$

We will use

$$\sigma_\pi = \sqrt{\dfrac{\pi(1-\pi)}{n}} = \sqrt{\dfrac{.5(.5)}{14}} = \sqrt{.01786} = .134$$

Since

$n\pi = 14(.5) = 7 > 5$ and $n(1-\pi) = 14(.5) = 7 > 5$

we use a normal distribution centered at π = .5. For a one-tailed test with α = .05, as shown in Figure 13.1, the critical value is z = 1.64. For the sample data, we find that

$$z = \dfrac{p - \pi}{\sigma_\pi}$$

$$= \dfrac{.714 - .5}{.134} = 1.60$$

and hence we cannot reject the null hypothesis. That is, the evidence is not sufficiently strong to indicate that the prices at Ace are consistently lower than those at Bestway at the 5% level of significance.

We now consider a more formal example involving the sign test.

FIGURE 13.1

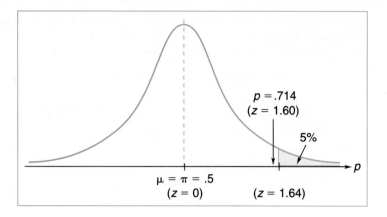

EXAMPLE 13.1 A college believes that a 1-hour review session can improve grades on a mathematics placement exam for incoming freshmen. To test this hypothesis, a placement test is administered to a group of students; then the students attend the mathematics review session and are given a second equivalent placement exam afterward. The results are as follows:

Before	After
22	21
26	29
17	15
20	20
28	26
31	32
23	25
13	14
19	19
25	27
28	27
24	25
27	27
18	20
20	23
14	16
24	26
15	20
19	20
18	17
27	29

On the basis of these results, can we conclude that most participants in the review session tend to improve their placement exam scores at the 5% level of significance?

SOLUTION The null hypothesis states

$$H_0: \text{There is no consistent improvement: } \pi = .5$$

while the alternate hypothesis is

$$H_a: \text{There is improvement: } \pi > .5$$

where π represents the proportion of students whose grades increase. We organize the given data in a table with x used as the first score and y as the second.

x	y	y − x	+/−
22	21	−1	−
26	29	3	+
17	15	−2	−
20	20	0	0
28	26	−2	−
31	32	1	+
23	25	2	+
13	14	1	+
19	19	0	0
25	27	2	+
28	27	−1	−
24	25	1	+
27	27	0	0
18	20	2	+
20	23	3	+
14	16	2	+
24	26	2	+
15	20	5	+
19	20	1	+
18	17	−1	−
27	29	2	+

Thus, out of the 21 students, there are 18 changes in score, so that $n = 18$. Further, of these changes, there are 13 plus signs, which represent the improvements. Therefore, the sample data we use are

$$n = 18$$
$$x = 13$$
$$p = \tfrac{13}{18} = .722$$

The assumption that there is no consistent difference implies that $\pi = .5$. Since

$$n\pi = n(1 - \pi)$$
$$= 18(.5)$$
$$= 9 > 5$$

we use a normal distribution with

$$\sigma_\pi = \sqrt{\frac{.5(1 - .5)}{18}}$$
$$= \sqrt{.01389}$$
$$= .118$$

As shown in Figure 13.2, we have a one-tailed test where the critical value corresponding to $\alpha = .05$ is $z = 1.64$. For the sample data, we find that

$$z = \frac{.722 - .5}{.118}$$
$$= 1.88$$

and therefore we reject the null hypothesis at the .05 level of significance. That is, taking the review session usually does improve placement test scores. ∎

The above approach is known as a *paired-sample sign test* since two sets of readings on the same individuals are compared. There are also instances where we take a sample from a single population and use it to test a hypothesis

FIGURE 13.2

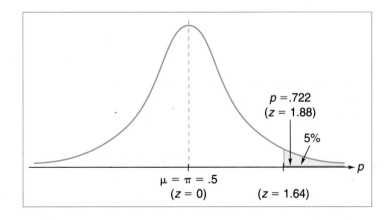

regarding the population median M of the underlying population. The methods used in such a *single-sample sign test* are virtually identical to what we did above. We illustrate such a case in the following example.

EXAMPLE 13.2 A real estate industry group claims that the median price for new homes in a certain region of the United States is $165,000. A study is conducted regarding the median price for new homes in one town in this region, and the following prices (in thousands of dollars) were paid:

142, 167, 145, 188, 165, 159, 179, 162, 139, 189, 219, 144, 159, 160, 138, 199, 159, 145, 160

Use these data to test whether the prices for new homes in this town are usually lower than the median for the region. Use $\alpha = .01$.

SOLUTION The null hypothesis states that

H_0: The median price in the town is $165,000

or equivalently that the proportion π of homes priced below the regional median is .5

H_0: $\pi = .5$

The alternate hypothesis is

H_a: The median price is below $165,000

or equivalently the proportion of houses priced below the median is greater than .5:

H_a: $\pi > .5$

Therefore, the hypothesis test is a one-tailed test on proportions with the rejection region in the tail to the right.

We begin by listing all the data values and comparing them to the assumed population median of 165,000, or 165 thousands. Thus, we obtain

x	+/−
142	−
167	+
145	−
188	+
165	0
159	−
179	+
162	−
139	−
189	+
219	+

x	+/−
144	−
159	−
160	−
138	−
199	+
159	−
145	−
160	−

We see that 12 houses cost less than the median, 6 houses cost more, and 1 house is precisely equal to the median. We ignore this one house price and consider the $n = 18$ remaining houses, of which $x = 12$ are below the median. The sample proportion is

$$p = \frac{12}{18} = .667$$

Since both $n\pi$ and $n(1 - \pi)$ are greater than 5, we can use a normal distribution with

$$\sigma_\pi = \sqrt{\frac{\pi(1 - \pi)}{n}}$$

$$= \sqrt{\frac{.5(.5)}{18}}$$

$$= \sqrt{.01389}$$

$$= .118$$

For a one-tailed test with $\alpha = .01$, the critical value is $z = 2.33$, as shown in Figure 13.3. For the sample data, we find that

FIGURE 13.3

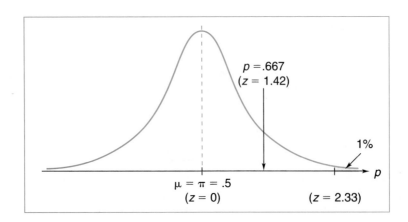

$$z = \frac{p - \pi}{\sigma_\pi}$$

$$= \frac{.667 - .5}{.118}$$

$$= 1.42$$

and so we cannot reject the null hypothesis at the $\alpha = .01$ level of significance. There is insufficient evidence to allow us to conclude that the prices for new homes in this town are lower than the median value for the entire region. ∎

Exercise Set 13.2

Mastering the Techniques

Test each of the following sets of paired data to see if one group is usually lower than the other at the $\alpha = .05$ level of significance:

1.
x	21	24	20	17	28	25	30	19	23	24	26
y	24	25	21	26	25	18	24	26	20	22	27

2.
x	14	18	12	10	15	17	13	13	21	15	12	16
y	15	17	18	15	18	20	15	12	17	13	17	18

3.
x	55	58	51	48	56	59	62	53	58	60	56	64	62	60
y	59	61	54	57	52	55	62	57	52	60	58	59	60	55

4.
x	120	154	135	142	165	140	176	139	147	160	150	165
y	124	148	127	146	156	134	167	132	147	154	143	155

Applying the Concepts

5. Two tellers in a bank are being compared to see whether one serves consistently more customers over the course of an hour. The head teller collects the following data on the number of customers each teller serves in the course of 1-hour periods:

Craig	13	15	18	11	14	16	24	16	13	8	13	16	12	13
Ken	11	11	14	15	14	12	16	14	19	12	15	13	11	14

Based on these data, does Craig usually serve more customers per hour than Ken at the $\alpha = .05$ level of significance?

6. Steve and Marian are neighbors and work around the corner from each other. Over a 3-week period, they decide to compare two different routes to work (A and B) in terms of which takes less time if they leave home at the same time.

They randomly choose who will drive each route each day. The commuting times, in minutes, they experience are as follows:

Route A	27	23	20	27	36	23	20	27	24	32	24	32	25	29	31
Route B	29	26	25	22	27	38	30	25	29	26	27	26	25	26	34

Can you conclude that route A is usually faster than route B at the $\alpha = .02$ level of significance?

7. An engineer is testing a new bobsled design for potential use in the next winter Olympics. Its times on the run are being compared to the median time, 43.41 seconds, on the same run for traditionally designed bobsleds. The new sled takes the following times to complete the course:

42.15, 43.04, 42.38, 42.17, 41.58, 42.40, 42.52, 43.36, 42.79, 42.53, 43.12, 42.87

Does the new sled usually complete the course faster than the median time at the 5% level of significance?

8. The engineer in Exercise 7 is now testing two bobsled designs for potential use in the Olympics. They are used alternately down a trial bobsled course with the following times to complete each run:

A	42.15	43.04	42.38	42.17	41.58	42.40	42.52	43.36	42.79	42.53	43.12	42.87
B	42.43	42.47	42.46	42.43	42.03	42.55	43.12	42.44	42.57	42.48	44.65	43.03

Does sled A usually complete the course faster than sled B at the 5% level of significance?

9. The cost for home insurance in a certain city has a median of $380 per year. An insurance broker has calculated the rates charged by a major insurance company for a number of his clients who request different types of coverage. He finds the following amounts:

255, 320, 352, 279, 439, 268, 375, 389, 285, 311, 290

Can he conclude that the company's rates are usually lower than the industry-wide rates at the 5% level of significance?

10. The insurance broker in Exercise 9 now wants to compare the rates charged by the company he used in Exercise 9 and by a second insurance company. He gets the following rate quotes:

A	255	320	352	279	439	268	375	389	285	311	290
B	247	309	336	295	446	259	360	344	292	297	276

Can he conclude that company B usually provides lower rates for comparable coverage at the 5% level of significance?

11. A company offering an SAT mathematics preparatory course administers a sample exam at the beginning of the course and another at the end to demonstrate

that the course makes a major difference to students. For one group of 80 students, the company finds that 58 achieved higher scores, 18 received lower scores, and 4 achieved the same score. Based on these data, can the company advertise that its course usually produces improved SAT scores at the $\alpha = .05$ level of significance?

12. A pharmaceutical research laboratory is conducting a series of tests involving a new drug which supposedly reverses hair loss in balding men. In performing the test, the number of hairs growing in a given (marked) region of the scalp is counted, the drug is administered, and the subject returns after 1 month for a hair count on the same region of the scalp. The results are that of 150 men involved in the experiment, 116 have higher counts, 22 have lower counts, and 12 have the same counts. Based on these results, can the company conclude that the drug usually has a positive effect at the $\alpha = .02$ level of significance?

13.3 The Rank-Sum Test

Some of the most important applications of inferential statistics involve comparisons between the means of two different groups to determine whether there is any difference between them. The difference-of-means test is probably the single most useful tool for studies comparing the means of independent samples. However, there are some limitations to its applicability. Its use is predicated on two assumptions—that the two underlying populations are roughly normal and their standard deviations are comparable in size. If either condition does not hold, then the difference-of-means procedure cannot be applied.

We now consider an alternative, nonparametric approach for dealing with such a situation known as the *rank-sum test*. It is also called the *Wilcoxin rank-sum test* and sometimes the *Mann-Whitney U-test*. The test involves studying the position or *ranking* in a single list of all the different data values from two independent random samples. We develop the ideas involved in the following situation.

Suppose a study is being conducted to compare the average time spent in space by U.S. astronauts and Russian cosmonauts. Random samples of the number of hours that various astronauts and cosmonauts spent in orbit are collected:

Astronauts: 9, 98, 331, 71, 147, 195, 216, 1427
Cosmonauts: 25, 119, 26, 73, 569, 143, 1632, 425, 223

The mean for the eight sample astronauts is 311.75 hours while the mean for the nine sample cosmonauts is 359.44 hours.

Unfortunately, based on these sample values, there is no reason to expect that either population is normal or that the standard deviations for the two populations are at all comparable. Therefore, we cannot use a difference-of-means test to compare the two sets.

The *rank-sum test* is based on combining these data into a single table or list in numerical order. Since there are a total of 17 entries, they are ranked from 1 to 17 when listed in either ascending or descending order. For this illustration, we arrange them in ascending order, as shown in Table 13.1, where the letters in the third column refer to astronauts (*A*) and cosmonauts (*C*). Notice that the eight U.S. times have ranks of 1, 4, 6, 9, 10, 11, 13 and 16 while the Soviet times have ranks of 2, 3, 5, 7, 8, 12, 14, 15 and 17.

TABLE 13.1

Rank	Time	Group
1	9	A
2	25	C
3	26	C
4	71	A
5	73	C
6	98	A
7	119	C
8	143	C
9	147	A
10	195	A
11	216	A
12	223	C
13	331	A
14	425	C
15	569	C
16	1427	A
17	1632	C

Since there are two sample groups, we denote the sample sizes by n_1 and n_2 with the understanding that n_1 always represents the size of the *smaller* group. Thus, in the above list, $n_1 = 8$ for the astronauts and $n_2 = 9$ for the cosmonauts. Of course, if both sample sizes are equal, then we can use either one as n_1.

Furthermore, the rank-sum test is based not on the rankings of the individual entries, but rather on the *sum of the rankings*. Since there are two groups, we use R_1 and R_2 to represent the sums of the rankings in each group. Thus, the sum of the rankings for the astronauts is

$$R_1 = 1 + 4 + 6 + 9 + 10 + 11 + 13 + 16 = 70$$

while that for the cosmonauts is

$$R_2 = 2 + 3 + 5 + 7 + 8 + 12 + 14 + 15 + 17 = 83$$

When n_1 and n_2 are both large enough (each equal to 8 or more), the distribution of the values for the sum of the rankings is approximately normal with mean

$$\mu_R = \tfrac{1}{2} n_1 (n_1 + n_2 + 1)$$

and standard deviation

$$\sigma_R = \sqrt{\frac{n_1 n_2 (n_1 + n_2 + 1)}{12}}$$

Moreover, in performing these calculations, it is sufficient to work only with the smaller group corresponding to n_1 and R_1. (Note that since we know the sum of all the rankings $1 + 2 + \cdots + 17 = 153$, it follows that $R_1 + R_2 = 153$ and so $R_2 = 153 - R_1$.)

For our data, $n_1 = 8$ and $n_2 = 9$, so that

$$\mu_R = \tfrac{1}{2}(8)(8 + 9 + 1)$$
$$= \tfrac{1}{2}(8)(18) = 72$$

and

$$\sigma_R = \sqrt{\frac{(8)(9)(8 + 9 + 1)}{12}}$$
$$= \sqrt{108}$$
$$= 10.39$$

The null hypothesis for this nonparametric test states

H_0: The populations of times spent in space are identical.

and the alternate hypothesis is

H_a: The populations are not identical.

Suppose we now conduct the test to see whether there is any difference in the populations at the $\alpha = .05$ level of significance. Since the R values are approximately normally distributed and we are considering a two-tailed test, we use critical values of $z = 1.96$ and $z = -1.96$, as shown in Figure 13.4. For the data listed, we have $R = R_1 = 70$, so that the test statistic is

$$z = \frac{R - \mu_R}{\sigma_R}$$
$$= \frac{70 - 72}{10.39}$$
$$= \frac{-2}{10.39}$$
$$= -.19$$

FIGURE 13.4

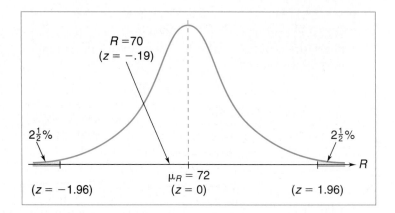

Consequently, we cannot reject the null hypothesis, and so we cannot conclude that there is any difference in the populations of time spent in space at the 5% level of significance.

Before we consider another example of the rank-sum test, several important points bear mentioning. First, notice that nowhere in the above analysis did we make any use of the means for the two groups. The rank-sum test does not use the parameters of the underlying populations; rather, it uses the nonparametric measure given by the rankings of the original data.

Second, in constructing the table of rankings, it is certainly possible that ties could occur for some ranks. In such a case, we simply assign the average of the ranks to the tied values. For instance, if the 3rd and 4th entries were the same, then both would be assigned rankings of $(3 + 4)/2 = 3.5$ and the following entry would remain in position number 5. Similarly, if the 7th, 8th and 9th entries were the same, all three would be assigned a ranking of $(7 + 8 + 9)/3 = 24/3 = 8$ so that the table would contain the following rankings:

$$1, 2, \ldots, 6, 8, 8, 8, 10, 11, \ldots$$

We now turn our attention to another example involving the rank-sum test.

EXAMPLE 13.3 A commuter has the choice of driving to work either by a highway (which is often clogged with traffic during morning rush hours) or by back roads. Over the course of several weeks, he tries both routes and times his trip to work. Based on the following set of data for the number of minutes the trip takes, test if there is any difference in the average time for his commute at the 5% level of significance.

A (highway)	34	28	46	42	56	85	48	25	37	49
B (back roads)	43	49	41	55	39	45	65	50	47	51

SOLUTION The null hypothesis for this problem states

H_0: There is no difference in times for the routes

while the alternate hypothesis states

H_a: There is a difference

Notice that the mean time taken by highway is 45 minutes while the mean time taken by the back roads is 48.5 minutes. However, if we cannot assume that the two sets of times are approximately normally distributed with equal variances, we cannot compare their means using the difference-of-means test. Consequently, we apply the rank-sum test.

We begin by organizing a table of rankings for these data.

Rank	Time	Group
1	25	A
2	28	A
3	34	A
4	37	A
5	39	B
6	41	B
7	42	A
8	43	B
9	45	B
10	46	A
11	47	B
12	48	A
13.5	49	A
13.5	49	B
15	50	B
16	51	B
17	55	B
18	56	A
19	65	B
20	85	A

Notice that there are two entries for 49 minutes; since they are in the 13th and 14th positions, we assign both a rank of 13.5. Further, since both groups involve 10 entries, we let $n_1 = 10$ represent the entries for group A. The sum of the ranks for group A is

$$R = 1 + 2 + 3 + 4 + 7 + 10 + 12 + 13.5 + 18 + 20$$
$$= 90.5$$

Since n_1 and n_2 are both greater than or equal to 8, the distribution of sample R values is approximately normal with mean

$$\mu_R = \tfrac{1}{2}n_1(n_1 + n_2 + 1)$$
$$= \tfrac{1}{2}(10)(10 + 10 + 1)$$
$$= 105$$

and standard deviation

$$\sigma_R = \sqrt{\frac{n_1 n_2(n_1 + n_2 + 1)}{12}}$$
$$= \sqrt{\frac{(10)(10)(10 + 10 + 1)}{12}}$$
$$= \sqrt{175}$$
$$= 13.23$$

Since this is a two-tailed test at the $\alpha = .05$ level of significance, the critical values are $z = \pm 1.96$, as shown in Figure 13.5. The test statistic for these data is

$$z = \frac{R - \mu_R}{\sigma_R}$$
$$= \frac{90.5 - 105}{13.23}$$
$$= -1.10$$

and hence we cannot reject the null hypothesis. There is insufficient evidence to conclude that there is any difference in the commuting times over the two routes. ∎

As we remarked earlier, the rank-sum test uses the normal distribution provided that both $n_1 \geq 8$ and $n_2 \geq 8$. If either or both of these conditions

FIGURE 13.5

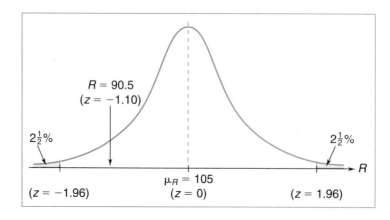

fail, it is necessary to use an exact distribution known as the Mann-Whitney U-distribution. Exact tables are available for performing this U-test, but we do not go into the details here.

Exercise Set 13.3

Mastering the Techniques

For the sets of data in Exercises 1 to 4, find the value of the rank sum R and the mean and standard deviation for the corresponding normal distribution:

1.
A	15	18	14	22	25	16	12	20
B	23	11	26	24	17	19	15	21

2.
A	43	38	39	44	53	42	55	47
B	41	40	52	48	46	51	57	45

3.
A	102	114	127	111	122	108	117	115	
B	105	114	120	124	132	118	125	125	123

4.
A	64	59	47	74	48	55	48	71	59	63	64
B	73	58	55	72	64	62	63	72	49	55	

Applying the Concepts

5. Two police cars patrol the same stretch of a major freeway. The officer in charge decides to compare the number of tickets issued daily by each officer over the course of a 2-week period. These are the results:

Officer McCarthy	32	14	26	37	45	28	32	36	25	30
Officer Abramowitz	44	37	24	33	27	31	41	29	25	34

Use the rank-sum test to determine whether there is any difference in the average number of tickets issued daily by the two officers at the $\alpha = .05$ level of significance.

6. A professional basketball recruiter is seeking to compare the scoring of two centers, one in a Division I school and the other in a Division II school. He collects the following sets of points scored by each player over the course of several games.

Player A	16	20	13	24	18	21	19	16
Player B	25	32	17	11	24	12	21	10

Use the rank-sum test to determine whether there is any difference in the average number of points scored by the two centers at the $\alpha = .10$ level of significance.

7. The manager of a stock brokerage office wants to compare the number of stock orders obtained each day by two brokers. These results are collected:

| Mr. Merrill | 42 | 36 | 58 | 27 | 48 | 85 | 38 | 44 | 62 |
| Ms. Hutton | 53 | 48 | 65 | 41 | 57 | 49 | 74 | 49 | 56 |

Determine whether there is any difference in the average number of orders obtained by the two brokers at the 5% level of significance.

8. The manager of the office in Exercise 7 also wants to determine if there is any difference in the total dollar amounts of the orders obtained by the two brokers. The results, in thousands of dollars per day, are as follows:

| Mr. Merrill | 92 | 85 | 110 | 76 | 116 | 131 | 163 | 90 | 118 |
| Ms. Hutton | 85 | 104 | 93 | 86 | 140 | 185 | 95 | 156 | 135 |

Determine whether there is any difference in the means of the total dollar amounts of the orders per day by the two brokers at the 5% level of significance.

9. A professor is charged with favoring the men in her class by giving them consistently higher grades on term papers. To test whether this complaint is true, the Dean collects the following sample set of grades for comparison:

| Men | 75 | 86 | 77 | 72 | 89 | 94 | 97 | 81 | 83 | 77 | 73 | 86 | 90 | 90 |
| Women | 77 | 83 | 72 | 67 | 84 | 91 | 82 | 73 | 65 | 72 | 70 | 72 | 65 | 88 |

Determine whether there is any difference in the average grades given to men and women at the $\alpha = .05$ level of significance.

10. Two supermarket clerks are being compared to see which one processes customers faster. The store manager records the number of customers served by each cashier per hour:

| Pam | 12 | 14 | 8 | 16 | 12 | 10 | 9 | 9 | 13 | 17 |
| Sam | 10 | 7 | 9 | 12 | 8 | 15 | 10 | 9 | 11 | |

Test whether there is any difference in the average number of customers served per hour by each cashier at the 2% level of significance.

11. One way that economists measure the health of the U.S. economy is by studying the number of jobs advertised in newspapers. In such a study, they might count the number of pages of ads in one year and compare it to the number of pages of ads in a succeeding year. Suppose the results for a random sample of days during a particular month are as follows:

| July 1992 | 12 | 10 | 25 | 18 | 14 | 18 | 20 | 24 | 20 |
| July 1993 | 21 | 16 | 26 | 17 | 20 | 28 | 19 | 20 | 18 |

Test whether there is any difference in the average number of pages of job ads in the newspaper between the two years at the 2% level of significance.

12. A college senior wants to determine if the average salary offered in her field is higher on the East Coast or the West Coast. To do this, she takes a random sample of job salaries listed in ads in a New York newspaper and compares them to the average salaries listed in a Los Angeles paper. These are the results (in thousands of dollars per year):

New York	28	31	26	28.5	32	27.5	30.5	36	32
Los Angeles	30	26	29	31.5	34.5	30	25	27.5	

Test whether there is any difference between the averages of salaries in her field between the two cities at the 5% level of significance.

13.4 The Spearman Rank Correlation Test

We have seen that regression and correlation analysis is an extremely powerful tool for studying situations involving two distinct variables. As an alternative to the usual correlation coefficient r which we discussed in Chapter 10, statisticians have developed a nonparametric method which can be applied when we cannot assume that the underlying populations are normal or in other situations where the correlation coefficient does not apply.

The *Spearman rank correlation test* uses the *Spearman rank correlation coefficient* to find an approximation to the correlation coefficient r. It is based on comparing the *rankings* of a set of paired data rather than the data values themselves. We illustrate its use in the following situation.

Suppose two sportswriters at a newspaper have made the following predictions for the order in which the six teams in the National League East will finish next season:

Team	Writer x	Writer y
Chicago Cubs	4	3
Montreal Expos	3	5
New York Mets	1	2
Philadelphia Phillies	5	6
Pittsburgh Pirates	2	1
St. Louis Cardinals	6	4

We want to know whether there is any correlation between these two sets of rankings or if they are essentially random. To do this, we introduce the *Spearman rank correlation coefficient* to compare their rankings. It is defined as

$$R_s = 1 - \frac{6\Sigma(x-y)^2}{n(n^2-1)}$$

where $\Sigma(x-y)^2$ represents the sum of the squares of the differences in the *rankings* and n is the number of pairs that have been ranked.

To calculate the Spearman rank correlation coefficient, we extend the above table as follows:

Team	Writer x	Writer y	x − y	(x − y)²
Chicago Cubs	4	3	1	1
Montreal Expos	3	5	−2	4
New York Mets	1	2	−1	1
Philadelphia Phillies	5	6	−1	1
Pittsburgh Pirates	2	1	1	1
St. Louis Cardinals	6	4	2	4
				$\Sigma(x-y)^2 = 12$

Therefore, the rank correlation coefficient is

$$R_s = 1 - \frac{6\Sigma(x-y)^2}{n(n^2-1)}$$

$$= 1 - \frac{(6)(12)}{6(6^2-1)}$$

$$= 1 - \frac{12}{35}$$

$$= 1 - .343$$

$$= .657$$

This value for the rank correlation coefficient is interpreted in much the same way as the usual correlation coefficient r. Its value always lies between -1 and 1. Values of R_s close to 1 indicate a high positive correlation between the rankings while values of R_s close to -1 indicate a high negative correlation. Further, values of R_s close to 0 indicate minimal correlation.

Moreover, as with the usual correlation coefficient r, the key to using R_s lies in being able to decide whether the correlation R_s between the rankings is significant. For a particular value of the rank correlation coefficient, is R_s large enough to infer a significant degree of correlation? To answer this question, we compare the calculated value for R_s to a critical value of the Spearman rank correlation coefficient given in Table IX. For the above example, the value we calculated was $R_s = .657$ based on $n = 6$ data pairs. If we wish to determine whether this is significant at the $\alpha = .05$ level of significance, then we must compare it to the appropriate entry in the table, $R_s = .886$. In order for the calculated value of R_s to be significant, it must be larger than .886 or smaller than $-.886$. In our case, the value of $R_s = .657$ is between the two critical values, and hence we conclude that there is no correlation between the two writers' rankings of the teams. The writers do not really agree in their predictions.

We note that the Spearman rank correlation coefficient need not be applied only when the set of data represents rankings, as we did above. It can also be calculated for any set of paired numerical data. All we have to do is rank the two sets of data individually. That is, we rank the x's and then rank the y's and apply the above procedure to the difference in the rankings rather than to the original data entries.

The above formula for R_s assumes that there are no ties in the rankings. However, if there are only a few ties, statisticians have found that the formula is still highly accurate. In such a case, we handle ties in precisely the same way as we did when working with the rank-sum test in Section 13.3—we simply average the rankings for the tied entries.

To illustrate this procedure, we again consider the data in the example from Section 10.1 on the relationship between high school average (HSA) and college GPA.

EXAMPLE 13.4 Determine the Spearman rank correlation coefficient based on the following set of data, and decide if there is any correlation at the $\alpha = .02$ level of significance.

x (HSA)	80	85	88	90	95	92	82	75	78	85
y (GPA)	2.4	2.8	3.3	3.1	3.7	3.0	2.5	2.3	2.8	3.1

SOLUTION We begin by setting up a table consisting of these entries as well as their respective rankings:

x	Rank of x	y	Rank of y
80	8	2.4	9
85	5.5	2.8	6.5
88	4	3.3	2
90	3	3.1	3.5
95	1	3.7	1
92	2	3.0	5
82	7	2.5	8
75	10	2.3	10
78	9	2.8	6.5
85	5.5	3.1	3.5

Notice that the x's and y's are listed in their original paired order. We must not reorganize the data to list the x's and the y's separately in ranked order. Also notice how we account for the duplications in each variable when the rankings are tied.

We now construct the associated table consisting of just the rankings and their differences:

Rank of x	Rank of y	Difference	(Difference)²
8	9	−1	1
5.5	6.5	−1	1
4	2	2	4
3	3.5	−.5	.25
1	1	0	0
2	5	−3	9
7	8	−1	1
10	10	0	0
9	6.5	2.5	6.25
5.5	3.5	2	4
			$\Sigma(\text{difference})^2 = 26.50$

From this, we find that

$$R_s = 1 - \frac{6(26.5)}{10(10^2 - 1)}$$

$$= 1 - .161$$

$$= .839$$

To decide whether this is significant at the $\alpha = .02$ level of significance, we consult Table IX and find that, corresponding to $n = 10$, the critical value for R_s is .745. Since our calculated value is larger than this, we conclude that there is indeed a correlation between HSA and GPA. ∎

For comparison, we note that the value we found for the usual correlation coefficient r in Section 10.2 for this same set of data was $r = .858$. (We recall that this also indicated a significant degree of correlation between the two sets of values.) Thus, in this instance, the Spearman rank correlation coefficient $R_s = .839$ is fairly close to the correlation coefficient r computed under the assumption that both high school averages and grade-point averages are normally distributed. Consequently, the Spearman rank correlation coefficient R_s provides a fairly accurate alternative to the correlation coefficient r.

The Spearman rank correlation coefficient can also detect a relationship between two sets of data which are not linearly related. In contrast, the usual correlation coefficient would show a low level of *linear* correlation in such a case.

In summary, if the original data consist of actual measurements and we can assume normality, then we can use a computer program such as Program 32: Linear Regression Analysis or a statistical calculator to obtain the correlation coefficient. But if normality is an unreasonable assumption or if the given data represent rankings rather than numerical measurements or if the apparent relationship is not linear, then the correlation coefficient r does not

Exercise Set 13.4

Mastering the Techniques

For each set of rankings in Exercises 1 and 2, calculate the Spearman rank correlation coefficient.

1.
x	y
1	2
2	4
3	1
4	5
5	3
6	6

2.
x	y
3	5
1	4
6	3
4	1
2	2
5	6

For each set of paired data in Exercises 3 through 6, calculate the Spearman rank correlation coefficient.

3.
x	11	24	31	36	42	50	60
y	27	44	38	53	75	67	72

4.
x	122	138	150	143	187	175	148
y	48	38	53	28	55	50	45

5.
x	16	11	30	25	38	40	35	44
y	88	77	65	54	61	47	42	40

6.
x	97	94	85	90	84	77	72	75	80
y	20	24	44	32	48	52	40	50	60

Applying the Concepts

7. Two movie critics have ranked five films (called, for convenience, A, B, C, D and E) nominated for the Academy Award for best picture of the year:

A:	2	4
B:	1	3
C:	5	2
D:	4	5
E:	3	1

Determine if there is any correlation between their rankings at the $\alpha = .05$ level of significance.

8. Two judges at a beauty contest have ranked the six finalists as follows:

Alison:	3	5
Beth:	6	4
Carmela:	2	3
Deborah:	1	2
Elvira:	5	6
Faith:	4	1

Determine whether there is any correlation between the judges' rankings at the 5% significance level.

Exercises 9 through 14 represent the same problems that you saw in Sections 10.1 and 10.2 with regression and correlation. You are now asked to calculate the Spearman rank coefficient coefficient R_s. You should also compare the result you obtain to the value you calculated for the usual correlation coefficient r in Section 10.2 and compare your conclusions in each case regarding whether there is any correlation.

9. A small company is interested in analyzing the effects of advertising on its sales. Over a 5-month period, it finds the following results:

x	5	8	10	15	22
y	6	15	20	30	39

Here x represents the money spent on advertising (in hundreds of dollars), and y represents the total sales (in thousands of dollars). Use these data to determine the Spearman rank correlation coefficient for sales as a function of advertising. Is there any correlation at the 5% level of significance?

10. A supermarket owner is studying the effects of the number x of checkout clerks working on the average waiting time y in minutes for customer checkout. The results are as follows:

x	3	4	5	5	6	7
y	9	7	6	5	5	4

Use these data to determine the rank correlation coefficient for waiting time as a function of number of clerks on duty. Is there any correlation at the 5% level of significance?

11. A college collects the following set of data on the number of credits C that a randomly selected group of students carry and the number of hours H that they work during the week.

H	20	25	30	50	20	23
C	12	13	12	15	16	16

Find the rank correlation coefficient based on these data, and determine whether there is any correlation at the $\alpha = .05$ level of significance.

12. The college in Exercise 11 uses the above information on the number of hours H that students work to study the relationship between time spent on a job with grade-point average G. The results are

H (hours)	20	25	30	50	20	23
G (GPA)	3.4	3.0	2.8	2.4	2.9	2.9

Find the rank correlation coefficient based on these data, and decide whether it implies a significant degree of correlation between time that students work on a job and their GPAs at the $\alpha = .05$ level of significance.

13. A government study on energy conservation is conducted to assess the relationship between the price of home heating oil P (in cents) and the mean number of gallons G used per month during January in a variety of different communities with similar climates. The results are:

P	75	80	86	90	95	98	106
G	120	125	114	110	112	106	97

Find the rank correlation coefficient relating oil use to its cost. Is there any correlation between them at the $\alpha = .05$ level of significance?

14. A TV network is concerned about the high cost of producing many of its programs. It therefore conducts a study to relate the production costs for 30 minutes of programming, C (in hundreds of thousands of dollars), to the ratings R that the program gets in a national ratings survey. The results are:

C	1.2	1.6	1.8	2.5	2.7	3.0	3.5	4.4
R	3.3	3.9	5.7	4.2	4.5	8.2	6.1	4.6

Find the rank correlation coefficient comparing the ratings to the production costs, and determine if there is any correlation at the $\alpha = .05$ significance level.

13.5 The Runs Test

A recurrent theme throughout statistics is the notion of *randomness*. We speak of *random samples*; we use *random numbers* to generate the random samples; we study *random processes* such as flipping coins, rolling dice, and drawing samples. Clearly randomness is an essential underlying necessity for most valid statistical analyses. We have indicated the need to use either a table of random numbers, such as Table I, or Program 2: Random Number Generator in selecting a random sample. We have tacitly assumed that all processes are random and that all mechanisms involved, such as the actual coins being flipped, are random.

However, we have not fully studied the question: How do we know that a set of outcomes is truly random? We addressed this question in part in a discussion involving flipping a coin. We considered the experiment of flipping a supposedly fair coin and focused on the number of heads and tails that came up. Suppose, for example, that such a coin is flipped 1000 times. We are able to conclude that any split close to 500-500 is an indication that the coin is fair. Thus, for instance, 510 heads and 490 tails seem reasonable (the probability is relatively large) while 600 heads and 400 tails are not (the probability of occurrence is incredibly small). Further, the goodness of fit test from Section 11.3 also provides a method for determining if the set of outcomes is indeed random.

Unfortunately, this is not adequate. For example, what if, out of the 1000 flips of the coin, the 510 heads and 490 tails arose in the following manner?

$$\underbrace{\text{HHH} \ldots \text{HHH}}_{510 \text{ H's}} \underbrace{\text{TTT} \ldots \text{TTT}}_{490 \text{ T's}}$$

Clearly, this *arrangement* does not seem in the least random, and if it occurred, we would have grave doubts about the randomness of the process or the fairness of the coin or the flipping ability of the person doing the flipping.

We therefore see that randomness is not always simply a matter of *counting the number of outcomes*, but may also involve noting the *order in which those outcomes arise*. To determine whether a series of outcomes is truly random, we introduce another nonparametric test known as the *runs test*. We develop it in the context of the following problem.

Suppose a coin is flipped 25 times with the following outcomes:

H H T H T T T H H T H H T T H T T H H H H T H H T

We want to determine if the *arrangement* of the H's and T's is random or if there is some underlying pattern to the position of the symbols.

A *run* is a succession of identical symbols (either letters or numbers) preceded by a different symbol and followed by a different symbol; a run can also occur at the beginning or end of a list.

In the above listing, the first two H's form a run of length 2, the next T forms a run of length 1, and so forth. In fact, this list consists of 14 different runs:

H H | T | H | T T T | H H | T | H H | T T | H | T T | H H H H | T | H H | T

Let's see how this number of runs can give an indication of whether the arrangement of the H's and T's seems random.

The smallest numbers of possible runs involving 25 coin tosses are either 1 (if all are heads or all are tails) or 2 (a run of one symbol followed by a run of the other). Clearly, these arrangements do not indicate randomness. In fact, if the number of runs is very small, it suggests some pattern. At the other extreme, the largest number of runs possible in this situation is 25 (the H's and T's strictly alternate). Again, this arrangement is certainly not random. Similarly, any arrangement involving a large number of runs (close to 25) suggests a nonrandom pattern. It is only when the number of runs is somewhere in the middle that we might conclude that the arrangement is apparently random.

Moreover, obtaining a reasonable number of runs also depends on the number of occurrences of each symbol. In the above list, there are 14 H's and 11 T's out of the 25 flips. We denote these numbers by n_1 and n_2. Thus, $n_1 = 14$ and $n_2 = 11$.

Notice that we have not mentioned the length of any of the runs, only the total number of runs in a list and the number of occurrences of each symbol. It turns out that the length of the runs is irrelevant to the question of randomness. All that we need to make a decision is the number of runs we observe in the list and the total number of occurrences of each symbol.

To make this decision at the $\alpha = .05$ level of significance, we use Table X. This table gives pairs of critical values—the lower limit and the upper limit on the number of runs we would expect if the outcomes were indeed random. To use the table, we first look for the larger of n_1 and n_2 ($n_1 = 14$) across the top of the table and for the smaller of the two ($n_2 = 11$) down the left-hand column. When we then read across, we find the box marked

$$\boxed{\begin{array}{c} 8 \\ 19 \end{array}}$$

As we indicated, these two numbers are the lower and upper limits on the *number of runs* which are likely if the arrangement is random. If the actual list contains a number of runs between 8 and 19 inclusive, then we conclude that the arrangement is random. If the list involves fewer than 8 runs or more than 19 runs, the arrangement is not random. In our example, we have 14 runs; hence we conclude that the arrangement is likely generated by a random process at the 5% significance level.

We illustrate this procedure with several examples.

EXAMPLE 13.5 A stock broker receives 23 orders from clients either to buy (B) stock or to sell (S) stock. The stock orders arrive in the following order:

B B B B B S S S S B B B B B S S S S B B B B B

Test whether the arrangement is random at the $\alpha = .05$ level of significance.

SOLUTION This is a hypothesis test with null hypothesis

H_0: The order is random

and alternate hypothesis

H_a: The order is not random.

The list of 23 stock orders consists of $n_1 = 15$ B's and $n_2 = 8$ S's. They are arranged in 5 runs:

B B B B B | S S S S | B B B B B | S S S S | B B B B B

Therefore, from Table X, we find that the critical values for the number of runs are

6 and 16

Since the list contains only 5 runs, we reject the null hypothesis and so conclude that the arrangement is likely not random at the 5% level of significance. ∎

The theory of runs can also be applied to many different situations provided that we can characterize each outcome by two possible cases (represented by two symbols). We show this in the next example.

EXAMPLE 13.6 A die is rolled 24 times with the following outcomes:

5 2 2 1 6 5 3 3 1 6 5 2 1 4 4 4 4 2 6 1

Test if the arrangement of 1's in this list is random at the $\alpha = .05$ level of significance.

SOLUTION The null and alternate hypotheses are

H_0: The arrangement of the 1's is random.

H_a: The arrangement is not random.

We use Y whenever the die comes up 1 and N for any other face. Therefore, the original list can be rewritten as

N N N Y N N N N Y N N N Y N N N N N N Y

There are $n_1 = 4$ occurrences of Y and $n_2 = 16$ occurrences of N. Moreover, there are a total of 8 runs in this list:

N N N | Y | N N N N | Y | N N N | Y | N N N N N | Y

From Table X, we see that the critical numbers of runs corresponding to the values of $n_1 = 4$ and $n_2 = 16$ are

4 and 10

Since we have 8 runs in the list, we cannot reject the null hypothesis and so conclude that the arrangement of the 1's may be random. ∎

We note that this only ensures that the arrangement of the 1's is random. It says nothing about the arrangement of any of the other outcomes 2, 3, . . . ,

6. Clearly, we can repeat the above analysis to study the arrangement of 2's, 3's, ..., 6's to check if each of them is arranged randomly. We will ask you to do some of this as part of Exercise Set 13.5.

We can also apply the theory of runs to test whether a list of numbers is random. As in Example 13.6, the key is to introduce a way of partitioning the list into two categories, and the simplest way of doing this is to use the median of the entries. With this approach, each entry is either above (A) or below (B) the median. Any entry which is equal to the median is discarded. We illustrate this procedure in the next example.

EXAMPLE 13.7 The number $\pi = 3.14159\ldots$ is an irrational number, so it can only be expressed as a nontermininating, nonrepeating decimal. The first 31 decimal places of this number are

1 4 1 5 9 2 6 5 3 5 8 9 7 9 3 2 3 8 4 6 2 6 4 3 3 8 3 2 7 9 5

Test whether this arrangement of digits seems random at the $\alpha = .05$ level of significance.

SOLUTION The null and alternate hypotheses are

H_0: The digits in π are random.

H_a: The digits in π are not random.

We first find the median for this set of numbers, and this involves putting them in numerical order. Thus:

1 1 2 2 2 2 3 3 3 3 3 4 4 4 5 5 5 5 5 6 6 6 7 7 8 8 8 9 9 9 9

Since there are 31 entries, the median is the middle or 16th entry and so is 5. Therefore, in the *original* list of digits for π, we replace each entry below 5 with a B and each entry above 5 with an A. Initially, for clarity, we also replace all entries equal to 5 with a dot. We will subsequently discard them altogether. The result is

B B B . A B A . B . A A A A B B B A B A B A B B B A B B A A .

After deleting the dots, we are left with

B B B A B A B A A A A B B B A B A B A B B B A B B A A

We therefore have a list of 27 entries composed of $n_1 = 12$ A's and $n_2 = 15$ B's. This list consists of 16 different runs:

B B B | A | B | A | B | A A A A | B B B | A | B | A | B | A | B B B | A | B B | A A

From Table X, we find that the critical numbers of runs are

8   and   20

Since this list contains 16 runs, we cannot reject the null hypothesis and so conclude that the arrangement of the first 31 digits in π seems random. ∎

You have undoubtedly noticed that Table X goes up to a maximum of $n_1 = n_2 = 20$. If either n_1 or n_2 is larger than 20, it turns out that we can approximate the distribution of runs by a normal distribution having mean

$$\mu = \frac{2n_1 n_2}{n_1 + n_2} + 1 \tag{13.1}$$

and standard deviation

$$\sigma = \sqrt{\frac{2n_1 n_2(2n_1 n_2 - n_1 - n_2)}{(n_1 + n_2)^2(n_1 + n_2 - 1)}} \tag{13.2}$$

We illustrate the use of this approach in the following example.

EXAMPLE 13.8 The first 50 entries in a table of "random" digits are:

04433 80674 24520 18222 10610 05794 37515 48611 62866 33963

Determine if the arrangement of the digits is random at the $\alpha = .05$ level of significance.

SOLUTION The null and alternate hypotheses are

H_0: The arrangement of the digits is random.

H_a: The arrangement is not random.

We first put the numbers in numerical order

000000 111111 222222 333333 444444 5555 6666666 777 8888 99

and so find that the median is the average of the 25th and 26th entries, or $(4 + 4)/2 = 4$. Therefore, in the original list, we label each entry as being either above (A) or below (B) the median. We thus obtain:

B..BB ABAA. B.ABB BABBB BBABB BAAA. BAABA .AABB ABAAA BBAAB

where we again initially use a dot to represent an entry equal to the median 4. When we delete the dots, we are left with a list of 44 entries composed of $n_1 = 24$ B's and $n_2 = 20$ A's. Further, this list contains 23 different runs:

BBB | A | B | AA | B | A | BBB | A | BBBBB | A | BBB |
AAA | B | AA | B | AAA | BB | A | B | AAA | BB | AA | B

Since both n_1 and n_2 are greater than 20, we cannot resort to the table, but rather have to utilize the normal approximation. The corresponding normal distribution has a mean given by

$$\mu = \frac{2n_1 n_2}{n_1 + n_2} + 1$$

$$= \frac{2(24)(20)}{24 + 20} + 1$$

$$= 22.82$$

and standard deviation given by

$$\sigma = \sqrt{\frac{2n_1 n_2 (2n_1 n_2 - n_1 - n_2)}{(n_1 + n_2)^2 (n_1 + n_2 - 1)}}$$

$$= \sqrt{\frac{2(24)(20)(2 \cdot 24 \cdot 20 - 24 - 20)}{(24 + 20)^2 (24 + 20 - 1)}}$$

$$= \sqrt{\frac{(960)(916)}{(44^2)(43)}}$$

$$= 3.25$$

The normal approximation to the runs test is a two-tailed test and, at the $\alpha = .05$ level of significance, the critical values for z are ± 1.96. The test statistic for z corresponding to the 23 different runs in the list is given by

$$z = \frac{23 - 22.82}{3.25}$$

$$= .055$$

and we therefore cannot reject the null hypothesis. That is, the number of runs is within acceptable limits and so the arrangement of the digits seems random. See Figure 13.6. ∎

FIGURE 13.6

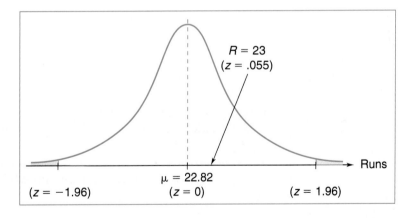

Computer Corner

We have included Program 40: Simulation of the Runs Test to allow you to investigate the sampling distribution associated with the runs test. To use the program, you must enter values for n_1 and n_2, the number of occurrences of each symbol, and the number of simulations you want. For example, suppose we select $n_1 = 20$ and $n_2 = 15$ and ask for 400 samples. The program then generates repeated random samples consisting of 20 A's and 15 B's and counts the number of runs in each sample. The resulting numbers of runs are then graphed, as shown in the typical output in Figure 13.7. In this case, the distribution is roughly normal in shape, and this fact therefore substantiates the statement we made earlier that the associated distribution is approximately normal provided that either n_1 or n_2 is greater than 20. You may also want to compare the values for the mean and standard deviation of the sample runs generated with the theoretical values for μ and σ predicted by formulas (13.1) and (13.2).

FIGURE 13.7

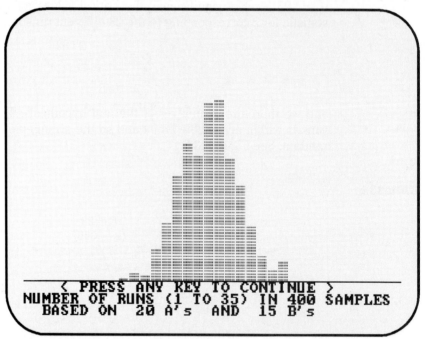

We suggest that you experiment with this program to see the effects of using:

1. Larger values for both n_1 and n_2
2. A larger value for one and a considerably smaller value for the other
3. Smaller values for both

Exercise Set 13.5

Applying the Concepts

1. Use the results on the dice rolling experiment in Example 13.6 to test whether the arrangement of 2's is random at the $\alpha = .05$ level of significance.
2. Repeat Exercise 1 for the 3's in the list.
3. Repeat Exercise 1 for the 4's in the list.
4. Repeat Exercise 1 for the 5's in the list.
5. A student guesses wildly at the answers to the 25 questions on a true-false test as follows:

   T T F F T F T F T T T F F T F T F F T T F F T T T

   Test whether the arrangement of the T's and F's is random at the $\alpha = .05$ level of significance.

6. A gambler is playing the colors (black and red) at a roulette table and observes the following arrangement of outcomes:

   B B B B B R R R R B B B B B B B R R R B B B B B B B R R R

   Based on this, can you conclude that the arrangement is random at the 5% level of significance?

7. The maître d' at a restaurant is seating people for lunch and observes the following pattern for whether customers are alone (A) or in a group (G):

   A A A G G A G A G G A A G G G A G A A A G G G A A A A G G

   Test whether the arrangement is random at the 5% level of significance.

8. An usher at a movie theater notices the order, in terms of their gender, in which customers are lined up waiting to purchase tickets:

   M F M F M F F F F M M M M F M F M F M F M M M M M F F F F

   Test whether the arrangement is random at the 5% level of significance.

9. A man wants to make a long-distance telephone call at a pay phone, but someone is already using it. While waiting, he notices that to place the call using his phone card, he must dial the following series of digits:

   1 5 1 1 3 2 2 4 1 1 6 5 7 8 0 0 6 7 5 3 3 1 1 2 5 8 8 0 6

   He begins to wonder if the arrangement of all these digits is random. Can you help him make this decision at the $\alpha = .05$ level of significance?

10. A basketball fan keeps track of the number of points that her school team scores each game and makes the following list:

    82, 73, 68, 72, 78, 75, 92, 84, 76, 85, 77, 85, 79, 84, 88, 80, 91, 86, 94, 88, 98, 94, 90, 83, 95, 97

    Test whether the arrangement of the points scored is random at the $\alpha = .05$ level of significance. (Alternatively, might it suggest improvement as the season progresses?)

Computer Application

11. Use Program 2: Random Number Generator to generate a set of at least 50 random integers between 0 and 1. Determine whether the arrangement of the 0's and 1's is random at the $\alpha = .05$ level of significance.
12. Use Program 2: Random Number Generator to generate a set of at least 50 random integers between 0 and 9. Determine whether the arrangement is random at the $\alpha = .05$ level of significance.

Student Projects

This project involves a statistical study relating to the controversy over global warming. Many scientists believe that the earth is experiencing an overall increase in temperatures due to a phenomenon called the greenhouse effect that may have a significant impact not only on our lifestyle but also on our very ability to maintain life on earth.

To conduct the study, you should pick a randomly selected date (say your birthday) during the past 12 months and a randomly selected year during the past hundred years, but at least 10 years in the past. From your library's microfilm records, you can find the national weather reports for your date, both for the past year and for the earlier year you selected. Pick a minimum of 30 randomly selected cities, either in the United States or around the world, and record the high temperatures on both days. Then apply the appropriate nonparametric procedure to determine if there is a consistent difference in the temperatures on the date you chose between the two years.

You should incorporate all your results into a formal statistical project report. The report should include

(a) A statement of the topic being studied including the particular populations being sampled
(b) The source of the data
(c) A discussion of how the data was collected and why you believe that it is a random sample
(d) A list of the data including possibly some tables and/or diagrams to display the data
(e) All the statistical analysis
(f) Your conclusions based on this analysis and a rationale for the level of significance you select
(g) A calculation of the average high temperatures on your date for the two years, if you find that there is a consistent difference in the high temperatures
(h) A discussion of any surprises you may have noted in connection with collecting and organizing the data or with your conclusions.

You may also want to compare your results with those of the other students in your class. In particular, what percentage of your classmates found a consistent difference in temperatures between the two years that they studied? Does this suggest a consistent pattern of warming, either nationally or globally?

CHAPTER 13 SUMMARY

Nonparametric statistical tests can be used as an alternative when some of the underlying conditions which are needed to apply the standard parametric tests, such as normality, cannot be assumed.

The **sign test** can be used to test whether there is a **consistent** difference between two sets of data. It is based on testing the proportions of pluses and minuses that arise when the two sets of data values are compared.

The **Wilcoxin rank-sum test** is a nonparametric alternative to the difference of means test. It is based on comparing the **rankings** of the entries in two sets of data to determine if there is any difference in the population means. If the two sample sizes n_1 and n_2 are both larger than 8, then the sums of the sample rankings R are approximately normally distributed with mean

$$\mu_R = \tfrac{1}{2} n_1 (n_1 + n_2 + 1)$$

and standard deviation

$$\sigma_R = \sqrt{\frac{n_1 n_2 (n_1 + n_2 + 1)}{12}}$$

The **Spearman rank correlation coefficient** R_s is used as a nonparametric alternative to the usual correlation coefficient r. It is found from

$$R_s = 1 - \frac{6 \Sigma (x - y)^2}{n(n^2 - 1)}$$

where $\Sigma (x - y)^2$ is the sum of the squares of the differences in the rankings. The value obtained for R_s is then compared to the critical values shown in Table IX to test whether there is any correlation between the two variables.

The **runs test** is used to test whether a given **arrangement** in a set of data is **random**. The data must be organized in such a way that the entries can be described by two symbols so that the number of **runs** can be calculated. If the number of runs that occur falls between two critical values found in Table X based on the number of occurrences n_1 and n_2 of each of the symbols, then the arrangement may be random.

If either n_1 or n_2 is greater than 20, then the distribution of sample runs is approximately normal with mean

$$\mu = \frac{2 n_1 n_2}{n_1 + n_2} + 1$$

and standard deviation

$$\sigma = \sqrt{\frac{2 n_1 n_2 (2 n_1 n_2 - n_1 - n_2)}{(n_1 + n_2)^2 (n_1 + n_2 - 1)}}$$

Review Exercises

1. A study is made comparing the number of passengers on a random sample of flights between New York and Chicago for two competing airlines on twelve successive days. The results are:

A:	232	265	249	250	255	236	270	266	249	240	257	239
B:	189	230	236	261	249	218	258	253	251	233	254	249

   Test whether airline A consistently carries more passengers than airline B on its flights at the 1% significance level.

2. The following are the weights in pounds of two random samples of turkeys fed two different diets:

A:	16.3	10.1	10.7	13.5	14.9	11.8	14.3	10.2	12.0	14.7	14.8
B:	21.3	23.8	15.4	19.6	12.0	13.9	18.8	19.2	15.3	20.1	18.9

   Test whether the second diet produces a greater weight at the alpha = 1% level.

3. A coin is being tested for fairness. The following results are obtained when it is flipped repeatedly:

   H H H T H T T H H H H T T H T H H H T H T T H T T H H H T H T

   Based on these results, can you conclude it is fair?

4. Two TV ratings services obtain the following sets of rankings for the top 10 TV programs:

Program	A	B	C	D	E	F	G	H	I	J
Service 1	2	5	7	4	1	6	10	3	9	8
Service 2	1	4	9	7	2	5	8	3	6	10

   Determine the rank correlation coefficient and test whether there is any correlation between the two sets of rankings at the 5% level.

5. A businessman claims the following number of lunches on his expense account for 24 consecutive months:

   6, 7, 5, 6, 8, 6, 8, 6, 6, 4, 3, 2, 4, 4, 3, 4, 7, 5, 6, 8, 6, 6, 3, 4

   An auditor wishes to determine if this is a random arrangement. What is his conclusion?

APPENDIX I BASEBALL STATISTICS DATA SET

American League Batting (based on 390 at-bats)

Player and Team	At-Bats	Runs	Hits	Doubles	Triples	Home Runs	Runs Batted In	Average
E. Martinez, Mariners	528	100	181	46	3	18	73	.343
Puckett, Twins	639	104	210	38	4	19	110	.329
Thomas, White Sox	573	108	185	46	2	24	115	.323
Molitor, Brewers	609	89	195	36	7	12	89	.320
Mack, Twins	600	101	189	31	6	16	75	.315
Baerga, Indians	657	92	205	32	1	20	105	.312
Alomar, Blue Jays	571	105	177	27	8	8	76	.310
Griffey, Mariners	565	83	174	39	4	27	103	.308
Harper, Twins	502	58	154	25	0	9	73	.307
Bordick, Athletics	504	62	151	19	4	3	48	.300
Hamilton, Brewers	470	67	140	19	7	5	62	.298
Knoblauch, Twins	600	104	178	19	6	2	56	.297
Raines, White Sox	551	102	162	22	9	7	54	.294
Vizquel, Mariners	483	49	142	20	4	0	21	.294
Listach, Brewers	579	93	168	19	6	1	47	.290
Winfield, Blue Jays	583	92	169	33	3	26	108	.290
Orsulak, Orioles	391	45	113	18	3	4	39	.289
Davis, Twins	444	63	128	27	2	12	66	.288
Mattingly, Yankees	640	89	184	40	0	14	86	.288
Polonia, Angels	577	83	165	17	4	0	35	.286
Brett, Royals	592	55	169	35	5	7	61	.285
Jefferies, Royals	604	66	172	36	3	10	75	.285
Lofton, Indians	576	96	164	15	8	5	42	.285
Olerud, Blue Jays	458	68	130	28	0	16	66	.284
Miller, Royals	416	57	118	24	4	4	38	.284
R. Henderson, Athletics	396	77	112	18	3	15	46	.283
Ventura, White Sox	592	85	167	38	1	16	93	.282
Hall, Yankees	583	67	163	36	3	15	81	.280
Johnson, White Sox	567	67	158	15	12	3	47	.279
Steinbach, Athletics	438	48	122	20	1	12	53	.279
Whitaker, Tigers	453	77	126	26	0	19	71	.278
Sierra, Rangers, Athletics	601	83	167	34	7	17	87	.278
G. Davis, Orioles	398	46	110	15	2	13	48	.276
Devereaux, Orioles	653	76	180	29	11	24	107	.276
Phillips, Tigers	606	114	167	32	3	10	64	.276
Zupcic, Red Sox	392	46	108	19	1	3	43	.276
R. Kelly, Yankees	580	81	158	31	2	10	66	.272
Maldonado, Blue Jays	489	64	133	25	4	20	66	.272
Velarde, Yankees	412	57	112	24	1	7	46	.272
Anderson, Orioles	623	100	169	28	10	21	80	.271
Seitzer, Brewers	540	74	146	35	1	5	71	.270
Munoz, Twins	418	44	113	16	3	12	71	.270
Wilson, Athletics	386	38	107	15	5	0	37	.270
Joyner, Royals	572	86	154	36	2	9	66	.269
Sorrento, Indians	458	52	123	24	1	18	60	.269
Palmeiro, Rangers	608	84	163	27	4	22	85	.268
McGwire, Athletics	467	87	125	22	0	42	104	.268
Stanklewicz, Yankees	400	52	107	22	2	2	25	.268
Reimer, Rangers	494	56	132	32	2	16	58	.267
Brunansky, Red Sox	458	47	122	31	3	15	74	.266
Tartabull, Yankees	421	72	112	19	0	25	85	.266
Fryman, Tigers	659	87	175	31	4	20	96	.266
Gomez, Orioles	468	62	124	24	0	17	64	.265
Lewis, Indians	413	44	109	21	0	5	30	.264
Yount, Brewers	557	71	147	40	3	8	77	.264
Carter, Blue Jays	622	97	164	30	7	34	119	.264

APPENDIX I BASEBALL STATISTICS DATA SET

American League Batting (based on 390 at-bats)

Player and Team	At-Bats	Runs	Hits	Doubles	Triples	Home Runs	Runs Batted In	Average
Lee, Blue Jays	396	49	104	10	1	3	39	.263
Lansford, Athletics	496	65	130	30	1	7	75	.262
Gonzalez, Rangers	584	77	152	24	2	43	109	.260
Belle, Indians	585	81	152	23	1	34	112	.260
Rodriguez, Rangers	420	39	109	16	1	8	37	.260
Boggs, Red Sox	514	62	133	22	4	7	50	.259
Curtis, Angels	441	59	114	16	2	10	46	.259
Hayes, Yankees	509	52	131	19	2	18	66	.257
T. Martinez, Mariners	460	53	118	19	2	16	66	.257
Bell, White Sox	627	74	160	27	0	25	112	.255
Gladden, Tigers	417	57	106	20	1	7	42	.254
Whiten, Indians	508	73	129	19	4	9	43	.254
Baines, Athletics	478	58	121	18	0	16	76	.253
Surhoff, Brewers	480	63	121	19	1	4	62	.252
C. Ripken, Orioles	637	73	160	29	1	14	72	.251
Leius, Twins	409	50	102	18	2	2	35	.249
White, Blue Jays	641	98	159	26	7	17	60	.248
Reed, Red Sox	550	64	136	27	1	3	40	.247
DiSarcina, Angels	518	48	128	19	0	3	42	.247
Deer, Tigers	393	66	97	20	1	32	64	.247
Reynolds, Mariners	458	55	113	23	3	3	33	.247
Gagne, Twins	439	53	108	23	0	7	39	.246
Felix, Angels	509	63	125	22	5	9	72	.246
Fielder, Tigers	594	80	145	22	0	35	124	.244
Canseco, Athletics, Rangers	439	74	107	15	0	26	87	.244
Hrbek, Twins	394	52	96	20	0	15	58	.244
Buhner, Mariners	543	69	132	16	3	25	79	.243
Borders, Blue Jays	480	47	116	26	2	13	53	.242
Pena, Red Sox	410	39	99	21	1	1	38	.241
Milligan, Orioles	462	71	111	21	1	11	53	.240
Tettleton, Tigers	525	82	125	25	0	32	83	.238
Sax, White Sox	567	74	134	26	4	4	47	.236
Macfarlane, Royals	402	51	94	28	3	17	48	.234
Palmer, Rangers	541	74	124	25	0	26	72	.229
Gruber, Blue Jays	446	42	102	16	3	11	43	.229
Vaughn, Brewers	501	77	114	18	2	23	78	.228
Gaetti, Angels	456	41	103	13	2	12	48	.226
McRae, Royals	533	63	119	23	5	4	52	.223
O'Brien, Mariners	396	40	88	15	1	14	52	.222

American League Pitching (minimum 100 innings)

Player and Team	Wins	Losses	Innings Pitched*	Hits	Bases on Balls	Strike Outs	Earned Run Average
Eldred, Brewers	11	2	100.1	76	23	62	1.79
D. Ward, Blue Jays	7	4	101.1	76	39	103	1.95
Clemens, Red Sox	18	11	246.2	203	62	208	2.41
Appier, Royals	15	8	208.1	167	68	150	2.46
Harris, Red Box	4	9	107.2	82	60	73	2.51
Mussina, Orioles	18	5	241.0	212	48	130	2.54
Mills, Orioles	10	4	103.1	78	54	60	2.61
Guzman, Blue Jays	16	5	180.2	135	72	165	2.64

*Standard newspaper usage indicates ⅓ inning by .1 and ⅔ inning by .2.

APPENDIX I   BASEBALL STATISTICS DATA SET

American League Pitching (minimum 100 innings)

Player and Team	Wins	Losses	Innings Pitched	Hits	Bases on Balls	Strike Outs	Earned Run Average
Meacham, Royals	10	4	101.2	88	21	64	2.74
Abbott, Angels	7	15	211.0	208	68	130	2.77
Perez, Yankees	13	16	247.2	212	93	218	2.87
Nagy, Indians	17	10	252.0	245	57	169	2.96
McDowell, White Sox	20	10	260.2	247	75	178	3.18
Wegman, Brewers	13	14	261.2	251	55	127	3.20
Smiley, Twins	16	9	241.0	205	65	163	3.21
Welch, Athletics	11	7	123.2	114	43	47	3.27
Brown, Rangers	21	11	265.2	262	76	173	3.32
Navarro, Brewers	17	11	246.0	224	64	100	3.33
Fleming, Mariners	17	10	228.1	225	60	112	3.39
Erickson, Twins	13	12	212.0	197	83	101	3.40
Viola, Red Sox	13	12	238.0	214	89	121	3.44
Key, Blue Jays	13	13	216.2	205	59	117	3.53
Bosio, Brewers	16	6	231.1	223	44	120	3.62
Langston, Angels	13	14	229.0	206	74	174	3.66
Guzman, Rangers	16	11	224.0	229	73	179	3.66
Stewart, Athletics	12	10	199.1	175	79	130	3.66
Darling, Athletics	15	10	206.1	198	72	99	3.66
Reed, Royals	3	7	100.1	105	20	49	3.68
Ryan, Rangers	5	9	157.1	138	69	157	3.72
Gubicza, Royals	7	6	111.1	110	36	81	3.72
Valera, Angels	8	11	188.0	188	64	113	3.73
Johnson, Mariners	12	14	210.1	154	144	241	3.77
Cook, Indians	5	7	158.0	156	50	96	3.82
Burns, Rangers	3	5	103.0	97	32	55	3.84
Doherty, Tigers	7	4	116.0	131	25	37	3.88
Hough, White Sox	7	12	176.1	160	66	76	3.93
Pichardo, Royals	9	6	143.2	148	49	59	3.95
Darwin, Red Sox	9	9	161.1	159	53	124	3.96
Finley, Angels	7	12	204.1	212	96	124	3.96
Tapani, Twins	16	11	220.0	226	48	138	3.97
Morris, Blue Jays	21	6	240.2	222	80	132	4.04
Dopson, Red Sox	7	11	141.1	159	38	55	4.08
Moore, Athletics	17	12	223.0	229	103	117	4.12
McCaskill, White Sox	12	13	209.0	193	95	109	4.18
Leiter, Tigers	8	5	112.0	116	43	75	4.18
McDonald, Orioles	13	13	227.0	213	74	158	4.24
Cadaret, Yankees	4	8	103.2	104	74	73	4.25
Fernandez, White Sox	8	11	187.2	199	50	95	4.27
Witt, Rangers, Athletics	10	14	193.0	183	114	125	4.29
Krueger, Twins	10	6	161.1	166	46	86	4.30
Gullickson, Tigers	14	13	221.2	228	50	64	4.34
Kamlenleckl, Yankees	6	14	188.0	193	74	88	4.36
Hesketh, Red Sox	8	9	148.2	162	58	104	4.36
Tanana, Tigers	13	11	186.2	188	90	91	4.39
Hibbard, White Sox	10	7	176.0	187	57	69	4.40
Sutcliffe, Orioles	16	15	237.1	251	74	109	4.47
Stottlemyre, Blue Jays	12	11	174.0	175	63	96	4.50
Nichols, Indians	4	3	105.1	114	31	56	4.53
Bones, Brewers	9	10	163.1	169	48	65	4.57
Gordon, Royals	6	10	117.2	116	55	96	4.59
Mesa, Orioles, Indians	7	12	160.2	169	70	62	4.59
Armstrong, Indians	6	15	166.2	176	67	114	4.64
Blyleven, Angels	8	12	133.0	150	29	70	4.74
Swan, Mariners	3	10	104.1	104	45	45	4.74

APPENDIX I BASEBALL STATISTICS DATA SET

American League Pitching (minimum 100 innings)

Player and Team	Wins	Losses	Innings Pitched	Hits	Bases on Balls	Strike Outs	Earned Run Average
Gardiner, Red Sox	4	10	130.2	126	56	79	4.75
Hanson, Mariners	8	17	186.2	209	57	112	4.82
Sanderson, Yankees	12	11	193.1	220	64	104	4.93
Terrell, Tigers	7	10	136.2	163	48	61	5.20
Alvarez, White Sox	5	3	100.1	103	65	68	5.20
Scudder, Indians	6	10	109.0	134	55	66	5.28
Leary, Yankees, Mariners	8	10	141.0	131	87	46	5.36
Wells, Blue Jays	7	9	120.0	138	36	62	5.40
Milacki, Orioles	6	8	115.2	140	44	51	5.84

National League Batting (based on 390 at-bats)

Player and Team	At-Bats	Runs	Hits	Doubles	Triples	Home Runs	Runs Batted In	Average
Sheffield, Padres	557	87	184	34	3	33	100	.330
Van Slyke, Pirates	614	103	199	45	12	14	89	.324
Kruk, Phillies	507	86	164	30	4	10	70	.323
Roberts, Reds	532	92	172	34	6	4	45	.323
Gwynn, Padres	520	77	165	27	3	6	41	.317
Pendleton, Braves	640	98	199	39	1	21	105	.311
Bonds, Pirates	473	109	147	36	5	34	103	.311
Butler, Dodgers	553	86	171	14	11	3	39	.309
Grace, Cubs	603	72	185	37	5	9	79	.307
Larkin, Reds	533	76	162	32	6	12	78	.304
Sandberg, Cubs	612	100	186	32	8	26	87	.304
Walker, Expos	528	85	159	31	4	23	93	.301
Clark, Giants	513	69	154	40	1	16	73	.300
McGee, Giants	474	56	141	20	2	1	36	.297
O. Smith, Cardinals	518	73	153	20	2	0	31	.295
Jose, Cardinals	509	62	150	22	3	14	75	.295
Caminiti, Astros	506	68	149	31	2	13	62	.294
Nixon, Braves	456	79	134	14	2	2	22	.294
Lankford, Cardinals	598	87	175	40	6	20	86	.293
DeShields, Expos	530	82	155	19	8	7	56	.292
Finley, Astros	607	84	177	29	13	5	55	.292
McGriff, Padres	531	79	152	30	4	35	104	.286
Biggio, Astros	613	96	170	32	3	6	39	.277
Dawson, Cubs	542	60	150	27	2	22	90	.277
Grissom, Expos	653	99	180	39	6	14	66	.276
Fernandez, Padres	622	84	171	32	4	4	37	.275
Bagwell, Astros	586	87	160	34	6	18	96	.273
Morris, Reds	395	41	107	21	3	6	53	.271
Daulton, Phillies	485	80	131	32	5	27	109	.270
Oliver, Reds	485	42	131	25	1	10	57	.270
Hollins, Phillies	586	104	158	28	4	27	93	.270
Snyder, Giants	390	48	105	22	2	14	57	.269
Bass, Giants, Mets	402	40	108	23	5	9	39	.269
Duncan, Phillies	574	71	153	40	3	8	50	.267
Morandini, Phillies	422	47	112	8	8	3	30	.265
Bell, Pirates	632	87	167	36	6	9	55	.264
Buechele, Pirates, Cubs	524	52	137	23	4	9	64	.261
Murray, Mets	551	64	144	37	2	16	93	.261

APPENDIX I BASEBALL STATISTICS DATA SET

National League Batting (based on 390 at-bats)

Player and Team	At-Bats	Runs	Hits	Doubles	Triples	Home Runs	Runs Batted In	Average
Offerman, Dodgers	534	67	139	20	8	1	30	.260
Thompson, Giants	443	54	115	25	1	14	49	.260
Gant, Braves	544	74	141	22	6	17	80	.259
Zeile, Cardinals	439	51	113	18	4	7	48	.257
Karros, Dodgers	545	63	140	30	1	20	88	.257
Justice, Braves	484	78	124	19	5	21	72	.256
Martinez, Reds	393	47	100	20	5	3	31	.254
Pagnozzi, Cardinals	485	33	121	26	3	7	44	.249
Bonilla, Mets	438	62	109	23	0	19	70	.249
Jackson, Padres	587	72	146	23	5	17	70	.249
Merced, Pirates	405	50	100	28	5	6	60	.247
O'Neill, Reds	496	59	122	19	1	14	66	.246
Clark, Padres	496	45	120	22	6	12	58	.242
Anthony, Astros	440	45	105	15	1	19	80	.239
Lind, Pirates	468	38	110	14	1	0	39	.235
King, Pirates	480	56	111	21	2	14	65	.231
Lemke, Braves	427	38	97	7	4	6	26	.227
Williams, Giants	529	58	120	13	5	20	66	.227
Wallach, Expos	537	53	120	29	1	9	59	.223
Schofield, Mets	420	52	86	18	2	4	36	.205

National League Pitching (minimum 100 innings)

Player and Team	Wins	Losses	Innings Pitched	Hits	Bases on Balls	Strike Outs	Earned Run Average
Rojas, Expos	7	1	100.2	71	34	70	1.43
D. Jones, Astros	11	8	111.2	96	17	93	1.85
Swift, Giants	10	4	164.2	144	43	77	2.08
Hernandez, Astros	9	1	111.0	81	42	96	2.11
Tewksbury, Cardinals	16	5	233.0	217	20	91	2.16
Maddux, Cubs	20	11	268.0	201	70	199	2.18
Schilling, Phillies	14	11	226.1	165	59	147	2.35
Martinez, Expos	16	11	226.1	172	60	147	2.47
Boever, Astros	3	6	111.1	103	45	67	2.51
Morgan, Cubs	16	8	240.0	203	79	123	2.55
Rijo, Reds	15	10	211.0	185	44	171	2.56
Portugal, Astros	6	3	101.1	76	41	62	2.66
Hill, Expos	16	9	218.0	187	75	150	2.68
Swindell, Reds	12	8	213.2	210	41	138	2.70
Fernandez, Mets	14	11	214.2	162	67	193	2.73
Glavine, Braves	20	8	225.0	197	70	129	2.76
Drabek, Pirates	15	11	256.2	218	54	177	2.77
Smoltz, Braves	15	12	248.2	206	80	215	2.85
Cone, Mets	13	7	196.2	162	82	214	2.88
Barnes, Expos	6	6	100.0	77	46	65	2.97
Candiotti, Dodgers	11	15	203.2	177	63	152	3.00
Smith, Pirates	8	8	141.0	138	19	56	3.06
Rivera, Phillies	7	4	117.1	99	45	77	3.07
K. Gross, Dodgers	8	13	204.2	182	77	158	3.17
Avery, Braves	11	11	233.2	216	71	129	3.20
Walk, Pirates	10	6	135.0	132	43	60	3.20
Nabhotz, Expos	11	12	195.0	176	74	130	3.32

APPENDIX I BASEBALL STATISTICS DATA SET

National League Pitching (minimum 100 innings)

Player and Team	Wins	Losses	Innings Pitched	Hits	Bases on Balls	Strike Outs	Earned Run Average
Benes, Padres	13	14	231.1	230	61	169	3.35
Leibrandt, Braves	15	7	193.0	191	42	104	3.36
Tomlin, Pirates	14	9	206.2	226	42	90	3.41
Castillo, Cubs	10	11	205.1	179	63	135	3.46
Ojeda, Dodgers	6	9	166.1	169	81	94	3.63
Schourek, Mets	6	8	136.0	137	44	60	3.64
Gooden, Mets	10	13	206.0	197	70	145	3.67
Hershiser, Dodgers	10	15	210.2	209	69	130	3.67
Cormier, Cardinals	10	10	186.0	194	33	117	3.68
Seminara, Padres	9	4	100.1	98	46	61	3.68
Lefferts, Padres	13	9	163.1	180	35	81	3.69
Hamisch, Expos	9	10	206.2	182	64	164	3.70
Osborne, Cardinals	11	9	179.0	193	38	104	3.77
Mulholland, Phillies	13	11	229.0	227	46	125	3.81
Olivares, Cardinals	9	9	197.0	189	63	124	3.84
Burkett, Giants	13	9	189.2	194	45	107	3.84
Jackson, Cubs, Pirates	8	13	201.1	211	77	97	3.84
Hurst, Padres	14	9	217.1	223	51	131	3.85
Belcher, Reds	15	14	227.2	201	80	149	3.91
Kile, Expos	5	10	125.1	124	63	90	3.95
Black, Giants	10	12	177.0	178	59	82	3.97
R. Martinez, Dodgers	8	11	150.2	141	69	101	4.00
Henry, Astros	6	9	165.2	185	41	96	4.02
J. Jones, Astros	10	6	139.1	135	39	69	4.07
Gr. Harris, Padres	4	8	118.0	113	35	66	4.12
Young, Mets	2	14	121.0	134	31	64	4.17
Wilson, Giants	8	14	154.0	152	64	88	4.21
Hammond, Reds	7	10	147.1	149	55	79	4.21
Gardner, Expos	12	10	179.2	179	60	132	4.36
DeLeon, Cardinals, Phillies	2	8	117.1	111	48	79	4.37
Clark, Cardinals	3	10	113.1	117	36	44	4.45
Abbott, Phillies	1	14	133.1	147	45	88	5.13

APPENDIX II   WEATHER DATA SET

Cities	Yesterday	Precip'n	Today		Cities	Yesterday	Precip'n	Today	
Albany	51/39	0	55/31	PC	Jackson	72/63	.16	75/57	PC
Albuquerque	72/43	0	68/44	PC	Jacksonville	83/54	0	80/56	PC
Anchorage	35/32	.44	35/26	C	Kansas City	66/35	0	62/42	PC
Atlanta	78/57	0	70/57	C	Key West	83/73	0	81/71	PC
Atlantic City	59/43	0	59/43	PC	Las Vegas	70/59	0	75/55	PC
Austin	86/62	0	87/55	S	Lexington	70/53	0	66/46	S
Baltimore	62/50	0	63/42	PC	Little Rock	70/62	.01	73/49	PC
Baton Rouge	79/62	.09	78/60	PC	Los Angeles	72/63	0	73/62	PC
Billings	53/39	0	54/28	S	Louisville	72/52	0	67/45	S
Birmingham	77/57	0	70/59	C	Lubbock	76/42	0	83/50	PC
Boise	63/42	0	64/42	PC	Memphis	70/63	18	72/51	PC
Boston	55/46	0	56/40	PC	Miami	84/68	0	82/67	PC
Bridgeport	56/46	0	58/40	PC	Milwaukee	61/37	0	55/38	PC
Buffalo	52/32	0	56/38	PC	Minn-St. Paul	60/30	0	47/33	C
Burlington	45/37	0	49/33	PC	Mobile	81/61	0	79/62	PC
Casper	63/41	0	50/30	C	Nashville	67/61	.02	69/48	PC
Charleston, WV	67/46	0	66/48	PC	Newark	60/46	0	61/44	PC
Charlotte	76/46	0	69/55	C	New Orleans	79/64	.01	79/62	PC
Chattanooga	69/50	0	69/53	PC	New York City	58/46	0	60/44	PC
Chicago	59/34	0	59/36	PC	Norfolk	63/50	0	67/50	C
Cincinnati	65/46	0	66/41	PC	Oklahoma City	72/45	0	72/47	PC
Cleveland	57/33	0	59/35	PC	Omaha	69/36	0	54/37	PC
Colo. Springs	69/39	0	52/37	C	Orlando	84/64	0	82/62	PC
Columbia	80/44	0	76/54	C	Philadelphia	58/48	0	59/43	PC
Columbus	63/44	0	65/37	PC	Phoenix	81/66	0	81/65	C
Concord	51/37	0	53/28	PC	Pittsburgh	61/33	0	65/36	PC
Dallas-Ft. Worth	79/59	0	79/53	PC	Portland, Me.	52/36	0	51/30	PC
Denver	71/45	0	51/38	Sr	Portland, Ore.	61/41	0	58/43	PC
Des Moines	64/30	0	56/39	PC	Providence	56/45	0	57/37	PC
Detroit	59/37	0	59/36	PC	Raleigh	76/48	0	69/52	C
El Paso	81/46	0	80/54	PC	Richmond	70/44	0	67/47	C
Fairbanks	32/19	0	22/15	Ss	Rochester	51/32	0	54/36	PC
Fargo	61/36	0	40/30	C	Sacramento	79/57	0	74/55	PC
Hartford	54/43	0	56/32	PC	St. Louis	64/43	0	66/41	PC
Honolulu	84/75	0	86/74	S	St. Thomas	94/72	0	92/78	PC
Houston	84/64	0	85/58	PC	Salt Lake City	68/46	0	64/48	Sr
Indianapolis	62/39	0	62/37	PC	San Antonio	87/61	0	87/57	S

APPENDIX II WEATHER DATA SET

Cities	Yesterday	Precip'n	Norm	Cities	Yesterday	Precip'n	Norm
Acapulco	92/77	0	88/73	Damascus	73/39	0	75/52
Amsterdam	47/41	.03	54/46	Dublin	45/43	.01	55/43
Athens	75/63	0	72/57	Edinburgh	46/39	trc	52/43
Bangkok	88/77	0	90/77	Edmonton	41/32	trc	45/21
Beijing	58/36	0	61/39	Geneva	54/39	.02	54/43
Berlin	37/30	.01	50/39	Guadalajara	75/61	0	77/54
Bermuda	78/68	0	77/70	Havana	84/70	0	86/70
Bonn	47/39	.01	55/43	Helsinki	25/21	trc	43/34
Brussels	*	*	55/43	Hong Kong	82/66	0	79/70
Budapest	49/43	.04	55/43	Istanbul	64/63	.02	64/54
Buenos Aires	69/68	.28	72/52	Jerusalem	70/52	0	77/57
Cairo	79/64	0	84/63	Johannesburg	82/45	0	77/54
Caracas	89/75	0	88/77	Kingston	84/77	0	88/75
Casablanca	70/55	0	73/55	Lima	63/61	.01	66/59
Copenhagen	45/32	0	50/41	Lisbon	68/61	.02	68/59
Dakar	87/73	0	88/75	London	52/43	.03	54/43
Madrid	63/52	0	64/46	Rio de Janeiro	*	*	77/66
Manila	*	*	88/73	Riyadh	88/63	0	90/59
Martinique	87/73	.51	88/73	Rome	69/59	trc	68/54
Merida	89/64	0	86/70	Seoul	61/39	0	61/39
Mexico City	*	0	70/48	Shanghai	*	*	68/54
Montego Bay	87/75	.28	88/75	Singapore	*	0	88/73
Monterrey	82/64	0	77/61	Stockholm	36/23	0	45/36
Montreal	41/37	trc	50/36	Sydney	69/59	.43	72/57
Moscow	27/25	0	41/32	Taipei	*	0	79/64
Nairobi	*	0	75/55	Tokyo	70/55	0	66/50
Nassau	83/64	0	82/72	Toronto	52/37	.04	54/36
New Delhi	88/61	0	90/59	Tunis	82/54	0	73/57
Nice	66/54	trc	68/52	Vancouver	49/36	0	54/41
Oslo	30/27	.02	43/34	Vienna	47/37	trc	54/39
Panama City	*	*	84/43	Warsaw	*	.20	50/39
Paris	52/45	1.10	55/45	Winnipeg	47/37	trc	45/23
Prague	41/32	.08	50/39				

SOURCE: National Weather Service observations, forecasts, and reports.
Weather conditions: C—cloudy; F—fog; H—haze; I—ice; PC—partly cloudy; R—rain; Sr—showers; S—sunny; Sn—snow; Ss—snow showers; T—thunderstorms; *—not available; trc—trace.

APPENDIX III STOCK EXCHANGE DATA SET

Stock	Sales 100s	Close	Change	Stock	Sales 100s	Close	Change	Stock	Sales 100s	Close	Change
AAR	537	11⅛	−⅛	ACyan	895	54⅝	+⅞	BaltGE	1253	22⅜	+⅛
AL Lab	155	23	+⅜	AEIPw	958	32⅝	+⅜	BncOne	1523	44⅛	+⅛
AbtLab	6669	28⅞	+¼	AmExp	8214	20⅜	−⅛	BncLat	167	20¾	−⅛
Abex	369	4⅞	−⅛	AGnCp	497	49	+⅜	BcpHw	213	42¾	−⅛
ACMln	449	11⅛	−⅛	AmGvl	203	9		Bandg	182	58	−⅛
ACM Sc	482	11⅛		AHltPr	451	21¾	−⅛	BkBost	1195	20	+¼
ACMSp	1262	9½	+⅛	AHome	4046	65½	+⅝	BkNY	468	43¾	+¼
ACM M	1668	10⅛	−¼	AlntGr	2098	109⅜	+2¾	BankAm	4129	42	+¼
ACMMM	463	9⅞	−¼	AmOil	371	14⅝	+⅛	BkA pfL	166	25⅛	−⅛
Acuson	655	15¾	+⅛	AREst	128	7⅛	+⅛	BankTr	315	63½	+¾
ADT	736	6¾	+⅛	AmStr	363	40	+¼	Bard	377	27⅛	
AMD	13769	12½	+⅝	ASIPII	125	15⅛	−⅛	Barnett	814	38¾	+⅛
AetnLf	596	42½	+⅞	AT&T	10994	42⅛	+⅜	Barold	823	6⅞	+⅛
AfilPb	274	11	+¼	Amrscb	1363	9⅝	−⅛	BatlMt	6924	5½	−½
AFLAC	255	30⅜	+⅜	Amrtch	1348	67⅜	+1⅜	Bausch	414	51⅜	+¾
AgMin	150	23¾	+⅜	Ameron	129	31⅝	+⅜	Baxter	5478	31⅝	+⅜
Ahmans	1759	13⅜	+¼	Amoco	2430	50¼	+⅛	BearSt	1160	14½	+¼
AirPrd	788	42⅞	+⅜	AMP	1039	59½		BectDk	387	75⅞	+¾
Airgas	374	32½	+¼	AMR	6687	55⅛	+⅛	BeldnH	168	32⅞	−1
Airlease	147	10¾	+¼	Amsco	887	25⅛	+⅛	BellAtl	5110	47	+¾
AlaP pfA	200	24¾		AmSth	169	27⅞	+¼	BellSo	3861	52⅞	+1⅜
AlskAir	780	15⅛	−⅛	Anadrk	1058	29⅛	+⅝	Bemis	486	24	+½
Albtsn	1441	43⅞	+1⅝	Analog	2275	10⅛	−¼	BenfCp	231	61½	+⅝
Alcan	2053	16	+¼	Anheus	1914	55½	+1	BestBy	1395	26⅞	+1⅛
Alcatel	146	23⅝	+¼	AnnTay	801	18	+⅝	BethStl	1720	11	+⅜
AlexBr	937	15⅜	+¼	Anthem	903	37¼	+¼	Bevrly	2133	9⅞	+¼
AlexAlx	201	25¾	+½	AonC	119	46½	+¼	BirStl	238	24¼	+⅝
vjAlexdr	618	32	+1⅛	AonCp	319	49	+1⅛	BJS	207	17¾	+¼
AllegCp	146	117½	−⅝	Apache	2127	19¼	+¼	BlackD	3864	16	+⅛
AlgLud	457	29⅞	−¼	Apex	2633	9⅝	+⅜	Blk 1998	767	10⅛	
AllgPw	469	46½		Arcadn	219	20¾	+⅛	Blk2001	779	9¾	
Alergn	801	22¼	+¾	ArchDn	2955	25⅝	+⅝	BlkAdv	272	10¼	−⅛
AllTch	203	26⅜	+⅛	Arkla	2093	9⅞	−⅛	BlkIT	1100	9½	+⅛
AldSgnl	3171	50⅛	+⅛	Armco	1018	5⅜	+⅛	Blk2008	158	15⅛	+⅛
AMIO2	676	8	+⅛	Armc	202	49⅜		BlkMT	459	10	
AMIO3	120	8¾	−⅛	ArmWl	1065	26⅝	+½	BlkIQT	412	10⅛	+⅛
AMIO	795	8	+⅛	ArowE	1817	25⅝	+⅝	BlkMTar	507	10	
AMIT	600	10¼	−⅛	Arvin	220	22⅞	−⅛	BlkNA	386	13⅝	−⅛
AMIT2	506	10½	+⅛	ASA	439	34⅛	−½	BlkStr	825	9⅞	
AMIT3	59	9⅝		Asarco	846	22⅝	−¼	BlkTT	2132	10¼	
AMPI	385	10½	+⅛	AshOil	266	26		BlckHR	2935	32⅞	
ALTEL	192	43¾	+⅜	AsiaPc	158	13⅝		BlockE	3169	13½	+⅛
Allwst	352	5½		AltEng	141	23		Boeing	3893	36⅝	+¼
Alcoa	2039	64⅞	+¾	AttRich	1629	117	+⅝	BoiseC	487	17½	
Alza s	9826	36¼	−1½	Attwood	919	9½		BordC	1124	14⅞	−¼
AmaxG	503	9¼	−⅜	Augat	207	11	+⅛	BordCh	486	14¾	−⅜
Amax	2255	16⅛	−¼	AutoDt	743	46⅛	+⅛	Borden	2844	27¼	
Ambac	486	36⅛	−⅛	Autozn	479	32¾	+1¼	BostEd	1784	25⅝	+¼
Amcast	142	12¾		Avalon	545	3⅝	+¾	BostSc	190	17¾	
AmHes	2316	47	+⅜	AveryD	142	25¼		Bowafr	754	18	−⅜
AAdj97	256	10		Avnet	354	29	+¼	BP Pru	147	31	−⅛
AAdj96	178	10⅛	+⅛	Avon	480	54¼	+⅛	Bradles	1072	16⅛	+⅛
AAdj98	476	10		BakrHu	1578	23¾	+1	Brazil	297	15⅛	+⅝
ABarck	2042	29⅝	−¾	Ball	518	28⅝	+⅜	BrazEF	205	10⅜	+⅝
ABrnd	1052	44⅜		BallyMf	683	4½	+⅛	BCAuto	3819	19½	−⅝
ACapln	532	7¾	−⅛	BaltBcp	418	6¼		Brinker	587	34¾	−1

APPENDIX III STOCK EXCHANGE DATA SET

Stock	Sales 100s	Close	Change	Stock	Sales 100s	Close	Change	Stock	Sales 100s	Close	Change
BrMySq	8199	62⅛	+⅜	Chiquta	486	17½		ConvHld	277	7⅞	+⅛
BritAir	157	49⅜	+¾	Chryslr	15429	22⅞	+¾	CvHd	207	11¾	+⅛
BritPt	4625	45⅛	+⅞	Chubb	1207	83½	+¾	Convex	1664	5⅜	+⅜
BritStl	352	10½		ChrDwt	133	30½	+½	Cooper	595	50⅞	
BritTel	290	62	+1¼	CIGNA	485	52¼	+½	CoopTr	1354	28⅛	+1⅛
Broadln	1656	23⅛	+¼	CIGHi	206	7⅝	−⅛	Cornin	1513	36⅞	+½
Broadln	43	26⅛	+⅛	CinnBel	214	16⅜	+⅛	CTF q	141	13⅛	−¼
BklyUG	264	33		CinGE	364	37		CntCrd	1222	23¾	+½
BmF B	157	82⅜	+⅜	CirCty	2474	34⅜	+⅜	CousPr	526	14	
BrwnFr	2243	23⅛	+⅜	Circuss	4027	50⅝	−⅞	CPC	1379	45½	+⅜
Bmwk	332	12⅝	+⅜	Citicorp	9127	15⅜	+½	CPI	466	15⅛	
BurlCt	244	16½	+⅛	CtzUt B	206	25¾		CrosTm	128	14½	−⅛
BurlIE	361	12⅞	+¼	CityNC	333	5⅜	−¼	CrayRs	321	23¼	+⅝
BurlNth	571	36⅜	+⅜	ClairSt	2097	9¾	+⅜	CRIIMI	328	9⅝	
BrlRsc	1435	40⅝	+½	ClarkE	110	18⅝	+⅜	CRI Liq	104	10⅝	+⅛
Cabltrn	967	63⅝	+1⅛	ClaytHs	640	23⅛	+½	CrmpK	130	18¼	+⅛
Cabot	186	47	−⅛	ClvClf	327	32¾	−¼	CwnCrk	586	35⅜	
Cadence	584	15⅜		Clorox	1415	41	+¼	CrysBd	292	2⅝	+⅛
Caesar	3681	33¾	−⅜	CMLs e	2049	27⅞	+1⅛	CSX	791	56½	+1¼
Calgon	1166	15⅞	−½	CMS Eng	646	17¼		CUC	816	20	
CalGolf	447	26½	−⅛	CNAFn	135	95¾	+2½	CumEn	838	62½	
CampSp	1469	41⅛	−¾	CoastSv	914	6⅞	+¼	Cumn	169	48¾	+¼
CapCits	113	433½	+4	Coastal	2254	28½	+⅝	CypSem	1587	8⅜	−⅛
CapHld	1085	60⅛	−⅝	CocaCl	15249	36⅞	−⅛	Cyprus	3652	27⅛	+⅜
CapRe	120	21½	−¼	CocaCE	227	11¾	−¼	CypM	581	54⅝	+⅝
CarnCr	425	26½	+⅛	ColgP	1451	58	+½	DalSem	3993	12⅜	+½
CaroPw	587	52	+⅜	ColHln	515	8½	+⅛	DameMr	626	15⅛	+⅜
CartWl	261	25¼	−¾	CollIn	128	11⅞	+⅛	DamnC	249	16¾	−⅛
CshAm	234	8⅝		CollHl	387	6¾	−⅛	DanaCp	469	36⅞	−¼
CatMkt	480	33½	+1¼	Coltec	1149	14½	+¼	Danher	186	21⅞	+⅝
Catelus	374	6½	+⅛	vjColGas	732	19⅝	+⅛	Daniel	1262	11⅝	−¼
Caterp	1931	49⅝	+1	Comdis	566	16⅛	+½	DataGn	3603	11½	+½
CBI	938	29	+⅜	Comdis	121	23¾	−⅛	DaytHd	1863	70⅛	+⅝
CBS	333	207⅝	−1⅞	Comeric	1071	60	+⅜	DeanFd	103	26½	
CCP	540	15⅝	+¼	Comdre	1467	7⅛	+⅛	DWGI	352	9¼	
CDI	187	6⅝	−⅛	CmwE	3268	22½	+⅛	Deere	1866	40⅝	+¾
CedrFr	201	22⅝	−¼	ComES	181	41⅛	+⅝	DelmPL	124	22⅞	+¼
Centel	1142	31⅝	+½	Comsat	170	41	+¼	DeltaAr	1991	54¾	−⅝
CentEn	2063	18⅛	−⅛	CPsyc	871	8⅞		DeltaA	475	53⅞	
Centex	474	25¾	+1¾	Compq	11595	36	−½	DeltaW	323	12	+⅛
CenSoW	2270	28½	+⅜	CmpAsc	1996	16	+⅜	Deluxe	536	41	+⅜
CenHud	708	30		CompSc	324	67¼	+½	Destec	172	15⅝	+⅜
CeMPw	197	22½		ConAgr	2121	30⅛	+⅜	DetEd	1671	31⅞	+⅜
CntyTl	607	36⅝	+⅛	ConeMl	270	14¾		Diagnst	907	16½	+¾
Ceridan	2082	14¼	+¼	ConnNG	129	25	+½	DialCp	330	38	
Chmpln	610	24⅛	+⅛	ConrPr	1324	19½	+⅝	DiaShm	151	18½	+⅜
ChartC	506	7		Conseco	1371	33¾	+⅞	Diasnc	111	13	
Chase	3408	21⅜	+⅜	ConEd	1154	31	+¼	DigtlCm	147	14¾	
Chse	105	29½	+⅛	CnsFt pfC	189	17⅛	−⅛	Digital	2010	36⅞	+1⅛
ChBk	118	25¼	−⅛	CnsFrt	156	13½	+¼	Dillard	1398	37¼	+¾
ChmBnk	3008	31¾	+¾	ConsNG	273	45⅝	+¼	Disney	7064	35⅜	+1⅛
ChWste	501	18½	+⅛	Conrail	1515	38	+⅞	Dole	958	29⅛	−½
Chspk	516	19	+⅝	CnStor	1171	15	+¼	DomRs	989	38	+¼
Chevrn	1681	73¾	+1⅛	CntlBk	564	16⅛	+¼	Donald	112	35¼	−⅛
Chile	291	29	+⅝	CntlCp	1500	25	+⅜	Donelly	877	29⅛	+⅜
China	449	11⅞	−⅛	CtlMed	216	16	+⅛	Dover	786	43½	+1

APPENDIX III STOCK EXCHANGE DATA SET

Stock	Sales 100s	Close	Change	Stock	Sales 100s	Close	Change	Stock	Sales 100s	Close	Change
DowCh	2112	54⅜	+⅞	Farah	172	5⅛		Gannett	665	46⅞	+⅞
DowJns	1436	28⅜	+1⅛	FayInc	248	7½	−⅛	Gap	4716	28⅝	
DPL	447	19⅜	+¼	Feders	718	4	+⅛	GenCr	1496	10¼	
DQE	572	31¼		FedExp	1432	38½	+¾	Genentc	231	33⅜	+⅛
Dressr	2989	18¾	+⅜	FdHmL	10494	39¾		GCinm	1175	28⅜	+1⅛
DryStG	221	12		FedMag	481	16½	+¼	GnDyn	870	93⅛	−⅜
DrySM	522	10		FedNM	4007	66	+1¼	GenEl	5339	75⅜	+⅜
DryStrt	1144	10½		FedPB	1443	26	+⅛	GnHost	306	9⅛	+¼
DuPont	3454	47¼	+1	FedP	371	50⅝	+¼	GnInst	172	17⅜	+½
DufPUtil	1722	10¼	−⅛	FedRlty	157	24⅛	−⅛	GnMills	1231	65⅞	+⅝
DukeP	884	35¼	+⅜	FedlSgn	352	18⅞	+⅜	GnMotr	17034	30¾	+1⅝
DukP pfS	265	27¾	−½	FedrDS	2513	14⅞	−⅛	GPU	1554	26⅜	−¼
DunBrd	1325	57½	+⅜	FidFn	564	20½		GenlRe	1100	108	+⅜
Duracl	429	30¾		Fidcrst	390	17	+⅛	GnSignl	469	53⅞	+¼
DutyF	1263	24¾	−¼	Finght	1401	25⅝	+¼	Gensco	285	7¼	+⅛
E Syst	307	37⅝	+1⅜	FtBrnd	114	24⅜	−¼	GenuPt	615	30⅛	−⅛
EastEn	138	25½	+⅛	FstChic	1953	31¼	+⅝	GaGulf	1438	17	
EastUtl	270	23⅜	+⅛	FtData	4754	26½	−½	GaPac	1581	50⅛	+½
EKodak	3890	42⅜	+¼	FFB	1089	36⅞	+¾	GrbPd	693	36⅛	+⅜
Echlin	845	19½	+⅛	FlFnMg	334	33½	+½	GerFd	206	11⅞	−⅛
Edward	737	19⅜	+¼	Flntste	341	40	+¾	GFC	109	22⅜	−⅛
EGG	284	21½	+¼	FtMiss	302	8⅛		Gillete	2146	57⅛	+1⅜
EKCO	256	8		FstAm	515	32⅛		Gitano	333	4¼	
ElPas	1184	27⅜	−1	FtPhil	255	12⅜	+⅛	Glaxo	6737	26½	+½
Elcor	168	12¾	−⅛	FstUC	3410	36¼	+⅜	GlbGvt	333	7½	+⅛
ElcAut	142	21¾		Ft USA	475	11⅝	−⅜	GlbHlt	321	11⅛	+⅛
ElfAquit	948	33⅝	+⅞	FtVaBk	140	32¼	+¼	GlncPl	379	9⅞	−⅛
Elscint	466	5⅝		Firstar	319	27⅛	+½	GlobNR	574	6¾	+¼
EMC	1207	20⅝	+⅜	FishPr	1009	22⅝	−⅛	GlobYld	286	8⅛	
EMTel	211	12½	+⅜	FltFln	685	29⅛	+⅛	GldWF	609	37	−¼
EmMex	173	16¼	+⅛	FltMdg	166	19⅝	+⅝	Gdrich	892	40¾	
EmrsEl	579	52½	+⅝	FleetEn	317	33⅛	+⅝	Goodyr	3409	62	+1⅞
EBP	3780	14	+2⅝	Flemng	230	28⅛		Grace	654	35¼	+¼
Endesa	553	27⅞	+¾	FlaPrg	972	32⅜		GrhmFl	168	6⅛	+⅛
Emplca	337	17⅝	−¼	Flower	182	18¼		Graingr	139	49⅝	+⅝
EnglCp	604	30¼	+⅜	Fluor	913	42½	+½	GrdMet	1718	25½	−⅛
Enhanc	1032	16⅞	+⅛	FMC	361	43	−⅜	GtAtPc	538	22¼	+⅝
EqStr	220	15	+⅜	FordH	155	24½	−¼	GtLkCh	726	68¼	+1⅝
Enron	1314	47½	+⅝	FordM	1753	68½	+2	GtWFn	1204	13⅞	+⅝
EnronL	159	26⅞		FordM	15477	36⅛	+1¼	GChina	328	12¾	+⅝
EnrOG	439	31⅝	+⅛	FostWh	1802	29½	+1	GrenTr	721	34¼	+¾
Ensrch	349	15⅜	+¼	FounH	1318	39	+⅛	GthSpn	393	7½	+⅛
Entergy	1269	31¼	+⅜	Foxmyr	148	12¾	−⅛	Grumn	683	19⅞	−¾
Entera	100	20⅜	+⅛	FPLGp	2300	35¾		GTE	7007	33⅞	+⅜
EnvEle	473	7⅜	+⅜	FrkPr	363	7⅞	+⅛	Gtech	319	26⅛	+¾
Equifx	662	14¾	−⅛	FrkQst	147	16⅜	−⅛	Gulfrd	194	19¼	−¼
Equmk	339	6½		FrkRs	436	29¾	+½	GlfStUt	5389	16⅜	+¼
EqtRes	180	47½	−⅛	FrkUnv	506	8⅜		Haemon	381	37⅝	+⅛
Esco	350	7⅛	+⅛	FrMeyer	258	26¼	+⅜	Halbtn	2239	32¼	+1⅝
Ethyl	884	26⅞	+⅜	FMCG	1379	19¾	−½	HancFb	241	9⅝	−⅛
EuroFd	125	12⅛	−¼	FrptMc	2306	18¾	−¼	Handlm	878	11⅛	
EurWtF	168	7⅜	−¼	FMRP	642	22¼	+⅛	HandH	111	12¼	+⅛
Exel	213	42¾		Frtrlns	137	33⅛	−¼	Hanson	3083	18¾	+⅜
Exxon	6755	61½	−⅛	Fuqua	248	10⅜	+⅜	Harlnd	143	23⅝	+⅞
FabCtr	618	10½	−⅜	GT Euro e	226	9½	−⅛	Harley	578	27⅝	+⅜
FamDir	533	18		Gabeli	1873	9⅝	+⅛	Harnish	421	16½	+⅜

APPENDIX III STOCK EXCHANGE DATA SET

Stock	Sales 100s	Close	Change	Stock	Sales 100s	Close	Change	Stock	Sales 100s	Close	Change
Harris	192	30⅜	+⅜	INCO	1460	22⅛	+⅛	KnghtR	920	55¼	
HrtfdSt	348	57	−⅜	Indres	707	7⅛		Knogo	163	6¾	−¼
Hartmx	309	3⅞		IngerRd	832	27¾	+1	Korea	487	12½	−⅛
HCA n	4838	17⅛	+¼	InldStl	331	16⅞		Kroger	1700	11¾	−⅛
HltRhb	677	12	+⅛	Integn	220	22		Kysor	198	12¾	−⅜
HCR n	524	24½	+⅜	InlMun n	278	15⅛	−⅛	LAGr	270	9¾	
HlthCP	1063	24		IntQlMu	225	15⅜	−¼	LQuMt	470	8¼	+⅛
HlthEq	231	9⅝		vjInterc	453	6⅞	−⅛	LACg	2080	6¼	−⅜
HltMg A	1037	22¾	−¼	IBM	8799	78½		LaidlwB	287	7⅝	+⅛
HltsthR	1251	18⅜	+½	IntFam	279	10⅜	−⅛	LkehdP	217	26⅛	+¼
Hlthtrst	1341	13	−⅜	IntFlav	161	106	+1⅝	LandsE	211	27¼	
HeclaM	841	8⅛	−⅝	IGame	4598	41	+2¼	LatADl	31	14⅞	+¼
Heilig	293	35	+⅛	IntMult	202	28	+¼	LatAm	709	12⅞	−⅛
Heinz	1876	39⅛	+⅝	IntPap	3521	61½	+⅞	Lawtins	150	13⅛	
HelFn pfA	116	24¼		IntlRec	173	12¼	+¼	LeeEnt	379	31¾	
Herculs	741	56¾		IntSpcl	495	11¼	−⅛	LegPlat	159	24½	+⅝
Hrshey	682	45¾		ITCrp	415	5	−⅛	Lennar	1998	23¾	+1¾
HewlPk	4706	52¾	+1⅛	IntpbG	612	31	+⅜	LGEEn	173	33⅞	+⅛
Hexcel	112	10¾	−¼	IntstBak	158	18⅝	+¼	LbtyAS	378	10⅛	+⅛
Hibern	394	4¾		IPTimb	123	26½	−¼	LbtTr	153	10⅝	+¼
Hilnco	1062	5¾	+⅛	Ipalco	163	34⅜	−¼	Lfetme	150	10⅝	+⅛
Hlncll	1030	5⅞		IRT	114	11½	+⅛	Lilly	3060	60½	+1
Hilnlll	378	6¾	−¼	Itel	493	18¾	−⅛	Limitd	3810	21¼	+¼
HiYld	210	7⅞		ITTCp	1100	64⅝		LincNtl	144	64⅜	+½
HiYdPl	313	7⅞	−¼	Jackpot	311	15¼	⅛	Litton	425	40⅛	+⅜
Hillnbd	328	37⅞	−1⅜	JRiver	393	17⅛	+⅛	LizClab	1782	37⅜	+½
Hilton	1017	42	−⅜	JpOTC	272	8½		Lockhd	232	47¼	+¾
HMO	174	16⅞	−¼	JarFCh	463	12⅛	+¼	Loews	368	118⅜	+½
HomeD	5589	53⅛	+1¼	JAlden	436	14¼	−⅜	Logicn	230	15⅜	+¼
HomeSh	1280	6	+⅛	JNuven	191	24	+⅜	LomFn	699	7⅝	+⅜
Hmstke	3167	12⅞	−⅝	JohnJn	7693	48⅛	+1⅜	LILCo	3758	24⅜	+⅜
Honda	209	20¾	+⅝	JohnCn	449	40¼	+½	Loral	325	39	+⅜
Honwel	1115	61¼	−1½	JonesAp	266	34⅜	−½	LaLand	160	36	+⅛
HK Tel	1362	31⅞	+⅝	Jostens	1231	26⅝	+⅜	LaPac	739	47	+1
HMann	422	26⅝	+¼	Jundt	244	12⅞	+½	Lowes	1523	18⅛	+⅛
HrzHlt	325	9⅝	−⅛	JWP	3437	2⅝	−⅛	LSI Lg	528	6¼	+¼
Horsh	314	8	−¼	Kmart s	3759	24¼	+⅜	vjLTV	455	⅝	+1/16
HospSt	811	6⅛	+⅜	KCtyPL	941	22¾	+⅛	Lubrzl	952	26¾	+⅛
HougM	365	30½	+1	Kasler	177	8½	+½	Lubys	301	15¾	−⅛
Holnt	142	10¾		Katyln	315	22⅜		Luxotc e	1059	24¾	−⅛
Houslnt	557	47¾	−¼	KaufBH	428	12¼	+½	MACOM	200	4½	+⅛
Houlnd	689	43⅛	+¼	Kellogg	1124	70	+1¼	MacFrug	908	12	
Hubel B	274	51⅛	−¼	Kellwd	117	28	+½	MagmC	735	11½	
Huffy	225	12⅜	+⅛	Kemper	248	24	+⅛	Magnal	1023	21⅛	−⅛
Human	1420	22½	+¼	KmpHl	289	9⅜	−⅛	Malaysa	122	13¼	+¼
HypTr	462	10⅛		KmplGv	439	9⅛	−⅛	MgdMu	377	12⅛	
Hyprn	225	11⅞		KmpMl	413	11⅛	+⅛	ManrCr	920	19¼	+1½
IBP	4245	16½	+⅜	KmpMu	531	12½		Manpwl	878	14	−¼
ICNPh	866	7		KmpStr	242	11⅝	−¼	MargFn	1038	12½	−¼
IdahoP	253	26⅝	−⅛	KentEl	135	20½		MarMD	5469	24⅜	−⅝
IlliCtr	822	20¼	+⅜	KerrMc	836	41¾		MrklV	152	12½	
IllPowr	378	21¼	−⅛	Keycp	829	33⅛		Maiot	601	58½	+⅜
ITW	291	65		KeyInt	168	22¼	+⅛	Marriot	2609	18¾	+⅛
IMCFrt	387	43⅞	+1¼	KimbCl	2059	50⅜	+¼	MrshMc	1270	90¾	+¾
Imcera	4078	33¼	−1¾	Kimco	208	28	+¼	Martch	232	9⅛	+⅛
Imolnd	618	8¼	−⅜	KingWd	110	25½	+¼	MartM	453	56½	+¾

APPENDIX III STOCK EXCHANGE DATA SET

Stock	Sales 100s	Close	Change
Marvel	943	33¾	+⅜
Masco	1357	22⅝	+⅛
MasPrt	169	8¼	
Mattel	5272	22¾	+⅝
Maxus	11224	7¼	+¼
MayDS	1266	64½	+⅝
Maytag	2097	13	+⅛
MBIA	459	54¾	
MBNA	612	40	+1⅜
McDerl	1458	22⅝	+½
McDonl	2796	43½	+½
McDnD	2213	40⅝	+¾
McGrH	345	59	+⅜
McKes	714	39	+1¼
MCN	354	26⅞	−⅝
Mead	1057	35⅜	
Mesrx	469	15⅝	
MedCrA	3611	18½	−⅜
Mediplx	146	14½	−⅛
Meditr	545	28⅝	+⅛
Medtrn	605	93⅝	+1
MEI	121	5⅛	
Mellon	766	44⅝	+⅝
Melvile	2608	47⅞	+¼
Merck	32135	41¾	+⅝
MerFn	203	13⅛	−⅛
MerLSP	336	10⅛	
MerLyn	3468	48	+1⅝
MeryG	1609	10¾	
MeryLd	281	12	+⅜
Mesa	1122	10⅛	
MetFn	561	13¼	−⅛
MexEqt	464	13½	+⅛
MexFd	732	19¾	+⅛
MCR	3124	10⅜	−⅛
MGF	1719	8	
MIN	4731	8⅛	
MMT	3349	8	−⅛
MFM	329	9	
MGIC	124	38¼	+½
MGMG	230	17	+⅛
MicrTc	1513	16⅛	−⅛
MAWst	491	14⅛	+⅜
MdwRs	172	17	+¼
Millipre	1093	31⅜	+⅜
MMM	2215	99⅜	+1½
MirRsrt	709	24¾	−¼
MNC	2386	10¾	+⅛
Mobil	3465	62	+⅜
MolBio	176	19½	+⅞
Monsan	2465	51	+1
MonPw	229	25	+¼
Moore	248	14⅞	+⅛
MorgGr	52	10½	−⅜
Morgan	2699	61	−¼
MorKeg	116	12⅛	+¼
MorgSt	2235	46⅝	+⅛
MorKnd	274	18⅜	+¼
Morton	705	55⅜	−⅛
Motorla	4294	89½	−½
Mueler	269	17⅜	+¾
MunHl	182	9¼	
MuniFd	351	12⅜	−¼
MunCA	33	14⅝	
MunFL	33	15⅛	−⅛
Muniyd	202	15¼	
MunIns	483	14⅞	−⅛
MunQll	177	15	
MunQl	242	14½	
MurpO	116	35	+⅛
MusicL	318	10⅛	−⅛
MutRisk	124	35	−¼
Mylan	1015	25⅛	+⅛
Nalco	610	31¼	+⅜
NtlCity	185	43¼	+¼
NatEdu	266	6¾	+⅜
NatFGs	256	25¼	+¼
NHltLb	2419	19⅝	−⅛
NII	211	12¼	
NMedEs	2085	10⅝	+⅜
NtSemi	4740	12⅛	−⅛
NtSvln	349	25⅝	−⅛
NtlWst	137	35¼	+¼
NatnsBk	3876	42⅛	+½
NatHP	135	30¼	+⅛
Navistr	4597	1⅞	+⅛
Nav	320	20⅜	+⅞
NBD	541	28¼	+¼
NtwkEq	418	13½	+⅝
NevPw	427	22	+¼
NewAm	472	4¼	−⅛
NEngEl	3202	34⅞	
NwGrm	204	11	−¼
NJRsc	196	22⅜	+⅛
NPlnRl	237	22	
NYSEG	239	31⅝	+⅛
Newell	860	37	+¾
Newhal	317	13⅜	−¼
NwmtG	495	40¼	−1⅝
NwtMg	911	45½	−1⅝
NewsC	861	34⅝	−⅛
NiaMP	2627	19¼	+⅛
NikeB	2939	77¾	+⅛
NIPSCO	1221	26	+⅛
NoblAf	657	18¾	−¼
NordRs	344	5	+¼
NflkSo	1770	54⅞	+1⅛
Norsk	1301	21¾	
Nortek	162	4⅜	+¼
NAmMt	288	14½	−⅛
NoestUt	2228	25⅝	−⅛
NoStPw	238	42¼	
NorTel	854	31	+¼
Norwst	1860	38⅛	+⅝
NovaCr	1014	16⅛	+⅜
Nucor	378	49⅝	−⅝
NUI	142	24	+½
Nutmeg	1604	11⅞	+¾
NvClQ	101	15⅝	
NvCMI	134	12⅜	+⅛
NvCPP	188	15½	
NvFL	261	16¼	+⅛
NvInQl	130	16⅛	+⅛
NvlQl	190	16½	
NvMAd	700	15¾	−⅛
NvMO	426	16¼	
NuvMu	922	11⅜	+⅛
NNYIQ	173	16¾	+⅛
NvNYQ	294	14⅝	
NuvPP	298	15⅝	−⅛
NvPMl	146	14⅜	
NuvPrm	238	14⅞	
NuvPl	946	16⅞	
NuvQin	383	14⅞	+⅛
NWNL	191	37¼	−½
Nynex	2289	84⅜	+2
OcciPet	7513	16½	
Oceaner	443	17⅜	−⅜
OEA s	952	18⅝	+½
OfcDpt	799	27¼	−⅛
OffshP	113	14⅜	−½
Ogden	951	17⅝	−⅛
OgdPr	213	14⅞	−⅜
OhioEd	585	21⅝	+⅛
OHMCp	140	6¼	−⅛
OklaGE	1053	33⅜	+⅛
OldRep	491	24	+½
Olin	406	40	
Olin	291	42½	−¼
OMI	656	4⅛	
Omnicm	1543	35⅞	+½
Oneida	118	12½	
ONEOK	437	16⅞	+⅛
OpnhCa	176	21¼	
OppMS	586	11½	
OranRk	2082	39¾	
OrngCo	160	6	−¼
OreStl	1011	17¼	−⅞
Oryx	580	23¼	+½
OutbdM	982	15⅞	+⅛
OvShip	130	13¾	+⅛
OwenC	235	30½	+¼
Ownlll	220	9	
Oxford	479	17⅛	+⅛
PacEnt	2532	19¼	+½
PacGE	2021	31½	+¼
PacTel	2147	44⅜	+1
PacifCp	1705	22⅞	+⅜

APPENDIX III STOCK EXCHANGE DATA SET

Stock	Sales 100s	Close	Change	Stock	Sales 100s	Close	Change	Stock	Sales 100s	Close	Change
PainWb	1312	17	+½	PugetP	183	26¼	−⅛	Safewy	1082	10⅜	−¼
PanEC	2056	18⅞	+⅝	PDIF	178	12⅝	−⅛	SallieM	689	70	+¼
ParCom	963	43¾	−½	PHYM	486	10	+⅛	SalmFd	259	13⅞	
ParkHn	365	27⅛	−¼	PIGIT	1110	9⅜		Salomn	2798	37	+1
PatrPr	231	10¼	−⅛	PMMI	396	10⅜	+⅛	SDieGs	437	23⅝	+¼
PatPRII	197	11⅞		PMIT	815	9¼		SJuanB	317	7⅛	
PatSel	270	17¾	−¼	PMIIT	834	8⅜	−⅛	SFER	1199	9½	+¼
Penney	1245	70⅜	+1⅜	PPrIT	2085	8½	+⅛	SFePC	1099	11⅝	+⅛
PaPL	1108	26⅞	+½	PTFHC	169	14⅛	+⅛	SntO pfB	349	25⅛	
Pennzol	388	52⅞	+¼	QMS	166	7¾		SaraLe a	1232	56⅞	+⅝
PeopEn	150	30	+⅛	QuakrO	867	60⅛	−⅜	SCANA	333	42¼	−⅛
PepBoy	1289	25⅛	+½	QuakSC	791	11	−⅛	SCEcp	1636	43½	−⅛
PepsiC	6603	37⅞	+⅝	Quanex	375	15½	−½	SchrPl	3908	59¼	+⅝
PerkEl	2551	29	−¼	Quanx	124	20½	−⅛	Schlmb	2081	68¼	+¾
Prmian	138	4⅝	+⅛	Quantm	470	12⅝	+⅛	Schwb	1115	17⅝	+⅛
Pet e	866	17¾	+⅞	QsfVC e	441	22⅞	+⅜	SciAtl	728	28½	−¼
Petrie	257	23		QstVI	130	13½		ScotP	311	36	
Pfizer	4758	71	+½	Quests	343	25¼	−¼	Scripps	154	23⅝	+¼
PhmRes	125	6⅜		QkReily	199	20	+⅛	Seagrm	658	25¾	+¼
PhelpD	2627	42⅛	+¼	ROCFd	263	8⅞	−⅛	Seagul	306	30⅜	+1¼
PhilaEl	1787	25⅞	+¼	RAC	146	12½		Sears	3512	42⅝	+⅝
PhilMr	11582	82¾	+1	RAC	694	21⅛	−¼	Sears	796	44⅝	+⅛
PhilGl	1777	13½	+⅛	RalsPu	750	43½	+¾	Sears	244	25⅞	
PhilPet	6962	24¼	+¼	Raycm	266	37¼	+⅞	Sensor	552	24⅞	+¼
PhlVH	476	24⅞	+⅜	RJamF	396	15⅞	+⅜	SvceCp	315	17⅜	+¼
Phlcorp	133	24		Rayonr	281	39¾	+¼	SvMer	1468	9⅝	−¼
PHM	617	23¼	+⅜	Raythn	2634	42⅛	+⅛	Svcmst	476	25⅞	+½
Pier	539	8¼		RdrDg A	682	52⅛	−⅛	Shaw	2933	21½	+⅜
PilPrm	358	9⅛		RdgBte	320	5⅝	−⅛	ShawNt	1209	15⅜	+⅝
PinWst	1793	19⅝	−⅛	REIT	119	13½	−⅛	Shrwin	633	27⅝	−⅛
PitnyBw	886	33⅜	+¼	Reebok	927	29	+⅜	Shoney	2020	17¾	+⅝
Pittstn	428	11⅞		RelElc	314	16⅛	−⅜	Shopko	209	15¼	+¼
PlacerD	5904	11⅜	−⅜	Repsol	436	21¼	+¼	SierPac	3596	19⅛	+¼
PNC	1250	52½	+⅝	RepNY s	950	43½	−½	SilcnGr	385	18½	−⅛
PogoPd	835	11½	+⅛	Revco	1325	8⅝	+⅜	Singer	114	25¼	
Polaroid	1786	32	+⅛	ReyRey	121	41½	−½	Sizzler	161	9¼	
Polygr	644	26⅛	−¼	ReyMtl	1714	49½	+¾	Skyline	148	16⅜	+⅝
PopeTal	155	13¾	+⅛	RPR s	321	46¼	+⅜	SthCor	295	6⅝	
PortGC	1296	19⅛		RiteAid	1204	22¾	+¾	Smithln	1048	9⅜	
Potash	188	20⅝	+¼	RJR Nab	8009	8¾	+⅜	SmithB eq wi	2648	38	+⅜
PotmEl	1249	25⅞	+½	RJR pfP	2062	10¼		Smckr B	142	26½	+⅜
PPG	2313	53⅞	+⅜	RckCtr	1391	8½	−½	Smckr	116	27¼	−½
Praxar	3148	15⅛	−⅛	Rockwl	1450	26¾	+¼	SnapOn	498	30⅛	+⅜
PrecCst	224	22⅞		RoHaas	155	54⅞	+⅜	Society	115	57	+½
PfdInc	168	19¾	+⅛	RollinE	497	11⅝	+⅛	Solctr	1183	24¼	+¾
PflOF	187	13¾		Rollins	152	32	+⅛	Sonat	399	39⅞	+½
Premrk	1676	38⅜	+⅛	RolLeas	348	12⅜	+¼	SonyCp	112	33	+¾
Primca	185	24⅞	−⅛	Rowan	6436	9	+¼	Sothbys	511	10⅜	
Primca	10595	42½	−¾	RoyApl	722	9⅜	+¼	SouthCo	1979	37	
ProctG	2950	49⅝	+⅜	RoylD	2625	86⅞	+1¾	SolnGs	102	32½	−⅜
Promu	382	37⅞	+¼	Royce	152	11⅜		SNETel	873	35¼	−1
ProsSt	488	4⅜		Rubmd	1149	31⅛	+⅝	SwAirl	482	23½	+¼
PSI	482	18¾	+⅛	RussBr	413	26¾	+¼	SwBell	3066	66⅜	+1⅞
PSyCol	162	28⅛	+⅛	Ryder	1082	22⅛	+⅛	SwEnr	315	36½	
PSvNM	1885	13	+¼	Saatchl	1327	7⅜	+⅛	SwtPS	822	32	
PSEG	1683	28⅛	+⅛	SaftKl	1775	28⅞	+1	Spain	209	8⅛	−⅜

APPENDIX III STOCK EXCHANGE DATA SET

Stock	Sales 100s	Close	Change	Stock	Sales 100s	Close	Change	Stock	Sales 100s	Close	Change
Sprint	1438	24¼	+½	Texaco	8131	61¼	+⅞	UKing	147	9¼	+⅛
SPSTec	154	20¼	+⅛	TexInst	4473	44⅝	+¾	UnTech	1321	47⅜	+1⅛
SPSTm	457	37½	+¾	TexUtil	797	41¼	+⅛	UnvFd	113	30¾	+½
StPaul	540	74⅜	+⅞	Textron	1111	35⅜	+½	UnvHR	130	15¾	+⅜
StFdBk	763	19⅜	+½	ThaiCF	431	9½		UnvHlt	158	12⅝	−⅛
StMotr	143	12¼	+¼	ThrmEl	191	42⅜	+¼	UnoRst	453	5⅞	+⅛
StdPac	956	4⅞	+¼	Thiokl	216	16	−⅛	Unocal	2807	25½	+⅝
StdPrd	170	26½		ThmBet	179	67	+1⅜	UNUM	253	44½	
StanlWk	566	35⅛	−¼	Tidwtr	1583	20	+⅝	Upjohn	2286	31½	+⅛
STI Gp	588	6½		Tifany	394	23⅝	−⅜	USShoe	2393	10½	+⅛
StoneC	2668	13¼	+½	TimeW	9913	23¾	+½	US Surg	4777	57⅝	−⅛
StopSh	123	15¼	−⅛	TmW	343	50¼		USWst	3368	38½	+⅝
StorEq	274	10	+¼	TmW	2426	53¾	+⅛	UsairG	2000	11¾	−¼
StorTch	4175	24¾	+2	TmMir	662	28⅞	+⅜	USFG	1009	10½	+⅛
StrGlb	222	14	+⅛	TNP	354	19½	+¼	USLIFE	143	45⅝	+¼
Stratus	1047	36	+⅛	Tokhem j	150	6¾	+½	UST	706	32¼	+¼
StridRt	1382	17⅝	+⅛	TollBro	360	8½	+⅛	USX	1462	23⅜	
SunCo	682	23¼	−⅛	Tommy	631	18	+¼	USXDel	715	14¼	−½
SunEng	246	8¾		Trchmk	199	50⅜	+⅞	USXMar	7454	16⅝	
SunOst	238	13⅜	−⅛	Tosco	633	19½	−¼	UtiliCo	156	27¼	+¼
Sunst	153	34¼	+½	Total	3111	21½	+⅜	VFCp	559	45½	+⅜
SunMed	374	22¼	+⅛	ToyRU	4003	37⅛	−⅜	Valero	206	24⅜	
SunTrst	713	39¼	−⅛	Transm	324	43⅝	+⅜	ValMer	257	10⅜	+⅜
Suplnds	465	24⅛	+⅛	TrnatH	1107	45½		VKMIG	219	15⅜	
SupValu	552	28		TrCda	198	14⅛	+⅛	VKMT q	707	7¾	−⅛
SurgAf	1037	20⅝	−⅛	Transco	2163	14⅞	+⅛	VKML	574	9⅜	−¼
Sybrn	248	17¾		TrMMx	170	8¾		VKmpM q	155	11⅝	
SyblTc	793	10½	−⅛	TMMxA	191	8¼	+¼	VKMMT q	549	15¾	
Synovs	223	23⅛	+¼	Tavl pf	323	26		Varco	482	5½	+⅛
Syntex	4885	26	+⅛	Travler	2844	22½	+⅛	Varian	389	34½	−⅛
Sysco	1121	25	+⅜	TriCn	3047	½	+1/32	Varty	710	20¼	+⅜
T2 Med	1830	19	−¼	TriCon	955	24¾	+⅛	Vencor	346	27¾	−¼
TadrnL	139	20	+⅜	Tribune	509	44⅜	+⅜	VentSt	538	27	−¾
Taiwan	358	18⅝	−⅛	Trimas	371	23⅜		VinPt	164	10⅝	−½
Tandem	744	11½	+⅜	Trinty	144	31	+⅛	Vishay	424	28¾	
Tandy	136	27⅞		Trinova	461	21	+⅛	Vitro	183	16⅛	−¼
Tandy	696	26½	+⅜	TritEng	1289	35⅞	−¼	Vons	342	23⅜	
TauNY	586	12½	−⅜	TRW	159	49⅜	+¼	VulcM	135	40	+⅜
TCBY	136	4⅛		TucsEP	837	4⅜	−⅛	Waban	1166	19	+½
TCW	445	8¾		TycoLb	431	34¼	+⅝	WbshNt	49	17⅜	−⅛
TchSym	145	10⅝	+¼	TycoTy	4272	13⅜	+⅛	Wachovia	355	61	+¾
TECO	140	40⅝	+⅛	TylrCbt	536	13¼	+⅛	WalMt	10722	58⅝	+⅝
Tektmx	1902	20¼		UALCp	1034	111½	−⅝	Walgm	1138	39⅛	+⅛
TeleNZ	972	23⅝	+¼	UDCHm	173	8⅜		Warnac	1004	34⅜	−⅛
Teldyn	179	17⅝	+⅛	UGI	581	21⅜		WarnL	1153	64¾	+1
Telef	2133	27¾	+⅛	UJB Fn	291	16⅝	+⅛	WashEn	300	21⅛	−¼
TelMex	5623	44⅝		Unifi	376	36¼	−¼	Waste	8260	38⅜	+½
Templl	1241	44⅝	+⅛	UnlNV	676	111⅛	+2⅝	WastMl	832	22⅛	+¼
TemplE	909	22⅜	−¼	UnCmp	3367	41⅞	+1½	Welcom	2997	16⅜	+¼
TpGGv	537	9		UCarb	2584	12⅜	−⅛	Wellmn	651	18⅜	−⅜
TmpGib	1729	9	−⅛	UnElec	978	36¾	+⅝	WellsF	1701	63¼	+1⅝
Tenn p	383	35⅛	−⅛	UnPac	1746	51¾	+1⅛	Wendys	2512	12⅜	+⅜
Tennco	3666	34⅝	+⅜	Unisys	9411	8¾		WCNA	466	6	+⅛
Tepco	206	22⅞	+⅛	UAM	192	26⅛	+¼	WDigitl	2242	6	
Terdyn	431	13⅝	+½	UnDomR	130	22⅜	+⅜	WtGR	204	27½	−⅜
Terra	513	4½	−⅛	UHltCr	1650	53⅝	+2¼	WstnRes	1399	29¾	

APPENDIX III STOCK EXCHANGE DATA SET

Stock	Sales 100s	Close	Change	Stock	Sales 100s	Close	Change	Stock	Sales 100s	Close	Change
WstgEl	4197	14¾	+⅛	Winnbg	299	6½	+⅛	Xerox	1058	78	+1½
Wstvco	446	35	+½	WiscEn	310	26¼	+⅛	Yarkin	808	33¼	−¼
Weyerh	1205	33⅜	+⅜	WiscPS	120	31¾	+⅜	ZenithE	361	5¾	
WhlTc	215	33¼	+¼	Witco	298	42¾	−¼	Zenix	345	7⅛	+⅛
Whrlpl	630	36¼	+⅝	WMS	451	18¼	−⅛	Zumin	121	30⅝	+½
Whitmn	299	12⅝	+⅛	Wolwth	1251	31⅞		Zweig	688	12⅜	−⅛
Williams	2453	36⅛	+¼	Wrigley	679	32¾	+½	ZweigTl	1900	10⅛	+⅛
WinDix	516	62⅞	+1⅜	WyleLb	1052	17⅞	+⅝				

TABLE I RANDOM NUMBERS

04433	80674	24520	18222	10610	05794	37515	48611	62866	33963	14045	79451	
60298	47829	72648	37414	75755	04717	29899	78812	03509	78673	73181	29973	
67884	59651	67533	68123	17730	95862	08034	19472	63971	37271	31445	49019	
89512	32155	51906	61662	64130	16688	37275	51266	11569	08697	91120	64156	
32653	01895	12506	88535	36553	23757	34209	55806	96275	26130	47949	14877	
95913	15405	13772	76638	48423	25018	99041	77527	81360	18180	97421	55541	
55864	21694	13122	44115	01601	50541	00147	77680	58788	33016	61173	93049	
35334	49810	91601	40617	72876	33967	73830	15404	96554	88265	34537	38526	
57729	32196	76487	11622	96297	24160	09903	14045	22917	60718	66487	46346	
86648	13697	63677	70119	94739	25875	38829	68376	43918	77653	04127	69930	
30574	47609	07967	32422	76791	39725	53711	93385	13421	67957	20384	58731	
81307	43694	83580	79974	45929	85113	72268	09858	52104	32014	53115	03727	
02410	54905	79007	54939	21410	86980	91772	93307	34116	49516	42148	57740	
18969	75274	52233	62319	08598	09066	95288	04794	01534	92058	03157	91758	
87863	82384	66860	62297	80198	19347	73234	86265	49096	97021	92582	61422	
68397	71708	15438	62311	72844	60203	46412	65943	79232	45702	67055	39024	
28529	54447	58729	10854	99058	18260	38765	90038	94209	04055	27393	61517	
44285	06372	15867	70418	57012	72122	36634	97283	95943	78363	36498	40662	
86299	83430	33571	23309	57040	29285	67870	21913	72958	75637	99936	58715	
84842	68668	90894	61658	15001	94055	36308	41161	37341	81838	19389	80336	
56970	83609	52098	04184	54967	72938	56834	23777	98392	31417	98547	92058	
83125	71257	60490	44369	66130	72936	69848	59973	08144	61070	73094	27059	
55503	52423	02464	26141	68779	66388	75242	82690	74099	77885	23813	10054	
47019	76273	33203	29608	54553	25971	69573	83854	24715	48866	65745	31131	
84828	32592	79526	29554	84580	37859	28504	61980	34997	41825	11623	07320	
68921	08141	79227	05748	51276	57143	31926	99915	45821	97702	87125	44488	
36458	96045	30424	98420	72925	40729	22337	48293	86847	43186	42951	37804	
95752	59445	36847	87729	81679	59126	59437	33225	31280	41232	34750	91097	
26768	47323	58454	56958	20575	76746	49878	06846	32828	24425	30249	78801	
42613	37056	43636	58085	06766	60227	96414	32671	45587	79620	84831	38156	
95457	30566	65482	25596	02678	54592	63607	82096	21913	75544	55228	89796	
95276	17894	63564	95958	39750	64379	46059	51666	10433	10945	55306	78562	
66954	52324	64776	92345	95110	59448	77249	54044	67942	24145	42294	27427	
17457	18481	14113	62462	02798	54977	48349	66738	60184	75679	38120	17640	
03704	36872	83214	59337	01695	60666	97410	55064	17427	89180	74018	44865	
21538	86497	33210	60337	27976	70661	08250	69599	60264	84549	78007	88450	
57178	67619	98310	70348	11317	71623	55510	64756	87759	92354	78694	63638	
31048	97558	94953	55866	96283	46620	52087	80817	74533	68407	55862	32476	
69799	55380	16498	80733	96422	58078	99643	39847	96884	84657	33697	39578	
90595	61867	59231	17772	67831	33317	00520	90401	41700	95510	61166	33757	
33570	04981	98939	78784	09977	29398	93896	78227	90110	81378	96659	37008	
15340	93460	57477	13898	48431	72936	78160	87240	52716	87697	79433	16336	
64079	42483	36512	56186	99098	48850	72527	08486	10951	26832	39763	02485	
63491	05546	67118	62063	74958	20946	28147	39338	32169	03713	93540	61244	
92003	63868	41034	28260	79708	00770	88643	21188	01850	69689	49426	49128	
52360	46658	66511	04172	73085	11795	52594	13287	82531	04388	64693	11934	
74622	12142	68355	65635	21828	39539	18988	53609	04001	19648	14053	49623	
04157	50079	61343	64315	70836	82857	35335	87900	36194	31567	53506	34304	
86003	60070	66241	32836	27573	11479	94114	81641	00496	36058	75899	46620	
41268	80187	20351	09636	84668	42486	71303	19512	50277	71508	20116	79520	

SOURCE: An excerpt from *Tables of 105,000 Random Decimal Digits*. Interstate Commerce Commission, Bureau of Transport Economics and Statistics, Washington, D.C.

TABLE II BINOMIAL PROBABILITIES

n	x	.05	.1	.2	.3	.4	π .5	.6	.7	.8	.9	.95
2	0	.902	.810	.640	.490	.360	.250	.160	.090	.040	.010	.002
	1	.095	.180	.320	.420	.480	.500	.480	.420	.320	.180	.095
	2	.002	.010	.040	.090	.160	.250	.360	.490	.640	.810	.902
3	0	.857	.729	.512	.343	.216	.125	.064	.027	.008	.001	
	1	.135	.243	.384	.441	.432	.375	.288	.189	.096	.027	.007
	2	.007	.027	.096	.189	.288	.375	.432	.441	.384	.243	.135
	3		.001	.008	.027	.064	.125	.216	.343	.512	.729	.857
4	0	.815	.656	.410	.240	.130	.062	.026	.008	.002		
	1	.171	.292	.410	.412	.346	.250	.154	.076	.026	.004	
	2	.014	.049	.154	.265	.346	.375	.346	.265	.154	.049	.014
	3		.004	.026	.076	.154	.250	.346	.412	.410	.292	.171
	4			.002	.008	.026	.062	.130	.240	.410	.656	.815
5	0	.774	.590	.328	.168	.078	.031	.010	.002			
	1	.204	.328	.410	.360	.259	.156	.077	.028	.006		
	2	.021	.073	.205	.309	.346	.312	.230	.132	.051	.008	.001
	3	.001	.008	.051	.132	.230	.312	.346	.309	.205	.073	.021
	4			.006	.028	.077	.156	.259	.360	.410	.328	.204
	5				.002	.010	.031	.078	.168	.328	.590	.774
6	0	.735	.531	.262	.118	.047	.016	.004	.001			
	1	.232	.354	.393	.303	.187	.094	.037	.010	.002		
	2	.031	.098	.246	.324	.311	.234	.138	.060	.015	.001	
	3	.002	.015	.082	.185	.276	.312	.276	.185	.082	.015	.002
	4		.001	.015	.060	.138	.234	.311	.324	.246	.098	.031
	5			.002	.010	.037	.094	.187	.303	.393	.354	.232
	6				.001	.004	.016	.047	.118	.262	.531	.735
7	0	.698	.478	.210	.082	.028	.008	.002				
	1	.257	.372	.367	.247	.131	.055	.017	.004			
	2	.041	.124	.275	.318	.261	.164	.077	.025	.004		
	3	.004	.023	.115	.227	.290	.273	.194	.097	.029	.003	
	4		.003	.029	.097	.194	.273	.290	.227	.115	.023	.004
	5			.004	.025	.077	.164	.261	.318	.275	.124	.041
	6				.004	.017	.055	.131	.247	.367	.372	.257
	7					.002	.008	.028	.082	.210	.478	.698
8	0	.663	.430	.168	.058	.017	.004	.001				
	1	.279	.383	.336	.198	.090	.031	.008	.001			
	2	.051	.149	.294	.296	.209	.109	.041	.010	.001		
	3	.005	.033	.147	.254	.279	.219	.124	.047	.009		
	4		.005	.046	.136	.232	.273	.232	.136	.046	.005	
	5			.009	.047	.124	.219	.279	.254	.147	.033	.005
	6			.001	.010	.041	.109	.209	.296	.294	.149	.051
	7				.001	.008	.031	.090	.198	.336	.383	.279
	8					.001	.004	.017	.058	.168	.430	.663
9	0	.630	.387	.134	.040	.010	.002					
	1	.299	.387	.302	.156	.060	.018	.004				
	2	.063	.172	.302	.267	.161	.070	.021	.004			
	3	.008	.045	.176	.267	.251	.164	.074	.021	.003		
	4	.001	.007	.066	.172	.251	.246	.167	.074	.017	.001	
	5		.001	.017	.074	.167	.246	.251	.172	.066	.007	.001
	6			.003	.021	.074	.164	.251	.267	.176	.045	.008
	7				.004	.021	.070	.161	.267	.302	.172	.063
	8					.004	.018	.060	.156	.302	.387	.299
	9						.002	.010	.040	.134	.387	.630

TABLE II BINOMIAL PROBABILITIES

n	x	.05	.1	.2	.3	.4	.5	.6	.7	.8	.9	.95
10	0	.599	.349	.107	.028	.006	.001					
	1	.315	.387	.268	.121	.040	.010	.002				
	2	.075	.194	.302	.233	.121	.044	.011	.001			
	3	.010	.057	.201	.267	.215	.117	.042	.009	.001		
	4	.001	.011	.088	.200	.251	.205	.111	.037	.006		
	5		.001	.026	.103	.201	.246	.201	.103	.026	.001	
	6			.006	.037	.111	.205	.251	.200	.088	.011	.001
	7			.001	.009	.042	.117	.215	.267	.201	.057	.010
	8				.001	.011	.044	.121	.233	.302	.194	.075
	9					.002	.010	.040	.121	.268	.387	.315
	10						.001	.006	.028	.107	.349	.599
11	0	.569	.314	.086	.020	.004						
	1	.329	.384	.236	.093	.027	.005	.001				
	2	.087	.213	.295	.200	.089	.027	.005	.001			
	3	.014	.071	.221	.257	.177	.081	.023	.004			
	4	.001	.016	.111	.220	.236	.161	.070	.017	.002		
	5		.002	.039	.132	.221	.226	.147	.057	.010		
	6			.010	.057	.147	.226	.221	.132	.039	.002	
	7			.002	.017	.070	.161	.236	.220	.111	.016	.001
	8				.004	.023	.081	.177	.257	.221	.071	.014
	9				.001	.005	.027	.089	.200	.295	.213	.087
	10					.001	.005	.027	.093	.236	.384	.329
	11							.004	.020	.086	.314	.569
12	0	.540	.282	.069	.014	.002						
	1	.341	.377	.206	.071	.017	.003					
	2	.099	.230	.283	.168	.064	.016	.002				
	3	.017	.085	.236	.240	.142	.054	.012	.001			
	4	.002	.021	.133	.231	.213	.121	.042	.008	.001		
	5		.004	.053	.158	.227	.193	.101	.029	.003		
	6			.016	.079	.177	.226	.177	.079	.016		
	7			.003	.029	.101	.193	.227	.158	.053	.004	
	8			.001	.008	.042	.121	.213	.231	.133	.021	.002
	9				.001	.012	.054	.142	.240	.236	.085	.017
	10					.002	.016	.064	.168	.283	.230	.099
	11						.003	.017	.071	.206	.377	.341
	12							.002	.014	.069	.282	.540
13	0	.513	.254	.055	.010	.001						
	1	.351	.367	.179	.054	.011	.002					
	2	.111	.245	.268	.139	.045	.010	.001				
	3	.021	.100	.246	.218	.111	.035	.006	.001			
	4	.003	.028	.154	.234	.184	.087	.024	.003			
	5		.006	.069	.180	.221	.157	.066	.014	.001		
	6		.001	.023	.103	.197	.209	.131	.044	.006		
	7			.006	.044	.131	.209	.197	.103	.023	.001	
	8			.001	.014	.066	.157	.221	.180	.069	.006	
	9				.003	.024	.087	.184	.234	.154	.028	.003
	10				.001	.006	.035	.111	.218	.246	.100	.021
	11					.001	.010	.045	.139	.268	.245	.111
	12						.002	.011	.054	.179	.367	.351
	13							.001	.010	.055	.254	.513

TABLE II BINOMIAL PROBABILITIES*

n	x	.05	.1	.2	.3	.4	.5	.6	.7	.8	.9	.95
14	0	.488	.229	.044	.007	.001						
	1	.359	.356	.154	.041	.007	.001					
	2	.123	.257	.250	.113	.032	.006	.001				
	3	.026	.114	.250	.194	.085	.022	.003				
	4	.004	.035	.172	.229	.155	.061	.014	.001			
	5		.008	.086	.196	.207	.122	.041	.007			
	6		.001	.032	.126	.207	.183	.092	.023	.002		
	7			.009	.062	.157	.209	.157	.062	.009		
	8			.002	.023	.092	.183	.207	.126	.032	.001	
	9				.007	.041	.122	.207	.196	.086	.008	
	10				.001	.014	.061	.155	.229	.172	.035	.004
	11					.003	.022	.085	.194	.250	.114	.026
	12					.001	.006	.032	.113	.250	.257	.123
	13						.001	.007	.041	.154	.356	.359
	14							.001	.007	.044	.229	.488
15	0	.463	.206	.035	.005							
	1	.366	.343	.132	.031	.005						
	2	.135	.267	.231	.092	.022	.003					
	3	.031	.129	.250	.170	.063	.014	.002				
	4	.005	.043	.188	.219	.127	.042	.007	.001			
	5	.001	.010	.103	.206	.186	.092	.024	.003			
	6		.002	.043	.147	.207	.153	.061	.012	.001		
	7			.014	.081	.177	.196	.118	.035	.003		
	8			.003	.035	.118	.196	.177	.081	.014		
	9			.001	.012	.061	.153	.207	.147	.043	.002	
	10				.003	.024	.092	.186	.206	.103	.010	.001
	11				.001	.007	.042	.127	.219	.188	.043	.005
	12					.002	.014	.063	.170	.250	.129	.031
	13						.003	.022	.092	.231	.267	.135
	14							.005	.031	.132	.343	.366
	15								.005	.035	.206	.463
20	0	.358	.122	.012	.001							
	1	.377	.270	.058	.007							
	2	.189	.285	.137	.028	.003						
	3	.060	.190	.205	.072	.012	.001					
	4	.013	.090	.218	.130	.035	.005					
	5	.002	.032	.175	.179	.075	.015	.001				
	6		.009	.109	.192	.124	.037	.005				
	7		.002	.055	.164	.166	.074	.015	.001			
	8			.022	.114	.180	.120	.035	.004			
	9			.007	.065	.160	.160	.071	.012			
	10			.002	.031	.117	.176	.117	.031	.002		
	11				.012	.071	.160	.160	.065	.007		
	12				.004	.035	.120	.180	.114	.022		
	13				.001	.015	.074	.166	.164	.055	.002	
	14					.005	.037	.124	.192	.109	.009	
	15					.001	.015	.075	.179	.175	.032	.002
	16						.005	.035	.130	.218	.090	.013
	17						.001	.012	.072	.205	.190	.060
	18							.003	.028	.137	.285	.189
	19								.007	.058	.270	.377
	20								.001	.012	.122	.358

*Values omitted in this table are .0005 or less.

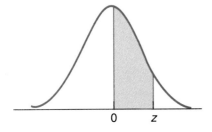

The entries in Table III are the probabilities that a random variable having the standard normal distribution takes on a value between 0 and z; they are equal to the area under the curve shaded in the figure.

TABLE III   AREAS UNDER THE STANDARD NORMAL CURVE

z	.00	.01	.02	.03	.04	.05	.06	.07	.08	.09	z
.0	.0000	.0040	.0080	.0120	.0160	.0199	.0239	.0279	.0319	.0359	.0
.1	.0398	.0438	.0478	.0517	.0557	.0596	.0636	.0675	.0714	.0753	.1
.2	.0793	.0832	.0871	.0910	.0948	.0987	.1026	.1064	.1103	.1141	.2
.3	.1179	.1217	.1255	.1293	.1331	.1368	.1406	.1443	.1480	.1517	.3
.4	.1554	.1591	.1628	.1664	.1700	.1736	.1772	.1808	.1844	.1879	.4
.5	.1915	.1950	.1985	.2019	.2054	.2088	.2123	.2157	.2190	.2224	.5
.6	.2257	.2291	.2324	.2357	.2389	.2422	.2454	.2486	.2517	.2549	.6
.7	.2580	.2611	.2642	.2673	.2704	.2734	.2764	.2794	.2823	.2852	.7
.8	.2881	.2910	.2939	.2967	.2995	.3023	.3051	.3078	.3106	.3133	.8
.9	.3159	.3186	.3212	.3238	.3264	.3289	.3315	.3340	.3365	.3389	.9
1.0	.3413	.3438	.3461	.3485	.3508	.3531	.3554	.3577	.3599	.3621	1.0
1.1	.3643	.3665	.3686	.3708	.3729	.3749	.3770	.3790	.3810	.3830	1.1
1.2	.3849	.3869	.3888	.3907	.3925	.3944	.3962	.3980	.3997	.4015	1.2
1.3	.4032	.4049	.4066	.4082	.4090	.4115	.4131	.4147	.4162	.4177	1.3
1.4	.4192	.4207	.4222	.4236	.4251	.4265	.4279	.4292	.4306	.4319	1.4
1.5	.4332	.4345	.4357	.4370	.4382	.4394	.4406	.4418	.4429	.4441	1.5
1.6	.4452	.4463	.4474	.4484	.4495	.4505	.4515	.4525	.4535	.4545	1.6
1.7	.4554	.4564	.4573	.4582	.4591	.4599	.4608	.4616	.4625	.4633	1.7
1.8	.4641	.4649	.4656	.4664	.4671	.4678	.4686	.4693	.4699	.4706	1.8
1.9	.4713	.4719	.4726	.4732	.4738	.4744	.4750	.4756	.4761	.4767	1.9
2.0	.4772	.4778	.4783	.4788	.4793	.4798	.4803	.4808	.4812	.4817	2.0
2.1	.4821	.4826	.4830	.4834	.4838	.4842	.4846	.4850	.4854	.4857	2.1
2.2	.4861	.4864	.4868	.4871	.4875	.4878	.4881	.4884	.4887	.4890	2.2
2.3	.4893	.4896	.4898	.4901	.4904	.4906	.4909	.4911	.4913	.4916	2.3
2.4	.4918	.4920	.4922	.4925	.4927	.4929	.4931	.4032	.4934	.4936	2.4
2.5	.4938	.4940	.4941	.4943	.4945	.4946	.4948	.4949	.4951	.4952	2.5
2.6	.4953	.4955	.4956	.4957	.4959	.4960	.4961	.4962	.4963	.4964	2.6
2.7	.4965	.4966	.4967	.4968	.4969	.4970	.4971	.4972	.4973	.4974	2.7
2.8	.4974	.4975	.4976	.4977	.4977	.4978	.4979	.4979	.4980	.4981	2.8
2.9	.4981	.4982	.4982	.4983	.4984	.4984	.4985	.4985	.4986	.4986	2.9
3.0	.4987	.4987	.4987	.4988	.4988	.4989	.4989	.4989	.4990	.4990	3.0
3.1	.4990	.4991	.4991	.4991	.4992	.4992	.4992	.4992	.4993	.4993	3.1
3.2	.4993	.4993	.4994	.4994	.4994	.4994	.4994	.4995	.4995	.4995	3.2
3.3	.4995	.4995	.4995	.4996	.4996	.4996	.4996	.4996	.4996	.4997	3.3
3.4	.4997	.4997	.4997	.4997	.4997	.4997	.4997	.4997	.4997	.4998	3.4
3.5	.4998	.4998	.4998	.4998	.4998	.4998	.4998	.4998	.4998	.4998	3.5
3.6	.4998	.4998	.4999	.4999	.4999	.4999	.4999	.4999	.4999	.4999	3.6
3.7	.4999	.4999	.4999	.4999	.4999	.4999	.4999	.4999	.4999	.4999	3.7
3.8	.4999	.4999	.4999	.4999	.4999	.4999	.4999	.4999	.4999	.4999	3.8
3.9	.5000										

For $z \geq 3.90$, the areas are .5000 to four decimal places.

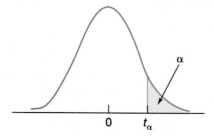

TABLE IV CRITICAL VALUES OF THE *t*-DISTRIBUTION

df	$t_{.05}$	$t_{.025}$	$t_{.01}$	$t_{.005}$	df
1	6.314	12.706	31.821	63.657	1
2	2.920	4.303	6.965	9.925	2
3	2.353	3.182	4.541	5.841	3
4	2.132	2.776	3.747	4.604	4
5	2.015	2.571	3.365	4.032	5
6	1.943	2.447	3.143	3.707	6
7	1.895	2.365	2.998	3.499	7
8	1.860	2.306	2.896	3.355	8
9	1.833	2.262	2.821	3.250	9
10	1.812	2.228	2.764	3.169	10
11	1.796	2.201	2.718	3.106	11
12	1.782	2.179	2.681	3.055	12
13	1.771	2.160	2.650	3.012	13
14	1.761	2.145	2.624	2.977	14
15	1.753	2.131	2.602	2.947	15
16	1.746	2.120	2.583	2.921	16
17	1.740	2.110	2.567	2.898	17
18	1.734	2.101	2.552	2.878	18
19	1.729	2.093	2.539	2.861	19
20	1.725	2.086	2.528	2.845	20
21	1.721	2.080	2.518	2.831	21
22	1.717	2.074	2.508	2.819	22
23	1.714	2.069	2.500	2.807	23
24	1.711	2.064	2.492	2.797	24
25	1.708	2.060	2.485	2.787	25
26	1.796	2.056	2.479	2.779	26
27	1.703	2.052	2.473	2.771	27
28	1.701	2.048	2.467	2.763	28
29	1.699	2.045	2.462	2.756	29
inf.	1.645	1.960	2.326	2.576	inf.

TABLE V CRITICAL VALUES OF r

n	$r_{.025}$	$r_{.005}$
3	.997	
4	.950	.999
5	.878	.959
6	.811	.917
7	.754	.875
8	.707	.834
9	.666	.798
10	.632	.765
11	.602	.735
12	.576	.708
13	.553	.684
14	.532	.661
15	.514	.641
16	.497	.623
17	.482	.606
18	.468	.590
19	.456	.575
20	.444	.561
21	.433	.549
22	.423	.537
27	.381	.487
32	.349	.449
37	.325	.418
42	.304	.393
47	.288	.372
52	.273	.354
62	.250	.325
72	.232	.302
82	.217	.283
92	.205	.267

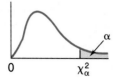

TABLE VI VALUES OF χ_α^2

df	$\chi^2_{.995}$	$\chi^2_{.99}$	$\chi^2_{.975}$	$\chi^2_{.95}$	$\chi^2_{.05}$	$\chi^2_{.025}$	$\chi^2_{.01}$	$\chi^2_{.005}$	df
1	.000	.000	.001	.004	3.841	5.024	6.635	7.879	1
2	.010	.020	.051	.103	5.991	7.378	9.210	10.597	2
3	.072	.115	.216	.352	7.815	9.348	11.345	12.838	3
4	.207	.297	.484	.711	9.488	11.143	13.277	14.860	4
5	.412	.554	.831	1.145	11.070	12.832	15.086	16.750	5
6	.676	.872	1.237	1.635	12.592	14.449	16.812	18.548	6
7	.989	1.239	1.690	2.167	14.067	16.013	18.475	20.278	7
8	1.344	1.646	2.180	2.733	15.507	17.535	20.090	21.955	8
9	1.735	2.088	2.700	3.325	16.919	19.023	21.666	23.589	9
10	2.156	2.558	3.247	3.940	18.307	20.483	23.209	25.188	10
11	2.603	3.053	3.816	4.575	19.675	21.920	24.725	26.757	11
12	3.074	3.571	4.404	5.226	21.026	23.337	26.217	28.300	12
13	3.565	4.107	5.009	5.892	22.362	24.736	27.688	29.819	13
14	4.075	4.660	5.629	6.571	23.685	26.119	29.141	31.319	14
15	4.601	5.229	6.262	7.261	24.996	27.488	30.578	32.801	15
16	5.142	5.812	6.908	7.962	26.296	28.845	32.000	34.267	16
17	5.697	6.408	7.564	8.672	27.587	30.191	33.409	35.718	17
18	6.265	7.015	8.231	9.390	28.869	31.526	34.805	37.156	18
19	6.844	7.633	8.907	10.117	30.144	32.852	36.191	38.582	19
20	7.434	8.260	9.591	10.851	31.410	34.170	37.566	39.997	20
21	8.034	8.897	10.283	11.591	32.671	35.479	38.932	41.401	21
22	8.643	9.542	10.982	12.338	33.924	36.781	40.289	42.796	22
23	9.260	10.196	11.689	13.091	35.172	38.076	41.638	44.181	23
24	9.886	10.856	12.401	13.848	36.415	39.364	42.980	45.558	24
25	10.520	11.524	13.120	14.611	37.652	40.646	44.314	46.928	25
26	11.160	12.198	13.844	15.379	38.885	41.923	45.642	48.290	26
27	11.808	12.879	14.573	16.151	40.113	43.194	46.963	49.645	27
28	12.461	13.565	15.308	16.928	41.337	44.461	48.278	50.993	28
29	13.121	14.256	16.047	17.708	42.557	45.722	49.588	52.336	29
30	13.787	14.953	16.791	18.493	43.773	46.979	50.892	53.672	30

TABLE VII CRITICAL VALUES FOR THE KOLMOGOROV-SMIRNOV TEST OF GOODNESS OF FIT

Sample Size (n)	Significance Level				
	.20	.15	.10	.05	.01
1	.900	.925	.950	.975	.995
2	.684	.726	.776	.842	.929
3	.565	.597	.642	.708	.829
4	.494	.525	.564	.624	.734
5	.446	.474	.510	.563	.669
6	.410	.436	.470	.521	.618
7	.381	.405	.438	.486	.577
8	.358	.381	.411	.457	.543
9	.339	.360	.388	.432	.514
10	.322	.342	.368	.409	.486
11	.307	.326	.352	.391	.468
12	.295	.313	.338	.375	.450
13	.284	.302	.325	.361	.433
14	.274	.292	.314	.349	.418
15	.266	.283	.304	.338	.404
16	.258	.274	.295	.328	.391
17	.250	.266	.286	.318	.380
18	.244	.259	.278	.309	.370
19	.237	.252	.272	.301	.361
20	.231	.246	.284	.294	.352
25	.21	.22	.24	.264	.32
30	.19	.20	.22	.242	.29
35	.18	.19	.21	.23	.27
40				.21	.25
50				.19	.23
60				.17	.21
70				.16	.19
80				.15	.18
90				.14	
100				.14	

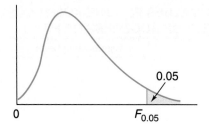

TABLE VIIIA CRITICAL VALUES OF THE F-DISTRIBUTION ($\alpha = 0.05$)

Degrees of Freedom for Denominator	Degrees of Freedom for Numerator									
	1	2	3	4	5	6	7	8	9	10
1	161	200	216	225	230	234	237	239	241	242
2	18.5	19.0	19.2	19.2	19.3	19.3	19.4	19.4	19.4	19.4
3	10.1	9.55	9.28	9.12	9.01	8.94	8.89	8.85	8.81	8.79
4	7.71	6.94	6.59	6.39	6.26	6.16	6.09	6.04	6.00	5.96
5	6.61	5.79	5.41	5.19	5.05	4.95	4.88	4.82	4.77	4.74
6	5.99	5.14	4.76	4.53	4.39	4.28	4.21	4.15	4.10	4.06
7	5.59	4.74	4.35	4.12	3.97	3.87	3.79	3.73	3.68	3.64
8	5.32	4.46	4.07	3.84	3.69	3.58	3.50	3.44	3.39	3.35
9	5.12	4.26	3.86	3.63	3.48	3.37	3.29	3.23	3.18	3.14
10	4.96	4.10	3.71	3.48	3.33	3.22	3.14	3.07	3.02	2.98
11	4.84	3.98	3.59	3.36	3.20	3.09	3.01	2.95	2.90	2.85
12	4.75	3.89	3.49	3.26	3.11	3.00	2.91	2.85	2.80	2.75
13	4.67	3.81	3.41	3.18	3.03	2.92	2.83	2.77	2.71	2.67
14	4.60	3.74	3.34	3.11	2.96	2.85	2.76	2.70	2.65	2.60
15	4.54	3.68	3.29	3.06	2.90	2.79	2.71	2.64	2.59	2.54
16	4.49	3.63	3.24	3.01	2.85	2.74	2.66	2.59	2.54	2.49
17	4.45	3.59	3.20	2.96	2.81	2.70	2.61	2.55	2.49	2.45
18	4.41	3.55	3.16	2.93	2.77	2.66	2.58	2.51	2.46	2.41
19	4.38	3.52	3.13	2.90	2.74	2.63	2.54	2.48	2.42	2.38
20	4.35	3.49	3.10	2.87	2.71	2.60	2.51	2.45	2.39	2.35
21	4.32	3.47	3.07	2.84	2.68	2.57	2.49	2.42	2.37	2.32
22	4.30	3.44	3.05	2.82	2.66	2.55	2.46	2.40	2.34	2.30
23	4.28	3.42	3.03	2.80	2.64	2.53	2.44	2.37	2.32	2.27
24	4.26	3.40	3.01	2.78	2.62	2.51	2.42	2.36	2.30	2.25
25	4.24	3.39	2.99	2.76	2.60	2.49	2.40	2.34	2.28	2.24
30	4.17	3.32	2.92	2.69	2.53	2.42	2.33	2.27	2.21	2.16
40	4.08	3.23	2.84	2.61	2.45	2.34	2.25	2.18	2.12	2.08
60	4.00	3.15	2.76	2.53	2.37	2.25	2.17	2.10	2.04	1.99
120	3.92	3.07	2.68	2.45	2.29	2.18	2.09	2.02	1.96	1.91
	3.84	3.00	2.60	2.37	2.21	2.10	2.01	1.94	1.88	1.83

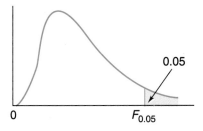

	Degrees of Freedom for Numerator								
	12	15	20	24	30	40	60	120	∞
1	244	246	248	249	250	251	252	253	254
2	19.4	19.4	19.4	19.5	19.5	19.5	19.5	19.5	19.5
3	8.74	8.70	8.66	8.64	8.62	8.59	8.57	8.55	8.53
4	5.91	5.86	5.80	5.77	5.75	5.72	5.69	5.66	5.63
5	4.68	4.62	4.56	4.53	4.50	4.46	4.43	4.40	4.37
6	4.00	3.94	3.87	3.84	3.81	3.77	3.74	3.70	3.67
7	3.57	3.51	3.44	3.41	3.38	3.34	3.30	3.27	3.23
8	3.28	3.22	3.15	3.12	3.08	3.04	3.01	2.97	2.93
9	3.07	3.01	2.94	2.90	2.86	2.83	2.79	2.75	2.71
10	2.91	2.85	2.77	2.74	2.70	2.66	2.62	2.58	2.54
11	2.79	2.72	2.65	2.61	2.57	2.53	2.49	2.45	2.40
12	2.69	2.62	2.54	2.51	2.47	2.43	2.38	2.34	2.30
13	2.60	2.53	2.46	2.42	2.38	2.34	2.30	2.25	2.21
14	2.53	2.46	2.39	2.35	2.31	2.27	2.22	2.18	2.13
15	2.48	2.40	2.33	2.29	2.25	2.20	2.16	2.11	2.07
16	2.42	2.35	2.28	2.24	2.19	2.15	2.11	2.06	2.01
17	2.38	2.31	2.23	2.19	2.15	2.10	2.06	2.01	1.96
18	2.34	2.27	2.19	2.15	2.11	2.06	2.02	1.97	1.92
19	2.31	2.23	2.16	2.11	2.07	2.03	1.98	1.93	1.88
20	2.28	2.20	2.12	2.08	2.04	1.99	1.95	1.90	1.84
21	2.25	2.18	2.10	2.05	2.01	1.96	1.92	1.87	1.81
22	2.23	2.15	2.07	2.03	1.98	1.94	1.89	1.84	1.78
23	2.20	2.13	2.05	2.01	1.96	1.91	1.86	1.81	1.76
24	2.18	2.11	2.03	1.98	1.94	1.89	1.84	1.79	1.73
25	2.16	2.09	2.01	1.96	1.92	1.87	1.82	1.77	1.71
30	2.09	2.01	1.93	1.89	1.84	1.79	1.74	1.68	1.62
40	2.00	1.92	1.84	1.79	1.74	1.69	1.64	1.58	1.51
60	1.92	1.84	1.75	1.70	1.65	1.59	1.53	1.47	1.39
120	1.83	1.75	1.66	1.61	1.55	1.50	1.43	1.35	1.25
∞	1.75	1.67	1.57	1.52	1.46	1.39	1.32	1.22	1.00

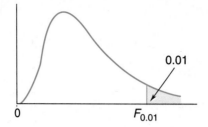

TABLE VIIIB CRITICAL VALUES OF THE F-DISTRIBUTION ($\alpha = 0.01$)

Degrees of Freedom for Denominator	Degrees of Freedom for Numerator									
	1	2	3	4	5	6	7	8	9	10
1	4,052	5,000	5,403	5,625	5,764	5,859	5,928	5,982	6,023	6,056
2	98.5	99.0	99.2	99.2	99.3	99.3	99.4	99.4	99.4	99.4
3	34.1	30.8	29.5	28.7	28.2	27.9	27.7	27.5	27.3	27.2
4	21.2	18.0	16.7	16.0	15.5	15.2	15.0	14.8	14.7	14.5
5	16.3	13.3	12.1	11.4	11.0	10.7	10.5	10.3	10.2	10.1
6	13.7	10.9	9.78	9.15	8.75	8.47	8.26	8.10	7.98	7.87
7	12.2	9.55	8.45	7.85	7.46	7.19	6.99	6.84	6.72	6.62
8	11.3	8.65	7.59	7.01	6.63	6.37	6.18	6.03	5.91	5.81
9	10.6	8.02	6.99	6.42	6.06	5.80	5.61	5.47	5.35	5.26
10	10.0	7.56	6.55	5.99	5.64	5.39	5.20	5.06	4.94	4.85
11	9.65	7.21	6.22	5.67	5.32	5.07	4.89	4.74	4.63	4.54
12	9.33	6.93	5.95	5.41	5.06	4.82	4.64	4.50	4.39	4.30
13	9.07	6.70	5.74	5.21	4.86	4.62	4.44	4.30	4.19	4.10
14	8.86	6.51	5.56	5.04	4.70	4.46	4.28	4.14	4.03	3.94
15	8.68	6.36	5.42	4.89	4.56	4.32	4.14	4.00	3.89	3.80
16	8.53	6.23	5.29	4.77	4.44	4.20	4.03	3.89	3.78	3.69
17	8.40	6.11	5.19	4.67	4.34	4.10	3.93	3.79	3.68	3.59
18	8.29	6.01	5.09	4.58	4.25	4.01	3.84	3.71	3.60	3.51
19	8.19	5.93	5.01	4.50	4.17	3.94	3.77	3.63	3.52	3.43
20	8.10	5.85	4.94	4.43	4.10	3.87	3.70	3.56	3.46	3.37
21	8.02	5.78	4.87	4.37	4.04	3.81	3.64	3.51	3.40	3.31
22	7.95	5.72	4.82	4.31	3.99	3.76	3.59	3.45	3.35	3.26
23	7.88	5.66	4.76	4.26	3.94	3.71	3.54	3.41	3.30	3.21
24	7.82	5.61	4.72	4.22	3.90	3.67	3.50	3.36	3.26	3.17
25	7.77	5.57	4.68	4.18	3.86	3.63	3.46	3.32	3.22	3.13
30	7.56	5.39	4.51	4.02	3.70	3.47	3.30	3.17	3.07	2.98
40	7.31	5.18	4.31	3.83	3.51	3.29	3.12	2.99	2.89	2.80
60	7.08	4.98	4.13	3.65	3.34	3.12	2.95	2.82	2.72	2.63
120	6.85	4.79	3.95	3.48	3.17	2.96	2.79	2.66	2.56	2.47
∞	6.63	4.61	3.78	3.32	3.02	2.80	2.64	2.51	2.41	2.32

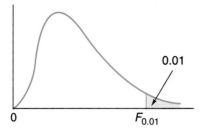

	Degrees of Freedom for Numerator								
	12	15	20	24	30	40	60	120	∞
1	6,106	6,157	6,209	6,235	6,261	6,287	6,313	6,339	6,366
2	99.4	99.4	99.4	99.5	99.5	99.5	99.5	99.5	99.5
3	27.1	26.9	26.7	26.6	26.5	26.4	26.3	26.2	26.1
4	14.4	14.2	14.0	13.9	13.8	13.7	13.7	13.6	13.5
5	9.89	9.72	9.55	9.47	9.38	9.29	9.20	9.11	9.02
6	7.72	7.56	7.40	7.31	7.23	7.14	7.06	6.97	6.88
7	6.47	6.31	6.16	6.07	5.99	5.91	5.82	5.74	5.65
8	5.67	5.52	5.36	5.28	5.20	5.12	5.03	4.95	4.86
9	5.11	4.96	4.81	4.73	4.65	4.57	4.48	4.40	4.31
10	4.71	4.56	4.41	4.33	4.25	4.17	4.08	4.00	3.91
11	4.40	4.25	4.10	4.02	3.94	3.86	3.78	3.69	3.60
12	4.16	4.01	3.86	3.78	3.70	3.62	3.54	3.45	3.36
13	3.96	3.82	3.66	3.59	3.51	3.43	3.34	3.25	3.17
14	3.80	3.66	3.51	3.43	3.35	3.27	3.18	3.09	3.00
15	3.67	3.52	3.37	3.29	3.21	3.13	3.05	2.96	2.87
16	3.55	3.41	3.26	3.18	3.10	3.02	2.93	2.84	2.75
17	3.46	3.31	3.16	3.08	3.00	2.92	2.83	2.75	2.65
18	3.37	3.23	3.08	3.00	2.92	2.84	2.75	2.66	2.57
19	3.30	3.15	3.00	2.92	2.84	2.76	2.67	2.58	2.49
20	3.23	3.09	2.94	2.86	2.78	2.69	2.61	2.52	2.42
21	3.17	3.03	2.88	2.80	2.72	2.64	2.55	2.46	2.36
22	3.12	2.98	2.83	2.75	2.67	2.58	2.50	2.40	2.31
23	3.07	2.93	2.78	2.70	2.62	2.54	2.45	2.35	2.26
24	3.03	2.89	2.74	2.66	2.58	2.49	2.40	2.31	2.21
25	2.99	2.85	2.70	2.62	2.53	2.45	2.36	2.27	2.17
30	2.84	2.70	2.55	2.47	2.39	2.30	2.21	2.11	2.01
40	2.66	2.52	2.37	2.29	2.20	2.11	2.02	1.92	1.80
60	2.50	2.35	2.20	2.12	2.03	1.94	1.84	1.73	1.60
120	2.34	2.10	2.03	1.95	1.86	1.76	1.66	1.53	1.38
∞	2.18	2.04	1.88	1.79	1.70	1.59	1.47	1.32	1.00

TABLE IX CRITICAL VALUE OF SPEARMAN'S RANK CORRELATION COEFFICIENT

	Level of Significance for Two-tailed Test			
n	0.10	0.05	0.02	0.01
5	0.900	1.000	1.000	—
6	0.829	0.886	0.943	1.000
7	0.714	0.786	0.893	0.929
8	0.643	0.738	0.833	0.881
9	0.600	0.683	0.783	0.833
10	0.564	0.648	0.745	0.794
11	0.523	0.623	0.736	0.818
12	0.497	0.591	0.703	0.780
13	0.475	0.566	0.673	0.745
14	0.457	0.545	0.646	0.716
15	0.441	0.525	0.623	0.689
16	0.425	0.507	0.601	0.666
17	0.412	0.490	0.582	0.645
18	0.399	0.476	0.564	0.625
19	0.388	0.462	0.549	0.608
20	0.377	0.450	0.534	0.591
21	0.368	0.438	0.521	0.576
22	0.359	0.428	0.508	0.562
23	0.351	0.418	0.496	0.549
24	0.343	0.409	0.485	0.537
25	0.336	0.400	0.475	0.526
26	0.329	0.392	0.465	0.515
27	0.323	0.385	0.456	0.505
28	0.317	0.377	0.448	0.496
29	0.311	0.370	0.440	0.487
30	0.305	0.364	0.432	0.478

TABLE X CRITICAL VALUES FOR TOTAL NUMBER OF RUNS*

The Smaller of n_1 and n_2	The Larger of n_1 and n_2															
	5	6	7	8	9	10	11	12	13	14	15	16	17	18	19	20
2								2 6	2 6	2 6	2 6	2 6	2 6	2 6	2 6	2 6
3		2 8	2 8	2 8	2 8	2 8	2 8	2 8	2 8	2 8	3 8	3 8	3 8	3 8	3 8	3 8
4	2 9	2 9	2 10	3 10	3 10	3 10	3 10	3 10	3 10	3 10	3 10	4 10	4 10	4 10	4 10	4 10
5	2 10	3 10	3 11	3 11	3 12	3 12	4 12	4 12	4 12	4 12	4 12	4 12	4 12	5 12	5 12	5 12
6		3 11	3 12	3 12	4 13	4 13	4 13	4 13	5 14	5 14	5 14	5 14	5 14	5 14	6 14	6 14
7			3 13	4 13	4 14	5 14	5 14	5 14	5 15	5 15	6 15	6 16	6 16	6 16	6 16	6 16
8				4 14	5 14	5 15	5 15	6 16	6 16	6 16	6 16	6 17	7 17	7 17	7 17	7 17
9					5 15	5 16	6 16	6 16	6 17	7 17	7 18	7 18	7 18	8 18	8 18	8 18
10						6 16	6 17	7 17	7 18	7 18	7 18	8 19	8 19	8 19	8 20	9 20
11							7 17	7 18	7 19	8 19	8 19	8 20	9 20	9 20	9 21	9 21
12								7 19	8 19	8 20	8 20	9 21	9 21	9 21	10 22	10 22
13									8 20	9 20	9 21	9 21	10 22	10 22	10 23	10 23
14										9 21	9 22	10 22	10 23	10 23	11 23	11 24
15											10 22	10 23	11 23	11 24	11 24	12 25
16												11 23	11 24	11 25	12 25	12 25
17													11 25	12 25	12 26	13 26
18														12 26	13 26	13 27
19															13 27	13 27
20																14 28

*Table shows critical values for two-tailed test at $\alpha = 0.05$.

Answers to Selected Exercises

Exercise Set 2.1, page 28

1.

Six Classes

Class	Class Limits	Frequency
1	50–124	3
2	125–199	5
3	200–274	7
4	275–349	5
5	350–424	3
6	425–499	2
Total		25

Nine Classes

Class	Class Limits	Frequency
1	50–99	2
2	100–149	2
3	150–199	4
4	200–249	4
5	250–299	5
6	300–349	3
7	350–399	3
8	400–449	0
9	450–499	2
Total		25

3.

Six Classes

Class	Class Limits	Frequency
1	36–43	3
2	44–51	6
3	52–59	6
4	60–67	8
5	68–75	3
6	76–84	4
Total		30

Ten Classes

Class	Class Limits	Frequency
1	36–40	1
2	41–45	3
3	46–50	4
4	51–55	3
5	56–60	7
6	61–65	4
7	66–70	2
8	71–75	2
9	76–80	3
10	81–84	1
Total		30

5.

Class	Class Limits	Frequency	Relative Frequency
1	50–124	3	0.12
2	125–199	5	0.20
3	200–274	7	0.28
4	275–349	5	0.20
5	350–424	3	0.12
6	425–499	2	0.08
Total		25	1.00

Class	Class Limits	Frequency	Cumulative Frequency
1	50–124	3	3
2	125–199	5	8
3	200–274	7	15
4	275–349	5	20
5	350–424	3	23
6	425–499	2	25
Total		25	

Answers to Selected Exercises **677**

Class	Class Limits	Frequency	Relative Frequency
1	50–99	2	0.08
2	100–149	2	0.08
3	150–199	4	0.16
4	200–249	4	0.16
5	250–299	5	0.20
6	300–349	3	0.12
7	350–399	3	0.12
8	400–449	0	0.00
9	450–499	2	0.08
Total		25	1.00

Class	Class Limits	Frequency	Cumulative Frequency
1	50–99	2	2
2	100–149	2	4
3	150–199	4	8
4	200–249	4	12
5	250–299	5	17
6	300–349	3	20
7	350–399	3	23
8	400–449	0	23
9	450–499	2	25
Total		25	

9. (a) .84 = 84%; (b) .72 = 72%.

11.
Class Limits	Frequency
750–849	3
850–949	9
950–1049	7
1050–1149	2
1150–1249	1
1250–1349	3
Total	25

13. **Six Classes**

Class Limits	Frequency
120–137	3
138–155	5
156–173	6
174–191	8
192–209	3
210–227	1
Total	26

Seven Classes

Class Limits	Frequency
120–134	2
135–149	5
150–164	3
165–179	8
180–194	5
195–209	2
210–225	1
Total	26

The percentage is 31%.

15.
Class Limits	Frequency	Relative Frequency
4.4–4.6	1	.05
4.7–4.9	6	.30
5.0–5.2	6	.30
5.3–5.5	4	.20
5.6 and over	3	.15

The percentage that cleared is .20 = 20%; the percentage that did not clear is .35 = 35%.

Exercise Set 2.2, page 35

1.
Interval	Frequency	Percent
Six Classes		
54–126.7	3	12%
126.67–199.3	5	20%
199.34–272.0	7	28%
272.01–344.6	5	20%
344.68–417.3	3	12%
417.35–490.0	2	8%
Nine Classes		
54–102.44	2	8%
102.44–150.8	2	8%
150.88–199.3	4	16%
199.32–247.7	4	16%
247.76–296.2	5	20%
296.20–344.64	3	12%
344.64–393.0	3	12%
393.08–441.5	0	0%
441.52–489.8	2	8%

3.

Interval	Frequency	Percent
Six Classes		
36–44	3	10%
44–52	6	20%
52–60	6	20%
60–68	8	26.7%
68–76	3	10%
76–84	4	13.3%
Ten Classes		
36–40.8	1	3.3%
40.8–45.6	3	10%
45.6–50.4	4	13.3%
50.4–55.2	3	10%
55.2–60	7	23.3%
60–64.8	4	13.3%
64.8–69.6	2	6.7%
69.6–74.4	0	0%
74.4–79.2	4	13.3%
79.2–84	2	6.7%

9.

Interval	Frequency	Percent
0–5	11	19%
5–10	15	25.9%
10–15	12	20.7%
15–20	7	12.1%
20–25	7	12.1%
25–30	3	5.2%
30–35	3	5.2%

Exercise Set 2.3, page 45

1.

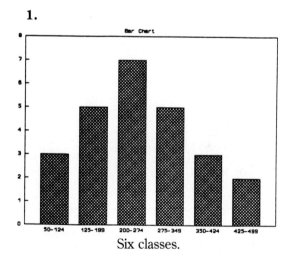

Six classes.

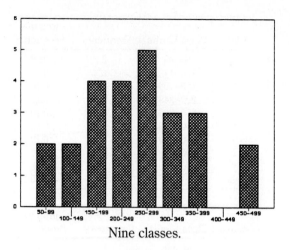

Nine classes.

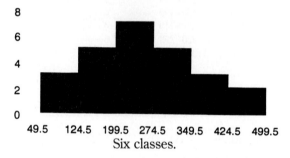

Six classes.

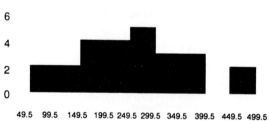

Nine classes.

Answers to Selected Exercises **679**

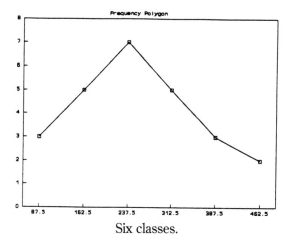

Six classes.

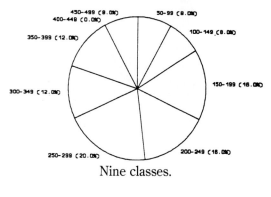

Nine classes.

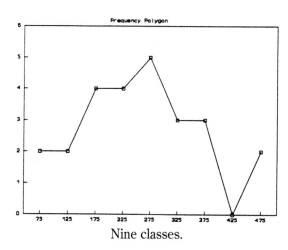

Nine classes.

5.

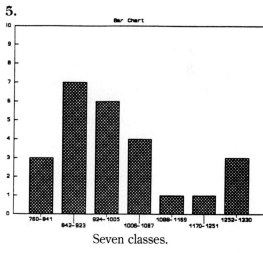

Seven classes.

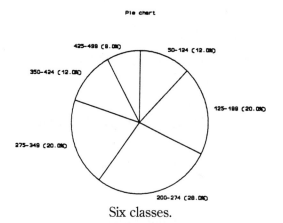

Six classes.

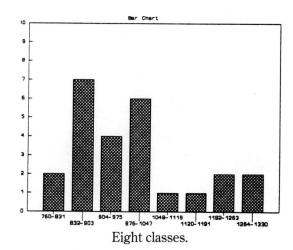

Eight classes.

680 Answers to Selected Exercises

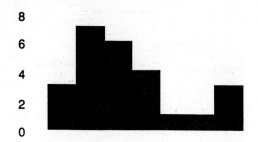

Seven classes.

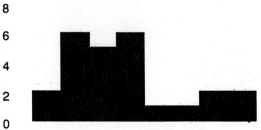

Eight classes.

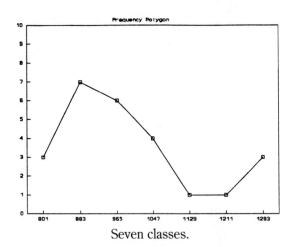

Seven classes.

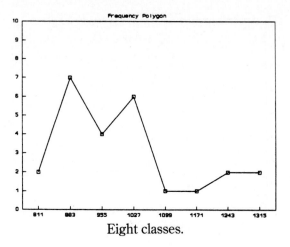

Eight classes.

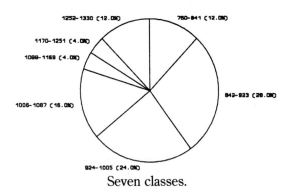

5a

Seven classes.

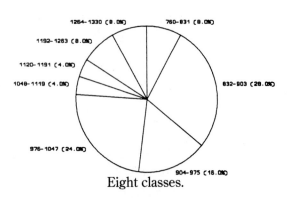

5b

Eight classes.

Answers to Selected Exercises **681**

9. 10% of the carrots are underweight.
11. 69% of the cars are speeding.

Exercise Set 2.4, page 50

1.
3	6
4	2246789
5	1357899
6	000333369
7	5578
8	04

3.
21	06
22	4
23	5
24	8
25	5
26	06
27	058
28	3578
29	038
30	5
31	6

5.
2*	09
2*	21, 22, 25
2*	42, 47, 50, 51, 53, 54, 59
2*	71, 78
2*	85, 88, 89, 90, 90, 95
3*	04, 11, 14, 15, 17

7.
2.	38
3.	056
4.	1349
5.	24778
6.	03368
7.	122
8.	022

9.
12	08
13	5
14	0788
15	468
16	566
17	24457
18	039
19	0256
20	6

Exercise Set 2.5, page 53

1. Ratio data.
3. Nominal data.
5. Ratio data.
7. Ratio data.
9. Ratio data.
11. Ordinal data.
13. Ratio data.

Exercise Set 2.6, page 59

1. Select two digits beginning in the upper left.
01   10   00   04   09
08   03   19   06   16
3. Start at upper left, go down columns.
245   519   522   587   335   520
332   304   368   584   436   332

Review Exercises, page 63

1.

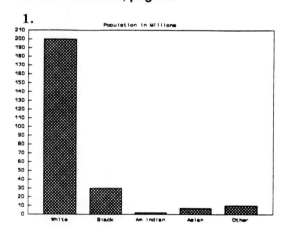

3. The data is nominal.

7.

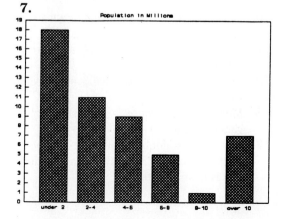

9. 75%.
11. 14%.

Exercise Set 3.1, page 72

1. Mean = 25; median = 24; there is no mode; midrange = 27.5
3. Mean = 59.4; median = 59.5; mode = 63; midrange = 60.
5. Mean = 986.4; median = 970; mode = 980; midrange = 1045.
7. Mean = 240.8; median = 244.5; mode = 190; midrange = 234.
9. The median indicates that half the students do better than 970; half do worse. This measure indicates the greatest need for improvement.
11. Mean = 167.8; median = 169; mode (there are 3) = 148, 166, 174.
13. Values for Exercise 1 are doubled.
15. Values for Exercise 3 are multiplied by 10.
17. Mean = 26.45; median = 27.
19. Mean = 5.544; median = 5.7.
21. Mean = 12.759; median = 12.
23. Mean = 27.567; median = 23.75.

Exercise Set 3.2, page 83

1. Population variance = 108.8; population standard deviation = 10.43.
3. Mean = 132; σ^2 = 346/10 = 34.6; σ = 5.88.
5. Mean = 5.72; population variance = .0456; population standard deviation = .214.
7. Sample variance = 120.9; sample standard deviation = 11.0.
9. Sample variance = 38.4; sample standard deviation = 6.2.
11. Sample variance = .057; sample standard deviation = .239.
13. Mean = 171; variance = 1733; standard deviation = 41.6.
15. Mean = 5.7; variance = 7.12; standard deviation = 2.67.
17. The mean is doubled (50), the variance is multiplied by 4 (435.2), the standard deviation is doubled (20.86).
19. Mean = 130; standard deviation = 48.5; variance = 2350.
21. Brand A: mean = 28; standard deviation = 4.2. Brand B: mean = 29, standard deviation = 11.4. Brand B lasts longer. Brand A is more consistent.
23. Mean = 26.46; s = 2.929; σ = 2.783.
25. Mean = 5.556; s = 1.728; σ = 1.654.
27. Mean = 25; s = 10.50; σ = 10.36.
29. Mean = 12.759; s = 8.71; σ = 8.56.

Exercise Set 3.3, page 89

1. Interval is (25, 35); interval is (20, 40).
3. (150, 170); (140, 180).
5. (280, 520); (160, 640).
7. (5.5, 8.5); (4, 10).
9. (8.7, 13.9); (6.1, 16.5).
11. (400, 550); (325, 625); (250, 700).
13. 68% is within one standard deviation of the mean. 34 inches to 56 inches.
15. 68%; 2.5%.
17. 65%; 97%.

Exercise Set 3.4, page 95

1. 1; 1.5; 2.3; 3.7; 0.7; −.9; 0.
3. 55; 62.5; 44.7; 60.6; 59.4; 20; 27.2.
5. Team A: μ = 78; s = 4.
   Team B: μ = 76; s = 3.
   Team A: z-scores of 1.5, .5, −.25, −.75 and −1.
   Team B: z-scores of 1.67, 0, 0, −.67, and −1.

Exercise Set 3.5, page 101

1. (a) Q_1 = .2565; Q_2 (Med) = .27; Q_3 = .295.
   (b) D_2 = .249; D_3 = .2594; D_7 = .2936.
   (c) P_{95} = .323.
3. (a) Q_1 = 55.25; Q_2 (Med) = 68; Q_3 = 79.
   (b) D_2 = 52; D_3 = 59; D_8 = 81.
   (c) P_{95} = 88.

Exercise Set 3.7, page 118

1. Mean = 18.5; variance = 52.25; standard deviation = 7.23.

Answers to Selected Exercises   **683**

3. Mean = 40; variance = 17; standard deviation = 4.12.
5. Mean = 522.4; standard deviation = 168.4.

Review Exercises, page 121

1. Mean = 5.722; median = 6.5; mode = 7, 8 (bimodal); σ^2 = 10.53, s^2 = 11.15; σ = 3.25, s = 3.34.
3. (a) 95%; (b) 1150 calories to 1750 calories.
5. Minimum 3; Q_1 = 5; median = 6; Q_3 = 6.5; maximum 8.
7. Nolan Ryan (z = 2.60) has the more outstanding accomplishment.

Exercise Set 4.1, page 139

1. 1/25.
3. 5/25.
5. 7/25.
7. 15/25.
9. 6/25.
11. (a) {BBB BBG BGB BGG GBB GBG GGB GGG}.
    (b) P(0 girls) = 1/8; P(1 girl) = 3/8; P(2 girls) = 3/8; P(3 girls) = 1/8.
    (c) 4/8. (d) 7/8. (e) 7/8.
13. (a) 2/15. (b) 6/15. (c) 2/11. (d) 6/11.
15. (a) 1/3. (b) 5/36. (c) 19/36. (d) 17/36.
    (e) 1/9. (f) 13/36.
17. (a) 12/19. (b) 15/19.
19. 5/12.
21. 2/3.

Exercise Set 4.2, page 143

1. (26)(26)(26)(10)(10)(10) = 17,576,000.
3. 10^7.
5. 8,000,000.
7. 27,600.
9. 800.
11. 360.
13. (a) 20. (b) 20. (c) 380.

Exercise Set 4.3, page 152

1. $_5P_2$ = 20.
3. $_8P_3$ = 336.

5. $_7P_1$ = 7.
7. $_7P_7$ = 7!
9. $_8C_4$ = 70.
11. $_8C_5$ = 56.
13. $_7C_0$ = 1.
15. $_{10}C_6$ = 210.
17. $_9P_5$ = 15120.
19. $_{16}C_3$ = 560.
21. $_9C_5$ = 126.
23. $_6C_4 \cdot {}_6C_4$ = 225.
25. $_{16}P_5$ = 524,160.
27. $_nP_0$ = 1.
29. $_nC_0$ = 1.

Exercise Set 4.4, page 164

1. (a) 39/52. (b) 28/52. (c) 32/52.
   (d) 15/52.
3. (a) {H1 H2 H3 H4 H5 H6 T1 T2 T3 T4 T5 T6}. (b) 1/12. (c) 7/12. (d) 3/12.
   (e) 9/12. (f) 6/12. (g) 3/12.
5. (a) 10/200. (b) 170/200. (c) 30/200.
7. 9/16; 6/16; 1/16.
9. (a) .81, .01. (b) .729, .001.
11. (a) .0225. (b) .7225.
13. (a) .0081. (b) .2401.
15. .729.
17. 1 − .9729 = .0271.
19. 1 − .9332 = .0668.

Exercise Set 4.5, page 170

1. 3/8.
3. 6/15.
5. 2/11.
7. 3/9.
9. 1/3.

Review Exercises, page 180

1. (a) 4/12. (b) 3/12. (c) 4/12.
3. (a) 8/36. (b) 6/36. (c) 10/36. (d) 20/36.
   (e) 27/36.
5. $_{12}C_4$ = 495.

Exercise Set 5.1, page 187

1. Discrete.
3. Continuous.

5. Discrete.
7. Continuous.
9. Continuous.
11. Probability distribution.
13. Not a probability distribution.
15. Not a probability distribution.
17. $p = .1$.
19. $p = .37$.
21. $\mu = 2.4; \sigma^2 = .84; \sigma = .92$.

23. | x | p(x) |
|---|---|
| 0 | .03125 |
| 1 | .15625 |
| 2 | .31250 |
| 3 | .31250 |
| 4 | .15625 |
| 5 | .03125 |

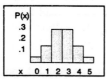

25. | x | p(x) |
|---|---|
| 1 | .25 |
| 2 | .25 |
| 3 | .25 |
| 4 | .25 |

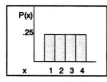

27. | x | p(x) |
|---|---|
| 0 | .81 |
| 1 | .18 |
| 2 | .01 |

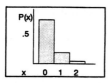

29. | x | p(x) |
|---|---|
| 0 | .0625 |
| 1 | .3750 |
| 2 | .5625 |

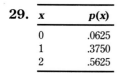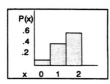

31. $\mu = 2.5; \sigma^2 = 1.25; \sigma = 1.12$.
33. $\mu = .20; \sigma^2 = .18; \sigma = .424$.

Exercise Set 5.2, page 198

1. $\mu = 50; \sigma^2 = 25; \sigma = 5$.
3. $\mu = 60; \sigma^2 = 40; \sigma = 6.32$.
5. $\mu = 200; \sigma^2 = 160; \sigma = 12.65$.
7. $\mu = 360; \sigma^2 = 36; \sigma = 6$.
9. $\mu = 30; \sigma^2 = 29.1; \sigma = 5.39$.
11. $\mu = 400; \sigma^2 = 398.4; \sigma = 19.96$.

Exercise Set 5.3, page 206

1. .09375.
3. .1094.
5. .3025.
7. .2753.
9. .094.
11. .109.
13. .303.
15. .275.
17. .259.
19. .117.
21. .057; probability of three or more burgers = .069.
23. .4437; probability of at least three packages arriving on time = .9014.
25. .158; probability that at most two saw the show = .253; probability that at most two saw the show = .128.
27. .000; .001; .011.
29. .082; .016; .002.
31. .017; .110; .067.

Exercise Set 5.4, page 211

1. .4242.
3. .4762.
5. .4167.
7. $\mu = 3.33; \sigma = .8409$.
9. .3874.
11. .2983.
13. .010.
15. .053.

Exercise Set 5.5, page 220

1. .2240.
3. .1404.
5. .1755.
7. 1.73.
9. .1044.
11. .00248.

Exercise Set 5.6, page 226

1. .12288.
3. .12.
5. .1647.

7. .1563.
9. .0354.
11. $p + q + r + s = 1$
$x + y + z + w = n$
$\dfrac{n!}{x!\, y!\, z!\, w!} p^x q^y r^z s^w$

Review Exercises, page 229

1. (a) Not a probability distribution.
   (b) Probability distribution.
3. $\mu = 1.95$; $\sigma^2 = 1.2675$; $\sigma = 1.13$.
5. This is a binomial problem; .248
7. .0564.

Exercise Set 6.1, page 242

1. .50.
3. .475.
5. .815.
7. .16.
9. .16.
11. .135.
13. .975.
15. .34.
17. .815.
19. .135.
21. .34.
23. .16.
25. .815.
27. .025.
29. .68.
31. .025.
33. 2.5%.

Exercise Set 6.2, page 256

1. .2019.
3. .3997.
5. .0516.
7. .4530.
9. .2577.
11. .2996.
13. .1293.
15. .4582.
17. .3083.
19. .0009.
21. .7486.
23. .1812.
25. .0038.
27. .3520.
29. .0918; .3707.
31. .0228.
33. 57.

Exercise Set 6.3, page 275

1. Yes.
3. No, 4.8 is less than 5.
5. Yes.
7. No, 4 is less than 5.
9. (a) .6757. (b) .0668. (c) .0336. (d) .5675.
   (e) .4325. (f) .2033. (g) .1052.
11. .121.
13. .7805.
15. .2810.
17. .2266.
19. .9286.

Review Exercises, page 280

1. .0336.
3. .2167.
5. 53.4.
7. .6436.

Exercise Set 7.1, page 295

1. $\mu_{\bar{x}} = 80$; $\sigma_{\bar{x}} = 6$. No conclusions about the shape of the distribution of $\bar{x}$ may be drawn. The distribution of x is unknown; $n < 30$.
3. $\mu_{\bar{x}} = 80$; $\sigma_{\bar{x}} = 3$. No conclusions about the distribution of $\bar{x}$ may be drawn. The distribution of x is unknown; $n < 30$.
5. $\mu_{\bar{x}} = 80$; $\sigma_{\bar{x}} = 2.4$. The distribution of $\bar{x}$ is approximately normally distributed with mean 80 and standard deviation 2.4.
7. $\mu_{\bar{x}} = 400$; $\sigma_{\bar{x}} = 7.5$. No conclusions about the shape of the distribution of $\bar{x}$ may be drawn.
9. $\mu_{\bar{x}} = 400$; $\sigma_{\bar{x}} = 2.73$. The distribution of $\bar{x}$ is approximately normally distributed.
11. $\mu_{\bar{x}} = 2.75$; $\sigma_{\bar{x}} = .12$.
13. $\mu_{\bar{x}} = 41$; $\sigma_{\bar{x}} = 1$; distribution of $\bar{x}$ is normal.
15. $\mu_{\bar{x}} = 1200$; $\sigma_{\bar{x}} = 8$.
17. $_{3600}C_2 = 6478200$.

Exercise Set 7.2, page 302

1. (a) $\mu_{\bar{x}} = 40$; $\sigma_{\bar{x}} = 2$. (b) .3413. (c) .4772. (d) .8185. (e) .6247. (f) .2417. (g) .0668. (h) .3085. (i) .7734.
3. (a) $\mu_{\bar{x}} = 350$; $\sigma_{\bar{x}} = 15$. (b) .4772. (c) .4525. (d) .5138. (e) .8156. (f) .0792. (g) .0918. (h) .9772. (i) .1922.
5. (a) .3790. (b) .5899. (c) .0475. (d) .2033.
7. .0475.
9. (a) .9938. (b) .0062. (c) .1056.
11. (a) .1056. (b) .1056. (c) .5328.
13. The sampling distribution of $\bar{x}$ is unknown.

Exercise Set 7.3, page 307

1. 5.75; 7.25; 3.75; 4.325.
3. .645; .968; −.968; −.226.
5. 3.8; 4.8; 2.5; 2.88.

Exercise Set 7.4, page 311

1. This is a binomial problem; .395.
3. This is a normal distribution problem; .2417.
5. This is a binomial problem; .136.
7. This is a normal distribution problem; .3674.
9. This is a binomial problem with large n; .0427.
11. This is a binomial problem; .1228.
13. This is a simple probability problem; 1/4.
15. This is a binomial problem; .0510.
17. This is a distribution of sample means problem; .3228.
19. This is a normal distribution problem; 68.79%.
21. This is a binomial problem with large sample size; .0582.

Exercise Set 7.5, page 323

1. If the underlying distribution is normal, the sample midranges appear to follow a normal distribution, especially as n increases. If the underlying distribution is skewed, the shape is flatter. Uniform and U-shaped distributions are "tighter" than the corresponding normal distribution.

Review Exercises, page 325

1. $\mu_{\bar{x}} = 25$; $\sigma_{\bar{x}} = .6$. The distribution of $\bar{x}$ is approximately normally distributed with mean 25 and standard deviation .6. No conclusions about the shape of the original distribution may be made.
3. $\mu_{\bar{x}} = 5.4$; $\sigma_{\bar{x}} = .47$; .1762. Almost zero.
4. (a) $\sigma = 2.4$. (b) $\mu_{\bar{x}} = 7.2$; $\sigma_{\bar{x}} = .4$. (c) The distribution is approximately normal with mean 7.2 and standard deviation .4. (d) .2266.

Exercise Set 8.1, page 344

1. Confidence level of 90%, $z = 1.64$; [236.88, 263.12]. Confidence level of 95%, $z = 1.96$; [234.32, 265.68]. Confidence level of 99%, $z = 2.58$; [229.36, 270.64].
3. [233.6, 266.4]; [230.4, 269.6]; [224.2, 275.8].
5. [116.8, 133.2].
7. [2.563, 3.437].
9. [83.1, 110.9].
11. [35.65, 40.35]
13. [17.5, 19.1].
15. [30.61, 33.39].
17. [20.91, 32.20].

Exercise Set 8.2, page 354

1. Confidence level of 90%, $t = 1.860$; [225.6, 374.4]. Confidence level of 95%, $t = 2.306$; [207.76, 392.24]. Confidence level of 99%, $t = 3.355$; [165.8, 434.2].
3. Confidence level of 90%, $t = 1.740$; [250.79, 349.21]. Confidence level of 95%, $t = 2.110$; [240.33, 359.67]. Confidence level of 99%, $t = 2.898$; [218.04, 381.96].
5. $t = 1.860$; [$15.14, $18.86].
7. $t = 3.106$; [23.53, 32.47].
9. $t = 1.740$; [$242190, $307810].
11. $t = 1.895$; [$132.40, $143.10].

Exercise Set 8.3, page 365

1. Confidence level of 90%: $z = 1.64$; [.18, .32] = [18%, 32%]. Confidence level of 95%: $z = 1.96$; [.17, .33] = [17%, 33%]. Confidence level of 99%: $z = 2.58$; [.14, .36] = [14%, 36%].

3. [.159, .241] = [15.9%, 24.1%]; [.151, .249] = [15.1%, 24.9%]; [.1355, .2645] = [13.55%, 26.45%].
5. [.204, .246] = [20.4%, 24.6%]; [.200, .250] = [20%, 25%]; [.191, .259] = [19.1%, 25.9%].
7. [.468, .632] = [46.8%, 63.2%].
9. (a) [.175, .525] = [17.5%, 52.5%]. (b) [.475, .825] = [47.5%, 82.5%].
11. [.314, .426] = [31.4%, 42.6%].
13. [.163, .197] = [16.3%, 19.7%].

Exercise Set 8.4, page 371

1. 44.
3. 87.
5. 202.
7. 1977.
9. 97
11. 485.

Exercise Set 8.5, page 373

1. [.428, .632] = [42.8%, 63.2%].
3. [.655, .905] = [65.5%, 90.5%].
5. [15.18, 16.82].
7. [4.422, 4.978]. The intervals do not overlap. This indicates that the treatment is effective.
9. $t = 1.860$; [90.84, 113.16].
11. [.069, .151] = [6.9%, 15.1%].
13. [$11.03, $12.97].
15. $t = 2.069$; [$63.52, $92.48].

Exercise Set 9.1, page 395

1. $z = 1$; critical $z = 1.64$; do not reject H_0.
3. $z = 2$; critical $z = 1.64$; reject H_0.
5. $z = -1.2$; critical $z = -1.64$; do not reject H_0.
7. $z = -1.2$; critical $z = \pm 1.96$; do not reject H_0.
9. $z = 1.33$; critical $z = 1.64$; do not reject H_0.
11. $z = 2.99$; critical $z = 1.64$; reject H_0.
13. $z = 1.67$; critical $z = \pm 1.96$; do not reject H_0.
15. $z = 2.07$; critical $z = 1.64$; reject H_0.
17. $z = 2.03$; critical $z = 1.28$; reject H_0.

Exercise Set 9.2, page 404

1. $P = .1587$.
3. $P = .0228$.
5. $P = .1151$.
7. .0456.
9. .0918.
11. .0041.
13. .0950.
15. .0192.
17. .0212.

Exercise Set 9.3, page 413

1. $t = 1.5$; critical $t = 1.753$; do not reject H_0.
3. $t = 1.88$; critical $t = 1.711$; reject H_0.
5. $t = -1.6$; critical $t = -1.860$; do not reject H_0.
7. $t = -1.625$; critical $t = \pm 2.160$; do not reject H_0.
13. $t = -4$; critical $t = -1.753$; reject H_0.
15. $t = -1.18$; critical $t = -2.201$; do not reject H_0.
17. $t = -2$; critical $t = -1.833$; reject H_0.
19. $t = -3.04$; critical $t = \pm 2.518$; reject H_0.
21. $t = -1.82$; critical $t = -1.895$; do not reject H_0.

Exercise Set 9.4, page 423

1. $z = 1$; critical $z = 1.64$; do not reject H_0.
3. $z = -1$; critical $z = -1.64$; do not reject H_0.
5. $z = -1.54$; critical $z = \pm 1.96$; do not reject H_0.
7. $P = .1587$
9. $z = 1.875$; critical $z = 1.64$; reject H_0.
11. $z = -3.65$; critical $z = -1.64$; reject H_0.
13. $z = -0.65$; critical $z = \pm 1.96$; do not reject H_0.
15. $z = -1.75$; critical $z = -1.64$; reject H_0.
17. $z = 1$; critical $z = 1.64$; do not reject H_0.
19. $z = -2.2$; $P = .0139$; reject H_0.

Exercise Set 9.5, page 440

1. $z = 1.25$; critical $z = \pm 1.96$; do not reject H_0.
3. $z = -1.65$; critical $z = \pm 1.96$; do not reject H_0.

5. $t = -1.60$; critical $t = -1.645$; do not reject H_0.
7. $z = -2.20$; critical $z = \pm 1.96$; reject H_0.
9. $z = 2.45$; critical $z = \pm 1.96$; reject H_0.
11. $z = 2.66$; critical $z = \pm 2.33$; reject H_0.
13. $t = 2.36$; critical $t = \pm 2.120$; reject H_0.
15. $[-1.8553, .0553]$.

Exercise Set 9.6, page 447

1. $t = 1.5$; critical $t = \pm 2.064$; do not reject H_0.
3. $t = 3.33$; critical $t = 1.711$; reject H_0.
5. $t = -1.897$; critical $t = -1.895$; reject H_0.
7. $t = -.56$; critical $t = \pm 2.447$; do not reject H_0.

Exercise Set 9.7, page 460

1. $z = .71$; critical $z = \pm 1.96$; do not reject H_0.
3. $z = -.0625$; critical $z = \pm 1.96$; do not reject H_0.
5. $z = -1.27$; critical $z = -1.64$; do not reject H_0.
7. $z = -.83$; critical $z = \pm 1.96$; do not reject H_0.
9. $z = -1.79$; critical $z = \pm 1.96$; do not reject H_0.
11. $z = 2.26$; critical $z = 2.33$; do not reject H_0.
13. $z = 1.36$; critical $z = \pm 2.33$; do not reject H_0.
15. $[-.1676, .0676]$.

Exercise Set 9.8, page 463

1. $z = 1.05$; critical $z = 1.64$; do not reject H_0.
3. $z = 1.18$; critical $z = 1.64$; do not reject H_0.
5. $z = 2.92$; critical $z = \pm 2.58$; reject H_0.
7. $z = 1.85$; critical $z = \pm 2.33$; do not reject H_0.
9. $t = 2.22$; critical $t = 2.624$; do not reject H_0.
11. $t = -1.17$; critical $t = \pm 2.447$; do not reject H_0.

Exercise Set 10.2, page 487

1. $r = .982$

3. $r = -.951$

5. $r = .982$; critical $r = .878$; a correlation between the two variables exists.
7. $r = -.951$; critical $r = .811$; a correlation between the two variables exists.
9. $r = .985$

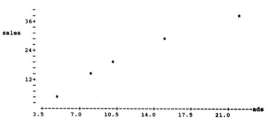

11. $r = .102$
13. $r = -.93$
17. $r = .849$; the relationship is significant.
19. $r = .745$; the relationship is significant.

Exercise Set 10.3, page 500

1. $y - 36 = 5.93(x - 7)$ or $y = 5.93x - 5.51$.
3. $y - 38 = -1.23(x - 12)$ or $y = -1.23x + 52.76$.

5. $y = 1.90x - .80$.
7. $y = .017x + 13.524$.
9. $y = -.80x + 184$.

Exercise Set 10.4, page 509

1. 4.96
3. 2.67
5. 2.59.
7. 2.11.
9. 3.70.
11. 11.88.
13. 8.555.

Exercise Set 10.5, page 516

1. $t = 8.79$; critical $t = \pm 3.182$; reject H_0; β_1 is nonzero.
3. $t = -6.15$; critical $t = \pm 2.776$; reject H_0; β_1 is nonzero.
5. $[-.25, 36.67]$; $[24.49, 59.37]$.
7. $[32.42, 48.58]$; $[19, 37.4]$.
9. $t = 10$; critical $t = \pm 3.182$; reject H_0; β_1 is nonzero.
11. $[0.85, 20.37]$; $[13.00, 31.02]$.
13. $t = 7.78$; critical $t = \pm 2.069$; the slope of the regression line is nonzero.
15. $t = 5.88$; critical $t = \pm 2.048$; the slope of the regression line is nonzero.

Review Exercises, page 525

1. $r = -.989$.
2. $r = -.989$; critical $r = .381$; a correlation between the two variables does exist.
3. $y = -.40x + 260.2$.
4. 1.31.
5. $t = -40$; critical $t = \pm 2.045$; reject H_0; β_1 is nonzero.
6. 3:40.2.
7. 2025.
8. 3:20.22.
9. 7:03.42.
10. 0:20.22.

Exercise Set 11.1, page 534

1.

20	50	70
60	150	210
80	200	280

3.

20 (27.55)	60 (52.04)	70 (70.41)	150
30 (27.55)	50 (52.04)	70 (70.41)	150
40 (34.90)	60 (65.92)	90 (89.18)	190
90	170	230	490

5.

70 (66)	80 (84)	150 (150)	300
40 (44)	60 (56)	100 (100)	200
110	140	250	500

7.

25 (38)	100 (85)	125 (127)	250
75 (62)	120 (136)	205 (203)	400
100	220	330	650

Exercise Set 11.2, page 547

1. (a) degrees of freedom $= 1$ (b) $\chi^2 = 0$.
3. df $= 4$; $\chi^2 = 4.882$.
5. $\chi^2 = 1.08$; critical value $\chi^2 = 5.991$; do not reject H_0; there is no difference in the proportion of drivers who prefer a particular brand based on gender.
7. $\chi^2 = 11.76$; critical value $\chi^2 = 9.210$; reject H_0; there is a difference in the proportions who are satisfied with the different cars.
9. $\chi^2 = 85.838$; critical value $\chi^2 = 11.345$; reject H_0; there is a difference in the proportions of men and women who are attracted to different colors on the dust jackets.

11. If the rows and columns are interchanged, $\chi^2 = 75.72$; critical value $\chi^2 = 15.507$; reject H_0; there is a difference in the proportions of people who express satisfaction living in each of the five cities.

Exercise Set 11.3, page 556

1. $\chi^2 = 2.8$; critical value $\chi^2 = 11.070$; the die is fair.
3. $\chi^2 = 28.42$; critical value $\chi^2 = 16.812$; there is a difference in the number of calls received based on the day of the week.
5. $\chi^2 = 3.91$; critical value $\chi^2 = 19.675$; the dice are fair.
7. $\chi^2 = 22.3$; critical value $\chi^2 = 7.815$; the given probabilities are not correct.

Exercise Set 11.4, page 565

1. $D_{10} = .409$; do not reject H_0.
3. $D_{30} = .242$; do not reject H_0.
5. $D_{30} = .264$; do not reject H_0.
7. $D_{24} = .264$; do not reject H_0.

Exercise Set 11.5, page 577

1. $\chi_L^2 = 5.629$; $\chi_R^2 = 26.119$.
3. $\chi_L^2 = 13.848$; $\chi_R^2 = 36.415$.
5. $\chi_L^2 = 12.401$; $\chi_R^2 = 39.364$.
7. [2.06, 3.85].
9. $\chi_R^2 = 42.557$; $\dfrac{[(n-1)s^2]}{\sigma^2} = 20.64$; do not reject H_0; there is insufficient evidence to support the claim.
11. (a) Confidence interval for the mean: [81.22, 88.78] (b) Confidence interval for the standard deviation: [4.98, 10.72]

Review Exercises, page 580

1. $\chi^2 = 10.305$; critical value $\chi^2 = 7.815$; reject H_0; there is a difference in the jeans color preferences for men and women.
3. $\chi^2 = 20.88$; critical value $\chi^2 = 9.488$; reject H_0; the probabilities are not correct.

Exercise Set 12.1, page 596

1. $F = 6.74$; critical $F = 5.14$; reject H_0.
3. $F = 1.42$; critical $F = 8.02$; do not reject H_0.
5. $F = 0.67$; critical $F = 3.49$; do not reject H_0.

Exercise Set 12.2, page 602

1. $F = 3.86$; critical $F = 4.26$; do not reject H_0.
3. $F = 4.70$; critical $F = 6.93$; do not reject H_0.
5. $F = 0.54$; critical $F = 3.49$; do not reject H_0.
7. The F-value is the same.

Exercise Set 13.2, page 617

1. $z = -.33$; critical $z = -1.64$; do not reject H_0.
3. $z = 0$; critical $z = -1.64$; do not reject H_0.
5. $z = .86$; critical $z = 1.64$; do not reject H_0.
7. $z = 3.57$; critical $z = 1.64$; reject H_0.
9. $z = -2.13$; critical $z = -1.64$; reject H_0.
11. $z = 4.33$; critical $z = 1.64$; reject H_0.

Exercise Set 13.3, page 625

1. $R = 60.5$; $\mu_R = 68$; $\sigma_R = 9.5$.
3. $R = 55.5$; $\mu_R = 72$; $\sigma_R = 10.4$.
5. $R = 100$; $z = -.38$; do not reject H_0.
7. $R = 71.5$; $z = -1.27$; do not reject H_0.
9. $R = 253.5$; $z = 2.30$; reject H_0.
11. $R = 74$; $z = -1.05$; do not reject H_0.

Exercise Set 13.4, page 631

1. $R_s = .6$.
3. $R_s = .86$.
5. $R_s = -.81$.
7. $R_s = -.1$.
9. $R_s = 1$; table value $= 1.00$; reject H_0.
11. $R_s = -.13$; table value $= .886$; do not reject H_0.
13. $R_s = -.93$; table value $= .786$, reject H_0.

Exercise Set 13.5, page 641

1. The data is random.
3. The data is not random.
5. The data is random.
7. The data is random.
9. Do not reject randomness.

Review Exercises, page 644

1. The problem is paired; sign test is used. $z = 1.79$; critical $z = 2.33$; reject H_0.
3. Sign test is used; $z = .78$; critical $z = \pm 2.58$; do not reject H_0.
5. The data is not random.

INDEX

Addition rule, 156
Alpha, α, 385
Alternate hypothesis, 380
Analysis of variance, 584
ANOVA, 584
   one-way, 600
   table for, 586
Average deviation, 76

Bar graph, 37
Before-and-after-type study, 442
Bell-shaped curve, 233
Bimodal, 69
Binomial coefficient, 150, 200
Binomial distribution, 192
   mean and standard deviation of, 193
   normal approximation to, 258
Binomial experiment, 190
Binomial probability formula, 200
Birthday problem, 163
Box-and-whisker plot, 99
Box plot, 99

$_nC_r$, 150
Central Limit Theorem, 291
Central tendency, measures of, 65
Chaos, 173
Chebyshev's Theorem, 89
Chi-square distribution, 537
Chi-square statistic, 533
Circle graph, 41
Class(es), 22
   boundaries, 24
   frequency, 22
   limits, 23
   mark, 39
   width, 23
Classical interpretation of probability, 128
Cluster sampling, 56
Coding, 74
Coefficient of linear correlation, 474
Combinations, 148
Complementary event, 158

Computer:
   IBM, 5
   Macintosh, 10
   programs (*see* Computer Program Index, page 685)
Conditional probability, 167
Confidence intervals:
   difference of means, 436
   difference of proportions, 457
   means (large samples), 329
   means (small samples), 345
   proportions, 355
Confidence level, 329
Consistent difference, 609
Contingency tables, 531
Continuity correction, 270
Continuous random variable, 182
Correlation coefficient, 474
Correlation, 470
   linear, 470
   negative, 471
   positive, 471
Critical region, 386
Critical values, 386
Cumulative frequency distribution, 25

Data:
   classification of, 51
   grouped, 112
   paired, 442
Decile, 98
Degree of confidence, 329, 551
Degrees of freedom, 305, 484, 537, 589
Dependent event, 167
Descriptive statistics, 21
Deviation:
   average, 76
   from the mean, 75
Difference of means, 424, 433
Difference of proportions, 449
Discrete random variable, 182
Distribution:
   bell-shaped, 233
   binomial, 192
   chi-square, 537

   of differences of sample means, 425
   of differences of sample proportions, 450
   F, 590
   frequency, 23
   Gaussian, 233
   hypergeometric, 207
   normal, 233
   Poisson, 213
   probability, 183
   of sample means, 285
   of sample medians, 313
   of sample midranges, 323
   of sample modes, 317
   of sample proportions, 356
   of sample variances, 566
   standard normal, 245
   Student's t, 304
   trinomial, 221
   uniform, 285, 555
Distribution-free methods, 608
Drunkard's walk, 171

Empirical Rule, 86
Equally likely events, 127
Error sum of squares, SS(Error), 586
ESP, 191, 227
Estimate, standard error of, 504
Estimation:
   of means, 328
   of proportions, 355
Event(s), 125
   complement of, 158
   equally likely, 127
   independent, 159
   mutually exclusive, 154
Expected frequency, 532
Expected value, 193
Experiment, 125

Factor level, 585
Factor sum of squares, SS(Factor), 586
Factorial notation, 146
F-distribution, 590

Index

Finite population correction factor, 293
F-ratio, 590
Frequency, 22
  cumulative, 25
  expected, 532
  observed, 531
  relative, 24
Frequency distribution, 23
Frequency polygon, 39
Fundamental counting principle, 141

Gaussian distribution, 233
Global warming study, 642
Goodness of fit, 550
Graph(s):
  bar, 37
  box-and-whisker plot, 99
  circle, 41
  frequency polygon, 39
  histogram, 38
  pie, 41
  stem-and-leaf, 47
Grouped data, 112

Hinges, 99
Histogram, 38
Hypergeometric distribution, 207
  mean and standard deviation of, 210
Hypothesis testing, 379

Independent events, 159
Inferential statistics, 21, 283
Interval(s):
  estimation, 329
  prediction, 514
Interval data, 52

Kolmogorov-Smirnov test, 560

Law of Large Numbers, 127
Least-squares line, 472
Level of significance, 385
Linear correlation, coefficient of, 474
Linear regression, 489

Mann-Whitney U-test, 619

Mean:
  population, 66
  of a probability distribution, 186
  sample, 67
  standard error of, 353
Mean squares, 590
Measures of central tendency, 65
Measures of variation, 75
Median, 67
Midrange, 69
Minitab, 13, 71, 83, 100, 109, 205, 353, 413, 438, 447, 480, 498, 521, 547, 597
Misuse of statistics, 71
Mode, 69
MS (Error), 590
MS (Factor), 590
Multivariate regression, 520
Multiplication rule for probability, 160
Mutually exclusive events, 154

Negative correlation, 471
Nominal data, 51
Nonlinear regression, 517
Nonparametric tests vs. parametric tests, 608
Normal distribution, 233
  applications of, 244
  approximation to the binomial distribution, 258
  standard, 245
Normality, testing for, 276, 558, 560
Null hypothesis, 380
Number of degrees of freedom, 305

Observed frequency, 531
One-sample sign test, 615
One-tailed test, 381
One-way ANOVA, 600
Ordinal data, 52
Outcomes, 125
Outlier(s), 277

$_nP_r$, 147
Paired-data test, 442
Paired-sample sign test, 614

Percentile, 96
Permutations, 146
Pie chart, 41
Point estimate, 284
Poisson distribution, 213
  approximation to binomial, 218
Population, 20
Population parameter, 66
Positive correlation, 471
Prediction intervals, 514
Probability, 125
  addition rule for, 156
  classical definition of, 128
  conditional, 167
  definition of, 126
  multiplication rule for, 160
  relative frequency definition of, 126
Probability distribution, 183
  mean and standard deviation, 186
Proportions:
  confidence intervals, 355
  tests concerning, 415
P-values, 397

Quartiles, 97

Random, 54
Random digits, table of, 54
Random numbers, 54
Random sample, 53
Random sampling, 56
Random variable, 182
  continuous, 182
  discrete, 182
Random walk, 171
Range, 22
Rank correlation coefficient, 628
Rank-sum test, 619
Ratio data, 53
Regression:
  linear, 489
  multivariate, 520
  nonlinear, 517
Regression line, 472
Rejection region, 386
Relative frequency distribution, 24
Risk level, 385

Run, 634
Runs test, 634

Sample(s), 20
Sample mean, 67
Sample proportion, 355
Sample size, determining, 366
Sample space, 125
Sample standard deviation, 79
Sample statistics, 66
Sample variance, 79
Sampling:
  cluster, 56
  stratified, 56
Sampling distribution:
  of the mean, 285
  of the median, 313
  of the midrange, 323
  of the mode, 317
  of proportions, 356
  of the variance, 566
Scatterplot, 107, 470
Sign test, 609
Significance level, 385
Skewed, 285
Skewness, 198
Spearman rank correlation coefficient, 627
Spearman rank correlation test, 627
SS (Error), 587
SS (Factor), 587
SS (Total), 588
Standard deviation, 76
  population, 76
  of a probability distribution, 186
  sample, 79
  of sample means, 353
Standard error of the estimate, 504
Standard error of the mean, 353
Standard normal distribution, 245
Standard scores, 92
Statistics:
  descriptive, 21
  inferential, 21, 283
  misuse of, 71
  nonparametric vs. parametric, 608
Stem-and-leaf plots, 47
Stratified sampling, 56
Student's t-distribution, 304
Sum of Squares:
  Error, 586
  Factor, 586
  Total, 586
Summation notation, 66

t-distribution, Student's, 304
t values, 304
Table of random digits (numbers), 54
Tail(s), 391

Tests:
  correlation coefficient, 484
  difference of means (large samples), 424
  difference of means (small samples), 433
  difference of proportions, 449
  means (large samples), 379
  means (small samples), 404
  paired data, 442
  proportions, 415
Total Sum of Squares, SS(Total), 588
Treatments, 585
Tree diagrams, 131
Trinomial distribution, 221
Two-tailed test, 381
Type I and II errors, 384

Uniform distribution, 285, 555

Variable, random, 182
Variance, 76
  analysis of, 584
  population, 76
  sample, 79
Venn diagram, 154

Waiting time, 176
Wilcoxon rank-sum test, 619

z-score or z-value, 92

Index of Computer Programs

1. Statistical Analysis of Data, 29, 44, 50, 71, 83, 100
2. Random Number Generator, 58
3. Comparing Two Sets of Data, 108
4. Grouped Data Analysis, 117
5. Coin Flipping Simulation, 128
6. Dice Rolling Simulation, 135
7. Law of Large Numbers, 138
8. Birthday Problem Simulation, 164
9. Drunkards (Random) Walk Simulation, 171
10. Patterns in Chaos, 174
11. The Binomial Distribution, 194
12. Binomial Simulation, 195
13. Binomial Probability, 204
14. The Hypergeometric Distribution, 211
15. The Poisson Distribution, 217
16. Trinomial Distribution Simulation, 224
17. Normal Distribution Probability, 254
18. Normal Distribution Simulation, 255
19. Normal Approximation to Binomial, 258, 274
20. Central Limit Theorem Simulation, 285
21. t-Distributions, 305
22. Distribution of Sample Medians, 313
23. Distribution of Sample Modes, 317
24. Distribution of Sample Midranges, 323
25. Confidence Intervals: Means/Props, 352, 361
26. Confidence Interval Simulation, 342
27. Distribution of Sample Proportions, 362
28. Hypothesis Tests, 411, 422
29. Hypothesis Testing Simulation, 411
30. Distribution of Difference of Means, 438
31. Distribution of Difference of Proportions, 459
32. Linear Regression Analysis, 478, 506
33. Correlation Simulation, 486
34. Regression Simulation, 506
35. Chi-Square Analysis, 544
36. Chi-Square Distributions, 547
37. Kolmogorov-Smirnov Test, 565
38. Distribution of Sample Variances, 571
39. Analysis of Variance, 597
40. Simulation of the Runs Test, 640